A MONOGRAPH
of the
FONTINALACEAE

A MONOGRAPH

of the

FONTINALACEAE

by

W<small>INONA</small> H. W<small>ELCH</small>, Ph.D.

Professor of Botany, DePauw University
Greencastle, Indiana

Springer-Science+Business Media, B.V. 1960

ISBN 978-94-017-5860-4   ISBN 978-94-017-6339-4 (eBook)
DOI 10.1007/978-94-017-6339-4

Dedicated to

T. G. YUNCKER

my first Teacher in Bryology

and to the Memory of

A. J. GROUT

and

H. N. DIXON

who suggested to me that a Monograph on

the Fontinalaceae be prepared

TABLE of CONTENTS

INTRODUCTION

In 1934, the author published a revision of the species of Fontinalaceae which occur in Canada and the United States. The study included three genera: *Fontinalis*, *Dichelyma*, and *Brachelyma*. The
Monographie des Fontinaleceés, by Jules Cardot, has been the basis
of treatises on this family of mosses since 1892. Cardot's monograph
included six genera in the Fontinalaceae: *Hydropogon*, *Hydropogonella*
(*Cryptangium*), *Fontinalis*, *Wardia*, *Brachelyma*, and *Dichelyma*. In
1943, a paper concerning the genera *Wardia*, *Hydropogon*, and *Hydropogonella* was published by the writer. These three groups of mosses
were removed from the Fontinalaceae because of the apparent absence
of a natural alliance between them and *Fontinalis*, *Dichelyma*, and
Brachelyma. The present monographic treatise is concerned only with
the genera *Fontinalis*, *Dichelyma*, and *Brachelyma*, the three groups
which, according to the judgment of the author, compose the family
of water mosses known as Fontinalaceae.

The diagnostic characteristics of the members of the Fontinalaceae
are: plants aquatic, attached at base, stems irregularly pinnately
divided, leaves tristichous, peristome double, the outer consisting of
16 teeth, the inner of 16 cilia, free or incompletely to completely
united by transverse strands into a cone-like trellis.

The manuscript for this monographic study was completed in
August 1949. A number of events causing the delay in publication were
unavoidable. The synonymy was brought up-to-date in January 1958.

EXPLANATION of TERMS and PROCEDURE

Type.– The type is considered to be the plant which was used as the basis of the original description. However, in the instance of mosses it is possible and probable that several plants of the same kind in one packet may have been used in the preparation of the original description. The author refers to these as the type or the type collection. These plants or this collection in the herbarium of the author of the original description, or the first author if more than one, with data on label corresponding to those given in their published description should be the type or the type collection, unless otherwise stated. Occasionally, but rarely, the word "type" occurs on the label or the packet. Frequently the abbreviation, sp. nov., etc. has been used. But, these indications are usually found in all herbaria to which the author of the species distributed duplicates. In the absence of definite indication as to the type, it has been necessary for the author to compare label and original description data in selecting the plant, plants, or collection which the author or authors may have used in writing the original description. These have been indicated by "selected to be the type," "selected to be the type collection," "assumed to be the type," or "assumed to be the type collection." The plants or packets of plants in other herbaria with labels bearing the same data as the type or selected type, apparently distributed by the author or authors of the original description, are indicated as being portions of the selected or assumed type or the selected or assumed type collection.

Macoun collections.– With reference to naming the collections of Fontinalaceae made by John Macoun by comparison with collection and exsiccati numbers in this monograph, it is recommended strongly that the plants in each packet be determined rather than relying upon agreement between numbers, dates, or locations.

Basis of classification.– Acknowledging the fact that characteristics of fruits are more stable than those of vegetative parts of plants, the author has used largely the gametophytic characteristics for separating the genera and for distinguishing the species and varieties within the genera because: first, the plants in this family frequently are not collected in fruit; second, the fruits of some species and varieties are unknown; and, third, differences in the sporophytes within the genera occur only in a few species and varieties. The anatomy of the stem, as observed in cross sections, is fundamentally in accord throughout the family, a central area of parenchymatous cells surrounded by several layers of sclerenchyma.

The plants of *Fontinalis* apparently are modified rather easily by environmental conditions. Those from still water, especially the younger growth, are frequently characterized by some degree of flaccidity, and by leaves rather distant; in some flaccid plants of the carinate-conduplicate group, plane cauline leaves are intermixed with the keeled-conduplicate blades; those bearing calcium deposit often have leaves and branches spreading instead of erect-spreading, the usual position.

On the same plant, the shape of the blade and the length-width ratio of the leaf and of the median cells frequently vary in leaves from younger portions of axes and those from the median locations; auricles may be distinct in mature leaves and lacking or slight in the blades on young branches and new growth of stems and branches; leaf bases may be decurrent in median cauline leaves and not decurrent in those on branches and younger portions of the axes; in the carinate-conduplicate group, frequently the younger leaves and occasionally the median branch leaves are plane and the median cauline blades are keeled-conduplicate, and the keels are distinctly curved in the mature leaves and straight in the younger blades; and, usually, according to size and (or) form, three different kinds of leaves occur, those roughly classified as near the base of the axes, those near the tips of the axes, and those in the middle or median portion of the stems and branches.

Upon the basis of the confusion of characteristics enumerated in the previous paragraph, two methods of classifying the plants in this family appear possible: one, the use of species or varieties for each variation which occurs, either genetic or ecological, and, second, the selection of one or more parts which seem to be rather stabilized throughout the Fontinalaceae and yet show sufficient differentiation for the recognition of the species and varieties. It appears to the writer that the former was chosen by many authors of forms, varieties, and species in *Fontinalis* and that the making of new species, varieties, or forms on plants or portions of plants which differ only because of the environmental conditions in which they are growing may serve to increase the confusion which already exists in the classification of plants belonging to this genus. The second method has been used in the present study in which the median cauline leaves in all instances, and the perichaetial leaves and sporophytes if present, have been used as the fundamental basis of determination.

Within a species or a variety some plants are more slender and others more robust, some more rigid and others more flaccid, than the majority of specimens within the unit. The form names indicative of robustness, slenderness, rigidity, and flaccidity have not been retained in this study. Neither have comparable varietal names been continued unless warranted by additional and distinctive characteristics of the median cauline leaves.

The observations by the writer of numerous plants which grew near the bottom of lakes, belonging to species of *Fontinalis*, are in general accord with those of Persson (1942 and 1944). In a given collection many transition forms have been noted as well as plants which are comparable to those of that species or variety growing elsewhere.

Procedure of examination.– All descriptions are based upon moistened or wet specimens. Observations and the larger measurements of the plant parts were made through a wide-field binocular, at 10 × and 23 × magnifications. The measurements of leaf length and width have been obtained through the wide-field binocular and by means of a glass ruler. Stem sections, selected leaves, peristomes, and spores were mounted for study through a calibrated microscope, at 100 × and 430 × magnifications. Cells, auricles, leaf apices, peristomes, and spores have been described as seen through the microscope. Calyptra, operculum, and urn measurements were made in length from base to apex and in width across the largest diameter, through the wide-field binocular and by means of a glass ruler. The peristome teeth and trellis were measured in length in a similar manner.

Plant size.– In consideration of size it is difficult to give measurements which will present the general appearance of plants. If distance across the foliated axis from leaf apex to leaf apex is given, the long, spreading, narrow leaves of a delicate plant may result in millimeters comparable to that of a very robust plant with large leaves somewhat appressed. A *Fontinalis* plant considered to be slender in the Fontinalaceae may appear robust in comparison with plants in another family of mosses. Although realizing the inefficiency of such relative terms as slender, delicate, medium, and robust, the author has used them in the absence of a more accurate means of classification and with reference to members of the Fontinalaceae, to indicate the size of a plant rather than a record of width in millimeters of the foliated stems and branches. Measurements of length of stems of individual plants and the diameter of the median portion of axes have been recorded.

Color.– Stem color in Fontinalaceae is not a criterion for recognition of species and varieties. The common transition in color from apex to base is yellowish green, through shades of red, to blackish. Older stems are characterized by the darker colors and are frequently denuded or nearly so. The stems of younger plants are generally yellowish green, reddish green, to reddish brown.

Leaf color can not be relied upon as a basis of specific and varietal recognition in the Fontinalaceae. The plants are often some shade of green but are frequently copper brown, golden brown, yellow brown, or brownish.

Surface gloss.– Plants or portions of same may be glossy or dull. This characteristic was determined by the examination of dried specimens, through the wide-field binocular and under the illumination of a microscope lamp.

Rigidity and flaccidity.– The rigidity or flaccidity of stems and branches frequently varies with the age of each, the older parts usually being more rigid than the younger. The author's data regarding the cells of the stem, as seen in transverse sections, are in accord with those of Leitgeb (1868) and Limpricht (1894). The inner portion of the axis is composed of parenchymatous tissue and is surrounded by a band of sclerenchymatous cells. The more rigid stems, or portions thereof, are characterized by a wider zone of sclerenchymatous tissue. The rigidity or flaccidity of the axes and the firmness or flaccidity of the leaves were determined when the plants were in water.

Branch and leaf positions.– The terms which indicate position of branches with relationship to the stem, and leaves with relationship to the axis, are applied as though the plants were upright. Erect branches are those extending somewhat parallel with the stem; erect-spreading branches form an angle with the stem of approximately 45°; spreading branches extend away from the stem in a direction almost perpendicular to it and form an approximate angle of 90°. The same interpretations are applicable to the leaves with reference to their position on the axis. The character of distichous branches is not regarded as a reliable diagnostic characteristic. If plants of *Fontinalis disticha* are in water for a brief time, the branches no longer appear distichous but are tristichous as in other species of *Fontinalis*.

Leaves:Dimorphism.– In this study the median cauline leaves

serve as the basis of contrast. The character of dimorphism used
previously by Cardot and also by the author is omitted. The size of
branch leaves and frequently their shape vary with the age of the
branch. A young branch may develop in the axil of a mature cauline
leaf and bear leaves which contrast distinctly with those of the stem.
A plant with older branches may show contrast in size and (or) shape
between median cauline and median branch leaves but less contrast
than would be evident in the instance of younger branches on old
stems. The branch leaves of plants growing in still water frequently
vary in form and flaccidity from branch leaves of the same species
growing in running water. Thus, it seems advisable not to use the
dimorphic characteristic.

Shape of blade if conduplicate.– Unless otherwise stated, the shape
of carinate-conduplicate leaves refers to the unfolded blades.

Terminal leaves.– The leaves at the ends of branches and stems
normally are subimbricate or imbricate, sometimes loosely, other
times tightly, causing the stem and branch tips to be obtuse, acute,
or attenuate.

Median leaves.– Although the adjective cauline may refer to both
stems and branches, in order to make specific references to leaves on
the main axis in comparison with those on the divisions of the chief
axis, the blades on the former are described in this study as cauline
leaves and those on the latter as branch leaves. The apical, basal, and
median leaves of the stems and branches on the same plant may vary
in size, shape, and position. The firmness and flaccidity of the leaves
may differ with age and habitat, as wet or dry, and still or running
water. Due to these conditions the writer has used the characteristics
of the mature median cauline and branch leaves unless otherwise
indicated. Median leaves of young branches usually appear different
from median cauline leaves, but the median leaves of older branches
generally resemble closely, except in size, the median cauline blades.
Thus, in the absence of the latter, one may be able to determine the
plants by means of the median branch blades.

Keel and median line.– If the leaf is definitely carinate-conduplicate
a keel extends through the center from base to apex. If the blade is
folded lengthwise because of being deeply concave or subconcave, the
axis of the fold is considered as the median line.

Basal and keel curvature.– The basal curve of a leaf has reference
to the curve of the base of the blade along the keel, median line, or
midrib immediately above the attachment of the leaf to the axis. As
to the curvature of the keel, reference is made to that portion between
the basal curve and the apical region.

Decurrent leaf base.– The absence or presence of decurrent leaf bases can not be relied upon as a single diagnostic characteristic in *Fontinalis*, because frequently on the same plant the leaf bases vary from scarcely decurrent to briefly or long decurrent. Occasionally a leaf base is decurrent on one side only.

Torn leaves.– Old leaves in *Fontinalis* frequently split or tear lengthwise through the center of the blade. This is especially true in the group in which the leaves are carinate-conduplicate. These leaf halves have lead to numerous incorrect determinations and occasionally have been the bases of new forms and varieties.

Leaf measurements.– In order that the records of leaf length and width may be comparable throughout the Fontinalaceae, the writer has measured medianly from base to apex for the length and across the leaf in the broadest portion for the width. The widths recorded in the conduplicate species and varieties are those of unfolded blades, computed by measuring one half of the conduplicate leaf and multiplying by two. The ratios recorded are to be interpreted, unless otherwise indicated, as length to width.

Cells: Median leaf cells.– The median leaf cells to which references are made are those located approximately half way between the middle of the leaf and the margin and between the base and the apex. The ratios recorded for cell lengths are to be interpreted, unless otherwise stated, as length to width. The ratio of length to width is expressed, for example, as 1–8 : 1, meaning 1–8 times as long as wide. Unless otherwise noted, the decrease in width of cells is very gradual from the median to the marginal.

Apical cells.– The measurements of the apical cells of the leaves are not recorded. They are commonly shorter and frequently broader than the median cells.

Alar cells.– The cells in the basal angles of the leaves which differ in size and (or) shape from the blade cells surrounding them are referred to as the alar cells. They may be essentially quadrate, rectangular, oval, or hexagonal. Since the perfect geometrical figures are not common, the author has abbreviated the descriptions by generally using the adjectives with the prefix, sub. If this alar group produces a definite outward curve in the outline of the base of the leaf, a slight outward curve, or no curve, the auricles are classified, respectively, as distinct, slight, or none. The outline of the alar cell group is generally suborbicular. If not, the variations are mentioned in the descriptions.

Perichaetial branch.– The term perichaetial branch has reference to the perichaetium together with the axis which supports it.

Trellis.– The trellis-like cone or inner peristome of *Fontinalis* is described as being either imperfect or perfect. A perfect or complete trellis is one in which transverse strands unite the cilia at regular intervals, from base to apex, into a cone-like trellis. In the incomplete or imperfect trellises, only the apices of the cilia are united by cross bars. In an imperfect trellis, the basal portion is frequently characterized by transverse strands broken or lacking. Occasionally one has difficulty in determining whether the cone is perfect or imperfect, because, first, the trellis may have fallen with the operculum, or it may have disappeared later, or it may have broken as the immature operculum became separated from the urn, and, second, a structure as fragile as the inner peristome of *Fontinalis*, subject to changes in habitat conditions such as rocks and torrents of water passing over the plants during and following heavy rains and melting snows, especially in mountain streams, may be very easily broken. If the transverse bars are not too badly damaged, one may ascertain whether or not they originally united the cilia into a complete lattice. If this condition can not be determined and if no perfect trellis has been found, the author of the original description states that the cone is imperfect. If the description of the type records the trellis as being perfect or imperfect, the writer assumes that the cone was originally complete and that those which were broken were described as the imperfect lattices. In a few instances in which the trellis is recorded in the description as being imperfect, the author has found upon examination of the type the base of the trellis or parts of it still attached to the urn and to be perfect instead of imperfect. When no trellises are exposed, frequently the operculum may be carefully removed leaving the trellis or parts of it intact so that one may discern whether or not the cross bars join the cilia at regular intervals from base to apex of the cone. However, if the capsule is immature it is sometimes difficult to remove a lattice in a condition such that completeness or incompleteness may be determined. Because of these facts, the author does not consider the imperfection or incompleteness of a trellis to be a reliable diagnostic characteristic within the genus *Fontinalis*.

Since the majority of the inner peristomes in *Fontinalis* are perfect, even some of those which were described originally as being imperfect, and since those trellises which are incomplete would have cross bars of the usual length if the broken segments were joined, the writer assumes that in *Fontinalis* the lattice cone is perfect or complete and becomes "imperfect" or "incomplete" presumably as a result of some mechanical factor. The problem can be solved only by availability of fruits in proper condition on all species and varieties. In *Brachelyma* and *Dichelyma* at the base of the cilia of imperfect trellises there are short projections but no indication that they ever united the cilia as they sometimes do at the apex of the cone.

ACKNOWLEDGMENTS

In the preparation of the Fontinalaceae of the United States and Canada for Grout's Moss Flora of North America by the writer, the need of a monographic treatment to include all known species of the Fontinalaceae became apparent, the latest comprehensive treatment of the family having been published by Cardot in 1892.

A grant from the Penrose Fund of the American Philosophical Society made possible this present study, April to September, 1938, in herbaria in the following institutions: Laboratoire de Cryptogamie, Muséum d'Histoire Naturelle, Paris, France; Rijksherbarium, Leiden, Holland; Jardin Botanique de l'État, Brussels, Belgium; Boissier and Delessert, Institut de Botanique Systématique de l'Université, Geneva, Switzerland; Botanisches Museum, Berlin-Dahlem, Germany; Botanisches Museum der Universität, Helsingfors, Finland; Naturhistoriska Riksmuseet, Stockholm, Sweden; Royal Botanic Garden, Edinburgh, Scotland; British Museum of Natural History and The Linnean Society of London, London, England; Royal Botanic Gardens, Kew, England; Oxford University, Oxford, England; and Cambridge University, Cambridge, England. Very sincere thanks to the directors and others in charge for the many courtesies extended to the author while working in these herbaria.

The publication of the monograph has been supported by a grant (G4782) from the National Science Foundation to DePauw University.

A grant from the Indiana Academy of Science secured the services of the artist. DePauw University grants defrayed certain expenses incidental to the preparation of the manuscript.

William D. Gray gave generously of his time in the preparation of the drawings. All drawings are original unless otherwise indicated. T. G. Yuncker liberally and sincerely gave counsel and criticism throughout the preparation of the manuscript. William C. Steere assisted in checking the validity of numerous obscure and questionable specific and varietal names, and gave helpful suggestions regarding the manuscript. H.S. Conard made valuable comments concerning the study. Herman Persson has given useful information and shared his collections pertaining to the transition forms from the bottom of lakes in Sweden. Frances Wynne Hillier has determined the specimens of *Drepanocladus* which have previously been recognized as *Dichelyma*. A. J. Grout, Hj. Möller, H. N. Dixon, C. Alb. Tärnlund, and J. B. Duncan shared with the author many of the Fontinalaceae in their herbaria, including some type materials. E. B. Bartram, Margaret Fulford, Mrs. Fay A. MacFadden, Cedric L. Porter, and Seville Flowers loaned the Fontinalaceae from their personal herbaria.

Librarians of numerous institutions made loans through the DePauw University Library or gave freely of their time in searching for obscure references. Many curators have made available for study the collections of Fontinalaceae in their herbaria.

Thanks are also due to Frans Verdoorn for his help in arranging for the publication of this monograph along the same lines as his former Annales Bryologici-Supplementary Volumes, also published by Messrs. Nijhoff, and for making the manuscript ready for press, with – as usual – the assistance of Mrs. Verdoorn.

The author regrets that the list of those who have assisted in various ways toward the completion of this monographic treatment is too long to be included in the acknowledgments. To the very large number who have contributed to this study in any way, the writer owes a debt of gratitude and extends very sincere appreciation and thanks.

ABBREVIATIONS OF NAMES OF HERBARIA CITED

The herbaria from which material has been available and from which specimens are cited in this study are indicated by the following abbreviations:[1]

ABS   Moss Herbarium of the American Bryological Society, Duke University Herbarium, Durham, North Carolina.
B   Herbarium of Botanisches Museum, Berlin-Dahlem, Germany.
BART   Herbarium of Edwin B. Bartram, Bushkill, Pennsylvania.
BG   Herbarium of Bergens Museum, Bergen, Norway.
BGSU   Herbarium of Bowling Green State University, Bowling Green, Ohio.
BM   Herbarium of Department of Botany of the British Museum of Natural History, London, England. (Includes Herbarium of H. N. Dixon, foreign collections.)
BOL   Bolus Herbarium, University of Cape Town, near Cape Town, South Africa.
BR   Herbarium of Jardin Botanique de l'État, Brussels, Belgium.
BUT   Herbarium of Butler University, Indianapolis, Indiana.
CAN   Herbarium of the National Museum of Canada, Ottawa, Ontario, Canada.
CAS   Herbarium of the California Academy of Sciences, The Science Museum, Golden Gate Park, San Francisco, California.
CGE   Botanical Museum and Herbarium of the University, Botany School, Cambridge, England.
CHI   Herbarium of the Chicago Natural History Museum, Chicago, Illinois. (Formerly the Field Museum of Natural History.)
CIN   Herbarium of the University of Cincinnati, Cincinnati, Ohio. (Includes Herbarium of Margaret Fulford.)
CM   Herbarium of the Carnegie Museum, The Carnegie Institute, Pittsburgh, Pennsylvania.
COLO   Herbarium of the University of Colorado, Boulder, Colorado.
CTM   Herbarium of the South African Museum, Cape Town, South Africa.
DPU   Herbarium of DePauw University, Greencastle, Indiana. (Includes Herbarium of Winona H. Welch.)
DS   Dudley Herbarium, Natural History Museum, Stanford University, Stanford, California.
DUKE   Herbarium of the Duke University, Durham, North Carolina. (Includes Herbarium of A. J. Grout.)
E   Herbarium of the Royal Botanic Garden, Edinburgh, Scotland.
FH   Farlow Herbarium of Cryptogamic Botany, Harvard University, Cambridge, Massachusetts.

---

[1] Lanjouw, J., On the Standardization of Abbreviations of herbaria for use in taxonomic publications. Chronica Botanica 3: 345–348. 1937; On the Standardization of Herbarium Abbreviations. Chronica Botanica 5: 142–150. 1939. With a few modifications, the abbreviations suggested by Lanjouw have been used.

FI   Istituto Botanico della Reale Università di Firenze, Firenze, Italy. Herbario Coloniale Florentino.

FLAS   Herbarium of Florida Agricultural Experiment Station, Gainsville, Florida.

FLSU   Herbarium of Florida State University, Tallahassee, Florida

G-BOIS   Institut de Botanique Systématique de l'Université, Herbier Boissier, Geneva, Switzerland.

G-DC   Institut de Botanique Systématique de l'Université, Herbier De-Candolle, Geneva, Switzerland.

G-DEL   Institut de Botanique Systématique de l'Université, Herbier Delessert, Geneva, Switzerland.

GRI   Herbarium of Grinnel College, Grinnell, Iowa. (Includes Herbarium of Henry S. Conard.)

H   Herbarium of Botanisches Museum der Universität, Helsingfors, Finland. (Includes the herbaria of V. F. Brotherus and S. O. Lindberg.)

HOL   Herbarium of Hollins College, Hollins College, Virginia.

ILL   Herbarium of the University of Illinois, Urbana, Illinois.

IOWA   Herbarium of the State University of Iowa, Iowa City, Iowa.

K   Herbarium of Royal Botanic Gardens, Kew, Surrey, England. (Includes H. N. Dixon Herbarium of British collections.)

L   Rijksherbarium, Leiden, Holland.

LINN   Herbarium of The Linnean Society of London, Piccadilly, London, England.

MACF   Herbarium of Mrs. Fay A. MacFadden, Los Angeles, California.

MD   Herbarium of the University of Maryland, College Park and Baltimore, Maryland.

MICH   Herbarium of the University of Michigan, Museums Building, Ann Arbor, Michigan.

MINN   Herbarium of the University of Minnesota, Minneapolis, Minnesota.

MO   Herbarium of the Missouri Botanical Garden, St. Louis, Missouri.

MT   Herbarium of the Institut de Botanique, Université de Montréal, Montréal, Quebec, Canada.

NEB   Herbarium of the University of Nebraska, Lincoln, Nebraska.

NJU   Herbarium of Rutgers University, New Brunswick, New Jersey.

NO   Herbarium of Tulane University, New Orleans, Louisiana.

NY   Herbarium of the New York Botanical Garden, New York, New York.

NYSM   Herbarium of University of the State of New York, New York State Museum, Albany, New York.

OS   Herbarium of The Ohio State University, Columbus, Ohio.

OXF   Oxford University Herbarium, Oxford, England. (Includes the Herbarium of J. J. Dillenius.)

PC   Herbarium, Laboratoire de Cryptogamie, Muséum d'Histoire Naturelle, Paris, France. (Includes the Herbarium of Jules Cardot.)

PENN   Herbarium of the University of Pennsylvania, Philadelphia, Pennsylvania.

PHIL   Herbarium of The Academy of Natural Sciences of Philadelphia, Philadelphia, Pennsylvania.

PRE   Cryptogamic Herbarium, Department of Agriculture, Pretoria, South Africa.

PSNH   Herbarium of the Portland Society of Natural History, Portland, Maine.

R   Herbarium of Museu Nacional Rio de Janeiro, Brazil, South America.

S   Herbarium of Naturhistoriska Riksmuseet, Stockholm, Sweden. (Includes the herbaria of N. C. Kindberg, Hj. Möller, and C. G. Myrin.)

SMU   Herbarium of Southern Methodist University, Dallas, Texas.

TENN   Herbarium of University of Tennessee, Knoxville, Tennessee.
TRT   Herbarium of University of Toronto, Toronto, Ontario, Canada.
TU   Herbarium of the University of Arizona, Tucson, Arizona.
UC   University of California Herbarium, Berkeley, California. (Includes Herbarium of Ira W. Clokey.)
UCCL   Herbarium of Ira W. Clokey.
UCLA   Herbarium of the University of California, Los Angeles, California.
UL   Herbarium of the University of Louisville, Louisville, Kentucky.
UMO   Herbarium of the University of Missouri, Columbia, Missouri.
UPS   Herbarium of Botaniska Museet, Uppsala University, Uppsala, Sweden.
US   United States National Herbarium, Smithsonian Institution, Washington, District of Columbia.
USC   Herbarium of the University of South Carolina, Columbia, South Carolina.
UT   Herbarium of the University of Utah, Salt Lake City, Utah. (Includes the Herbarium of Seville Flowers.)
V   Herbarium of Provincial Museum of Natural History, Victoria, British Columbia.
WASH   Herbarium of the University of Washington, Seattle, Washington. (Includes Herbarium of T. C. Frye.)
WELC   Herbarium of Wellesley College, Wellesley, Massachusetts.
WIS   Herbarium of the University of Wisconsin, Madison, Wisconsin. (Includes the herbaria of C. R. Barnes, L. S. Cheney, and L. M. Umbach.)
WJC   Herbarium of William Jewell College, Liberty, Missouri.
WVA   Herbarium of West Virginia University, Morgantown, West Virginia.
WYO   Rocky Mountain Herbarium, Laramie, Wyoming. (Herbarium of the University of Wyoming.) (Includes Herbarium of C. L. Porter.)
YALE   Herbarium of Yale University, Osborn Botanical Laboratory, New Haven, Connecticut.

Fontinalaceae Schimp.

Plants amphibious, normally submerged but frequently exposed by recession of the water, attached at base of stems by rhizoids, filiform to robust in size, yellowish green, green, olive green, grayish green, brownish green, blackish green, brownish yellow, brown, golden brown, reddish brown, copper brown, copper color, or golden; stems with central strand lacking, flaccid, subflaccid, subrigid, or rigid, short to much elongated, up to 90 cm in length, occasionally up to 150 cm long, 0.2–0.5 mm in diameter, occasionally 0.75 mm in diameter, often denuded at or near the base with age, regularly or irregularly branched, sometimes subpinnately, pinnately, bipinnately, or tripinnately divided; branches few to numerous, erect-ascending, erect-spreading, or spreading, appearing distichous in some species of *Fontinalis* in herbarium specimens, up to 15 cm in length, occasionally up to 30 cm long, distant to close, occasionally appearing subfasciculate to fasciculate, ends of foliated stems and branches acuminate, acute, or obtuse, in some species conspicuously three-angled, frequently curved to uncinate in *Dichelyma*; leaves tristichous, median cauline imbricate to distant, bases 0.25–2 mm apart, occasionally up to 3 mm apart, blades flaccid to firm, erect, erect-spreading, or spreading, often secund, subfalcate, falcate, or uncinate in *Dichelyma*, plane, subconcave, concave, subcanaliculate, or canaliculate, irregularly longitudinally folded, faintly keeled, subcarinate, or carinate-conduplicate, keel straight or slightly to strongly curved above basal curve, ecostate in *Fontinalis*, costate in *Dichelyma* and *Brachelyma* with the single midrib subpercurrent, percurrent, briefly excurrent, or long excurrent, plane or unfolded blades subulate, lanceolate-subulate, sublanceolate, lanceolate, oblong-lanceolate, ovate-lanceolate, oval-lanceolate, subelliptic-lanceolate, subovate, ovate, suboval, oval, suborbicular, subrhomboidal, or rhomboidal, apices subulate, short to long acuminate, subacute, acute, subobtuse, or obtuse, sometimes subcucullate to cucullate, occasionally subcymbiform, entire to sinuolate or serrulate, margins of apices sometimes narrowly to broadly involute, blades 2–8.5 mm long, occasionally up to 10 mm in length, 0.35–6.5 mm wide, sometimes up to 8.5 mm in width; median cells of leaves subrhombic, subrhomboidal, subhexagonal, linear-rhomboidal, linear-rhombic, or linear, ends obtuse or attenuate, frequently cells subflexuous or flexuous; marginal cells occasionally forming an indistinct to distinct border; alar cells not enlarged to much enlarged, subquadrate, subrectangular, suboval, or subhexagonal, walls hyaline, subhyaline, greenish, yellowish, yellowish brown, golden, golden brown, reddish brown, or brown; auricles distinct, slight, or none; leaf bases not decurrent, briefly decurrent, or long decurrent, occasionally subclasping or clasping; median branch leaves similar to median cauline except smaller in size; dioecious generally; male plants similar to the female in *Dichelyma*, similar or more slender in *Brachelyma*, and generally more slender in *Fontinalis*; reproductive organs borne on stems and principal branches; antheridial clusters axillary, small, numerous, bud-like in appearance, subsessile to sessile in *Fontinalis*, sessile in *Dichelyma* and *Brachelyma*; perigonial leaves ecostate, concave, oval, oval-lanceolate, or oblong, obtuse, briefly acuminate, or acuminate; archegonial clusters usually on upper portion of stem in *Brachelyma* and *Dichelyma* and on the lower part in *Fontinalis*; antheridia and archegonia numerous; paraphyses present; perichaetial branch 3–10 mm in length; perichaetium suboval, oval, oblong, subcylindrical, narrowly cylindrical, or cylindrical, 0.25–2.5 mm in diameter; perichaetial leaves imbricate, concave, ecostate, linear, narrowly lanceolate, narrowly ovate-lanceolate, broadly ovate, broadly oval-lanceolate, suboval, oval, suborbicular, or orbicular, convolute to tubular and twisted in *Dichelyma*, apices truncate, subobtuse, obtuse, subacute, acute, subapiculate, abruptly apiculate, or long acuminate, entire, subserrulate, or serrulate, lacerate with age; calyptra long conical or mitriform, dimidiate in *Dichelyma* and *Brachelyma*, in *Dichelyma* clasping seta when young; capsule usually erect, occasionally oblique in *Dichelyma*, sessile, subsessile, or on short to moderately long seta, completely immersed in perichaetial leaves or partially to completely emergent, brown, brownish yellow, or brownish orange when mature, neck, annulus, and stomata lacking; operculum obtuse conical in *Fontinalis*, rostrate in *Dichelyma* and *Brachelyma*, beak straight, oblique, or curved; seta 0.1–1.5 mm long in *Fontinalis* and *Brachelyma*, 3–21.5 mm in length in *Dichelyma*; urn erect, suboval, oval, oval-oblong, oblong-cylindrical, subcylindrical, or cylindrical, 0.65–3 mm long, 0.35–2 mm in diameter; peristome double, outer composed of teeth, 16, yellowish, brownish yellow, or brownish orange, long acuminate, subulate, linear, or linear-lanceolate, sometimes cleft or perforated along the divisural line, free or united in pairs at the apices, 0.26–1.5 mm long, smooth, finely muricate, or muricate, lamellae 8–23 in *Dichelyma* and *Brachelyma*, 11–50 in *Fontinalis*, inner of cilia, 16, alternating with the teeth, yellowish, brownish yellow, or brownish orange, free or incompletely to completely united by transverse strands into a conical trellis open at the apex, 0.4–1 mm long, finely muricate, subspinulose or spinulose, transverse bars smooth, muricate, nodulose, subappendiculate, appendiculate, subspinulose or spinulose; spores yellowish green, green, olive green, yellowish brown, or brown, smooth, submuricate, or muricate, 10–35 μ in diameter, ripe in summer.

Plants growing attached to soil, rock, stones, boulders, weeds, wood, logs, sticks, stumps, roots, and low branches of trees and shrubs, usually partially or completely submerged, in clear, muddy, fresh, stagnant, still, slowly flowing, swiftly running, shallow, or deep water, in creeks, ditches, streams, mountain rivulets, rivers, ravines, gorges, canals, lake inlets and outlets, in lakes, bays, bayous, coves, pools, ponds, sloughs, swamps, bogs, bog outlets, culverts, tanks, quarries, wells, springs, and waterfalls, and also in creek beds, bogs, marshes, pools, and ponds which are dry during a part of the year. A few collections made at edge of brackish water have been examined. The known habitats of the species of *Fontinalis*, *Dichelyma*, and *Brachelyma* are, in general, located in the temperate regions.

The common name of this family, Water Mosses, is very suitable, as the plants are either submerged all the time or through the larger portion of the year.

The Fontinalaceae are classified as pleurocarpous mosses, and more specifically, by some authors, as cladocarpous. The sporophyte terminates a short, lateral, special, fertile branch known as the perichaetial branch.

KEY TO THE GENERA

1. Leaves ecostate, from plane to concave, tubular, or carinate-conduplicate; perichaetium oval, oblong, or cylindrical; perichaetial leaves oblong, ovate, oval, oval-lanceolate, or orbicular; calyptra conical; capsule immersed in perichaetial leaves or emergent . . . . 1. *Fontinalis* (p. 18)
1. Leaves costate, always carinate-conduplicate; perichaetium always cylindrical; perichaetial leaves ovate-lanceolate, elliptic-lanceolate, or linear-lanceolate; calyptra dimidiate.
   2. Ends of foliated stems and branches conspicuously three-angled, not curved or uncinate; leaves not secund; keel straight to moderately curved, frequently abruptly curved near apex; calyptra covering operculum only; capsule completely immersed; seta 0.75–1.5 mm long . . . . 2. *Brachelyma* (p. 196)
   2. Ends of foliated stems and branches not conspicuously three-angled, commonly slightly to distinctly curved, frequently uncinate; leaves often secund and falcate; calyptra enveloping capsule; capsule emergent to surpassing perichaetium; seta 3–21.5 mm long . . . . 3. *Dichelyma* (p. 204)

Fontinalis

Fontinalis [Dill.] Hedw., Sp. Musc., p. 298. 1801.

Pilotrichum, Sect. 2. *Fontinalis* C. Müll., Syn. Musc. 2: 148. 1850 (excl. *P. fontinaloides* C. Müll. and *P. gymnostomum* C. Müll.)

Plants normally aquatic, floating, occasionally exposed by recession of the water, also on tree trunks and shrub axes in very humid atmosphere; stems up to 90 cm long, 0.2–0.5 mm in diameter; leaves tristichous, plane, subconcave, concave, subcanaliculate, canaliculate, subtubular, convolute-tubulose, subcarinate, subcarinate-conduplicate, carinate-conduplicate, keels straight to strongly curved, plane leaves sometimes asymmetrically longitudinally folded, occasionally with faint vertical ridges, ecostate, plane or unfolded blades subulate, sublanceolate, lanceolate, oblong-lanceolate, ovate-lanceolate, oval-lanceolate, subelliptic-lanceolate, subovate, ovate, suboval, oval, suborbicular, subrhomboidal, or rhomboidal, margins occasionally narrowly to broadly involute, apices short to long acuminate, subacute, acute, subobtuse, or obtuse, sometimes subcucullate to cucullate, occasionally subcymbiform, entire, sinuolate, subserrulate, or serrulate, margins of apices sometimes narrowly to broadly involute, blades 2–8 mm long, occasionally up to 10 mm in length, 0.35–6.5 mm wide, occasionally up to 8.5 mm in width, 1–8 : 1, sometimes 1–10 : 1, uncommonly 1 : 1–2; median cells of leaves subrhomboidal, rhomboidal, linear-rhomboidal, linear, ends attenuate or obtuse, frequently cells subflexuous or flexuous; alar cells enlarged, subquadrate, subrectangular, or subhexagonal; auricles none, slight, or evident; leaf bases not decurrent to long decurrent, occasionally subclasping or clasping; dioecious commonly, synoecious or monoecious occasionally; perichaetial branch 3–6 mm in length; perichaetium suboval, oval, oblong, subcylindrical, or cylindrical, 0.5–2.5 mm in diameter; perichaetial leaves oblong, ovate, suboval, oval, oval-lanceolate, suborbicular, or orbicular, apices acute, subapiculate, apiculate, subobtuse, obtuse, or truncate; calyptra long conical, 0.75–2.5 mm long; capsule sessile or subsessile, immersed in perichaetial leaves or emergent; operculum long conical, obtuse; seta short,

0.1–1 mm in length; urn suboval, oval, oval-oblong, oblong, oblong-cylindrical, subcylindrical, or cylindrical, 1.25–3 mm long, 0.35–2 mm in diameter; peristome teeth 0.45–1.5 mm in length, lamellae 11–50; cilia incompletely to completely united by transverse strands into a conical trellis, 0.4–1 mm long; spores 10–35μ in diameter. [The word, *Fontinalis*, is assumed to have had its origin in the Latin words, *fons, fontis*, meaning spring or fountain, having reference to the aquatic habitat of the plants of this genus. The Latin adjective, *fontinalis*, means of or belonging to a spring or fountain.]

KEY TO THE GROUPS AND SPECIES OF FONTINALIS

1. Leaves [1] usually carinate or carinate-conduplicate . . . .Group 1[2] (p. 22)
   2. Leaves with keels or median lines predominantly curved above basal curve.
      3. Leaves generally wider than long or as wide as long, asymmetrical, curvature of margins of blade halves if longitudinally folded rarely parallels curvature of keel, greatest curvature of keel being approximately at the middle and that of margins below the middle . . . . .
      . . . . 5. *F. antipyretica* var. *Heldreichii* (p. 71)
      3. Leaves generally longer than wide, symmetrical, curvature of margins of blade halves if longitudinally folded usually parallels curvature of keel.
         4. Ends of foliated stems and branches conspicuously elongated, triangular pyramidal in shape; keels slightly curved: leaf apices acute, entire; perichaetial leaves apiculate . . . .
         . . . . 7. *F. neo-mexicana* (p. 77)
         4. Ends of foliated stems and branches not conspicuously elongated triangular pyramidal in shape; keels moderately to strongly curved; leaf apices subobtuse to broadly obtuse; perichaetial leaves obtuse.
            5. Leaves broadly ovate or ovate-lanceolate, oval, or suborbicular, 1–1.6 : 1; keels moderately curved to almost semicircular.
               6. Plants robust; leaves 4–8 mm long, 3–6.5 mm wide, greatest width in basal half . . . .
               . . . . . . 3. *F. antipyretica* var. *gigantea* (p. 55)
               6. Plants slender; leaves 3.5–5 mm long, 2.5–5 mm wide, greatest width in middle portion . . . .
               . . . . 4. *F. antipyretica* var. *mollis*(p. 67)

---

[1] All references concerning leaves are applicable to the majority of median cauline blades unless otherwise stated. The leaf shapes are those of unfolded blades. Statements in keys refer to the usual or most common conditions, exceptions and variations being recorded in the description.

[2] Cardot in his monograph on Fontinalaceae arranged the species of *Fontinalis* under six sections: Section 1. *Tropidophyllae* Card., Section 2. *Heterophyllae* Card., Section 3. *Lepidophyllae* Card., Section 4. *Malacophyllae* Card., Section 5. *Stenophyllae* Card., and Section 6. *Solenophyllae* Card. In the opinion of the writer these groups do not appear to merit sectional rank. But, having been established, the author used them in a previous publication. However, because of the overlapping of sectional characteristics, necessitating the placement of some species in more than one section, it is now considered better to replace Cardot's sections by the groupings as used in this study.

  5. Leaves ovate- to oval-lanceolate, 1.5–3.5 : 1; keels slightly to moderately curved.
   6. Plants medium in size; leaves 3–8.5 mm long, 2–4 mm wide, 1.5–3.5 : 1; median cells frequently linear-rhomboidal . . .
   . . . .   1. *F. antipyretica* (p. 22)
   6. Plants slender; leaves 3–4 mm long, 1.5–2.2 mm wide, 1.7–2 : 1; medium cells frequently linear and flexuose . . . .
   . . . . 2. *F. antipyretica* var. *gracilis* (p. 46)
2. Leaves with keels or median lines predominantly straight above basal curve.
  3. Plants medium to robust; leaves ovate- to broadly ovate-lanceolate, 1.5–3 : 1.
   4. All leaves conduplicate, 5–7 mm long, 1.5–3.5 mm wide; keels sometimes slightly concave; apices long and narrowly acuminate
   . . . .   9. *F. Howellii* (p. 89)
   4. Some leaves carinate and open, or concave, 4–6 mm long, 1.5–2.5 mm wide; keels frequently abruptly curved at apex; apices briefly and broadly acuminate, often subconcave to plane . . .
   . . . .   8. *F. patula* (p. 84)
  3. Plants slender to medium; leaves narrowly ovate-lanceolate to lanceolate, 2–8 : 1.
   4. Leaves sublanceolate, 3.5–6 mm long, 0.75–2 mm wide, 2–4.5 : 1, occasional blades antipyretica-like in shape and size; apices obtuse . . . . .   6. *F. antipyretica* var. *oreganensis* (p. 73)
   4. Leaves narrowly lanceolate, 4–5.5 mm long, 0.5–1 mm wide, 6–8 : 1, antipyretica-like leaves absent; apices long acuminate
   . . . .   10. *F. chrysophylla* (p. 95)
1. Leaves usually concave . . . .   Group 2. (p. 98)
  2. Apices sometimes subcarinate to carinate.
   3. Apices frequently subcarinate to carinate, broadly obtuse . . . . .
   . . . .   22. *F. Allenii* (p. 152)
   3. Apices occasionally subcarinate to carinate near one margin, sometimes one side narrowly and briefly involute, now and then one margin plane and oblique, frequently twisted, acuminate to acute
   . . . .   21. *F. Mac-Millanii* (p. 150)
  2. Apices plane, concave, canaliculate, or tubular.
   3. Marginal cells abruptly narrower than adjacent cells and with thicker walls, forming indistinct to distinct border in some portion of mature blades.
    4. Apices of perichaetial leaves broadly obtuse to truncate . . .
    . . . .   11. *F. squamosa* (p. 98)
    4. Apices of perichaetial leaves apiculate or acute . . . .
    . . . .   12. *F. squamosa* var. *Curnowii* (p. 105)
   3. Marginal cells gradually narrower than adjacent cells and with thickness of walls approximately the same as in adjoining cells.
    4. Leaves erect to slightly spreading, appearing to be appressed.
     5. Apices of perichaetial leaves acute or apiculate . . . . . .
     . . . .   13. *F. dalecarlica* (p. 106)
     5. Apices of perichaetial leaves broadly obtuse . . . .
     . . . . . .   14. *F. dalecarlica* var. *Macounii* (p. 121)
    4. Leaves erect-spreading to spreading, not appearing to be appressed.
     5. Margins frequently involute.
      6. Margins in apical portions commonly narrowly involute.
       7. Blades usually firm, ovate-lanceolate, 2.5–4 mm long, 1–2 mm wide . . .   15. *F. novae-angliae* (p. 121)
       7. Blades usually flaccid, broadly ovate-lanceolate, 4–7.5 mm long, 1.5–3.5 mm wide . . . .
       . . . .   17. *F. novae-angliae* var. *latifolia* (p. 140)

6. Margins in apical portions frequently broadly involute, the broad involution sometimes extending to base of blade, occasionally margins convolute and blades subtubular to tu-' bular.

 7. Apices long acuminate; apical cells rhombic, rhomboidal, quadrate, rectangular, or hexagonal; blades narrowly ovate-lanceolate. 23. *F. biformis* (aestival leaves) (p. 155)

 7. Apices generally obtuse; apical cells linear or linear-rhomboidal.

  8. Apices obtuse to truncate, margins gradually narrowing to tips; blades ovate- to oblong-lanceolate or lanceolate. . . 16. *F. novae-angliae* var. *cymbifolia* (p. 136)

  8. Apices obtuse or abruptly narrowed into acute or obtuse tips; blades narrowly lanceolate. . . . . 24. *F. Langloisii* (p. 158)

5. Margins not involute.

 6. Apical cells rhomboidal, rhombic, quadrate, rectangular, or hexagonal; apices narrowly to broadly obtuse; blades broadly ovate-lanceolate or lanceolate . . . . 23. *F. biformis* (vernal leaves) (p. 155)

 6. Apical cells linear or linear-rhomboidal.

  7. Leaves generally obtuse.

   8. Apices broadly obtuse to truncate, sometimes incurved; blades ovate-, oblong-, or subelliptic-lanceolate . . . . 19. *F. bogotensis* (p. 145)

   8. Apices broadly obtuse or abruptly narrowed into acute or obtuse tips; blades narrowly lanceolate . . . . 24. *F. Langloisii* (p. 158)

  7. Leaves acute, acuminate, or subulate.

   8. Apices acuminate or acute, usually entire; alar group often rectangular in outline, parallel with leaf margin; alar cells in 4 or 5 vertical rows. 18. *F. Bryhnii* (p. 141)

   8. Apices acuminate, frequently serrulate; alar group subcircular in outline.

    9. Stems flaccid; leaves usually subflaccid, lanceolate to ovate-lanceolate . 20. *F. missourica* (p. 147)

    9. Stems rigid; leaves firm, narrowly lanceolate.

     10. Blades concave throughout or subconcave at base and plane above, occasionally canaliculate to subtubular; apices narrowly acuminate; leaves 0.75–1.5 mm wide . . . . 25. *F. disticha* (p. 161)

     10. Blades concave or deeply concave to convolute tubulose; apices long acuminate or subulate; leaves 0.35–0.5 mm wide . . . . 26. *F. filiformis* (p. 166)

1. Leaves usually plane . . . . Group 3 (p. 170)

2. Leaves generally broadly ovate-lanceolate or oval-lanceolate, 1–2.5 mm wide; margins tapering from approximate middle into apex; apices short and broadly acuminate; auricles frequent . . . 28. *F. Duriaei* (p. 178)

2. Leaves generally narrowly ovate-lanceolate or lanceolate, 0.5–1.75 mm wide; margins tapering from basal fourth or half into apex; apices long, acuminate.

 3. Apices gradually narrowed; tips commonly acute and entire, auricles usually none, occasionally very slight . . 27. *F. hypnoides* (p. 170)

 3. Apices frequently abruptly narrowed; tips commonly obtuse to truncate and serrulate; auricles very conspicuous . . . . 29. *F. flaccida* (p. 191)

GROUP 1: LEAVES CARINATE-CONDUPLICATE

Leaves generally carinate-conduplicate, but intermixed with these blades may be those which are subcarinate-conduplicate, carinate throughout but not conduplicate, subcarinate, deeply concave, irregularly folded, or plane.

1. **Fontinalis antipyretica** Hedw., Sp. Musc., p. 298. 1801.

Fontinalis trifaria Voit, Hist. Musc., p. 125. 1812. [On basis of description; type not seen.]
Pilotrichum antipyreticum C. Müll., Syn. Musc. 2: 148. 1850.
Fontinalis californica Sull., in Senate Rep. U.S. Pacific R.R. Expl. and Surv. 4: 189. 1856.
Fontinalis antipyretica var. *crassa* Mol., in Lorentz, Beitr. Biol. u. Geogr. Laubm., p. 21. 1860. [On basis of inadequate description; type not seen.]
Fontinalis antipyretica var. *montana* H. Müll., Westf. Laubm. No. 378, [brief description printed on label]. Year ?; H. Müller, in Verh. Nat. Ver. Rheinl. u. Westf. 24 Jahrg. 3 Folge, 4: 138. 1867.
Fontinalis antipyretica var. *alpestris* Milde, Bry. Siles., p. 276. 1869. [On basis of inadequate description; type not seen.]
Fontinalis laxa Milde, Bry. Siles., p. 276. 1869. [1] [On basis of description; type not seen.]
Fontinalis androgyna Ruthe, in Hedwigia 11: 166. 1872.
Fontinalis abyssinica Schimp., Syn. Musc., p. 556. 1876 (nomen nudum).
Fontinalis longifolia Jens., in Bot. Not., p. 83. 1885. [On basis of description and fragments of type.]
Fontinalis arvernica Ren., in Rev. Bryol. 15: 69. 1888.
Fontinalis antipyretica var. *californica* (Sull.) Lesq., in Cardot, in Rev. Bryol. 18: 82. 1891.
Fontinalis antipyretica subsp. *arvernica* (Ren.) Card., in Rev. Bryol. 18: 82. 1891.[2]

---

[1] Some authors give this citation as *Fontinalis laxa* Milde; others use *F. antipyretica* var. *laxa* Milde. It is difficult to ascertain Milde's meaning. He refers to a third remarkable form from Sagan and from Hamburg. It may be assumed that the word, form, in this instance, is a general reference to the plants and is not of taxonomic significance. Milde suggests that this form may become a species after the fruits are found and proposes for it the name of *Fontinalis laxa*, indicating specific rank of *laxa*, but does not number it as is his custom with regard to species. Also, one may infer from the position of the description and from his introductory statement that *laxa* might be a variety of *F. antipyretica*. The author interprets *F. laxa* in this reference to indicate specific rank.

[2] Cardot, in Rev. Bryol. 18: 82–83. 1891, ranked *F. Kindbergii*, *F. arvernica*, *F. neo-mexicana*, *F. columbica*, *F. Delamarei*, *F. dalecarlica*, *F. Cardoti*, *F. nitida*, and *F. tenella* as subspecies. In the same publication, pp. 84–86, these plants were treated as species. Perhaps he regarded it unnecessary to repeat

Fontinalis hypnoides C. J. Hartm., f. *androgyna* (Ruthe) Card., in Rev. Bryol. 18: 83. 1891.
Fontinalis islandica Card., in Rev. Bryol. 18: 84. 1891 [in descriptive key]; Cardot, Mon. Font., p. 70. 1892; Cardot, in Rev. Bryol. 18: 82. 1891 (nomen nudum).
Fontinalis antipyretica f. *tenuis* Card.,.Mon. Font., p. 50. 1892.
Fontinalis antipyretica f. *diffusa* Card., Mon. Font., p. 51. 1892.
Fontinalis antipyretica f. *imbricata* Card., Mon. Font., p. 51. 1892.
Fontinalis antipyretica var. *pseudosquamosa* Card., Mon. Font., p. 52. 1892.
Fontinalis antipyretica f. *dunensis* Card., Mon. Font., p. 146. 1892.
Fontinalis antipyretica var. *arvernica* (Ren.) Husn., Musc. Gall., p. 286. 1892.
Fontinalis antipyretica var. *laxa* (Milde) Limpr., Laubm. 2: 655. 1894.
Fontinalis Jacquini L. herb., according to Limpricht, Laubm. 2: 656. 1894 (nomen nudum).
Fontinalis antipyretica f. *laxior* Schimp., according to Limpricht, Laubm. 2: 656. 1894 (nomen nudum).
Fontinalis minor L. herb., according to Limpricht, Laubm. 2: 656. 1894 (nomen nudum).
Fontinalis antipyretica f. *minor* Schimp., according to Limpricht, Laubm. 2: 656. 1894 (nomen nudum).
Fontinalis antipyretica subsp. *californica* (Sull.) Kindb., in Can. Rec. Sci. 6: 75. 1894.
Fontinalis hypnoides C. J. Hartm. subsp. *longifolia* (Jens.) Kindb., Sp. Eur. and N. Am. Bryin., Part 1: 147. 1896.
Fontinalis dolosa Card., in Rev. Bryol. 23: 68. 1896.
Fontinalis antipyretica subsp. *dolosa* (Card.) Dixon, Stud. Handb., p. 355. 1896.
Fontinalis cavifolia Warnst. and Fleisch., in Fleischer and Warnstorf, in Bot. Centralbl. 65: 300. 1896.
Fontinalis thulensis Jens., in Bot. Tid. 20: 110. 1896. [On basis of description and fragments of type.]
Fontinalis antipyretica subsp. *islandica* (Card.) Kindb., Sp. Eur. and N. Am. Bryin., Part 1: 149. 1896.
Fontinalis antipyretica subsp. *androgyna* (Ruthe) Kindb., Sp. Eur. and N. Am. Bryin., Part 1: 149. 1896.
Fontinalis antipyretica var. *azorica* Card., in Ann. Rep. Mo. Bot. Gard. 8: 66. 1897.
Fontinalis antipyretica var. *laxa* (Milde) Limpr. f. *robustior* Fleisch. and Warnst., in Hedwigia (Beibl.) 36: 74. 1897 (nomen nudum).
Fontinalis antipyretica var. *ligurica* Fleisch., in Fleischer and Warnstorf, in Hedwigia (Beibl.) 37: 141. 1898.
Fontinalis antipyretica var. *cymbifolia* Nicholson, in Journ. Bot. 39: 427. 1901.
Fontinalis stagnalis Kaal., in Nyt. Mag. Nat. 40: 259. 1902.
Fontinalis antipyretica var. *pseudohypnoides* Velenov., in Rozpr. České Akad. Císaře Františka Josefa p. Vědy, Slovesn. Umění 12: 13. 1903. [On basis of description; type not seen.]
Fontinalis gracilis Lindb. var. *Grebeana* Roth, Eur. Laubm. 2: 282. 1904.
Fontinalis antipyretica f. *alpina* Card., in Warnstorf, Laubm., p. 626. 1905.
Fontinalis antipyretica var. *yezoana* Card., in Bull. Soc. Bot. Genève (Sér. 2) 1: 131. 1909; in Brotherus, in Hedwigia 38: 225. 1899 (nomen nudum).

---

the indication of subspecies. In Mon. Font., pp. 23–24. 1892, Cardot listed the members of *Fontinalis* in places of first, second, third, and fourth order. Since those listed above were not in first rank, one could assume that this use of subspecies was indicative of subordinate position. However, he listed others in lower rank which were not designated as subspecies in Rev. Bryol. 18: 82–83. 1891. In Cardot's Monographie des Fontinalacées, in 1892, all of the above subspecies were treated as species with the exception of *F. columbica* which was reduced to a variety of *F. neo-mexicana*. In consideration of the above data, the author has given Cardot's subspecies a literal interpretation.

24 THE FONTINALACEAE

Fontinalis cavifolia Warnst. and Fleisch. var. *rhenana* Roth, in Hedwigia 49: 221. 1910.

Fontinalis Lachenaudi Card., in Coppey, in Rev. Bryol. 38: 119. 1911.

Fontinalis fasciculata Lindb. var. *danubica* Card., in Familler, Denkschr. König.-Bayer. Bot. Gesell. Regensburg 11: 2. 1911.

Fontinalis antipyretica var. *pseudo-kindbergii* Card., in Zahlbruckner, Ann. K. K. Naturhist. Hofmus. 27: 278. 1913.

Fontinalis antipyretica f. *vulgaris* Mönkem., in Pascher, Süsswasserfl. 14: 102. 1914.

Fontinalis antipyretica f. *laxa* (Milde) Mönkem., in Pascher, Süsswasserfl. 14: 103. 1914.

Fontinalis antipyretica f. *cymbifolia* (Nicholson) Mönkem., in Pascher, Süsswasserfl. 14: 104. 1914.

Fontinalis antipyretica f. *alpestris* (Milde) Mönkem., in Pascher, Süsswasserfl. 14: 105. 1914.

Fontinalis antipyretica f. *montana* (H. Müll.) Mönkem., in Pascher, Süsswasserfl. 14: 105. 1914.

Fontinalis antipyretica f. *pseudosquamosa* (Card.) Mönkem., in Pascher, Süsswasserfl. 14: 105. 1914.

Fontinalis antipyretica var. *laxa* (Milde) Limpr. f. *paroica* Mönkem., in Pascher, Süsswasserfl. 14: 107. 1914.[1] [On basis of description; type not seen.]

Fontinalis antipyretica var. *mollissima* Warnst., in Bryol. Zeitschr. 1 : 38. 1916. [On basis of description; type not seen.]

Fontinalis antipyretica var. *danubica* (Card.) Warnst., in Bryol. Zeitschr. 1: 40. 1916.

Fontinalis antipyretica var. *Lachenaudii* (Card.) Warnst., in Bryol. Zeitschr. 1: 41. 1916.

Fontinalis fasciculata Lind. f. *danubica* (Card.) Familler, in Krypt. Forsch. 3: 167. 1918.

Fontinalis antipyretica f. *funiculata* Hj. Möll., in Ark. Bot. 17 (14): 24. 1922.

Fontinalis antipyretica f. *latifolia-tenuis* Fuchs., in Internat. Rev. Hydrobiol. u. Hydrogr. 12: 201. 1924. [On basis of description; type not seen.]

Fontinalis antipyretica f. *tenuis-latifolia* Fuchs., in Internat. Rev. Hydrobiol. u. Hydrogr. 12: 201. 1924. [On basis of description; type not seen.]

Fontinalis antipyretica f. *tenuis-vulgaris* Fuchs., in Internat. Rev. Hydrobiol. u. Hydrogr. 12: 202. 1924. [On basis of description; type not seen.]

Fontinalis antipyretica f. *lacustris* Fuchs., in Internat. Rev. Hydrobiol. u. Hydrogr. 12: 202. 1924. [On basis of description; type not seen.]

Fontinalis hypnoides C. J. Hartm. f. *obtusifolia* Thér., in E. Bauer, Musci Eur. et Am. Exsic., Ser. 39, No. 1932. 1927 [printed description on label]; E. Bauer, Sched. u. Bemerk. z. 39 Ser., p. 9. 1927.

Fontinalis ligurica (Fleisch.) Mönkem., Laubm. Eur. Erg.-Bd. 4: 660. 1927.

Fontinalis antipyretica f. *fasciculata* (Lindb.) Mönkem., Laubm. Eur. Erg.-Bd. 4: 665. 1927.

Fontinalis antipyretica f. *flagellacea* Podp., in Zprávy komise na přírodovědecký výzkum Moravy a Slezska. Oddělení botanické Č. 9: 16. 1932. [On basis of description; type not seen.]

Fontinalis antipyretica var. *erythraea* Tong. f. *robusta* Tong., in Nuov. Giorn. Bot. Ital. (N.S.) 45: 400. 1938.

Fontinalis antipyretica var. *erythraea* Tong. f. *laxa* Tong., in Nuov. Giorn. Bot. Ital. (N.S.) 45: 400. 1938.

Fontinalis gothica Card. and Arn. var. *stagnalis* (Kaal.) Jens., Skand. Bladmossfl., p. 379. 1939.

Fontinalis antipyretica var. *ambigua* Biz. and Hill., in Rev. Bryol. et Lichénol. (Nouv. Sér.) 15: 71. 1945.

---

[1] See page 22, note 1.

Fontinalis antipyretica subsp. *vulgaris* (Mönkem.) Giacom., in Atti Ist. Bot. Univ., Pavia, Lab. Crittog. (Ser. 5) 4 (2): 250. 1947.

Fontinalis antipyretica subsp. *vulgaris* (Mönkem.) Giacom. var. *montana* H. Müll., in Giacomini, in Atti Ist. Bot. Univ., Pavia, Lab. Crittog. (Ser. 5) 4 (2) 250. 1947.

Fontinalis pseudosquamosa (Card.) Giacom., in Atti Ist. Bot. Univ., Pavia, Lab. Crittog. (Ser. 5) 4 (2): 250. 1947.

Fontinalis Kindbergii Ren. and Card. subsp. *dolosa* (Card.) Giacom., in Atti Ist. Bot. Univ., Pavia, Lab. Crittog. (Ser. 5) 4 (2): 250. 1947.

Fontinalis squamosa Hedw. var. *corsica* Card., in Giacomini, in Atti Ist. Bot. Univ., Pavia, Lab. Crittog. (Ser. 5) 4 (2): 250. 1947 [1] (nomen nudum).

Fontinalis antipyretica subsp. *vulgaris* (Mönkem.) Giacom. f. *diffusa* Card., in Podpěra, Consp. Musc. Eur., p. 505. 1954.

Fontinalis antipyretica subsp. *vulgaris* (Mönkem.) Giacom. f. *flagellacea* Podp., Consp. Musc. Eur., p. 505. 1954.

Fontinalis antipyretica subsp. *vulgaris* (Mönkem.) Giacom. f. *funiculata* Hj. Möll., in Podpěra, Consp. Musc. Eur., p. 505. 1954.

Fontinalis antipyretica subsp. *vulgaris* (Mönkem.) Giacom. f. *imbricata* Card., in Podpěra, Consp. Musc. Eur., p. 505. 1954.

Fontinalis antipyretica subsp. *vulgaris* (Mönkem.) Giacom. f. *lacustris* Fuchs., in Podpěra, Consp. Musc. Eur., p. 505. 1954.

Fontinalis antipyretica subsp. *vulgaris* (Mönkem.) Giacom. f. *laxa* Mönkem., in Podpěra, Consp. Musc. Eur., p. 505. 1954.

Fontinalis antipyretica subsp. *vulgaris* (Mönkem.) Giacom. f. *mollissima* (Warnst.) Podp., Consp. Musc. Eur., p. 505. 1954.

Fontinalis antipyretica subsp. *vulgaris* (Mönkem.) Giacom. f. *pseudohypnoides* (Velenov.) Podp., Consp. Musc. Eur., p. 505. 1954.

Fontinalis antipyretica subsp. *vulgaris* (Mönkem.) Giacom. f. *tenuis* Card., in Podpěra, Consp. Musc. Eur., p. 505. 1954. [Cited erroneously as var. *tenuis* Card. and f. *tenuis* Mönkem.]

Fontinalis antipyretica subsp. *vulgaris* (Mönkem.) Giacom. var. *alpestris* Milde, in Podpěra, Consp. Musc. Eur., p. 505. 1954.

Fontinalis antipyretica subsp. *vulgaris* (Mönkem.) Giacom. var. *montana* H. Müll., Podpěra, Consp. Musc. Eur., p. 505. 1954.

Fontinalis antipyretica subsp. *vulgaris* (Mönkem.) Giacom. var. *pseudosquamosa* Card., in Podpěra, Consp. Musc. Eur., p. 505. 1954.

Fontinalis antipyretica subsp. *vulgaris* (Mönkem.) Giacom. var. *cymbifolia* Nicholson, in Podpěra, Consp. Musc. Eur., p. 506. 1954.

Fontinalis antipyretica subsp. *vulgaris* (Mönkem.) Giacom. var. *danubica* (Card.) Warnst., in Podpěra, Consp. Musc. Eur., p. 506. 1954.

Fontinalis antipyretica subsp. *vulgaris* (Mönkem.) Giacom. var. *ligurica* Fleisch., in Podpěra, Consp. Musc. Eur., p. 506. 1954.

Fontinalis antipyretica subsp. *vulgaris* (Mönkem.) Giacom. var. *arvernica* (Ren.) Husn., in Podpěra, Consp. Musc. Eur., p. 506. 1954.

Fontinalis antipyretica subsp. *vulgaris* (Mönkem.) Giacom. var. *ambigua* Biz. and Hill., in Podpěra, Consp. Musc. Eur., p. 506. 1954.

Fontinalis antipyretica var. *ambigua-Lachenaudi* Dismier, in Podpěra, Consp. Musc. Eur., p. 506. 1954 (nomen nudum). [Name on label.]

Fontinalis antipyretica subsp. *gracilis* (Lindb.) Kind. var. *Grebeana* Roth, in Podpěra, Consp. Musc. Eur., p. 506. 1954.

Fontinalis antipyretica subsp. *gothica* (Card. and Arn.) Podp. f. *stagnalis* (Kaal.) Podp., Consp. Musc. Eur., p. 507. 1954.

Fontinalis antipyretica subsp. *cavifolia* (Warnst. and Fleisch.) Podp., Consp. Musc. Eur., p. 507. 1954.

---

[1] The author saw a manuscript description of var. *corsica* Card. in Cardot's Herbarium on a label with plants which are *F. antipyretica* Hedw.

Fontinalis antipyretica subsp. *Lachenaudi* (Card.) Podp., Consp. Musc. Eur., p. 507. 1954.
Fontinalis pseudo-Durieui Card., in Jelenc, Musc. Afrique Nord, p. 117. 1955.
Fontinalis constantinica Card., in Jelenc, Musc. Afrique Nord, p. 118. 1955.
Fontinalis antipyretica var. *constantinica* Card., in Jelenc, Musc. Afrique Nord, p. 118. 1955 (nomen nudum). [From herbarium label.]
Fontinalis antipyretica var. *occidentalis* Card., in Gaume, Cat. Musc. Bretagne, p. 71. 1956 (nomen nudum).

Plants generally medium in size but varying from slender to slightly robust, usually dark green, sometimes pale green, yellowish green, olive green, pale brownish green, copper brown, or golden brown, either glossy or dull when dry; stems usually slightly rigid but occasionally somewhat flaccid, up to 80 cm in length, 0.3–0.5 mm in diameter, denuded and blackish at base with age, irregularly pinnately branching; branches erect, erect-spreading, or spreading, up to 15 cm in length, varying from close to distant, ends of foliated stems and branches obtuse, acute, or acuminate, with leaves loosely to tightly erect-imbricate; bases of median cauline leaves usually rather distant and 1–2 mm apart but occasionally up to 2.5 mm distant, sometimes closely imbricate and up to 0.5 mm apart, blades commonly firm but occasionally flaccid, generally erect-spreading but sometimes spreading, carinate-conduplicate with keels usually moderately curved but at times varying from straight to slightly curved to moderately curved on same plant, sometimes plane or asymmetrically longitudinally folded or faintly keeled if flaccid, occasionally split along the keel, ovate-lanceolate or oval-lanceolate, margins on one side frequently reflexed near base, sometimes both sides reflexed, blades symmetrical with curvature of the margins of the leaf halves when folded lengthwise usually paralleling the curvature of the keels; apices short to long acuminate, leaf tips generally subobtuse to obtuse, sometimes acute, usually serrulate, occasionally entire; median cauline blades 3–8.5 mm long, 2–4 mm wide, occasionally up to 4.5 mm in width, 1.5–3.5 : 1; median cells of leaves usually linear-rhomboidal, occasionally rhomboidal, or linear with ends attenuate or obtuse, 10–20.5 μ wide, 4–15 : 1; alar cells enlarged, subrectangular to subquadrate or subhexagonal, vertical rows of cells in alar group occasionally up to 13, walls yellowish, yellowish green, yellowish brown, or brown, generally forming distinct auricles, leaf bases varying from not decurrent to long decurrent on same axis; median branch leaves comparable with median cauline except smaller in size and sometimes varying from carinate-conduplicate to concave to plane; perichaetial branch 4.5–5 mm in length, perichaetium oval, suboval, or subcylindrical, 0.75–2 mm in diameter; upper perichaetial leaves oval to suborbicular, obtuse, entire, usually lacerate with age; calyptra long conical, 1.5 mm in length, 0.75–1 mm in diameter; operculum short conical, apex obtuse, 0.7–1.25 mm in length, 0.75–1.5 mm in diameter; seta 0.25–0.5 mm long; urn usually immersed but occasionally slightly to half emergent, suboval to subcylindrical, usually contracted beneath mouth when dry, 1.5–3 mm long, 0.75–2 mm in diameter, 1.6–2.6 : 1; peristome teeth brownish

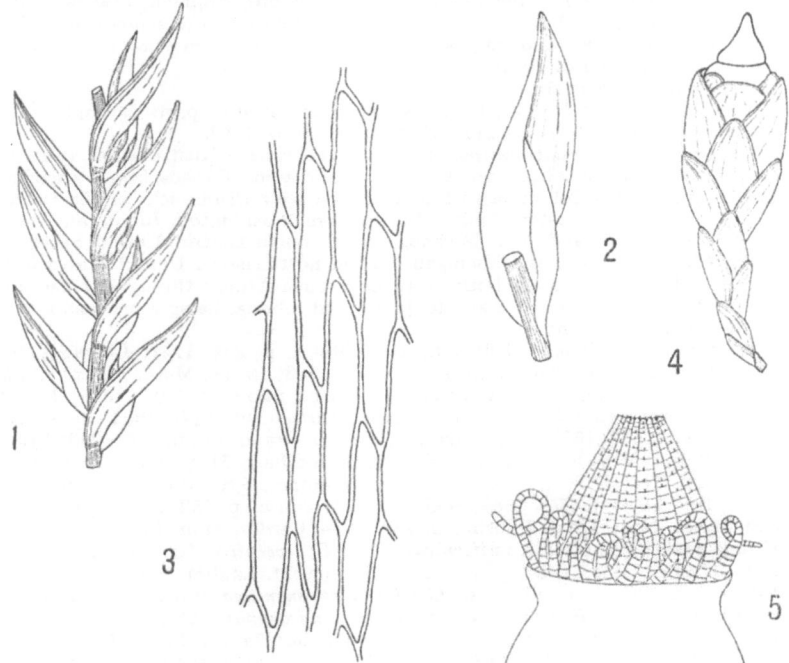

FIG. 1*. – *Fontinalis antipyretica* Hedw. – 1.Portion of stem with median cauline leaves. 2.Median cauline leaf. 3.Median cells of leaf. 4.Perichaetial branch and portion of capsule. 5. Portion of urn with inner and outer peristomes. Nos. 1, 2, 4, and 5 drawn after Bruch and Schimper, Bry. Eur. pl. 429; no. 3 from A. Nelson 7708, Wyoming.

orange, linear, long acuminate, frequently united in pairs at apex, 0.75–1 mm in length, muricate, lamellae 25–35; trellis of inner peristome perfect, brownish orange, 0.75–1 mm long, muricate, transverse bars subappendiculate to appendiculate; spores pale green, olive green, or yellowish green, submuricate to muricate, 13.5–18 μ in diameter, ripe in summer. (According to Bryologia Europaea, the specific name, *antipyretica*, given to this moss by Linnaeus, owes

* The portions of the plants illustrated in the figures are shown enlarged.

its origin to the usage by the peasants in Sweden of surrounding the chimneys of their houses with this plant to prevent fire. In Die Laubmoose, Limpricht states that the name, *antipyretica* (against fire), has reference to the old Germanic custom of filling the cracks of the log houses with this moss, whereby, according to popular belief, conflagration was supposed to be prevented.

Type: Not previously designated. The specimens in the J. J. Dillenius Collection in the Oxford University Herbarium, Oxford, England, first described and illustrated by Dillenius as *Fontinalis triangularis major complicata, e foliorum alis capsulifera,* and later known as *Fontinalis antipyretica* Hedw., have been assumed to be the type.

Type collector: Not known.

Type locality: "In Brooks, Rivulets and Ponds in most parts of England, all Seasons," according to Dillenius, Hist. Musc., p. 6. 1763.

Distribution: Fontinalis antipyretica has a wide range of distribution. It occurs in North America, in Alaska and Greenland, and across Canada from Vancouver Island and British Columbia to Labrador and New Brunswick; in the United States, throughout western United States from Washington, Idaho, and Montana to California and New Mexico, also in north central United States in Minnesota, Wisconsin, and Michigan, and in northeastern United States from Maine to Pennsylvania: in both north and south Africa; throughout Europe; and in Asia from Asiatic Turkey to Japan and Siberia, being more common in the northwestern countries.

Additional descriptions: Dillenius, Hist. Musc., p. 254. 1741; Linnaeus, Sp. Pl. 2: 1107. 1753; Linnaeus, Sp. Pl. 2: 1571. 1763; Bridel, Musc. Recent. 2 (3): 157. 1803; Hedwig, Schwaegrichen, Sp. Musc., Suppl. 1: 307. 1816; Bridel, Musc. Recent. Suppl. 3: 107. 1817; Sturm, Deutsch. Fl. 4 (2): 15. 1818; Bridel, Bry. Univ. 2: 655. 1827; C. J. Hartman, Handb. Skand. Fl., p. 434. 1843; Bruch and Schimper, Bry. Eur., Fasc. 16, 5: 4. 1842, and Fasc. 31, 5: 5. 1846; Sullivant, Musci and Hep. U.S., p. 654 (54). 1856; Schimper, Syn. Musc., p. 456. 1860; Milde, Bry. Siles., p. 275. 1869; Schimper, Syn. Musc., p. 552. 1876; Lesquereux and James, Man. Moss. N. Am., p. 268. 1884; Cardot, Mon. Font., pp. 48 (as *F. antipyretica*), 54 (as var. *californica*), 57 (as *F. arvernica*), 106 (as *F. longifolia*), 1892; Husnot, Musc. Gall., p. 285. 1892; Limpricht, Laubm. 2: 652 (as *F. antipyretica*), 654 (as var. *alpestris*), 655 (as vars. *montana* and *laxa*), 656 (as *F. arvernica*), 657 (as *F. islandica*), 662 (as *F. androgyna*), 665 (as *F. longifolia*). 1894; Fleischer and Warnstorf, in Hedwigia 36: 74 (*as F. cavifolia*). 1897; Limpricht, Laubm. 3: 802 (as *F. dolosa*), 803 (as *F. stagnalis* and *F. cavifolia*). 1903; Roth, Eur. Laubm. 2: 277 (as *F. antipyretica*), 278 (as vars. *alpestris* and *montana*), 279 (as var. *laxa* and as *F. arvernica*), 280 (as *F. dolosa*), 281 (as *F. islandica*), 284 (as *F. stagnalis*), 285 (as *F. androgyna*), 286 (as *F. longifolia*), 288 (as *F. cavifolia*), 289 (as *F. thulensis*). 1904; Braithwaite, Br. Moss-Fl. 3: 210 (as *F. antipyretica*), 211 (as var. *cymbifolia*), 212 (as *F. dolosa*). 1905; Warnstorf, Laubm. 2: 631 (as *F. laxa*). 1905; Mönkemeyer, in Pascher, Süsswasserfl. 14: 102 (as *F. antipyretica* and as *F. tenuis*), 105 (as *F. cavifolia* and as *F. cavifolia* var. *rhenana*), 106 (as *F. androgyna*). 1914; Hj. Möller, in Ark. Bot. 17 (14): 21 (as *F. antipyretica*), 32 (as var. *alpestris*), 33 (as var. *montana*), 35 (as var. *pseudosquamosa*), 68 (as *F. androgyna*). 1922; Brotherus, Laubm. Fenn., pp. 394 (as *F. antipyretica*), 395 (as vars. *alpestris, montana,* and *pseudosquamosa*), 397 (as *F. androgyna*). 1923; Jensen, Danm. Moss., pp. 192 (as *F. antipyretica*), 193 (as vars. *alpestris* and *laxa*). 1923; Dixon, Handb. Br. Moss., pp. 390 (as *F. antipyretica* and as var. *cymbifolia*), 391 (as *F. dolosa*). 1924; Luisier, Musci Salmant., p. 198 (as *F. antipyretica* and as var. *alpestris*). 1924; Mönkemeyer, Laubm. Eur. Erg.-Bd. 4: 655 (as *F. antipyretica*), 657 (as f. *tenuis,* f. *vulgaris,* f. *laxa,* and f. *cymbifolia*), 658 (as f. *alpestris,* f. *montana,* f. *pseudosquamosa,* and

F. arvernica), 659 (as *F. androgyna*), 660 (as var. *laxa*, f. *paroica*, *F. cavifolia* var. *rhenana*, *F. cavifolia*, *F. ligurica*, *F. dolosa*, and *F. islandica*), 665 (as *F. longifolia*). 1927; Welch, in Grout, Moss Fl. N. Am. 3: 235. 1934; Jensen, Skand. Bladmossfl., pp. 377 (as *F. antipyretica*, as f. *tenuis*, f. *imbricata*, f. *diffusa*, f. *alpestris*, f. *montana*, f. *pseudosquamosa*, and f. *laxa*), 380 (as *F. androgyna*). 1939; Jennings, Man. Moss. W. Penn. and Adj. Reg., p. 174. 1951.

Illustrations: Dillenius, Hist. Musc., pl. 33, fig. 1. 1741 and 1763; Sturm, Deutsch. Fl. 4 (2): pl. 24. 1818; Bruch and Schimper, Bry. Eur., Fasc. 16, 5: pl. 429. 1842; Sullivant, Musci and Hep. U.S., pl. 18. 1856; Schimper, Syn. Musc., pl. 5. 1860; Lesquereux and James, Man. Moss. N. Am., pl. 4. 1884; Husnot, Musc. Gall., pl. 80 (as *F. antipyretica* and as var. *arvernica*). 1892; Limpricht, Laubm. 2: figs. 325 and 326. 1894; Braithwaite, Br. Moss Fl., pl. 122 (as *F. antipyretica*), pl. 123 (as *F. dolosa*). 1905; Brotherus, in Engler and Prantl, Nat. Pflanz.: Bryophyta 2: fig. 544. 1905; Roth, in Hedwigia 53: pl. 3 (as *F. Lachenaudi*). 1913; Mönkemeyer, in Pascher, Süsswasserfl. 14: fig. 32. 1914; Hj. Möller, in Ark. Bot. 17 (14); figs. 1–2 (as *F. antipyretica*), fig. 3 and pl. 1² (as var. *alpestris*), fig. 4 and pl. 2³ (as var. *montana*), fig. 5 and pl. 2⁴ (as var. *pseudosquamosa*), figs. 29–31 and pl. 8¹⁶ (as *F. androgyna*). 1922; Brotherus, Laubm. Fenn., fig. 68. 1923; Dixon, Handb. Br. Moss., pl. 48c (as *F. antipyretica*), pl. 48d (as *F. dolosa*). 1924; Brotherus, in Engler and Prantl, Nat. Pflanz., Musci 11 (2): fig. 474. 1925; Mönkemeyer, Laubm. Eur. Erg.-Bd. 4: fig. 143a (as *F. antipyretica* and as f. *vulgaris*), fig. 145b (as f. *fasciculata*). 1927; Welch, in Grout, Moss Fl. N. Am. 3: pl. 74. 1934; Tongiorgi, in Nuov. Giorn. Bot. Ital. (N. S.) 45: 401, fig. 1 (as var. *erythraea*, f. *laxa*), fig. 2 (as var. *erythraea* f. *robusta*). 1938. – FIGURE 1.

Specimens examined: AFRICA: ABYSSINIA: Schimper 1055, Nov. 13, 1852, Débra Càna, alt. 2790 m (BM, H, PC); Steudner, Jan. 15, 1862, near Ghaba (H).

ALGERIA: Delestre in 1845, Tiaret (BM); Durieu (BM); Durieu, LaCalle (PC); Durieu, May 1840, Constantine (K); Flagey, Constantine (PC); Gandoger, Feb. 1880, Kabylie, Fl. Alger. Exsic. 1801 (BM); Trabut, Kabylie (PC); Trabut, Constantine (PC).

CAPE COLONY: Barnard 35628, July 1927, Stellenbosch, Jonkers Hoek (CTM, DPU); Wicht, Feb. 2, 1944, Stellenbosch, Jonkers Hoek (PRE).

ERITREA: Terracciano and Pappi 375, Feb. 11, 1893, Amasen, Sciumma-Negus, alt. 2200 m (TYPE of *F. antipyretica* var. *erythraea* f. *laxa*, FI; DPU); Terracciano and Pappi 4663, Feb. 18, 1893, Amasen, Amba Derò, alt. 2400 m (TYPE of *F. antipyretica* var. *erythraea* f. *robusta*, FI; DPU).

MOROCCO: Gattefossé, July 1932, alt. 2400 m (DPU).

TUNISIA: Labbé, May 5, 1945 (DPU); Pitard, Feb. 1907, Plantae Tunetanae 55, Tunis (PC); Pitard, Feb. 1907, alt. 800 m, Plantae Tunetanae 56 (FH, PC).

ASIA: ASIA MINOR: Bornmüller, June 6, 1899, Anatolia, Phrygia, Iter Anatolicum III. 3613 (G-DEL. PHIL); Schekovnikov, June 1916, Kurdistan (H); Von Handel-Mazzetti 821, in 1907, (Sandschak) Trapezunti, alt. 600–800 m (H).

JAPAN: Faurie 3104, in 1904 (H, NY); Miyabe 263, Aug. 9, 1891, Ishikari, Yezo (Assumed to be TYPE of *F. antipyretica* var. *yezoana*, PC; FH, H, K); Okamura 805, 1–10–1909, Kanazawa (H).

PERSIA: Bornmüller, Feb. 3, 1892, Teheran, alt. 1500 m, Iter Persico-turcicum 4477 (G-BOIS, G-DEL, K, PC, PHIL); Stapf, May 23, 1885, South Persia (H, K).

SYRIA: Haradjian 3916, July 1911, Amanus, alt. 700–1200 m (K); Postian, July 1891, Amanus (G-BOIS); Postian, July 1891, Amanus, Plantae montium Syriae borealis 141 (PC).

UNION OF SOVIET SOCIALIST REPUBLICS: Alexandrov, Sept. 2, 1911, Siberia (H); Arnell, June 30, 1876, Siberia, Yenisei (H, PC); Arnell, July 12, 1876, Siberia, Yenisei (PC); V. F. Brotherus, May 29, 1896, Turkestan (H); Fedtschenko, Mar. 7, 1869, Turkestan, Samarkand (K); Fedtschenko 3081, Mar. 7, 1869, Turkestan, Samarkand (H); Fedtschenko, Aug. 23, 1902, Tien-Shan (FH, H); Kuschakevitsch 3090, Apr. 25, 1878, Turkestan (H); Kuschakevitsch 3097, June 5, 1878, Turkestan (H); Kuznetsov 1357, July 24, 1910, Siberia (H); Lazarenko, June

1928, Tien-Shan, alt. 1500 m, Verdoorn, Musci Selecti et Critici 23 (CM, DUKE, MICH, NY, UC, WASH, WIS, YALE); Martianov, Aug. 1888, Siberia, Yenisei R. (H); Martianov, Aug. 20, 1888, Siberia (H); Martianov, 1894–98, Siberia (H); Regel 3289, in 1876, Turkestan, alt. 1200–1800 m (H); Regel 3339, Aug. 1876, Turkestan (H); Regel 3356, July 4, 1877, Turkestan, alt. 2400–2700 m (H); Regel 3354, Aug. 6, 1877, Turkestan, alt. 600–900 m (H); Regel, Mar. 1878, Turkestan, alt. 1500 m (FH, H, K); Regel 3352, Mar. 13, 1878, Turkestan, alt. 1500 m (H); Schinschkin, Sept. 5, 1924, Siberia (H); Schipchinsky, June 1, 1914, Semipalatinsk (H); Timofeyev, Aug. 20, 1907, Siberia, Yenisei (H); Tolmatschev, July 16, 1908, Siberia (H).

EUROPE: ALBANIA: Baldacci 164, Aug. 9, 1894 (CHI, G–DEL, K, PC).

AUSTRIA: Diettrich-Kolkhoff, Oct. 1911, southern Tirol, alt. 90 m (PC); Eiben 34 (NY); Heimerl 73, Aug. 22, 1934 (WASH); Loitlesberger, Dec. 1906, Küstenland, Bauer, Musci Eur. Exsic. 551 (BART, DPU, PC); L. Müller, Aug. 1931, Lunz, alt. 600 m (E); K. and L. Rechinger, Styria, near Aussee, submerged about 3 m, Krypt. Exsic. 2091 (G–BOIS, K, NY, PC, US); Richter, lower Austria, near Gloggnitz, alt. 400 m, Kerner, Fl. Exsic. Aust.-Hungar. 1110 (FH, G–BOIS, K, MINN, PC, US).

AZORES: V. and P. Allorge, July 12, 1937, Ribeira da Cruz (DPU, PC); V. and P. Allorge, July 22, 1937, Ribeira, alt. 600 m (DPU, PC); Godman, Flores (NY); Godman, in 1865, Flores (K); Trelease, Aug. 8, 1894 (NY); Trelease 1367, Aug. 8, 1894, Flores (Assumed to be the TYPE of *F. antipyretica* var. *azorica*, PC; MO); C. Watson, in 1842 (NY).

BELGIUM: Bouly de Lesdain, Sept. 1, 1912 (WASH); Cardot (PC); Cardot, Mar. 8, 1884 (PC); Cornet, July 1899, bed of the Lomme (BR); Cornet, Apr. 1910, Theux (BR); Demaret 1546a, Mehaigne R. at Moha (BR, DPU); Douret, Apr. 21, 1884, Rouge-Cloitre (BR); Gravet, June 1868, Namur, Louette-Saint-Pierre, Ardennes (BR); Gravet, Apr. 1872 (BR); Gravet, Aug. 25, 1872 (G–DEL, NY); Gravet, June 1874, Namur, Louette-Saint-Pierre (H); Gravet, June 1874, Namur, Louette-Saint-Pierre, Bryoth. Belg. 283 (BR, K, NY); Gravet, May 6, 1886, Namur, Louette-Saint-Pierre (BR, L); Gravet, May 19, 1886, Louette-Saint-Pierre (WIS); Henrotay, Sept. 11, 1863 (PC); Mansion, Oct. 10, 1890, Fléron (BR); Mansion , Oct. 13, 1890, Fléron (BR); Mansion, June 28, 1893 (BR); Mansion 306, May 28, 1901, Amblève Sougnez (BR); Michot, near Mons (BR, FH, K, NY); Mitten, in 1877, Amblève R. (NY); Mosseray and LeBrun, Aug. 10, 1933 (BR); Naveau, Sept. 29, 1908, Anvers (ILL); Piré, May 17, 1885 (BR, FH); Troch, June 1885 and 1886 (BR); Vandenbroeck, Sept. 28, 1881, Anvers (L); Vandenbroeck, in 1882, Anvers (PC); Vandenbroeck, Apr. 12, 1882, Anvers (BR); Vandenbroeck, Sept. 15, 1882, Anvers (BR).

BULGARIA: Podpěra, July 16, 1908, near Bystrica (MT); Podpěra, July 17, 1908, alt. 1800 m (S).

CANARY ISLANDS: Wells, in 1906 or 1907, La Palma, alt. 900–1800 m (BM, DPU).

CORSICA: Camus, May 23, 1901, Foce de Vizzavona, alt. 1000–1100 m (PC); Camus, June 3, 1901, Foce de Vizzavona, alt. 1200 m (PC); Kralik, July 29, 1849 (G–DEL, K).

CZECHOSLOVAKIA: D. Bauer, Aug. 27, 1889, Bohemia, Flora von Böhmen 88 (S); D. Bauer, July 18, 1890, Bohemia, Flora von Böhmen 125 (S); E. Bauer, Oct. 21, 1897, Bohemia, alt. 700 m (PC, S); E. Bauer, in 1898, Bohemia (PC); E. Bauer, Sept., 1898, Bohemia, alt. 500 m, Bryoth. Bohem. 343 (FH, PC, WASH); E. Bauer, June 14, 1903, Bohemia, alt. 1400 m, Bryoth. Bohem. 341 (Assumed to be the TYPE of *F. antipyretica* f. *alpina*, PC; S. WASH); E. Bauer, Sept. 3, 1911, Bohemia, Riesengebirge, alt. 1400 m, Musci Eur. Exsic. 1153 (BART, L, S, WASH); Deschner, Nov. 1899, Bohemia, Bauer, Bryoth. Bohem. 142 (PC, S, WASH); Matouschek, northern Bohemia, near Machendorf, in the "Jeschkengebirge" (Assumed to be the TYPE of *F. antipyretica* var. *pseudo-kindbergii*, PC); Matouschek, northern Bohemia, near Machendorf, in the "Jeschkengebirge,"

Krypt. Exsic. 2092 (K, NY, PC, S, US); Petrak, Oct. 1911, Bohemia, alt. 300 m, Fl. Bohem. et Morav. Exsic. 32 (BR); Pilous, Oct. 8, 1933, near Obecnice, alt. 570 m, Kavina and Hilitzer, Crypt. Čechoslov. Exsic. 193 (NY); Podpěra, June 1908, Petrak, Fl. Bohem. et Morav. Exsic. 33 (BR, FH, S); Prager, July 16, 1908, Bohemia, Riesengebirge, alt. 1400 m, E. Bauer, Musci Eur. Exsic. 560 (BART, PC); Wilms, July 24, 1882, Tatra (DPU, S).

DENMARK: Holmberg, Aug. 9, 1896, Bornholm Isl. (S); Jensen, Aug. 1884 (PC); Kopsch, Aug. 1913, Bornholm Isl. (WASH); Leth, Sept. 22, 1870, Jylland (UC); F. Möller, Aug. 20, 1863, Jylland (S); Mönkemeyer, July 1910, Bornholm Isl. (DPU, S); Nyman, Sept. 6, 1886, Bornholm Isl. (S); Nyman, July 21, 1888, Bornholm Isl. (S); Pedersen, July 4, 1888 (S).

ENGLAND: J. G. Baker, Sept. 1870 (US); Barker, July 7, 1899, Derbyshire (UC); Bellerby, May 27, 1926, Yorkshire, Malham Cove (H); Binstead, Feb. 26, 1896, Herefordshire, R. Wye (DPU, K); Boswell, in Thames R. near Oxford (FH); Braithwaite, Hebden Valley (E); Braithwaite, Oct. 1891, Yorkshire (BM); Brocas, in 1851, Huntingdonshire, Winchester (BR); Brocas, in 1853, Surrey (BR); Catcheside, June 1924, Surrey Co., Mitcham, Wandle R. (CGE, DPU); Catcheside, Aug. 13, 1924, Devon Co., Doone Valley (CGE, DPU); Catcheside, May 16, 1925, Surrey Co., Mortlake, bank of Thames R. (CGE, DPU); Catcheside, Sept. 25, 1926 (WASH); Catcheside, Sept. 4, 1927, Yorkshire, Malham Tarn, alt. about 225 m (CGE, DPU); Catcheside, Sept. 24, 1927, Yorkshire, Malham, Gordale Beck, alt. 225 m (CGE, DPU); Catcheside, Aug. 28, 1940, Westmorland Co., near Grasmere, Rothay R. (CGE, DPU); Catcheside, Sept. 6, 1940, Cumberland Co., Loweswater (CGE, DPU); H. N. Dixon, Yorkshire, Malham Cove (BM); H. N. Dixon, Aug. 17, 1895, Cumberland (DUKE); H. N. Dixon, July 23, 1896, Yorkshire, Malham Cove (BM, DPU, DUKE); H. N. Dixon, Aug. 1901, Huntingdonshire, R. Ouse at Hemingford Grey (BM, E, K, S); Dyer, July 30, 1867, Twickenham (K); Wm. Green, Aug. 1878, Gloster (OS); Hamilton, Shrewsbury, in small stone cistern (K); Hamilton, Aug. 5, 1897, Monkmoor (K); Hamilton, May 12, 1898, Owestry (K); Hamilton, May 12, 1898, Shrewsbury (K); Hamilton, July 1898, Monkmoor (K); Hamilton, in 1904 (K); Hobson, Musci Britain (K); Hunt, May 17, 1803 (H); Hunt, May 18, 1803, Cheshire (H); Hunt, in 1869, Derbyshire, Repton (BM); Ingham, Mar. 6, 1901, Yorkshire, Ackworth (E); Mosely, Oct. 4, 1909, Test R. (BM); Nicholson, Aug. 1900, Sussex, Ouse R. near Lewes (BM, CGE); Nicholson, Aug. 19, 1900, Sussex Co., near Hamsey, R. Ouse, "Vide Journal of Botany, December 1901" (Assumed to be the TYPE of F. antipyretica var. cymbifolia, CGE; DPU); Nicholson, Nov. 1901, Sussex, R. Ouse near Lewes (FH, H, S, WASH); Nicholson, Aug. 1910, East Sussex, R. Ouse near Barcombe Mills (DPU); Nowell, May 31, 1856, Yorkshire (BM); Rhodes, June 1909, North West Yorkshire, alt. 360 m (BM); Saunders, "In aqua stagnante, ad ligna submersa," Limbury, Bedfordshire ("Fontinalis dolosa Cardot sp. nova," with fruit, assumed to be the TYPE of F. dolosa, PC); Saunders, Limbury Ponds, South Bedfordshire ("Ex herb. H. N. Dixon" as F. dolosa Card., S); Saunders, submerged logs, Limbury Ponds, South Bedfordshire, ex herb. H. N. Dixon and part of type of F. dolosa Card., according to H. N. Dixon, DPU); Saunders, in 1882, Limbury Beds ("Original Specimen" of F. dolosa Card., K); Saunders, in 1882, submerged logs, Limbury ponds, Bedfordshire (" 'Type specimen' of F. dolosa Card.," with fruit, BM); Saunders, in 1882, submerged logs, Limbury, South Bedfordshire, ponds ("Original specimen ex herb. H. N. Dixon" of F. dolosa Card., E); Saunders, Aug. 6, 1882, Limbury, Bedfordshire (From herb. H. N. Dixon as F. dolosa Card., DUKE); [1] Saunders, in 1886, Limbury (BM); Saunders, May 1896, Limbury, Bedfordshire (NY); Shacklette 2316, Feb. 14,

---

[1] The author assumes that the specimens from the herbarium of H. N. Dixon, those indicated as "Original Specimen," as "Type Specimen," and as "sp. nova" are portions of the type, as Cardot states with the original description, "Angleterre: In aqua stagnante ad ligna submersa. Limbury, Bedfordshire. (Leg. James Saunders; comm. H. N. Dixon)."

1943, Devon Co., near Weston Manor and Honiton, near R. Otter (DPU); Sherrin, July 21, 1919, Dorsetshire, Morton Heath (S); Sutton, Aug. 9, 1917, Thames R. near Egham (BM); E. C. Wallace, Oct. 13, 1926, Surrey (H); E. C. Wallace, Sept. 17, 1938, North Wiltshire (BART, DUKE, MICH); E. C. Wallace, Apr. 1, 1948, Derbyshire, Lathkill Dale (DPU); W. Watson 451d, Oct. 1931, West Monkton, Somerset (BART, MICH); Welch 5568, Aug. 17, 1938, lawn of South London Botanical Institute (CHI, DPU); W. Wilson, July 1842, Mere, Musci Britannici 442 (BR, MO, NY, YALE); W. Wilson, in 1826, Cheshire (K); C. A. Wright, Sept. 1883 (BM).

ESTONIA: Niclasen, June 6, 1911, near Jäätmaa, Mikutowicz, Bryoth. Balt. 294a (S, US).

FAEROE ISLANDS: Jensen, May 8, 1896, Süderö Isl. (H); Simmons 536, Aug. 2, 1895, Süderö Isl., alt. 50 m (S); Simmons 471, Aug. 19, 1895, Osterö Isl. alt. 25 m (MO, PC).

FINLAND: Blom and Rengriod, Jan. 1, 1876, Tavastland (H); Blom and Renquist, Sept. 1, 1876, Tavastland (H, S); Bomansson, in 1864, Alandia (S); V. F. Brotherus, in 1868, near Helsinki (NY); V. F. Brotherus, June 1871, Nylandia (NY); V. F. Brotherus, July 5, 1871, Nylandia (S); V. F. Brotherus, July 16, 1871, near Helsinki, Musci Fenn. Exsic. 22 (K, NY); V. F. Brotherus, July 1872, Lapponia (DPU); V. F. Brotherus, Aug. 1872, Lapponia (PC); V. F. Brotherus 182, Aug. 1872, Lapponia (H); V. F. Brotherus, in 1887, Lapponia murmanica, Kola Expeditionen (H); V. F. Brotherus, in 1887, Kola Expeditionen, Musci Fenn. Exsic. 133 (H); V. F. Brotherus, Aug. 4, 1887, Lapponia, Kola Expeditionen (H); V. F. Brotherus, Aug. 4, 1887, Lapponia murmanica, Kola Expeditionen (H); Buch, July 21, 1911, Nylandia, Brotherus, Bryoth. Fenn. 358a (CHI, S, YALE); Collander, Sept. 1938, Snappertuna, South Finland, in brackish water (approximately 0.25% salt) (DPU); Collin, July 13, 1878, Nylandia (S); Erlandsson, July 9, 1936, Tetsamo (S); Kihlman, July 5, 1887, Lapponia murmanica, Kola Expeditionen (S); Kihlman, July 25, 1887, Lapponia imandrensis, Kola Expeditionen (S); Kihlman 522, Aug. 6, 1892, Lapponia imandrensis (S); Lang, Oct. 22, 1898, Tavastland (S); H. Lindberg, Nov. 5, 1900, Helsinki (S); H. Lindberg, Nov. 5, 1900, Helsinki, E. Bauer, Musci Eur. Exsic. 555 (BART, H, PC, S); S. O. Lindberg, June 30, 1868, Helsinki (K, US); S. O. Lindberg, July 12, 1868, Nylandia (H. MINN); Palmén, Aug. 1865, Tavastland (K); Renquist, July 28, 1875, Tavastland (H, MO, S); Wegelius, Aug. 12, 1913, Tavastia australis, Brotherus, Bryoth. Fenn. 358b (DPU, DUKE, H, YALE); Welch 5660, July 8, 1938, near Helsinki (CHI, DPU, MICH, NY).

FRANCE: P. Allorge, July 7, 1929, Parc-Marsinvast, Cherbourg, Manche (L); Beauvard, Mar. 20, 1904, near Ferney (G-BOIS); Beauvard, Sept. 21, 1912, Haute-Saône (G-DEL); Becker, July 8, 1866 (PC); Becker, May 30, 1867, Haut-Rhin (G-DEL); Bernet, Sept. 19, 1884, Haute-Savoie, alt. 2000 m (G-DEL); Bimont 4304, Apr. 1922, Manche, Saint-Michel-aux-Loups (MT, UCLA); Binstead, Apr. 1930, Cavalaire (BART); Bizot, June 27, 1929, Côte-d'Or, La Courtavaux, Dismier, Bryoth. Gall. 309 (Assumed to be a portion of TYPE of F. antipyretica var. ambigua Biz. and Hill., BART, CM, YALE); Bizot, July 1934, Côte-d'Or (MICH); Blot, July 1907, Lac Pavin, Puy-de-Dôme (S); Bonpland, Fontainebleau (PC); Bouly de Lesdain, Aug. 1895, Nord (H); Bouly de Lesdain, Aug. 1896, Nord, near Dunkerque (PC); Bouly de Lesdain, Aug. 1896, Nord, near Dunkerque, Bauer, Musci Eur. Exsic. 562 (BART, L, PC, S); Bouly de Lesdain, Aug. 24, 1896 (DUKE); Bouly de Lesdain, Apr. 19, 1897 (PC); Bouly de Lesdain, Aug. 12, 1899, Hautes-Pyrénées (PC); Bouly de Lesdain, Sept. 20, 1910, Nord (PC); Bouvet, Maine-et-Loire, Angers, Husnot, Musci Gall. 87 (CHI, DUKE, FH, K, NY, PC, WASH, WIS); Briquet, Sept. 6, 1906, Flore des Alpes Lémaniennes (G-DEL); Brunard, Sept. 1909, Lac Pavin (S); Bureau and Camus, Apr. 14, 1892, Loire-Inférieure (PC); Bureau and Camus, Mar. 25, 1894, Loire-Inférieure (PC); Camus, Sept. 10, 1890 (PC); Camus, Apr. 16, 1892, Loire-Inférieure (CHI, PC); Camus, Mar. 4, 1894, Loire-Inférieure (PC); Camus, Apr. 1, 1894 (PC); Camus, in 1895 (PC); Camus, July 21, 1900 (PC); Camus, Nov. 8, 1904, Belle-Ile-en-Mer (PC); Camus, Dec. 14, 1904, Ile d'Yen (PC); Chassagne, Aug. 1912, Auvergne, Lac Pavin (DUKE, G-DEL, L, MT); Chassagne, July 1913, Lac Pavin (S); Coppey, Aug. 17, 1909, Haute-Saône (PC); Coppey, Sept. 1910, Haute-Savoie (DUKE); Coppey, Sept. 14, 1910, Haute-Saône (H, PC); Corbière, Haute-Vienne (PC); Corbière, Sept. 15, 1886, Cherbourg (PC); Crozals, Mar. 1902, central France (PC); De Brébisson 51, Normandy (BR, G-DEL); Déséglise, July 30, 1855, forest of Allogny (G-DEL); Desmazières, in 1825, Plantes Cryptogames de France 97 (G-DEL, K, NY, PC); Dismier, in 1897 (PC); Dismier, June 2, 1927, Borne, Bryoth. Gall. 299 (BART, WASH, YALE); Dufour, Landes, Saint Sever (PC); Dufour, Pyrénées (H); Dufour, in 1914 (PC); Flagey, June 1880 (PC); Gaudefroy, May 6, 1869, near Mandreuse, Seine and Oise (NY); Giraudias, May 1884, Charente-Inférieure, Aulnay (CHI, MT); Girod, Apr. 1893 (G-DEL); Girod, Sept. 1901 (G-DEL); Girod, Aug. 1903, Ain, Hauteville, Albarine R., alt. 800 m (PC); Girod 727, Aug. 1903, Ain (G-DEL); Girod 727b, Aug. 1903, Albarine R., near Hauteville (G-DEL); Guinet, July 31, 1906, Sabaudia, alt. 1900 m (G-DEL); Guinet, Aug. 2, 1906, Sabaudia, alt. 2100 m (G-DEL); Héribaud, in 1887, Puy-de-Dôme, Lake Pavin (NY); Héribaud, Sept. 5, 1887, Puy-de-Dôme, Lake Pavin (MO); Héribaud, Sept. 25, 1887, Puy-de-Dôme, Lake Pavin, alt. 1200 m (G-BOIS, MINN, PC); Héribaud, in 1888, Auvergne, Lake Pavin (K); Héribaud, Sept. 1892, Puy-de-Dôme, Lake Pavin (PC); Héribaud, Sept. 8, 1896, Lake Pavin (PC); Héribaud, Sept. 1902, Puy-de-Dôme, Lake Pavin (NY); Héribaud, Sept. 7, 1906, Puy-de-Dôme, Lake Pavin (NY, S); Héribaud, Biélawski, and Gonod d'Arte-mare, Aug. 3, 1891, Puy-de-Dôme (PC); Hommey, in 1889, Orne, Sees (G-BOIS, MT); Husnot, Gen. Musc. Eur. Exsic. 74 (CHI); Husnot, Gard, Musci Gall. 832 (CHI, FH, K, PC, WASH, WIS); Jeanpert, Sept. 24, 1893, near Dol (CHI); Jeanpert, Dec. 16, 1894, St. Maur, in Marne R. (CHI); Jeanpert, May 14, 1896 (CHI); Jeanpert, Oct. 7, 1896, Emerainville (CHI); Jeanpert, Mar. 19, 1899, forest of Bougeaux (CHI); Jeanpert, Aug. 2, 1901 (CHI); Lachenaud, July 29, 1898, Haute-Vienne (PC); Lachenaud, Aug. 17, 1898, Haute-Vienne, forest of La-vergne, near St. Priest-Ligoure (Assumed to be the TYPE of F. Lachenaudi, PC; H, S); Langeron and Sullerot, June 1898, Côte-d'Or, Exsic. Musc. Côte-d'Or 526 (PC); Libert, in 1814, Charente Inférieure, St. Bonnet (BR); Madiot, Mar. 31, 1882, Haute-Saône (BR); Maire, Loire (CHI); Maire, in 1848, Loiret (CHI); Mougeot and Nestler, in 1812, Stirpes Cryptogamae Vogeso-Rhenanae 238 (CHI, FH, G-DEL, K, NY); Renauld, Arvernia, Lake Pavin, alt. 1200 m (WIS); Renauld, Arvernia, Lake Pavin, Renauld and Cardot, Musci Eur. Exsic. 30 (CHI, DUKE, FH, MINN, PC, UC, WIS, YALE); Renauld, Sept. 1886, Puy-de-Dôme, Lake Pavin (PC); Renauld, Oct. 1886, Puy-de-Dôme, Auvergne, Lake Pavin, alt. 1200 m (Assumed to be the TYPE of F. arvernica, PC); Renauld and Paillot, Nov. 6, 1873, "Arcier (D)," alt. 250 m, Vendrely, Fl. Crypt. Sequaniae Exsic., 1895, 138 (PC); Renou, in 1857 (PC); Roussel, in 1817 (PC); Roze and Bescherelle 93 (G-DEL); Schimper, "in rivulis Vogesi" (H, PC); Schimper, in 1840, "Vogesi" (NY); Schimper, in 1844, "in rivulis Vogesi" (G-BOIS); Schimper, in 1845, "in rivulis Vogesi" (PC); Thériot, May 11, 1889 (G-BOIS, PC); Thériot, May 12, 1889, Sarthe (G-BOIS, PC); Vigineix, July 22, 1855, near Vichy (G-DEL); Walker, in 1884, Nord (PC).

GERMANY: Andres, June 1921 and 1922, Westerwald, alt. 435 m, Wirtgen, Flora Rhenanae 32 (BR); E. Bauer, June 11, 1903, Riesengebirge (PC); Baum-gartner, July 6, 1911 (FH); W. Baur, July 15, 1891 (G-BOIS); Bergner, July 1921, Rhön (BART, MT, UC, WASH); Bertram, Aug. 1874, Westfalen (NY); Brendel, in 1844, Erlangen (FH); Breutel 41, Harz Mts. (UC); Breutel 184, Königshayn and Berthelsdorf (NY, PC, UC); Cassebeer, in 1832 and 1841, Wetterauische Laub-moose 40 (BR, NY); Chen, July 30, 1937, Ilsenburg, Harz (DPU); DeHaller 195, Hanover, (DeHaller, 1758–1828) (G-DEL); Dietrich (TRT); Familler, Aug. 1911, lower Bavaria, near Neuburg, Danube R. (PC); Familler, Aug. 1911, lower Bavaria, near Neustadt, alt. 350 m, Fl. Exsic. Bavar: Bry. 26b (Assumed to be a portion of the TYPE of *F. fasciculata* var. *danubica*, PC, S); Familler, Sept. 1911, lower Bavaria, near Neustadt, alt. 350 m, Bauer, Musci Eur. Exsic. 1155 (BART, G-DEL, L, S, WASH); Fischer, in 1902, Mossflora von Bamberg, Regnitz-Altung 587a (CAS); Fleiden and Berneek, Apr. 25, 1843 (NY); Freiberg, Sept. 4, 1909, Westpreusen (BR); Freiberg, May 29, 1910, Westpreussen, near Waltersdorf (H); Freiberg, Aug. 17, 1910, Westpreussen, near Wahlendorf (H); Freiberg, Aug. 20, 1910, Westpreussen (PC); Freiberg, Aug. 20, 1910, Westpreussen, near Neustadt (H); Freiberg, Aug. 1912, Ostpreussen, Allenstein (MT, S); Freiberg, Mar. 28, 1913, Ostpreussen, Allenstein, in water 2–3 m deep, Bauer, Musci Eur. Exsic. 1154 (BART, G-DEL, L, S, WASH); Funck, Bavaria, Cryptogamische Gewächse des Fichtelgebirges 76 (NY); Garrigus, July 22, 1849, near Ostiode (CHI, MICH); Gäumann, Aug. 1907, Rheinland, Ostenland, near Paderborn (PC); Geheeb, Rhön (PC); Geheeb, in 1872, Rhöngebirge (BR, PC); Glowacki, Sept. 10, 1897, Steirmark, near Römerbad (DPU); Goldmann, Rhein (PC); Goldmann, May 1906, Prussia, Ostenland, near Paderborn (S); Goldmann, July 1907, Hessen (DPU); Goldmann, Sept. 1908, near Biebrich, Rhine R. (L); Goldmann, Aug. 1908 and 1909, near Biebrich (S); Goldmann, Sept. 1908, Hessen, Biebrich on the Rhein, Bauer, Musci Eur. Exsic. 557 (BART, H, PC); Goldmann, June 1910, "an der Schwimmschule zu Biebrich am Rhein" (DUKE, MICH, S); Goldmann, June 1910, Nassau, Biebrich (Assumed to be portion of TYPE of *F. Rhenana*, "*F. Rhenana* G. Roth. TYPUS. Det. G. Roth.", S; DPU); Gräbner, Aug. 1893, Pommerania (MINN); Grebe, Reinhardswald (PC); Grebe, Oct. 1899, Reinhardswald (Assumed to be a portion of the TYPE of *F. gracilis* var. *Grebeana*, S); Grebe, Aug. 1900, Reinhardswald (PC, S); Grebe, July 1904 (H, S); Grebe, Aug. 1904, Hessen, Forsthaus Mariendorf, Bauer, Musci Eur. Exsic. 1159 (BART, L, WASH); Hallier, Apr. 9, 1892, near München (L); Heimerl, July 22, 1934 (FH); Hellsing, May 27, 1916, Taunus Mts., Musci Germanici 160 (S); Hellsing, July 21, 1917, Musci Germanici 466 (S); Hespe 242a, Aug. 31, 1903, Oldenburg (WIS); Holler 350, Aug. 11, 1883 (FH); Jonas, May 1930, Flora Rhenanae 534 (BR); Jungling, 1868–70 (CHI); Kaulfuss, Mar. 27, 1891, Bäyern, near Nürnberg (S); Kern, July 1886, Riesengebirge, alt. 1300 m (PC); Kopsch, Sept. 1921, Saxony, Erzgebirge, near Tellerhäuser, alt. 930 m, Bryoth. Sax. 258 (BM, MICH); Kopsch, July 1922, Erzgebirge, near Tellerhäuser, alt. 850 m, Bryoth. Sax. 257 (BM, MICH); Kopsch, July 1925, Leipzig, near Dölzig, Bryoth. Sax. 256 (BM, MICH); Krause, Aug. 1861, Dresden (MO); Krause, Sept. 1863 (MO); Leiner, in 1858, Constanz, Leiner and Stizenberger, Kryptogamen Badens 236 (CHI, K, NY, YALE); Limpricht, May 6, 1868, Silesia, near Bunzlau, Bryoth. Siles. 234 (NY, PC); Loeske, May 19, 1901, Flora Marchica (MINN); Luse, Aug. 21, 1903, Waldeck, Pyrmont (DPU, S); Luse, Aug. 26, 1906, Bayern, Berchtesgaden (DPU, S); Luse, Sept. 4, 1911 (DPU, S); Molendo, Franconia (ILL); Molendo, Oct. 12, 1867, Monte-Ponifero, Ruhlberg, near Bedwiz (FH); Mönkemeyer, July 1902, Fichtelgebirge (PC); Mönkemeyer, July 1903, Baiern, Fichtelgebirge, Bauer, Musci Eur. Exsic. 490 (BART, G-DEL, PC); Mönkemeyer, July 1905, Rhön (PC); Mönkemeyer, July 1906, Rhön (BART, DUKE, MT, PC); Mönkemeyer, July 1906, Hessen, Rhön, Bauer, Musci Eur. Exsic. 496 (BART, PC); H. Müller, Lippstadt, Westfalens Laubmoose 27 (K, NY, PC); H. Müller, near Niedersfeld, Westfalens Laubmoose 378 (Assumed to be a portion of the TYPE of *F. antipyretica* var. *montana*, FH, L, S); Prager, June 1886, Potsdam (R); Prager, June 23, 1887, Potsdam (MINN); Prasch, in 1920, in Rhine R. (DPU, DUKE); Progel, June 1851, upper Bavaria, München (NY); Rauschenberg, May 1861, near Ostheim, Rabenhorst, Bryoth. Eur. 431 (NY, PC); Reinhardt and Fried, June 10, 1862, Vogelsgebirge (K); Reinsch, Fichtelgebirge (K); Reiss, July 1881, Münster (BR); Roemer, Rhein-Provinz, Eupen (NY); Roeper, in 1829, Megapolitano (NY); Röll, July 4, 1882, Rhön (G-DEL); Roth, Flora Rhenana, "an der Schwimmschule bei Biebrich am Rhein" (G-BOIS); Roth, Jan. 9, 1882, Hessen, near Laubach (DPU, S); Roth, July 5, 1883, Hessen, near Laubach (CHI, DPU, S); Roth, Aug. 1891, Hessen, near Laubach, on basalt, alt. 275 m (S); Roth, Aug. 1902, Hessen, near Laubach, Bauer, Musci Eur. Exsic. 554 (BART, PC); Roth, July 1903, Hessen, near Laubach, on basalt, alt. 250 m (S); Roth, July 1907, Hessen, near Laubach, on basalt (S); Roth, July 1908, Hessen, near Laubach (DPU, S); Roth, July 1908, Hessen, near Laubach, on basalt, alt. 250 m (CHI, DPU, S); Ruthe, in 1872, Oderwiesen, near Bärwalde, Neumark (NY); Ruthe, June 1872, Flora Neomarchiae, Oderwiesen near Bärwalde (Assumed to be a portion of the TYPE of *F. androgyna*, S; Ruthe's Type has not been available.); Ruthe, June to Sept. 1872, Oderwiesen, near Bärwalde, Neomarchiae (S); Ruthe, Sept. 1872, Oderwiesen, near Bärwalde, Neomarchiae (B); Ruthe, Sept. 1872, Oderwiesen, near Bärwalde, Neomarchiae, Rabenhorst, Bryoth. Eur. 1292 (FH, G-BOIS, NY, PC); Ruthe, in 1873, near Bärwalde, Neomarchiae (K); Ruthe, July 1873, Oderwiesen, near Bärwalde (PC); Ruthe, Aug. 1874, Flora Neomarchiae, near Bärwalde, Oder R. (R); Ruthe, Aug. 1874, Oderwiesen, near Bärwalde, Neomarchiae (FH, G-BOIS, S, UCLA); Ruthe, Sept. 1874, Oderwiesen, near Bärwalde, Neomarchiae (CHI, S); Ruthe, July 1876, Oderwiesen, near Bärwalde, Neomarchiae (K, S); Ruthe, in 1879, near Bärwalde, Neomarchiae (H); Ruthe, July 1879, Oderwiesen, Bärwalde, Neomarchiae (H); Ruthe, Aug. 7, 1879, Oderwiesen, Bärwalde, Neomarchiae (CHI, FH, PC, S, US); Ruthe, Sept. 1879, Oderwiesen, near Bärwalde, Neomarchiae (S); Schellenberg, Mar. 28, 1908, Oberbayern (DPU, S); Schemmann, Sept. 15, 1895, Niedersfeld (PC, S); Schemmann, May 7, 1909, Westfalen (G-DEL); Schenk, July 1931, Mähr-Schönberg, alt. approximately 900 m (CAS); Schultz 238, July 1898, Sommerfeld (FH); Schultz 240, July 1898 (FH); Schultz, July 28 or Aug. 28, 1899, Sommerfeld (FH); Schulze, Apr. 22, 1878, Silesia (NY); Schwab and Familler, Sept. 1900, and Nov. 2, 1900, *a* collected by Schwab near Unterlind in the Fichtelgebirge, alt. 580 m, and *b* collected by Familler in the Donau near Regensburg, alt. 320 m, Fl. Exsic. Bav.: Bry. 26 (S); Hans Sydow, Rhön Mts, Gersfeld, Krypt. Exsic. 2094 (K, NY, US); P. Sydow, July 1903, Bavaria, Fichtelgebirge (S); Thiry, in 1859, Zähringen, Leiner, and Stizenberger, Kryptogamen Badens 236 (CHI, K, NY, YALE); Timm, Mar. 29, 1905, Hamburg (DUKE); Timm, Oct. 1905, Hamburg (S); Toepffer, May 13, 1894, Flora Megapolitana: Schwerin 4085 (PC); Trautmann, May 1918, Bayern, Oberstorf (DPU, S); Vigener, in 1859, Westfalen (NY); Voigt, Aug. 1874, Dresden (CHI); Walther, Franconia (ILL); C. Warnstorf, Sept. 1873, Neu-Ruppin, Märkische Laubmoose 188 (CAS); C. Warnstorf, June 1880, Neu-Ruppin, Deutsche Laubmoose Flora von Brandenburg (BR, G-BOIS); C. Warnstorf, July 1883, Harz Mts. (BR); C. Warnstorf, Aug. 1883, Brandenburg (G-BOIS); C. Warnstorf, Nov. 1899, Hamburg (PC); Wehrhahn, in 1887, Bry. Exsic. Hannov. 102 (Fasc. 3) (E); Wehrhahn, May 31, 1898, Bry. Exsic. Hannov. 103 (Fasc. 3) (BM); E. Winter, Aug. 27, 1922, Saxony, near Bautzen, A. Kopsch, Bryoth. Sax. 255 (BM, MICH); Wirtgen, Ruhr R., near Essen (FH); Wirtgen, in 1858 (MINN); Zeuker and Dietrich, Thuringia, Musci Thuringici 4, Fasc. 1, Jena, 1821 (NY).

GIGLIO: Bottini, in 1887 (PC); Bottini, Apr. 18, 1887 (H).

GREECE: Morsé 1365 (G-DEL).

HOLLAND: Lacoste, near Callantsoog (Assumed to be the TYPE of *F. antipyretica* f. *dunensis*, PC); Lacoste, near Dordrecht (H, PC); Lacoste, 1842–1843, near Dordrecht (H, S); Lacoste, Sept. 1846 (H).

HUNGARY: Boros, June 15, 1919, alt. 100 m, Pl. Hungar. Exsic. 1 (S); Boros, Apr. 11, 1920, alt. 120 m, Pl. Hungar. Exsic. 2 (S); Boros, July 6, 1920, alt. 100 m, Pl. Hungar. Exsic. 3 (S); Boros, May 17, 1925, near Kis Körös (WASH); Boros, Oct. 14, 1928, near Ocsa (WASH); Branik, July 1, 1876 (FH); Hazslinszky (G-BOIS).

ICELAND: Davidsson, June 5, 1898 (S); Davidsson, July 6, 1903 (DPU, S); Feddersen, July 25, 1884, Helgå (Assumed to be portion of TYPE of *F. longifolia*, PC, S); Feddersen, July 21, 1886, Laxá (Hjardarholt), (Assumed to be portion of TYPE of *F. thulensis*, PC, S); Gaimard and Robert, in 1835 (PC); Grønlund, Aug. 14, 1876 (S); Hesselbo, Aug. 1, 1909 (H); Hesselbo, June 4, 1912 (S); Hesselbo, June 12, 1912 (S); Hesselbo, July 14, 1912 (H, S); E. Jardin, May 1865,

Bay of Faxkrudfjord (TYPE of *F. islandica*, PC; assumed to be portion of Type, s); Jonsson (L); Robert, in 1835 (PC).

IRELAND: Ardle (BM); Bradwell (K); Bradwell, in 1798 (K); Conard, Aug. 7, 1935, Cloghane, Kerry Co. (DPU, FH); Conard, Aug. 8, 1935, Kerry Co., Cloghane (DPU); W. H. Harvey, in 1831, in spring flowing into Suir R., near Caher (BM, K); Mitten, July, 1880, Lake Killarney (NY); Taylor (K).

ITALY: Arcangeli, Aug. 14, 1874, Apennines (FH); Artaria (PC); Artaria, Mar. 19, 1895 (PC); Artaria, Feb. 3, 1896 (E); Artaria, Aug. 2, 1897, alt. 1300 m (WASH); Artaria, Aug. 2, 1898, Comensis (DPU); Artaria, June 4, 1899, Novara (S); Artaria, Aug. 27, 1899, Comensis (PC); Artaria, Aug. 28, 1899, Comensis, alt. 1350 m (S); Artaria, Oct. 7, 1903, Comensis, Bauer, Musci Eur. Exsic. 563 (BART, BR, H, PC); Artaria, Dec. 1903 (DUKE); Artaria, Dec. 7, 1903, Comensis (DUKE, PC, UC, WASH); Artaria, Oct. 1922 (PC); Artaria, June 20, 1925, Novara (WASH); Artaria, June 20, 1925, Novara, Bauer, Musci Eur. et Am. Exsic. 1929 (BART, NY, S); Bottini, in 1879, Apennines, alt. 1620 m (PC); Bottini, Apr. 17, 1887 (PC); De Notaris, in 1838 (PC); Fleischer, in 1895, near Pompeii (PC); Fleischer, in 1896, Liguria (K, PC); Fleischer, Aug. 1896, Liguria, near Rapallo (PC); Fleischer, Aug. 1896, Liguria, Rapallo, Fleischer and Warnstorf, Bryoth. Eur. Merid. 172 (H, PC); Fleischer, Aug. 4, 1896, Liguria, near Rapallo (PC, WASH); Fleischer, Aug. 4, 1896, Liguria, near Rapallo, Fleischer and Warnstorf, Bryoth. Eur. Merid. 171 (Assumed to be a portion of the TYPE of *F. antipyretica* var. *ligurica*, FH, PC); Fleischer, Aug. 25, 1896, Liguria, Rapallo (PC); Fleischer, Sept. 1896, Liguria (PC); Fleischer, Sept. 5, 1896, Liguria, Rapallo (S, WASH); Fleischer, Sept. 25, 1896, Liguria, Rapallo, Fleischer and Warnstorf, Bryoth. Eur. Merid. 173 (H, PC); Fleischer, Sept. 27, 1896, Liguria, Rapallo, Migula, Krypt. Ger., Aust., et Helv. 435 (Fasc. 55), (CHI, G-BOIS, ILL, NY); Fleischer, Oct. 1896, Liguria, Rapallo (PC); Fleischer, Oct. 5, 1896, Liguria (PC); Gennari, in 1851, near Genoa, Erbar. Crittog. Ital. 5 (1005) (NY, WASH); Gresino, July 31, 1922, Cuneo (WASH); Gresino, Aug. 11, 1924, Bologna (DPU, WASH); Gresino, Aug. 1924, Cuneo, near S. Rocco di Bernezzo ("Les feuilles sont nettement obtuses, et non aigues." "Prof. I. Thériot, in lit." Assumed to be a portion of the TYPE of *F. hypnoides* f. *obtusifolia*, PC); Gresino, Aug. 21, 1924, Cuneo, near S. Rocco di Bernezzo, Bauer, Musci Eur. et Am. Exsic. 1932 ("Les feuilles sont nettement obtuses, et non aigues." "Prof. I. Thériot in lit." Assumed to be a portion of the TYPE of *F. hypnoides* f. *obtusifolia*, BART, L, NY, PC, S); Gresino, Apr. 26, 1925 (DPU, WASH); Gresino, Oct. 4, 1926, Vararre, Lake Donato (UC); Gresino, June 26, 1929, "Celle Ligure" (BART); Grigoletto, Aug. 23, 1923 (UC, WASH); Loser, Aug. 11, 1860, Trieste (K); Loser, Aug. 11, 1860, Muggia (H); Sbarbaro, Apr. 28, 1921, Liguria, Alba (H); Sbarbaro, Apr. 1924, near Bologna (BART); Sbarbaro, June 1924, Cuneo (BART).

LITHUANIA: Minkevičius, May 22, 1926, Obeliai, Flora Lithuanica 200 (H).

MADEIRA ISLAND: Johnson, in 1859 (NY); Lowe, Oct. 6, 1858 (K); Lowe, June 6, 1863 (K).

NORWAY: Arnell, Aug. 22, 1891, Tromsø Amt (L); Blytt, Akershus Amt, Christiania (S); Blytt, in 1824 (DPU); Blytt, in 1824, Jarlsberg und Larviks Amt (S); Blytt, in 1826, Smaalenenes Amt (S); Cedergren, July 16, 1920 (S); Greve, July 1890, Søndre Bergenhus Amt (S); Hagen, June 9, 1889 (US); Hagen, July 20, 1891, Nedenes Amt (PC, S); Hagen, Aug. 11, 1891, Nedenes Amt (CAN); Kaalaas, July 16, 1881, Lister und Mandals Amt, Mandal (S); Kaalaas, Aug. 10, 1894, Nodre Nordlands Amt (S); Kaalaas, July 19, 1896, Stavanger Amt, near Sand in the Suldalslaagen R. (Assumed to be a portion of the TYPE of *F. stagnalis*, S); Kaalaas, July 21, 1900, Nordre Bergenhus Amt (S); Kaalaas, July 13, 1902, Nordlands Amt (S); Kaalaas, July 23, 1909, Nordre Trondhjems Amt (S); Kaalaas, July 22, 1911, Nordre Trondhjems Amt (S); Kaalaas, Aug. 6, 1916, Tromsø Amt (S); Kiaer, July 28, 1884, Akershus Amt (S); Kiaer, July 29, 1884, Jarlsberg und Larviks Amt (S); Kidr 194, Aug. 1866 (NY); Lillefosse, Aug. 10, 1911, Nordlands Amt (S); Lörenson, June 11, 1918, Lister und Mandals Amt (s); Nyman, July 13, 1893, Stavanger Amt (s); Öhrstedt, July 26, 1930 (US); Oldberg, Herjedalen (US); Ryan, June 1886, Smaalenenes Amt (s); Ryan, Nov. 1886, Smaalenenes Amt (s); Ryan, Sept. 3, 1887, Smaalenenes Amt (s); Ryan, July 8, 1888, Smaalenenes Amt (L, s); Ryan, Oct. 1888, Smaalenenes Amt (s); Wulfsberg, Sept. 4, 1867, Nordre Bergenhus Amt (DPU, s).

POLAND: Zmuda, Aug. 22, 1912, Bryoth. Pol. 132 (PC, US).

PORTUGAL: Amann, Aug. 12, 1878 (H); Luisier, Apr. 1909, alt. 1300 m, Bryoth. Lus. 9 (K); Nicholson and H. N. Dixon, May 10, 1911, Algarve, Monchique (BM, DPU); Solms, in 1866, Flora Lusitanica Algarve 99 (NY); Welwitsch 7, Algarve, Serra de Monchique (NY); Welwitsch, in 1847, Algarve, Serra de Monchique (G-BOIS, PC); Welwitsch, June 1847, Algarve, alt. 600 m (H).

RUMANIA: Barth, Aug. 10, 1874, Flora Transsilvanica (NY); Péterfi, April 7, 1916, Transylvania, alt. 600 m, Fl. Rom. Exsic. 1132 (PC, US).

SARDINIA: Fleischer, in 1894, near Isili (s); Fleischer, Mar. 23, 1894, near Isili, alt. 200 m (WASH); Fleischer, Mar. 23, 1894, near Isili, alt. 300 m, Fleischer and Warnstorf, Bryoth. Eur. Merid. 73 (Assumed to be a portion of the TYPE collection of F. cavifolia, FH, PC); Fleischer, Apr. 8, 1894, Fleischer and Warnstorf, Bryoth. Eur. Merid. 72 (PC, WASH); Marcucci, in 1866 (G-BOIS, NY, PC).

SCOTLAND: Barker, July 1868, Aberdeen (UC); Bell, Sept. 1860, Ben Lawers (E); Catcheside, May 27, 1929, Perthshire, Ben Lawers, alt. 480 m (CGE, DPU, WASH); Croall 569, July 1856, Aberdeenshire, Braemar (K); G. Dixon (CHI); G. Dixon, May 1852 (CHI); W. Evans, Apr. 12, 1902 (E); Ferguson, about 1846 (CM); Fleming, Aberdeenshire (CIN); Gardiner, July 1844, Glen Callater (BM); Gardiner, July 1844, Dundee (K, NY); G. Gardner, in 1835, near Glasgow (K); Greville (NY, US); Hunt, July 8, 1869 (K); Johnston, Aug. 23, 1912, Orkney (BM, K); McAndrew, Sept. 28, 1904, Water of Leith at Juniper Green (E); Meldrum, Apr. 1887, Perthshire (E); Meldrum, Nov. 1903 (E); Sadler, Aug. 1866, Ben Lawers (E); Shacklette 2411, Aug. 17, 1943, Perthshire, Aberfeldy (DPU); W. Smith, July 1886, Clova (E); Stirling, May 19, 1902, Ayrshire (BM); M. E. Taylor, in 1896 (BM); James Thompson, June 1843, R. Skill (E); G. T. West, July 18, 1904, Inverness (E); G. T. West, Sept. 14, 1904, Loch Ness (E).

SPAIN: Casares-Gil 24 (H); Eliás 12, May 2, 1905 (PC); Fahr (s); Levier, Aug. 12, 1878 (G-BOIS); Levier, Aug. 12, 1878, alt. 2100 m, Fleischer and Warnstorf, Bryoth. Eur. Merid. 364 (H); Levier, July 23, 1879 (G-BOIS); Levier, Aug. 12, 1879, alt. 2000–2200 m, Musci Hispanici 50 (s).

SWEDEN: Åberg, Jämtland (s); Åberg 137, Aug. 6, 1912, Jämtland (s); Åberg 138, Sept. 7, 1912, Jämtland (s); Åberg 466, Aug. 20, 1913, Jämtland (s); Åberg 689, Sept. 23, 1914, Jämtland (s); Åberg 843, Sept. 7, 1916, Jämtland (s); Åberg 1596, July 15, 1920 (s); Åberg 1597, Aug. 23, 1920, Jämtland (s); Åberg 1906, Sept. 1, 1921, Värmland (s); Åberg 1910, Sept. 20, 1921, Västergötland (s); Åberg 1908, Sept. 21, 1921, Dalsland (s); Åberg 1907, Sept. 27, 1921, Värmland (s); Åberg, Sept. 9, 1923, Värmland (s); Adlerz, June 25, 1886, Östergötland (s); Ahrling, July 1855 (CHI); Angström, Uppsala (MINN); Arnell, June 18, 1884, Småland (s); Arnell, July 1887, Västergötland (PC); Arnell, July 7, 1887, (PC); Arnell, July 7, 1887, Västergötland, Bauer, Musci Eur. Exsic. 556 (BART, H, PC, s); Arvén, July 26, 1893, Småland (FH); Arvén, May 14, 1913, Småland (s); Arvén, Sept. 10, 1919, Småland (s); Arvén, May 20, 1920, Småland (s); Arvén, Aug. 2, 1920, Hälsingland (s); Arwidsson, Apr. 24, 1921, Uppland (L); Behm, Aug. 15, 1878, Jämtland (DPU, s); Berggren, July 12, 1886, Skåne (s); S. and C. Bergström, July 6, 1913, Dalsland (s); S. and C. Bergström, June 27, 1915, Dalsland (s); Carlson 116, June 29, 1901, Småland (s); Cedergren, Apr. 30, 1910, Uppland (s); Ekstrand, Småland (NY); Eleinke, June 1900 (DPU, s); Elmqvist, July 10, 1874, Gottland (s); Florin, July 22, 1923, Uppland (DPU, s); Florin, June 21, 1927, Småland (DPU, s); Florin, Oct. 9, 1927, Södermanland (DPU, s); Florin, Sept. 22, 1929, Södermanland (DPU, s); Florin, Aug. 27, 1933, Södermanland (DPU, s); Halle, May 20, 1925, Uppland (DPU, s); Halle, July 31, 1928, Dalarna (s); Hardin, Apr. 1833, Dalsland (s); C. Hartman, Uppland (TRT);

38   THE FONTINALACEAE

C. Hartman, July 24, 1872, Närke (H); Robert Hartman, in 1858, Bry. Scand. Exsic. 83, Gästrikland (K, S); Hasslow, May 1, 1934, Skåne (S); Hellsing, June, 1895, Uppland (S); Hellsing 397, June 22, 1915, Bohuslän (S); Hellsing, July 6, 1918, Musci Scandinavici 72 (S); Hellsing, July 6, 1918, Hälsingland, Musci Scandinavici 74 (S); Hellsing 110, Sept. 13, 1920, Dalsland (S); Hjärne, Feb. 1900, Göteborg (DPU); Hjärne, June 2, 1901, Göteborg (DPU, H, S, WASH); Hjärne, Apr. 15, 1928, Västergötland (S); Hjärne, May 14, 1930, Göteborg (UC); Hjärne, Apr. 16, 1934, Göteborg (CAS, DPU, S, UC); Hjertman, May 1912, Göteborg (DPU, S, US); Hovgard, July 14, 1929, Skåne (S); Hovgard, May 20, 1934, Skåne (CAS); Hovgard, May 30, 1934, Skåne (S); Hovgard, May 20, 1935, Skåne (S, US); Hovgard, June 7, 1936, Skåne (CAS); Hovgard, May 25, 1938, Skåne (CAS); Hülphers, July 15, 1920, Uppland (S); Hülphers, May 20, 1923, Västergötland (S); Hülphers, Oct. 1923, Västergötland (DUKE); Hülphers, July 1, 1927, Västergötland (S); Hülphers, Aug. 15, 1927, Västergötland (WASH); K. Johansson, July 8, 1900, Dalarna (S); E. Jonsson, July 1906, Göteborg (DPU, S, WASH); Larsson, Mar. 11, 1921, Dalsland (WASH); Laurent, Oct. 1920, Uppland (DPU, S); Lénström, June 1870, Uppland (S); Lillieroth, Aug. 6, 1934, Småland (UC); S. O. Lindberg, May 1854, Stockholm (CAN, H); Lindgren (PC); Löfvander, May 1905, Småland (S); S. Medelius, July 1911, Öland (S); S. Medelius, July 15, 1913, Småland (DPU, H, S); Hj. Möller, Sept. 14, 1906, Småland (DPU, S); Hj. Möller, May 15, 1907, Öland (DPU, S); Hj. Möller, Oct. 4, 1908, Dalarna (DPU, S); Hj. Möller, June 11, 1909, Dalarna (Assumed to be the TYPE of F. antipyretica f. funiculata, S; DPU, H); Hj. Möller, July 14, 1909, Dalarna (DPU, S); Hj. Möller, July 21, 1909, Dalarna (DPU, S); Hj. Möller, June 18, 1910, Dalarna (DPU, S); Hj. Möller, Aug. 15, 1911, Gästrikland (DPU, S); Hj. Möller, July 13, 1912, Torne Lappmark (DPU, H, S); Hj. Möller, May 26, 1913, Gästrikland (DPU, S); Hj. Möller, July 29, 1913, Skåne (DPU, S); Hj. Möller, May 1914, Dalarna (DPU, S); Hj. Möller, May 22, 1914, Dalarna (DPU, S); Hj. Möller, May 28, 1914, Dalarna (DPU, S); Hj. Möller, Oct. 11, 1914, Småland (DPU, S); Hj. Möller, May 13, 1915, Småland (DPU, S, US); Hj. Möller, May 25, 1916, Småland (DPU, S); Hj. Möller, May 29, 1916, Öland (DUKE); Hj. Möller, May 19, 1919, Skåne (DPU, S); Hj. Möller, May 26, 1919, Skåne (DPU, S); Hj. Möller, July 1919, Lule Lappmark, alt. 300 m (DPU, S); Hj. Möller, Apr. 12, 1921, Stockholm (DPU, S); Hj. Möller, May 17, 1921, Södermanland (H); Hj. Möller, June 12, 1924, Uppland (DPU, S); Hj. Möller, Oct. 19, 1924, Södermanland (DPU, S); Nolander, Mar. 1903, Göteborg (DPU, S); Nordenström, Aug. 1885, Östergötland (DPU, S); O. Nordstedt, June 3, 1899, Skåne (S); O. Nordstedt, Aug. 1900, Västergötland (S); Olsson, June 1857, Götland (S); Östman 159, July 9, 1871, Småland (S); Östman, Nov. 9, 1895, Götland (DPU); J. E. Palmer, July 1915, Bohuslän (S); J. Persson, Apr. 21, 1913, Småland (DPU, S); C. Sandberg, July 13, 1936, Bohuslän (S); Scheutz, Aug. 1878, Blekinge (S); Seth 27, Aug. 1872, Småland (S); Sillén, Västmanland, Musci Frond. Scand. Exsic. 118 (S); Sillén, June 24, 1836, Västmanland (S); Sillén, June 9, 1884, Västmanland (DPU, S); Stenholm, Aug. 29, 1920, Halland (S); Stenholm, Aug. 9, 1922, Åsele Lappmark (DUKE, S); Stenholm, Aug. 11, 1923, Torne Lappmark (DUKE); Stenholm, Oct. 14, 1931 (DUKE); Stenholm, June 26, 1932 (DUKE); Svedberz, Aug. 3, 1921, Dalarna (DPU, S); Svenson, Nov. 6, 1920, Halland (S); Tärnlund, June 15, 1919, Västmanland (DPU, S); Tärnlund, June 19, 1919, Västmanland (S); Tärnlund, Aug. 17, 1927, Västmanland (DPU, S); Tärnlund, Sept. 4–6, 1929, Småland (S); Tärnlund, Sept. 18, 1933, Västmanland (DPU, S); Tärnlund, Aug. 24, 1937, Södermanland (CHI, DPU, S); K. F. Thedenius, Aug. 1834, Gästrikland (DPU, S); K. F. Thedenius, in 1844, Gästrikland (K); Tolf, July 1891 (FH); Tordal 429, Oct. 1944, Dalarna (DPU); Trulsson, Mar. 11, 1918, Skåne (S); Trulsson, Mar. 19, 1918, Skåne (S); Tufvesson, May 21, 1933, Skåne (CHI, S); Vesterlund, June 1916, Dalarna (S); Von Post, Oct. 16, 1865, Östergötland (DPU, S); Wahlstedt, July 1869, Västergötland (CHI); Wennersten, July 1892, Götland (S); Westling, Aug. 1865, Småland (CHI); Zetterstedt 176, July 10, 1873, Skåne (PC, S).

SWITZERLAND: Bernet, in 1872, Rhône R. (G-BOIS); Bernet, Mar. 30, 1886, Rhône R., near Geneva (FH, S); Bernet, May 10, 1888, Salvan (G-BOIS); Bernet, Apr. 16, 1895, Valais (G-BOIS); Chenevard, Aug. 3, 1911, Bergamasque Alps (G-DEL); Conti, Sept. 1894 (G-BOIS); Courbon, in 1846, Neuchâtel (G-DEL); Guinet, Apr. 1, 1906, alt. 431 m (G-DEL); Guinet, Sept. 4, 1917, Rhône R. (G-DEL); Mitten, Wassen (NY); Moricand, Rhône R. (G-DEL); J. Müller, in 1857 (G-DEL); Payot and Bernet, Sept. 1884, Flora Helvetiae and Sabaudiae (PC). Reuter, June 1847 (G-DEL); Rome, Jan. 1886, Geneva (G-DEL); Weber, June 17, 1878 (G-DEL).

THRACE: Tedd K 47, July 23, 1933, alt. 450 m (K); Tedd K 206, Feb. 21, 1937 (K).

UNION OF SOVIET SOCIALIST REPUBLICS: Borshchov 206, Aug. 1857, Russia, Ural (H); A. H. Brotherus and V. F. Brotherus, May 1881, Caucasia (K); A. H. Brotherus and V. F. Brotherus 44, May 30, 1881, Caucasia (H); V. F. Brotherus, in 1877, Caucasia (H); V. F. Brotherus, May 15, 1877, Caucasia (PC); V. F. Brotherus, June 1877, Caucasia (H); V. F. Brotherus, June 15, 1877, Caucasia (H); V. F. Brotherus, June 25, 1877, Caucasia (S); V. F. Brotherus, Aug. 1877, Caucasia (H); V. F. Brotherus 490, Aug. 1877, Georgia (H); V. F. Brotherus, May 30, 1881, Caucasia (H); V. F. Brotherus, July 20, 1881, Caucasia (H); Chernayev, Oct. 1836, Russia (H); Fedtschenko 899, July 31, 1893, Russia (H); Haussknecht, in 1869, Georgia (BM, K); Haussknecht, Jan. 1869, Georgia (BM); Kihlman, July 17, 1891, Russia (H); Kursky 70, June 26, 1909, Altai (H); S. O. Lindberg, June 24, 1874, Karelia (H); Navaschin, Aug. 1888, Russia (PC); Sommier and Levier 246, Aug. 16, 1890, Caucasia, alt. 1200–1300 m (G-BOIS, H); Sommier and Levier 662, Sept. 14, 1890, Caucasia, alt. 900 m (H); Tomin, June 15, 1910, Altai (H); Voronov, in 1923, Caucasia (H); Voronov, Aug. 19, 1923, Caucasia (H).

WALES: Catcheside, Aug. 23, 1932, Carnarvonshire, near Llanberis (CGE, DPU); Welch 5566, Aug. 31, 1938, Ceiriog (DPU); Welch 5667, Aug. 31, 1938, Ceiriog (DPU); W. Wilson, Sept. 1828, Carnarvonshire, near Aber (MICH); W. Wilson, July 1844 (E); W. Wilson, July 6, 1844, Merionethshire, Dyffryn (DPU); W. Wilson, Mar. 24, 1856 (BM).

YUGOSLAVIA: Baenitz, Apr. 18, 1897, Dalmatia (E); Baenitz, May 30, 1897, Mostar, alt. 1000 m (US); Baenitz, May 17, 1898, Gravosa (E, US); Baumgartner, Apr. 5, 1909, Dalmatia, Bauer, Musci Eur. Exsic. 553 (BART, PC).

NORTH AMERICA: ALASKA: Eyerdam 582, Aug. 1, 1932, Unimak. Isl. (BART, MICH); Harrington, June 28, 1872, Popoff Isl. (MICH); Mehner, Oct. 15, 1906, Juneau (DPU); Trelease 2368, June 16, 1899, Sitka (MO, NY, PC); Mr. and Mrs. E. P. Walker 902, Aug. 3, 1915, Tongas Village, Portland Canal (DPU, FH, NY, WYO).

CANADA: ANTICOSTI ISLAND: Macoun, Aug. 16, 1883 (CAN).

BRITISH COLUMBIA: Boyd M 24, Aug. 19, 1946, Meade Creek, Cowichan Lake (DPU); Brinkman 442, Aug. 8, 1908, Shuswap Lake (MICH); T. C. Frye, Aug. 9, 1923, Jordan R., Tugwell Creek (DPU, WASH); A. J. Hill, New Westminster (PENN); A. J. Hill 789, May 1902, New Westminster (DUKE, FH, K, MO, NY, PC, S, UC, US, WIS, WYO); A. J. Hill, Mar. 10, 1903, Glen Brook (ABS, NY, PC); A. J. Hill, July 10, 1903, New Westminster (ABS, DUKE); A. J. Hill, Oct. 1903, New Westminster (ABS, FH, H, PC); A. J. Hill, Apr. 1904, New Westminster (NY); Macoun, in 1889, Selkirk Range (PC); Macoun 165, June 23, 1909 (S); Macoun, Mar. 29, 1916, Sidney (V); Noble, in 1923, Vancouver, Point Grey (MACF).

CAPE BRETON ISLAND: G. E. Nichols, in 1914 (NY); G. E. Nichols 682, July-Aug. 1914, South Ingonish (YALE); G. E. Nichols, in 1915 (NY).

LABRADOR: Waghorne, July 28, 1894, Forteau (PC); Wickes 64, July 16, 1940, Gannet Isl. (DPU); Wickes 29, July 24, 1940, Grady Isl. (DPU); Wickes 48, July 24, 1940, Grady Isl. (DPU).

NEW BRUNSWICK: James Fowler, July 30, 1873, Bass R., Bathurst (MO).

ONTARIO: Lawson, June 7, 1859, Loughboro Lake, Macoun, Can. Musci 995

(CAN); Lepage 6194, June 23, 1944, Moosonee (DPU); Macoun, Apr. 30, 1867, Can. Musci 227 p.p. (MO, NY, WIS); Macoun, July 8, 1884, Nipigon R. (CAN); Macoun 829, July 25, 1900, Algonquin Park (S); McCall, June 1925, Owen Sound (DUKE); Moxley ,July 18, 1925, Owen Sound (MICH); Moxley, Aug. 16, 1926, Grey Co., Jones Falls (DPU, PENN, WYO).

PRINCE EDWARD ISLAND: Macoun, July 11, 1888, Can. Moss. 201 p.p. (CAN, US).

QUEBEC: Dupret, June 12, 1905, Mt. St. Hilaire, near Montreal (DPU); Gauthier 2220, Aug. 7, 1934, St. Alexis des Monts (DPU, MT); Marie-Anselme 3784, July 1942, St. Francois R., Disraeli (DPU); Marie-Anselme 4076, July 17, 1943, Beauceville (DPU); Marie-Jean-Eudes 3021, Aug. 13, 1944, Montcalm Co. (DPU); Marie-Victorin, Sept. 1906, Mt. Béloeil, near Montreal (PC); Marr M348, July 8, 1939, Cairn Isl., Richmond Gulf, Hudson Bay (DPU, MICH); Urban, June 1858, R. Rouge (CAN).

VANCOUVER ISLAND: James Anderson, July 11, 1897, Shawnigan (NY); Lyall, in 1858 (K); MacFadden, Aug. 16, 1926, near Victoria (H, NY, UC, WASH); MacFadden, Sept. 1926, near Victoria (DPU, MACF); Macoun, Can. Moss. 203 p.p., Nanaimo (CAN, US); Macoun, May 9, 1875 (ILL); Macoun, May 8, 1893 and May 15, 1893, near Victoria, Can. Moss. 208 p.p. (MT); Macoun, May 15, 1893, Victoria, Can. Musci 598d p.p. (US); Macoun, May 22, 1893, near Victoria (PC); Macoun, May 26, 1893, Shawnigan Lake, Can. Moss. 209 (BM, CAN, FH, K, MO, MT, NY, US); Macoun, May 26, 1893, Shawnigan Lake, Can. Musci 598e p.p. (PC, US); Macoun, July 1, 1893, Shawnigan Lake, Can. Moss. 209 (CAN, US); Macoun, July 1, 1893, Can. Musci 3 (CAN); Macoun, July 1, 1893, Comox, Can. Musci 4 (CAN); Macoun, May 26, 1894, Shawnigan Lake, Can. Musci 4 (CAN); Waldron, July 12, 1901, along Straits Juan de Fuca (WYO).

Canada, without locality: Macoun, Aug. 26, 1885, Can. Musci 228 p.p. (NY, PC, US, YALE); Macoun, Apr. 30, 1887, Can. Musci 227 p.p. (UC).

GREENLAND: Breutel (G-BOIS, PC); Vahl, Sept. 1829, Fredericksdal (H).

UNITED STATES: ARIZONA: Haring, 3259, Mar. 11, 1945, Pima Co., Santa Catalina Mts., Romero Canyon, alt. 1425 m (DPU, TU).

CALIFORNIA: C. F. Baker 3000, May 12, 1902, near Lakeport (NY); C. F. Baker 3000, May 12, 1903 (PC); M. S. Baker, May 31, 1894, Burney Valley (BART, UC); Baker and Nutting, May 18, 1894, Cow Creek (CAS, UC); Baker and Nutting, May 31, 1894, Burney Valley (CAS, NY, UC); Bigelow, Apr. 11, 1854, Tamalpais (TYPE of F. californica, FH; NY, WIS); Bloomer 66, Contra Costa Co., near Oakland (NY); Bolander, Marin Co. (CHI); Bolander 952 (NY, US); Bolander, in 186–, Marin Co., Paper Mill Creek (FH); Bolander, in 1863 (H, K); Bolander, Apr. 14, 1863, Marin Co., Sullivant and Lesquereux, Musci Bor.-Am. Exsic. 333 (CHI, CM, FH, G-BOIS, IOWA, K, MICH, NY, PC, PHIL, UC, WIS, YALE); Bolander, 1864–1870 (CAS, YALE); Bolander, in 1866, Marin Co. (G-DEL); Brewer 952, Apr. 5–7, 1862, Marin Co. (BART, DPU,UC); H. E. Brown 432, June 25, 1897, Mt. Shasta (US); Byxbee, July 1895, Mariposa Co., Fish Creek (NY, UC); California State Survey 4603 (UC); Catcheside 4759, July 8, 1947, Tuolumne Co., Mather (CGE, DPU); Catcheside 4760, July 9, 1947, Mariposa Co., Yosemite Valley, Mirror Lake (CGE, DPU); Davy and Blasdale 5225, June 3, 1899, Mendocino Co., Sherwood Valley (BART, UC); Degener 16920, June 7, 1942, Sequoia Nat. Park (DPU, FH, NY); Doty 4256, Mar. 26, 1942, Tuolumne Co., near Strawberry (DPU); DuVall, July 20, 1935, Redwood Forest, near Crescent City (DPU); Alice Eastwood, Feb. 9, 1896, Marin Co., Mt. Tamalpais (NY, UC); N. L. Gardner, Aug. 4, 1927, Yosemite Nat. Park, Merced R. (BART, UC); C. C. Hall 41, June 1921, Eldorado Co., alt. 1050 m (NY); Jepson 67, Apr. 1, 1893, near Olema (NY); Lehnert, Dec. 1884 (US); Lemmon, in 1874, Sierra Valley (MICH, NY); MacFadden, July 1934, Nevada Co. (DPU, MACF, NY); MacFadden, July 3, 1940, Yosemite Nat. Park, Mirror Lake (DPU, MACF, NY); Mason, Oct. 25, 1930, Monterey Co., Arroyo Seco, Santa Lucia Mts. (BART, CAS, MICH, NY, UC, US, WASH); Nutting, May 26, 1894, Goose Valley (BART, UC); S. B. Parish 3425, June 21,

1894, San Bernardino Co., San Bernardino Mts., Bluff Lake, alt. 2220 m (NY, US); S. B. and W. F. Parish 1684, July 1881, San Jacinto Mts. (FH, UC, WIS); Parks 2842, May 10, 1925, Mendocino Co., South Fork, Eel R., near Garberville (UC); Pendleton 23, June 20, 1909, Sisson (WASH); Lewis S. Rose 36282, May 10, 1936, Santa Lucia Park (WASH); Wiggins, July 31, 1938, Tuolumne Co., Niagara Creek, near the Sonora Pass Road, alt. 1860 m (DPU).

COLORADO: C. F. Baker 41, March to May 1895, Larimer Co., near Fort Collins, alt. 1500–2100 m (CIN, COL, FH, MINN, NY, WASH); C. F. Baker 3, May 4, 1896, Fort Collins (PC); Brandegee (NY); Giles, Aug. 21, 1939, Jackson Co. (CHI); Grout, July 1914, Tolland, alt. 2700–3000 m (ABS, DUKE, US); Grout, July 28, 1914, Tolland (DPU, DUKE, US, UT); E. Hall (MICH); Purpus 749, Aug. 1892–1893 (PC); Purpus, May 1893 (CHI); Sayre 677, Aug. 10, 1938, Boulder Co., Little Royal Gorge, University Camp, alt. 2850 m (COL, DPU).

IDAHO: H. E. Brown 32, Aug. 1896, alt. 1050 m (US); Daubenmire 45230, July 1, 1945, Idaho Co., near Frenchman Butte (DPU); A. A. and E. Gertrude Heller 3450, July 16, 1896, Nez Perces Co., alt. 1050 m (DUKE, MINN, NY, PC, US); G. N. Jones 2074, May 26, 1928, Latah Co., Moscow Mt. (WASH); Leiberg 114 = 230 p.p., Kootenai Co., edges of depressions in granite between Rathdrum and Lake Pend d'Oreille (NY); MacFadden, Sept. 14, 1942, Elmore Co., Boise Nat. Forest (DPU, MACF, NY); J. H. Sandberg, July 1892 (PC); Svihla, May 23, 1931, Moscow Mt., near Viola (DPU, WASH).

MAINE: J. A. Allen, July 19, 1876, Hebron (NY); Bacon 348, May 22, 1932, Norway (NY); Cléonique-Joseph 5989, Aug. 1933, Alfred (DPU, MT).

MASSACHUSETTS: Conard, Sept. 1932, Hemlock Grove, north of Deerfield (DPU).

MICHIGAN: Dodge, June 5, 1917, Berrien Co., Rush Lake (MICH); Dodge, June 5, 1917, Marquette Co., near Huron Mt. (NY); Gleason 2409, Aug. 23, 1939, Delta Co. (DPU); Nichols and Steere, Aug. 20–27, 1935, Ontonagon Co., Porcupine Mts. (DPU, MICH, YALE).

MINNESOTA: Moyle 3621, June 24, 1940, Lake Co., Lower Knife R. (MINN); Moyle 3625, June 25, 1940, Lake Co., Knife R. (MINN); Moyle 3636, July 12, 1940, Lake Co., Rock Creek (MINN); Moyle 3620, July 26, 1940, Lake Co., Knife R. (MINN); Moyle 3619, Aug. 14, 1940, Lake Co. (MINN); Moyle 3644, Aug. 15, 1940, Lake Co., Rock Creek (MINN).

MONTANA: T. C. Frye, July 16, 1934, Glacier Nat. Park, Cut Bank Creek (DPU, WASH); W. P. Harris 494, June 28, 1901, Weed Creek, alt. 810 m (NY); Maguire 414, in 1912, Lake Josephine (UC); Maguire and Piranian 5331, July 4, 1934, Glacier Nat. Park, Divide Lake (UC); Maguire and Piranian 5326, Aug. 4, 1934, Glacier Nat. Park, Snyder Lake (UC); Umbach, in 1903, Midvale (PC); Umbach 58, June 16, 1903, Midvale (CHI, FH, ILL, NY, US, WIS); Umbach 867, Sept. 4, 1903, Lake McDonald (CHI, ILL, NY, US, WIS); Whitham and Wetzel 317, Gallatin Co., Gallatin Forest (MICH); R. S. Williams, Aug. 17, 1886, Belt R. (YALE); R. S. Williams 417, July 24, 1897, North Fork of Cut Bank Creek (CAS); T. G. and E. C. Yuncker 7120, July 26, 1937, Glacier Nat. Park, Fish Lake, alt. 1200 m (DPU).

NEVADA: Brandegee, Aug. 1885, Fish Creek (FH); S. Watson 1452, Sept. 1868, Ruby Valley, alt. 1800 m (US, YALE).

NEW HAMPSHIRE: Frost, in 1858, Cheshire Co., Mt. Wantastiquet (US, YALE); James 59, July 1852, White Mts., Crawford Notch (US).

NEW MEXICO: A. K. Fisher, Oct. 28, 1918, San Mateo Mts., Datel Nat. Forest (BART, US); Wright (H, PC); Wright, Sullivant and Lesquereux, Musci Bor.-Am. Exsic. 334 p.p. (FH).

NEW YORK: E. G. Britton, Feb. 11, 1900, Seaton's Falls (DUKE); Vasey (CHI).

OREGON: Coville and Leiberg 229, Aug. 3, 1896, Quartz Valley (US); Doty 5260, Sept. 20, 1942, Lane Co., Salt Creek Falls (DPU); DuVall, June 21, 1935, Emigrant State Park (DPU); A. S. Foster, Jan. 2, 1905, Portland (PC); A. S. Foster 1430 p.p., July 20, 1910, near Silverton (YALE); A. S. Foster, Aug. 20,

1910, near Silverton (WASH); Henderson 12067, Mar. 9, 1930, Lane Co., Willamette R., Jasper (DUKE); Howell, Oct. 1885, Sauvie's Isl. (CHI); Ingram (CHI); Ingram, U.S. Forest Service 32501, Wolf Creek, Ochoco Nat. Forest (NY); N. L. T. Nelson 2828, July 20, 1910 (H); Nevius, Portland (US); Savage, Cameron, and Lenocker, Aug. 1898, Portland (IOWA); Rev. and Mrs. R. W. Summers, in 1863, Yamhill Valley (FH); Mrs. R. W. Summers 1863, mountain streams, near Braley woods (UC); Mrs. R. W. Summers 1863, in 1878, McMinnville (YALE); Mrs. R. W. Summers, May 1878, near McMinnville (FH).

PENNSYLVANIA: Wolle, Sept. 1874, Stony Creek (US).

UTAH: Flowers 806, Aug. 12, 1924, Duchesne Co., Granddaddy Lakes, Uintah Mts., alt. 3090 m (DPU, UT); Flowers 2323, Aug. 10, 1938, Summit Co., near Mirror Lake, Uintah Mts., alt. 3090 m (DPU, UT); Flowers 2526, Aug. 12, 1938, Summit Co., in Mirror Lake, Uintah Mts., alt. 3090 m (DPU, UT); Flowers 2522, Aug. 4, 1940, Summit Co., Lily Lake, near Bald Mt., Uintah Mts., alt. 2700 m (DPU, UT); Graham 7981, May 31, 1933, Duchesne Co., Uintah Basin, alt. 2250 m (BART, CM); B. F. Harrison 10765, Aug. 30, 1944, Summit Co., Uintah Mts. (UT); Kelley, July 28, 1926, Weber Co., North Ogden Canyon, Liberty (DPU, NJU, NY); T. C. Porter, July 28, 1873, Uintah Mts. (PHIL, US); S. Watson 1452, July 1869, Evanston, alt. 1800 m (FH, NY).

WASHINGTON: J. A. Allen, May 11, 1898, upper Nesqually Valley (NY); J. W. Bailey, July 15, 1905, Seattle (NY); Brandegee, in 1883, Yakima Co. (NY); DuVall, Apr. 10, 1935, Bothel (CHI, DPU); Flett, Oct. 11, 1902, Tacoma (NY, UC); A. S. Foster 1098, Feb. 10, 1905, Hamilton (WASH); A. S. Foster, Apr. 5, 1905, Hamilton (WASH); A. S. Foster, Dec. 31, 1905, North Yakima (WASH); A. S. Foster, Apr. 20, 1907, Cathlamet (DPU, WASH); A. S. Foster 883, May 27, 1908, Quinault R., Indian Agency (NY); A. S. Foster, June 6, 1908, head of Sequalitchew Lake, Tacoma (WASH); A. S. Foster 883 or 906b, May 14, 1909, Chehalis Co., Quinault R. (DPU, WASH); A. S. Foster, Nov. 9, 1909, Sincoe Mts., Goldendale (DPU, WASH); A. S. Foster, Nov. 25, 1910, Copalis Crossing (H); A. S. Foster 1428, Nov. 26, 1910, Copalis Crossing (DPU, FH, WASH, YALE); A. S. Foster, Aug. 11, 1911, Lake Crescent, Port Angeles, Olympic Forest Reserve (WASH); A. S. Foster 1821 p.p., Aug. 11, 1911, Clallam Co., Lake Crescent (DPU, DUKE, NY, PC, US, WASH); A. S. Foster, June 1912, Ronald Bog, near Seattle (WASH); A. S. Foster 2171, June 1912, Ronald Bog, near Seattle (NY); Freye, Aug. 15, 1904, Nesqually, Mt. Rainier (PC); T. C. Frye, July 8, 1904, San Juan Co., Orcas Isl., alt. 660 m (WASH); T. C. Frye, Aug. 7, 1907, Eelwha R. valley, Olympic Mts., alt. 750 m (DPU, WASH); T. C. Frye, Apr. 13, 1908, Long Beach (WASH); T. C. Frye, Dec. 22, 1917, in tank, Union Bay of Lake Washington, Seattle (DPU, WASH); T. C. Frye, Aug. 20, 1921, Lake Kechelas (DPU, WASH); N. L. Gardner, Snoqualmie Falls (NY); N. L. Gardner 26, Snoqualmie Falls (UC); N. L. Gardner 122, Seattle (NY) Nelson 2523, in 1919, Montesaw (US); Grant, in 1919, Elma (MO); Grant, Jan. 1922, Langley (BART); Grant, Aug. 1922, Marysville (MICH); Grant, June 1923, Langley (BR, OS, WASH); Grant, Aug. 1932, Marysville (WASH); E. T. and S. A. Harper, July 1906, San Juan Co., San Juan Isl. (CHI, DPU, FH, NY); Lyall, 1858–1859, Oregon Boundary Commission, 49° North Latitude, Sumass Prairie (FH, PC, NY, WIS); M. L. Miller, Apr. 10, 1935, King Co., Bothel (UCCL); N. L. T. Nelson 2523, Nov. 9, 1909, Goldendale (S); E. E. Nichols 1428, Nov. 26, 1910, Pacific Beach (PC); Patchin June 1908, Blue Mts. (DUKE, NY); Piper, Oct. 1892, Seattle (WASH); Piper, Aug. 1895, Olympic Mts. (WASH); Piper, Aug. 1895, upper Skokomish R., Olympic Mts., Washington Flora 224 (NY); Suksdorf, May 31, 1889, Spokane Co. (NY); Svihla 706, Aug. 15, 1930, mouth of Sekin R., Olympic Peninsula (MICH); Svihla 2076, Aug. 2, 1932, Spirit Lake, region of Mt. St. Helens (DPU, WASH); Svihla, Sept. 17, 1939, McCoy Creek, Sultan (DPU, WASH); W. A. Weber 2653, Nov. 16, 1944, near Hemlock (DPU); W. A. Weber, Feb. 1945, Carson (DPU).

WISCONSIN: Cheney 5943 p.p., Ashland Co., Stockton Isl. (WIS); Cheney 1927, July 19, 1893, Oneida Co., near Rhinelander (DPU, WIS); Cheney, July 1896

(PC); Cheney 4835, July 15, 1896, Ashland Co., near Lake Superior (DPU, WIS); Fasset 8770, Aug. 27, 1929, Vilas Co., Crystal Lake (NY, WIS); Fasset 8772, Aug. 29, 1929, Vilas Co., Crystal Lake, in 19 m of water (WIS); Fasset 8773, Aug. 29, 1929, Vilas Co., Crystal Lake, in 14 m of water (WIS); Fasset 8775, Aug. 29, 1929, Vilas Co., Crystal Lake, in 15 m of water (WIS); Goessl 7507, July 6, 1917, Washburn Co., Minong (WIS); T. J. Hale, in 1861, Baraboo (MO); Schnooberger, Aug. 27, 1929, Vilas Co., Lake Constance, in 4.5 m of water (WIS); L. R. Wilson 100, May 12, 1927, Douglas Co., near Amnicon Falls (WIS).

WYOMING: Dennings 1091, July 17, 1941, Centennial (DPU, NY, OS); A. Nelson 2371 p.p., July 16, 1896, Little Goose Creek (UC, WYO, YALE); A. Nelson 70, Aug. 12, 1896, Lincoln Gulch (PC); A. Nelson 7708, July 26, 1900, North Fork, Centennial (PC); A. Nelson 8001, Aug. 7, 1900, Albany Co., Middle Fork, Little Laramie, Centennial (CHI, CM, FH, K, MINN, MO, NY, PC, US, WYO); A. Nelson 8521 p.p., July 28, 1901, Big Horn Mts. (CHI); A. Nelson 9669, Aug. 1912, Albany Co., Medicine Bow Mts. (DPU, MO, WYO); C. L. Porter 1437, Oct. 28, 1933, Albany Co., Medicine Bow Mts., North French Creek, alt. 3000 m (DPU, FH, NY, WYO); C. L. Porter 1704, June 27, 1934, Carbon Co., Mill Creek, alt. 2100 m (WYO); Richardson, Aug. 23–26, without year, Shoshone Geyser Basin, "Near a boiling spring on the edge of the creek. The water too hot to get the plant out with the hand.", alt. 2340 m (US); Röll 1529, Sept. 2, 1888, Yellowstone Nat. Park, alt. 2100 m (PC); Solheim, July 22, 1939, Albany Co., Snowy Range (WYO).

United States, without locality: Drummond, Musci Am. 232 (CAN, G-DEL, K, MICH, NY, PHIL).

Fontinalis antipyretica is apparently one of the most widely distributed and one of the most frequently collected species in the Fontinalaceae. In the examination of a large number of plants belonging to this species, the author has found great variation on the same plant in shape and size of cauline and branch leaves, degree of curvature of keel from almost straight to strongly curved, degree of conduplication from plane to concave, subcarinate, or definitely conduplicate, flaccidity and firmness of leaves, flaccidity and rigidity of stems and branches, position of leaves and branches on the stem from erect to spreading, length of decurrent leaf bases from not decurrent to long decurrent, leaf distance from close to distant, auricles none to distinct, length and width ratio of median cells from smaller proportions in younger leaves to larger in the older, and leaf apices from acute to rather broadly obtuse. Under such conditions it has become necessary to select a definite basis for making comparisons and contrasts. The characteristics of the median cauline leaves have been chosen to serve as that basis.

Fontinalis antipyretica may be distinguished from other species and varieties in the carinate-conduplicate group by the following combination of characteristics: plants medium in size, median cauline leaves with majority of keels moderately to strongly curved above basal curve, blades symmetrical with the curvature of margins of leaf halves folded lengthwise usually parallel with curvature of keels, ovate-lanceolate or oval-lanceolate, 3–8.5 mm long, 2–4 mm wide, occasionally 4.5 mm in width, 1.5–3.5 : 1, apices short to long acuminate, majority of median cells linear-rhomboidal, and upper perichaetial leaves oval to suborbicular with obtuse apices.

Fontinalis antipyretica var. *alpestris* has presented a problem because of insufficient data. Milde (1869) described var. *alpestris* as being different from *F. antipyretica* on the basis of gloss and color: "Pflanze glänzend; gelb, grün und roth gescheckt." Milde gave no other data in his original description, and his type of var. *alpestris* has not been available. Since the author has not found gloss and color of plants to be reliable characteristics for distinguishing species and varieties in *Fontinalis*, var. *alpestris* has been reduced to the synonymy of *F. antipyretica*.

Fontinalis antipyretica var. *laxa* or *F. laxa* [1] has been placed in the synonymy of *F. antipyretica* because the writer believes that the flaccidity of the stems and leaves and the mixture of carinate-conduplicate, subcarinate, and almost plane leaves on the same axis are conditions resulting from habitat factors or are indicative of young plants of *antipyretica*. On plants formerly determined as *laxa*, the more firm leaves and the carinate-conduplicate leaves agree in all details with those of *F. antipyretica*.

The author's reduction of *F. androgyna* to the synonymy of *F. antipyretica* is on the bases of: some leaves firm, carinate-conduplicate, with alar cells much enlarged, forming slight auricles; lack of evidence that the occurrence of antheridia and archegonia separately or together in the same cluster is sufficient cause for the formation of a species;[2] the study of a very large number of collections of *F. antipyretica* in which young plants or young portions of plants, intermixed with older and more mature plants or plant portions, have been found lacking, in themselves alone, distinct diagnostic characteristics; and the examination of numerous packets of *F. antipyretica* in which the habitat had greatly influenced the spacing of leaves, the rigidity of stems, the firmness of leaves, the degree of carinate-conduplication, the curvature of the keel or median line, etc. On one label, observed in Kew Herbarium, Ruthe recorded the water as being stagnant or still in which he collected: "*F. androgyna* Ruthe sp. nova! in Hedwigia 1872, pag. 166. Germania: in stagnis prope Bärwalde, Neomarchiae. G. R. Ruthe 1873. Com. A. Geheeb 12/74." The writer can not agree with Cardot's suggestion that *F. androgyna* is a form

---

[1] See page 22, note 1.

[2] In observation of the synoecious condition in *F. androgyna*, numerous clusters of reproductive structures have been examined. The following results were recorded: the majority of the clusters contained both antheridia and archegonia, but some had only antheridia and others had only archegonia.

of *F. hypnoides*, because some leaves are definitely carinate-conduplicate as in *F. antipyretica*, and leaf form, alar cells, and auricles are
not comparable to those of *F. hypnoides* but to *F. antipyretica*. It was
of interest to the author when studying the Fontinalaceae in the
herbarium of S. O. Lindberg to find *F. androgyna* included in his
folder with *F. antipyretica* and a note on the packet, written by
Lindberg, Sept. 30, 1884, "*F. antipyretica* forma."

Jensen, in the original description, stated that the leaves of *F.
longifolia* are carinate, with the exception of the youngest, and are
often split along the keel, and compared the plants with *F. gracilis*.
On the basis of carinate leaves, *F. longifolia* should be assigned to
the carinate-conduplicate group. Jensen referred to the stems and
leaves in his description and discussion as being flaccid. This characteristic was used by Cardot, apparently, as a basis for placing the
plants in the *Malacophyllae*. The author has seen leaves in this latter
group of Fontinalaceae, the group with the majority of leaves plane,
folded lengthwise either symmetrically or unsymmetrically, but does
not consider them to be carinate. Also, the writer has found various
degrees of flaccidity and firmness in stems and leaves of plants which
are definitely *F. antipyretica*, as well as leaves ranging from plane to
longitudinally folded, subcarinate, or carinate in this species. The
type material of *F. longifolia* available to the author is very fragmentary and accurate determination is difficult. The stem portion examined is rigid and the shredded blade bases, remnants of cauline
leaves, are firm. Some of the branch leaves are split longitudinally
through the center, a condition which is seen sometimes in cauline
and branch leaves of plants of *antipyretica*. The branch leaves are
rather firm and not as flaccid as is common in Cardot's *Malacophyllae*
with which he associates *F. longifolia*. Some of the apices of the leaves
on a very short branch, the only complete blades available, are comparable with those of *F. hypnoides* and others resemble those of *F. antipyretica*. The characteristics of the median cells have given no clue
for determination. The groups of alar cells of the leaf remnants, on
the small portion of stem available, are suborbicular in outline, and
the cells are numerous, in contrast with the alar group of *hypnoides*
which is subrectangular in outline, parallel with leaf margin, and in
which the cells are usually few in number. The measurements and
characteristics cited by Jensen are more applicable to *F. antipyretica*
than to var. *gracilis* as he suggested. On the basis of Jensen's description and the fragments of type material at hand, the author
assumes that *F. longifolia* may be *F. antipyretica*.

In distinguishing *F. islandica* from other species in the *Tropido-*

phyllae, Cardot emphasized the irregularity of median cells, both in width of cells and in thickness of walls. The writer has observed similar conditions in leaves of other species, and does not regard this cell variation as a characteristic of sufficient diagnostic importance for the retention of *F. islandica* as a species.

2. Fontinalis antipyretica Hedw. var. gracilis (Lindb.) Schimp., Syn. Musc., p. 552. 1876.

Fontinalis antipyretica var. *minor* Brid., Bry. Univ. 2: 657. 1827. [On basis of brief description; type not seen.]
Fontinalis gracilis Lindb., in Hedwigia 6: 39. 1867; in Not. Sällsk. Fauna et Fl. Fenn. 9: 274. 1868.
Fontinalis antipyretica var. *denudata* Saelan in MS., according to Lindberg, in Not. Sällsk. Fauna et Fl. Fenn. 9: 274. 1868.
Fontinalis antipyretica subsp. *gracilis* (Lindb.) Kindb., in Bih. K. Svensk. Vet.-Akad. Handl. 7 (9): 50. 1883.
Fontinalis dichelymoides Nordst., in Kat. Växt. Lunds Bot. För., Höstterminen, p. 11, 1888, according to Hj. Möller, in Ark. Bot. 17 (14): 41. 1922. [Not *F. dichelymoides* Lindb.]
Fontinalis gothica Card. and Arn., in Cardot, in Rev. Bryol. 18: 84 [in descriptive key], 87. 1891.
Fontinalis subglobosa Wils. MS. in herb. 1869, according to Cardot, Mon. Font., p. 56. 1892.
Fontinalis squamosa Hedw. var. *laxa* Erbar. Crittog. Ital. 6 (1006), [printed Latin description on label]. [Cardot refers to this collection in Mon. Font., p. 102. 1892.]
Fontinalis sparsifolia Limpr., Laubm. 2: 659. 1894.
Fontinalis antipyretica subsp. *sparsifolia* (Limpr.) Kindb., in Sp. Eur. and N. Am. Bryin., Part 1: 149. 1896.
Fontinalis amblyphylla Card., in Rev. Bryol. 24: 36. 1897.
Fontinalis gracilis Lindb. var. *tenella* Warnst. in litt., according to Warnstorf, Laubm. p. 628. 1905.
Fontinalis gracilis Lindb. var. *patens* Bryhn, in Nyt Mag. for Naturvid. 45: 123. 1907.
Fontinalis antipyretica var. *subgracilis* Card., in Bull. Soc. Bot. Genève (Sér. 2) 1: 131. 1909; in Brotherus, in Hedwigia 38: 226. 1899 (nomen nudum).
Fontinalis perfida Card., in Bull. Soc. Bot. Genève (Sér. 2) 1: 131. 1909.
Fontinalis seriata Lindb. var. *dentata* Roth and Von Bock, in Roth, in Hedwigia 49: 220. 1910.
Fontinalis seriata Lindb. var. *penicillata* Roth and Von Bock, in Roth, in Hedwigia 49: 221. 1910.
Fontinalis gothica Card. and Arn. var. *dimorphophylla* Hj. Möll., in Ark. Bot. 17 (14): 44. 1922.
Fontinalis antipyretica f. *gothica* (Card. and Arn.) Perss., in Bot. Not. 1942, p. 322. 1942.
Fontinalis antipyretica subsp. *gracilis* (Lindb.) Kindb. var. *patens* Bryhn, in Podpěra, Consp. Musc. Eur., p. 506. 1954.
Fontinalis antipyretica subsp. *gothica* (Card. and Arn.) Podp., Consp. Musc. Eur., p. 507. 1954.
Fontinalis antipyretica subsp. *gothica* (Card. and Arn.) Podp. f. *dimorphophylla* (Hj. Möll.) Podp., Consp. Musc. Eur., p. 507. 1954.

Plants very slender, light green, yellowish green, dark green, greenish yellow, reddish brown, golden brown, copper color, frequently glossy, especially younger portions, occasionally very glossy, sometimes dull; stems subrigid to rigid, up to 40 cm in length, 0.3–0.5 mm in diameter, denuded at base with age, irregularly pinnately branching; branches numerous, erect-ascending to erect-spreading, occasionally spreading, generally close, frequently appearing fasciculate, subflaccid to flaccid when young or if long, up to 9.5 cm in length, ends of imbricately foliated stems and branches acuminate or acute, subpyramidal to pyramidal in outline and subtriangular to triangular in cross section; median cauline leaves erect-imbricate to erect-spreading, occasionally spreading, bases usually 0.5–1 mm apart, sometimes up to 1.5 mm apart, blades firm, carinate-conduplicate with keel slightly curved to moderately curved above basal curve, occasionally almost straight, leaves generally subovate, ovate, or ovate-lanceolate, frequently margins at base reflexed on one side, rarely both sides, blades symmetrical with curvature of the margins of the leaf halves when folded lengthwise usually paralleling the curvature of the keels; apices commonly broadly obtuse, occasionally narrowly obtuse to acute, leaf tips generally entire, sometimes subserrulate to serrulate; median cauline leaves 3–4 mm long, occasionally up to 5.5 mm in length, 1.5–2.2 mm wide, sometimes up to 3 mm in width, 1.7–2 : 1, the occasional larger leaves frequently resemble leaves of *antipyretica*; median cells of leaves usually linear with ends obtuse or attenuate, occasionally narrowly rhomboidal, commonly subflexuous or flexuous, 8.5–17 μ wide, 5.5–12 : 1; alar cells enlarged, subquadrate to quadrate, rectangular, or hexagonal, walls yellowish, yellowish brown, golden brown, reddish brown, or brown, commonly forming distinct auricles, some leaf bases long decurrent, extending 1/2 – 3/4 of distance to base of next leaf below, frequently 0.5 mm, some briefly decurrent, others not decurrent; median branch leaves usually comparable to median cauline except smaller in size, occasionally keels straight above basal curve and width of blades gradually decreasing from base into acuminate apices; perichaetial branch 5–5.5 mm long, perichaetium suboval, approximately 2 mm in diameter, upper perichaetial leaves oval or suborbicular, apices round, entire, lacerate with age; calyptra long conical, 1.75 mm long, 0.6 mm in diameter; urn and operculum sometimes entirely immersed, operculum short, obtuse, conical, 0.5–1 mm in length, 0.75–1.25 mm in diameter, urn immersed or very slightly emergent, frequently contracted beneath mouth when dry, suboval to oval, 2–3 mm long, 1–1.5 mm in diameter, 1.6–2.5 : 1; seta 0.3 mm long; peristome teeth brownish orange, linear-acuminate, freqently united in pairs at the apex, muricate, 0.5–1.14 mm long, lamellae 25–38; trellis brownish orange, perfect, 0.75–1 mm long, muricate, transverse bars appendiculate; spores yellowish green, very muricate to almost smooth, 13.6–20.4 μ in diameter, occasionally up to 25.5 μ in diameter, ripe in summer. [The varietal name, *gracilis*, meaning slender, has reference to the foliated stems and branches being very slender.]

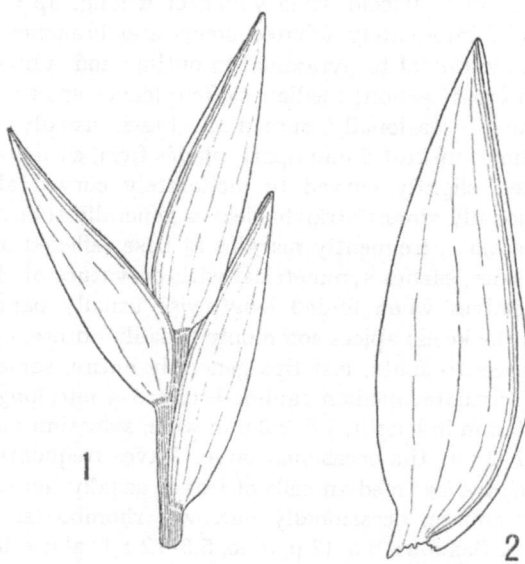

FIG. 2. – *Fontinalis antipyretica* Hedw. var. *gracilis* (Lindb.) Schimp. – 1.Portion of stem with median cauline leaves. 2.Median cauline leaf. Drawn from E. Åhrling, June 1857 and 1861, Sweden.

Type: Not previously designated. In the absence of material so marked, the writer has assumed to be the type the plants in the herbarium of S. O. Lindberg in the herbarium of the Botanisches Museum der Universität, Helsingfors, Finland, under the label, "*Fontinalis gracilis* Lindb., Finland: Helsingfors, Tali-å, inter *Fontinalis antipyreticum*, etc., 30 Junii 1868. S. O. Lindberg." The material is abundant and with fruit. This collection is cited in Lindberg's list of specimens examined which is included with his original description.

Type collector: S. O. Lindberg, June 30, 1868.

Type locality: Finland: Helsingfors, Tali-å.

Distribution: Throughout Europe, in Asiatic parts of Union of Soviet Socialist Republics, including Siberia, in Korea, and in Japan.

Additional descriptions: Milde, Bry. Siles., p. 276 (as *F. gracilis*). 1869; Cardot, Mon. Font., pp. 56 (as var. *gracilis*), 71 (as *F. gothica*). 1892; Husnot, Musc. Gall., p. 286 (as var. *gracilis*). 1892; Limpricht, Laubm. 2: 658 (as *F. gracilis*), 659 (as *F. sparsifolia*), 665 (as *F. gothica*). 1894; Kindberg, Sp. Eur. and N. Am. Bryin., Part 1: 149 (as subsp. *gracilis*). 1896; Roth, Eur. Laubm. 2: 281 (as *F. gracilis*), 282 (as *F. sparsifolia*), 283 (as *F. gothica*). 1904; Braithwaite, Br. Moss-Fl. 3: 211 (as *F. gracilis*). 1905; Mönkemeyer, in Pascher, Süsswasserfl. 14: 102

(as var. *gracilis*). 1914; Hj. Möller, in Ark. Bot. 17 (14): 38 (as var. *gracilis*), 41 (as *F. gothica*), 45 (as *F. sparsifolia*). 1922; Brotherus, Laubm. Fenn., pp. 395 (as var. *gracilis*), 396 (as *F. sparsifolia* and as *F. gothica*), 397 (as *F. gothica* var. *dimorphophylla*). 1923; Jensen, Danm. Moss., pp. 193 (as var. *gracilis*), 194 (as *F. gothica*), 195 (as *F. sparsifolia*). 1923; Dixon, Handb. Br. Moss., p. 390 (as var. *gracilis*). 1924; Mönkemeyer, Laubm. Eur. Erg.-Bd. 4: 657 (as var. *gracilis* and as *F. sparsifolia*), 661 (as *F. gothica* and as *F. gothica* var. *dimorphophylla*). 1927; Jensen, Skand. Bladmossfl., pp. 378 (as *F. sparsifolia*), 379 (as *F. gothica* and as *F. gothica* var. *dimorphophylla*). 1939.

Illustrations: Husnot, Musc. Gall., pl. 80. 1892; Roth, Eur. Laubm. 2: pl. 30 (as *F. gracilis*), pl. 31 (as *F. gothica*). 1904; Braithwaite, Br. Moss Fl. 3: pl. 123a. 1905; Roth, in Hedwigia 49: pl. 8 (as *F. seriata* var. *dentata*). 1909; Mönkemeyer, in Pascher, Süsswasserfl. 14: fig. 31. 1914; Hj. Möller, in Ark. Bot. 17 (14): figs. 7–8 and pl. 4[7] (as var. *gracilis*), figs. 9–10 and pl. 4[8] (as *F. gothica*), fig. 11 and pl. 5[9] (as *F. gothica* var. *dimorphophylla*), figs. 12–14 and pl. 5[10] (as *F. sparsifolia*). 1922; Jensen, Danm. Moss., pl. 9 (as *F. gothica*). 1923; Mönkemeyer, Laubm. Eur. Erg.-Bd. 4: fig. 144 (as var. *gracilis*), fig. 145f (as *F. gothica*). 1927. – FIGURE 2.

Specimens examined: ASIA: JAPAN: Faurie, in 1892, Lake Kushiro (s); Faurie 8621,[1] Aug. 25, 1892, Lake Kushiro (Assumed to be TYPE of *F. amblyphylla*, PC; BM, S); Faurie 8623,[1] Aug. 25, 1892, Lake Kushiro (Assumed to be Co-TYPE of *F. amblyphylla* (PC); Miyabe 359, July 15, 1891, Yezo, Oshima, Shiriuchi (Assumed to be TYPE of *F. antipyretica* var. *subgracilis*, PC; FH, K); Nakamura 992, Aug. 26, 1908 (H); Toba, 8–5–1927, Kikutyu, Nisiyama (H); Yasuda 493, June 30, 1918, Shinano, Mt. Yatsugatake (H); Yasuda 661, Sept. 17, 1918, Iwashire (H).

KOREA: Faurie 374, in 1906, Tjyang-Tjyen (Assumed to be the TYPE of *F. perfida*,[2] PC; G-BOIS, H, NY).

UNION OF SOVIET SOCIALIST REPUBLICS: Abramon, July 8, 1910, Siberia, Amur (H); Arnell, July 27, 1876, Siberia (H); Blagovestschensky and Poplav-skoyo, June 15, 1909, Transbaikalia (H); Doctvovsky, July 11, 1908, Siberia, Amur (H); Fhurarlet, Aug. 5, 1912, Siberia (H); Gorodkov, Aug. 14, 1913,

---

[1] The plants of Faurie 8621 and 8623, Aug. 25, 1892, Japan, are difficult to determine with accuracy because of the absence of complete median cauline leaves. The leaf fragments are badly lacerated. The median branch leaves compare favorably in all characteristics with those of *F. antipyretica* var. *gracilis*. Occasional leaves are carinate-conduplicate and others are concave or subconcave, which is frequently true in plants known to belong to var. *gracilis*. On the basis of these comparisons the author assumes that these plants may be *F. antipyretica* var. *gracilis*.

[2] The type of *F. perfida* should be in Cardot's herbarium in the Paris Museum. The author was unable to locate it in the herbarium in 1938. Faurie 374 occurs in the herbaria of Thériot in the Paris Museum, Brotherus in the Botanisches Museum, Helsingfors, Boissier in Geneva, and the New York Botanical Garden. All, according to the labels, are from "Herb. J. Cardot." In the absence of a designated type, the writer has assumed to be the TYPE the plants in Cardot's herbarium in the Paris Museum under the label: "*Fontinalis perfida* Cardot. Corée: Tjyang-Tjyen, ruisseaux. 1906. Faurie 374." In a letter, Aug. 12, 1947, Mrs. Pierre Allorge stated that *F. perfida* does not exist in Cardot's Herbarium in the Paris Museum; that it is possible that Cardot's type was lost during World War I, 1914–1918, and did not reach the Paris Museum Herbarium; and that Cardot and Thériot corresponded and exchanged mosses, and in that way *F. perfida* may have been deposited in the Paris Museum in Thériot's Herbarium. In case the assumed type is not located, the writer has selected to substitute for the type the plants under a comparable label in the I. Thériot Herbarium in the Paris Museum, France.

Siberia (H); Keller 356, July 18, 1908, Altai (H); Krascheninnikov 31, June 7,
1909, Transbaikalia (H); Kuzeneva, in 1909, Siberia, Amur (H); Kuzeneva,
July 27, 1910, Siberia (PC); Kuzeneva, Aug. 27, 1910, Siberia, Amur (H);
Kuzeneva, in 1914, Siberia, Amur (H); Kuzeneva, July 17, 1914, Siberia, Amur
(H); Kuzeneva, July 31, 1914, Siberia, Amur (H); Kwaschir-Samarin, June 2,
1910, Siberia, Amur (H); Levisky, in 1909, Siberia, Amur (H); Lokolov, in 1909,
Siberia (H); Novopokrovsky, in 1908, Siberia (H); Smirnov, in 1911, Siberia
(H); Timofeyev, Aug. 20, 1907, Siberia, Yeniseisk (H); Turyev 284, Aug. 10, 1910,
Siberia (H).

EUROPE: AUSTRIA: Heimerl and Conard, Aug. 25, 1935, Styria, near Krieg-
lach (DUKE, FH, WASH); L. Müller, Aug. 1931, Lunz (DPU); Venturi, in 1882,
Tyrol (PC); Wagner, near Randegg, alt. 600 m, Kerner, Fl. Exsic. Aust.-
Hungar. 1921 (G-BOIS, K, MINN, MO, PC).

BELGIUM: Cornet, July 1904 (BR); Cornet, Aug. 1921, Amblève R. (BR).

CZECHOSLOVAKIA: E. Bauer, Sept. 29, 1898, Bohemia, Bryoth. Bohem. 342
(PC, WASH); Milde, in 1866 (H); Podpěra, Aug. 1906, Moravia (DUKE).

ENGLAND: Braithwaite and Beesley, Aug. 5, 1903, Cumberland, R. Kent
(BM); Braithwaite and Beesley, Aug. 5, 1903, Cumberland, near Keswick (BM);
Burrell, in 1921, Yorkshire (BM); Burrell and C.A.C., Aug. 1920 (FH); Cheetham,
Jan. 1912, Marham Cove (BM); Dixon, Aug. 1895, Cumberland (PC); Gorum
(US); Hunt, July 1869 (FH, S).

FAEROE ISLANDS: Jensen, May 24, 1896 (H); Simmons 509, Iter Faeroense
1895, Aug. 23, 1895, Osterö Isl. (MO, MT, PC).

FINLAND: V. F. Brotherus, Helsingfors (NY, PC); V. F. Brotherus, July 9,
1869, Tavastia (NY); V. F. Brotherus, July 5, 1871, Nylandia, Musci Fenn.
Exsic. 23b (K, NY); V. F. Brotherus, July 16, 1871, Nylandia, in R. Tali-å, near
Helsingfors (K, NY); V. F. Brotherus, July 16, 1871, R. Tali-å, near Helsingfors,
Musci Fenn. Exsic. 23a (K, NY); V. F. Brotherus, July 30, 1871, Helsinki (NY);
V. F. Brotherus, Oct. 1871, near Kajana (PC); V. F. Brotherus, July 1872,
Lapponia (L, PC); V. F. Brotherus 180, July 1872, Lapponia (H); V. F. Brothe-
rus, June 17, 1887, Lapponia murmanica, Kola Expedition 1887 (H); V. F.
Brotherus, July 17, 1887, Lapponia murmanica, Kola Expedition 1887 (H);
V. F. Brotherus, July 7, 1916, Nylandia, Bryoth. Fenn. 361 (CHI, DUKE, L);
Hällström, June 15, 1905, Ostrobothnia (S); Hult and Kihlman 229, Aug. 4,
1880, Lapponia inarensis (H); Lackström, Savolex (MICH); Lackström, June 30,
1868, near Helsingfors, R. Tali-å (WASH); Lackström, July 30, 1868, near
Helsingfors, R. Tali-å (NY); H. Lindberg, June 30, 1881, Nylandia, Helsingfors
(S); H. Lindberg, in 1884, Helsingfors (DPU); H. Lindberg, July 19, 1894,
Karelia (S); H. Lindberg, Nov. 5, 1900, Helsingfors, R. Tali-å, E. Bauer, Musci
Eur. Exsic. 494 (BART, PC); S. O. Lindberg, Helsingfors, R. Tali-å (NY); S. O.
Lindberg, Oct. 14, 1866 (H); S. O. Lindberg, in 1868, Helsingfors (MINN, NY);
S. O. Lindberg, June 30, 1868, Helsingfors, R. Tali-å (Assumed to be the TYPE,
H; CAN, H, K, L, MINN, MO, NY, PC, S, US, WIS, YALE); S. O. Lindberg, in 1873,
Helsingfors, R. Tali-å, R. Hartman, Bry. Scand. Exsic. 411 (BM, K); Lundström
and Palmén, July and August 1865, Savonia borealis (K); Malmberg, June 8,
1866 (H); Nyberg, July 2, 1864, Uleåborg (S, YALE); Palmén, Aug. 1865, Ta-
vastia (K); Rancken, June 19, 1907, Ostrobothnia (UC).

FRANCE: Camus, Aug. 24, 1884 (PC); Camus, Aug. 28, 1884 (PC); Camus, Aug.
24, 1889 (PC); Camus, in 1892, Seine R., Paris (PC); Camus, Oct. 6, 1910 (PC);
Gasilien, Aug. 18, 1901, Haute-Savoie (FH); Girod, Sept. 10, 1907, Rhône
(G-DEL); Lachenaud, July 14, 1898, Vézère R. (PC); Monguillas, June 1888
(PC); Philibert, July 1877 (PC).

GERMANY: Bertram, Flora Hercynica (FH); Bertram, in 1874, Prussia (NY);
Bertram, Aug. 1874, Westfalen (NY); Brückner, May 10, 1903, Bavaria, Rhöngebirge, alt. 800 m, Familler, Fl. Exsic. Bavar.: Bry. 285 (PC); Fischer, in 1902,
Moosflora von Bamberg 583d (CAS); Geheeb, Rhön, alt. 800 m (FH, PC); Geheeb,
in 1871, Rhöngebirge, alt. 800 m (CHI, NY); Geheeb, Aug. 30, 1871, Rhöngebirge

(L); Geheeb, in 1872, Rhöngebirge (K); Geheeb, in 1872, Rhöngebirge, alt. 800 m, Rabenhorst, Bryoth. Eur. 1229 (FH, MICH, NY, UC); Geheeb, Aug. 13, 1872, Rhöngebirge, alt. 780 m (BM, K, WASH); Geheeb, in 1873, Rhöngebirge (H, K); Geheeb, July 14, 1873, Rhöngebirge (G-BOIS, G-DEL, K, NY, US); Geheeb, Aug. 1873 (S); Geheeb, Aug. 13, 1873 (FH); Geheeb, in 1875, Rhöngebirge (K, PC); Geheeb, Aug. 16, 1875, Rhöngebirge, alt. 800 m (K); Geheeb, Aug. 2, 1877, Rhöngebirge, 860 m (H); Geheeb, July 14, 1878, Rhöngebirge, alt. 780 m (L); Geheeb, Aug. 12, 1886, Rhöngebirge (PC); Grebe, June 23, 1895, Westfalen, alt. 800 m (S); Kalmus, July 1, 1906, near Helisberg (DUKE, PC, S); Kern, July 31, 1886, Riesengebirge, alt. 1400 m (FH, G-BOIS, G-DEL, NY, WIS); Kurz (MO); Limpricht, Silesia (MINN); Limpricht, in 1870, Silesia, Bryoth. Siles. 38 (PC); Limpricht, July 27, 1870, Riesengebirge, Bryoth. Siles. 336 (G-BOIS, NY, YALE); Milde, Breslau (H); Mönkemeyer, Rhön (PC); Mönkemeyer, July 1904, Erzgebirge (PC); Mönkemeyer, July 1906, Rhön, in cistern, near Römmers (PC, S); Mönkemeyer, July 1906, Hessen, Rhön, Bauer, Musci Eur. Exsic. 495 (BART, PC); Römer, July 1879, on granite, Flora des Unterhasses (BR); Römer, in 1880, Prussia, Harz (NY, PC); Roth, July 1889, Hessen, near Laubach, on basalt (Assumed to be a portion of TYPE [1] of *F. antipyretica* var. *minor* Roth, S); Schimper, in 1837, near Münster (PC); Schultz 274, Aug. 13, 1899 (FH); Warnstorf, Feb. 1, 1906, Prussia (DUKE); Zametzer, Familler, Fl. Exsic. Bavar.: Bry. 794 (PC).

IRELAND: Barlow, June 1844, near Dublin (K); McGaw 453, Aug. 1935, Derry Co., R. Bann, almost at sea level (DPU).

LATVIA (LIVLAND): Malts, Aug. 13, 1918 (S); Mikutowicz, July 27, 1906, near Tuckum, attached to granite, Bryoth. Balt. 37 (B, H, US); Mikutowicz, May 18, 1908, near Riga, attached to granite, Bryoth. Balt. 37a (B, E, US); Mikutowicz, May 18, 1908, near Riga, attached to granite, Bryoth. Balt. 295 (H, US); Roth (PC); Treboux, Aug. 20, 1907, near Pernau, Mikutowicz, Bryoth. Balt. 37b (B, E, US); Von Bock, Kersel (PC); Von Bock, Aug. 1907, near Rahezama, alt. 130 m (PC); Von Bock 357, Aug. 1907, near Rahezama, alt. 130 m (S); Von Bock 434, 480, Aug. 1907, Rahezama (Kersel), alt. 130 m (S); [2] Von Bock, Aug. 1908, Rahezama (Kersel), alt. 130 m (G-BOIS, K, PC, S, WASH); [3] Von Bock 908, Aug.

---

[1] The type, or, if not so designated, the assumed type, should be in Roth's Herbarium. It is the author's understanding from conversation with Hj. Möller in 1938 that many bryologists shared their types with him for his research and publications concerning mosses. There are numerous packets in Möller's Herbarium from Roth's Herbarium. Not being able to locate Roth's types, and under the assumption that Roth sent a portion of his type material to Hj. Möller, the writer has assumed that this is a portion of Roth's type. The information on the label is in accord with that in Roth's original description. In case Roth's type is non-existent, this portion in Hj. Möller's Herbarium has been chosen to substitute for the type.

[2] The type, or, if not so designated, the assumed type, of *F. seriata* var. *dentata* should be in Roth's Herbarium but has not been available. The author has studied several collections made by Von Bock in the type locality, Rahezama not far from Fellin in Livland, in August 1907, August 1908, and August 1911, from the Herbarium of Georg Roth in other herbaria. According to the original description, the type was collected in September 1907. All specimens examined have been determined to be *F. antipyretica* var. *gracilis*. There are no conflicting characteristics between those of var. *gracilis* and those given in the original description of *F. seriata* var. *dentata*.

[3] The author is not certain that the type of *F. seriata* var. *penicillata* has been available. The original description does not give the date of the type collection but states that the var. *penicillata* was collected in the same vicinity as var. *dentata*, Rahezama not far from Fellin in Livland. The specimens examined by the writer bear habitat data comparable to those of var. *dentata*, were collected

1908, near Rahezama (s); Von Bock, Aug. 1908, near Rahezama, alt. 130 m, E. Bauer, Musci Eur. Exsic. 497 (BART, PC); Von Bock, July 1909, Rahezama (Kersel), near Fellin, alt. 130 m (s); Von Bock, Aug. 1909, Rahezama (Kersel), alt. 120 m (s); Von Bock, July and Aug. 1909, Rahezama (Kersel), alt. 130 m, E. Bauer, Musci Eur. Exsic. 552 (BART, PC, s); Von Bock, Aug. 1910, Rahezama (Kersel), (CHI, US); Von Bock, Aug. 1911, near Rahezama, Fellin, attached to granite, alt. 130 m (DUKE, L, s); Von Bock, July 1913, near Kersel and Fellin, alt. 120 m (G-BOIS).

NORWAY: A. Blytt, Aug. 12, 1881 (s); A. Blytt, in 1885, Akershus Amt (s); Bryhn, July 1879, Alpes Jetunfjeldene, Gjendesheim (Assumed to be portion of TYPE of *F. sparsifolia*, BG, H, PC, s); [1] Bryhn, Aug. 2, 1887, Hedemarkens Amt (s); Bryhn, July 1900, Buskeruds Amt (H, s); Bryhn, July 1906, near Strømbuen, as *F. gracilis* var. *patens* Bryhn, PC); Bryhn, Aug. 1906, alt. 800 m, near Strømbuen (as *F. gracilis* var. *patens* Bryhn, PC, s); Bryhn, Aug. 4, 1906, Hedemarkens Amt (s); Bryhn, Aug. 6, 1906, alt. 800 m, near Strømbuen (as *F. gracilis* var. *patens* Bryhn, s); Conradi, July 14, 1899, Hedemarkens Amt (s); Fristedt and Looin, July 1853 (H); Kiaer, Aug. 1, 1890, Buskeruds Amt (s); Ryan, July 20, 1896, Søndre Trondhjems Amt (s).

PORTUGAL: P. Allorge, May 30, 1931, Gerez Mts., attached to granite, alt. 750 m (PC).

SARDINIA: Gennari, in 1862 (NY); Gennari, in 1862, Erbar. Crittog. Ital. 6 (1006) as *F. squamosa laxa* (NY, UC).

SCOTLAND: J. C. Adams, Sept. 4, 1915, Eddleston Water, near Eddleston (E); J. C. Adams, Jan. 15, 1916, Eddleston Water, near Nether Fala (E); M. L. Anderson, in 1868, West Brechin (BM); M. L. Anderson, in 1868, Forfarshire (BM); M. L. Anderson, May 8, 1868, Forfarshire (BM); M. L. Anderson, June 1869, Forfarshire (PC); M. L. Anderson, July 1869, Noran Water, Forfarshire (BM, H, K); M. L. Anderson and Fergusson, in 1871, Noran Water, Forfarshire (E); Catcheside, May 25, 1929, Perthshire, Ben Lawers (CGE, DPU); Croall, July 1853, Forfarshire (BM); Croall, July 1854, Aberdeenshire, Plants of Braemar 470 (K); J. Dickson 25 p.p. (E); H. N. Dixon, Oct. 14, 1905, North Esk R., Pentlands (E); J. B. Duncan, Sept. 1902 (K); William Evans, Apr. 1902, in Logan Water (E); J. Ferguson, in 1876, Forfarshire, R. Noran, near Fern (H); J. Ferguson, Aug. 1880, Forfarshire, near Fern (E); Gardiner, July 1845, Benna-Bourd (BM); Howie, in 1870, Cog Fife (BM); Hunt, July 1869, Forfarshire, Clova (US); Kidston, Apr. 27, 1896, Stirlingshire, near Kippan (BM); Kidston, June 1896, Stirlingshire (CAS, NY); Kidston, Aug. 8, 1896, Stirlingshire (BM); O'Loughlin, Sept. 2, 1925 (BM); James Thompson, July 1843 (CHI); Waddell, July 1898, Trossachs (BM).

SPAIN: Fleischer, Apr. 1908, Sierra Guadarrama, near Escorial, alt. 660 m, Fleischer and Warnstorf, Bryoth. Eur. Merid. 365 (H).

SWEDEN: Åberg, Sept. 8, 1916, Jämtland (s); Åberg 842, Sept. 8, 1916, Jämtland (s); Åberg, Aug. 9, 1919, Jämtland (s); Åberg 1395, Aug. 9, 1919, Jämtland (s); Åberg 1396, Aug. 9, 1919, Jämtland (s); Åberg 1598, July 13, 1920, Jämtland (s); Åberg 3330, July 14, 1926, Härjedalen (s); Adlerz, July 18, 1904, Närke (s); Åhrling, in 1857, Stockholm (H); Åhrling, June 1857 and 1861, Södermanland (s); Alm, Aug. 14, 1919, Torne Lappmark (s); Ångström, Lycksele

---

by Von Bock, August 1908, and July 1913, and were distributed by Georg Roth. All specimens examined have been determined to be *F. antipyretica* var. *gracilis*. There are no conflicting characteristics between those of var. *gracilis* and those given in the original description of *F. seriata* var. *penicillata*.

[1] The type has not been available. The writer has assumed to be the type the collection in the Limpricht Herbarium, bearing the label herein cited. If not in the Herbarium of Limpricht, the collection with comparable data in the N. Bryhn Herbarium in the Bergens Museum, Norway, has been chosen to substitute for the type.

Lappmark (DUKE, s); Arnell, Aug. 17, 1886, Medelpad (s); Carlson, in 1901, Småland (s); Cedergren, Aug. 6, 1916, Härjedalen (s); Cleve, July 1862, Dalarna (s); K. F. Dusén, June 7, 1872, Östergötland (G-DEL, H, PC, s); P. Dusén, July 30, 1889, Bohuslän (FH, NY); Falk, Oct. 1920, Östergötland (s); Grape, June 28, 1900, Jämtland (s); Hässler, Sept. 18, 1921, Västergötland (s); Hellsing, June 1895, Uppland (s, UCLA); Hellsing, Aug. 10, 1912, Västergötland, Musci Scand. 268 (s); Hjärne, Mar. 1900, Göteborg (WASH); Holmgren, Östergötland (H); Hovgard, July 14, 1929, Skåne (FH); Hovgard, July 14, 1930, Skåne (CAS); Hovgard, May 18, 1934, Skåne (s); Hovgard, May 30, 1934, Skåne (s); Hovgard, June 6, 1934, Blekinge (CAS); Hovgard, Oct. 4, 1934, Skane (CAS); Hülphers, Sept. 10, 1917, Västergötland (s); Hülphers, Sept. 3, 1919, Västergötland (s); Hülphers, Aug. 1921, Västergötland (s); Hülphers, Sept. 10, 1921, Västergötland (DUKE, FH, WASH); Hülphers, Aug. 29, 1922, Västergötland (s); Hülphers, Sept. 1925, Västergötland (WASH); Hülphers, Oct. 10, 1926, Västergötland (DUKE); Hülphers, Aug. 25, 1927, Västergötland (FH, s); Hülphers, Sept. 1, 1927, Västergötland (FH); Jäderholm, July 25, 1893, Dalarna (s); Jäderholm, July 13, 1911, Torne Lappmark (s); Jäderholm, Aug. 11, 1911, Torne Lappmark (s); Jäderholm, Oct. 1920, Östergötland (DPU); Jensen and Arnell, July 30, 1902, Lule Lappmark (H, s); Kindberg, in 1869, Östergötland (US); Kindberg, in 1878, Östergötland (s); Larsson, Sept. 3, 1920, Dalsland (s); Liljedahl, in 1920, Hälsingland (s); Löddeström, July 1873, Skåne (DPU); Lundqvist, July 1873, Skåne (s); Magnusson, July 18, 1908, Uppland (s); L. Medelius, July 27, 1920, Blekinge (s); L. Medelius, July 30, 1920, Blekinge (s); Hj. Möller, June 23, 1897, Skåne (DPU); Hj. Möller, July 2, 1909, Dalarna (DPU, H, US); Hj. Möller, July 3, 1909, Dalarna (s); Hj. Möller, June 29, 1911, Västergötland (H); Hj. Möller, July 1912, Torne Lappmark (DPU, DUKE, MICH); Hj. Möller, July 8, 1912, Torne Lappmark (DPU, FH, H, s); Hj. Möller, Sept. 10, 1920, Östergötland (DPU); Hj. Möller, Oct. 1920, Östergötland (CAS, DPU, DUKE, FH, H, MICH); Hj. Möller, Oct. 15, 1920, Östergötland, Bauer, Musci Eur. et Am. Exsic. 1744 (BART, NY, WASH); Hj. Möller, July 5, 1921, Lule Lappmark, alt. 350 m (H); Hj. Möller, July 23, 1922, Bohuslän (DPU); Nordstedt, Västergötland (K, US); Nordstedt, Västergötland, Renauld and Cardot, Musci Eur. Exsic. 185 (CHI, DUKE, FH, PC, YALE); Nordstedt, Västergötland, Krypt. Exsic. 297 (K, MICH, MT, NY, US); Nordstedt, Aug. 30, 1869, Östergötland (CHI, K, NY, s); Nordstedt, Aug. 1887, Västergötland (PC, s); Nordstedt, Aug. 1888, Västergötland, Sandhem, Sjöbacksjön (Assumed to be the TYPE of *F. gothica*, PC; DPU); Nordstedt, Aug. 20, 1888, Västergötland, alt. ½–2 m (BR, CAN, CHI, DPU, DUKE, H, NY, S, US); Nordstedt, July 1891, Västergötland (s); Nordstedt, July 21, 1891, Västergötland (s); Nordstedt, Aug. 5, 1895, Västergötland (CAS, DUKE, MT, PC, s); Nordstedt, Aug. 5, 1895, Västergötland, Bauer, Musci Eur. Exsic. 493 (BART, PC, s); Nordstedt, Aug. 1898, Västergötland, alt. 2–4 m (CAS, DPU, FH, PC, s); Nordstedt, Aug. 1898, Västergötland, Husnot, Musci Gall. 932 (CHI, DPU, FH, K, PC, WIS); Nordström, Sept. 26, 1915, Härjedalen (s); Nyman, Oct. 19, 1884, Östergötland (s); Nyman, Aug. 1885, Östergötland (s); Nyman, Aug. 10, 1888, Östergötland (s); Nyman, Aug. 13, 1888, Östergötland (DPU); Olsson, July 2, 1873, Östergötland (CHI); Samuelson, July 19, 1917, Dalarna (s); Scheutz, Skåne (s); Scheutz, in 1885, Småland (s); Sillén, Aug. 1875, Gästrikland (NY); Sillén 50 p.p., Aug. 1875, Gästrikland (NY); Sillén 45, June and Aug. 1880, Västmanland (DPU, s); Sillén 46, Aug. 1880, Västmanland (H, s); Sillén, Aug. 4, 1886, Västmanland (WASH); Simmons, June 23, 1897, Skåne (CHI, FH, G-DEL, H, s); Simmons, June 23, 1897, Skåne, Krypt. Exsic. 2093 (K, NY, PC, S, US); Stenholm, Västergötland (DUKE); Swartz, in 1804, Dalarna (H); Tärnlund, Aug. 1889, Stockholm (UC, US); Tärnlund, Sept. 10, 1918, Västmanland (DPU); Tärnlund, June 1, 1921, Västmanland (s); Tärnlund, June 27, 1935, Västmanland (DPU); Tärnlund, July 10, 1936, Västmanland (DPU); H. Thedenius, Västergötland (DPU); Tolf, June 26, 1885 (PC); Vesterlund, Aug. 14, 1910, Lule Lappmark (Assumed to be the TYPE of *F. gothica* var. *dimorphophylla*, s); Vester-

lund, July 1912, Dalarna (s); Vesterlund, June 17, 1915, Dalarna (s); Vetter-
hall, June 5, 1874, Östergötland (DPU, H, MINN, S, WASH); Westling, Aug. 1866,
Småland (DPU, S, US); Zetterstedt, Uppland, Uppsala (PC); Zetterstedt, July 16,
1859, Uppland, Uppsala (G-DEL, S); Zetterstedt, July 10, 1873, Skåne (FH);
Zetterstedt, July 30, 1879, Uppland, Uppsala (S).
    SWITZERLAND: Bernet, in 1883, Rhône and Arve Rivers, at Geneva (PC);
Bernet 729, Mar. 1886, Rhône R. near Geneva (G-DEL); Bernet, Apr. 1886
(G-BOIS).
    UNION OF SOVIET SOCIALIST REPUBLICS: Enwald, Aug. 5, 1880, Lapponia
rossica (H); Gordiagin 74, July 4, 1907, Russia (H); Krilov, Aug. 1876, Russia
(PC); Fr. Nylander, in 1844, Lapponia rossica (H); Sahlberg, June 24, 1870,
Lapponia rossica (H); Tomin, June 21, 1908, Russia, Institutum Cryptogami-
cum Horti Botanici Petropolitani 378 (H); Zickendrath 1150, June 13, 1895,
Russia (H); Zickendrath, July 5, 1895, Russia (H).
    WALES: Painter, June 15, 1902, Breconshire (K).
    YUGOSLAVIA: Pilous, July 1934, western Slovenia, alt. 500 m (CAS).

The very slender habit, the small carinate-conduplicate median
cauline leaves, and the keels commonly curved to some degree form
a combination of characteristics which are usually indicative of var.
gracilis. The median cauline leaves are commonly 3–4 mm long by
1.5–2.2 mm wide, 1.7–2 : 1. Occasionally, median cauline leaves are
up to 5.5 mm in length by 3 mm in width and are also comparable in
form as well as in size to smaller median cauline leaves sometimes
found on plants of *F. antipyretica*, so that it is difficult to decide
whether the plants in question are *F. antipyretica* or var. *gracilis*. In
these cases of similar leaf shape overlapping size, the frequent linear
and subflexuous or flexuous median cells in var. *gracilis* will separate,
sometimes but not always, the plants from *F. antipyretica* in which
the majority of median cells are rhomboidal or linear-rhomboidal,
and occasional cells are linear-attenuate and subflexuous or flexuous.
In spite of these characters which seem sufficiently apparent to the
author for consideration of var. *gracilis* as a species, the limits of
gracilis become confusing because these diagnostic characteristics are
not constant and in examining numerous collections of *F. antipyretica*
and of var. *gracilis* the writer has found many plants, apparently in
transition between the two, which were difficult to determine as
being either the species or the variety. Sometimes detached branches
of *F. antipyretica* are not readily distinguished as belonging to *F.
antipyretica* or to var. *gracilis*.

Because of the very slender habit the plants of var. *gracilis* are
occasionally determined incorrectly as *F. dalecarlica* but may be
easily distinguished from the latter by its carinate-conduplicate
leaves. Glossy plants of var. *gracilis* and especially plants with leaves
split along the keel are sometimes determined, in error, as *F. squamosa*, from which var. *gracilis* is quickly distinguished by its carinate-
conduplicate leaves.

The author regards the plants formerly determined as *F. gothica*, *F. gothica* var. *dimorphophylla*, *F. sparsifolia*, and *F. perfida* to be immature plants or immature portions of plants of var. *gracilis*, or plants of var. *gracilis* somewhat changed by environmental conditions. On the stems of type plants of each, mature leaves occur which are comparable to the median cauline leaves of *F. antipyretica* var. *gracilis*. On plants which are definitely var. *gracilis*, there frequently occur branches on which all leaves are comparable to those on the plans selected to represent the type of *F. sparsifolia*. Leaves narrowly lanceolate with keels straight above basal curve and auricles usually absent or very slight are frequent in var. *gracilis* near the ends of stems and branches, on young plants, and on the younger growth of old plants. These leaves, according to the opinion of the writer, have caused confusion, occasionally, in recognizing var. *gracilis*. This emphasizes the necessity of relying upon mature median cauline and branch leaves in determining species and varieties of *Fontinalis*.

Cardot was also of the opinion that *F. sparsifolia* should be in the synonymy of *F. antipyretica* var. *gracilis*, because he annotated the packet with the label, "*Fontinalis sparsifolia* Limpricht. Norwegia: Jul. 1879. Bryhn. Com. Hagen," to the effect that *F. sparsifolia* is only a mere form of *F. antipyretica* var. *gracilis*. Cardot was also of the judgment that *F. seriata* var. *penicillata* is a synonym of var. *gracilis*. On a packet in the herbarium of the Paris Museum, the author found his notation of November 1916. The label reads, "*Fontinalis seriata* var. *penicillata* Roth and Von Bock. Livland, Rahezama (Kersel). August 1908. Leg. Von Bock." Cardot annotated the plants as being a form very near to those of *F. antipyretica* var. *gracilis*.

3. **Fontinalis antipyretica** Hedw. var. **gigantea** (Sull.) Sull., Icon. Musc., p. 106. 1864.

Fontinalis gigantea Sull., Musci and Hep. U.S., p. 104, as an addition to p. 54. 1856.
Fontinalis antipyretica var. *latifolia* Milde, Bry. Siles. p. 276. 1869. [On basis of description; type not seen.]
Fontinalis antipyretica var. *robusta* Card., in Rev. Bryol. 9: 88. 1882.
Fontinalis antipyretica f. *robusta* (Card.) Card. Mon. Font., p. 51. 1892.
Fontinalis antipyretica var. *rufescens* Besch., in Cardot, Mon. Font., p. 53. 1892; Bescherelle, Cat. Mouss. d'Algerie, p. 30. 1882 (nomen nudum).
Fontinalis antipyretica subsp. *gigantea* (Sull.) Kindb., in Can. Rec. Sci. 6: 75. 1894.
Fontinalis antipyretica var. *monensis* Card. and Simm., in Simmons, in Bot. Not., p. 222. 1896.
Fontinalis antipyretica var. *macrophylla* Warnst., Laubm., p. 626. 1905. [On basis of description; type not seen.]
Fontinalis livonica Roth and Von Bock, in Roth, in Hedwigia 49: 221. 1910.

Fontinalis mollis C. Müll. var. *livonica* Roth in litt. as syn. of *F. livonica* Roth and Von Bock, according to Roth, in Hedwigia 49: 221. 1910.
Fontinalis antipyretica f. *latifolia* (Milde) Mönkem., in Pascher, Süsswasserfl. 14: 104. 1914.
Fontinalis antipyretica f. *gigantea* (Sull.) Mönkem., in Pascher, Süsswasserfl. 14: 105. 1914.
Fontinalis antipyretica f. *livonica* (Roth and Von Bock) Mönkem., in Pascher, Süsswasserfl. 14: 105. 1914.
Fontinalis antipyretica var. *livonica* (Roth and Von Bock) Mönkem. in litt., according to Hj. Möller, in Ark. Bot. 17 (14): 35. 1922.
Fontinalis antipyretica f. *latifolia-vulgaris* Fuchs., in Internat. Rev. Hydrobiol. u. Hydrogr. 12: 202. 1924. [On basis of description; type not seen.]
Fontinalis antipyretica f. *monensis* (Card. and Simm.) Jens., Skand. Bladmossfl., p. 377. 1939.
Fontinalis antipyretica var. *robusta* Card. f. *subsecunda* Biz., in Rev. Bryol. and Lichénol. (N.S.) 15: 166. 1946. [On basis of description; type not seen.]
Fontinalis antipyretica subsp. *vulgaris* (Mönkem.) Giacom. var. *rufescens* Besch., in Giacomini, in Atti Ist. Bot. Univ., Pavia, Lab. Crittog. (Ser. 5) 4 (2): 250. 1947.
Fontinalis sardoa Herzog, in Giacomini, in Atti Ist. Bot. Univ., Pavia, Lab. Crittog. (Ser. 5) 4 (2): 250. 1947 (nomen nudum).
Fontinalis antipyretica subsp. *vulgaris* (Mönkem.) Giacom. var. *latifolia* Milde, in Giacomini, in Atti Ist. Bot. Univ., Pavia, Lab. Crittog. (Ser. 5)4 (2): 250. 1947.
Fontinalis antipyretica subsp. *vulgaris* (Mönkem.) Giacom. f. *monensis* (Card. and Simm.) Jens., in Podpěra, Consp. Musc. Eur., p. 505. 1954.
Fontinalis antipyretica subsp. *vulgaris* (Mönkem.) Giacom. var. *livonica* (Roth and Von Bock) Mönkem., in Podpěra, Consp. Musc. Eur., p. 505. 1954.
Fontinalis antipyretica subsp. *vulgaris* (Mönkem.) Giacom. var. *robusta* Card., in Podpěra, Consp. Musc. Eur., p. 505. 1954.
Fontinalis antipyretica subsp. *vulgaris* (Mönkem.) Giacom. var. *gigantea* (Sull.) Sull., in Podpěra, Consp. Musc. Eur., p. 506. 1954.
Fontinalis antipyretica subsp. *vulgaris* (Mönkem.) Giacom. var. *gigantea* (Sull.) Sull., f. *macrophylla* (Warnst.) Podp., Consp., Musc. Eur., p. 506. 1954.

Plants moderately robust to very robust, glossy, green, brownish green, golden green, copper brown, or golden yellow; stems subrigid to rigid, up to 150 cm in length, 0.3–0.5 mm in diameter, denuded at base with age, irregularly pinnately branching; branches usually erect to erect-spreading, occasionally spreading, varying in length, commonly elongate, up to 70 cm long, ends of foliated stems and branches acute, frequently conspicuously three-angled, elongated pyramidal in outline and triangular in cross section; median cauline leaves slightly distant, subimbricate or imbricate with bases commonly 0.5–1.5 mm apart but occasionally up to 3 mm apart, blades usually erect-ascending or erect-spreading but occasionally spreading, firm, carinate-conduplicate with keel moderately curved to almost semicircular above basal curve, ovate, broadly ovate, ovate-lanceolate, oval, or suborbicular, commonly broadest in basal half or basal third, width often decreasing rapidly from median portion or lower two-thirds into narrow apex, frequently margins on one side reflexed near base, blades symmetrical with curvature of the margins of the leaf halves when folded lengthwise usually paralleling the curvature of the keels; apices gradually to rather abruptly narrowed, commonly narrowly to broadly obtuse, sometimes acute, leaf tips entire to serrulate; median cauline blades large, 4–8 mm long, occasionally up to 10 mm in length, usually 3–6.5 mm wide, sometimes up to 8.5 mm in width, approximately 1–1.5 : 1, at times up to 1.9 : 1; median cells of leaves generally linear-rhomboidal, sometimes linear with

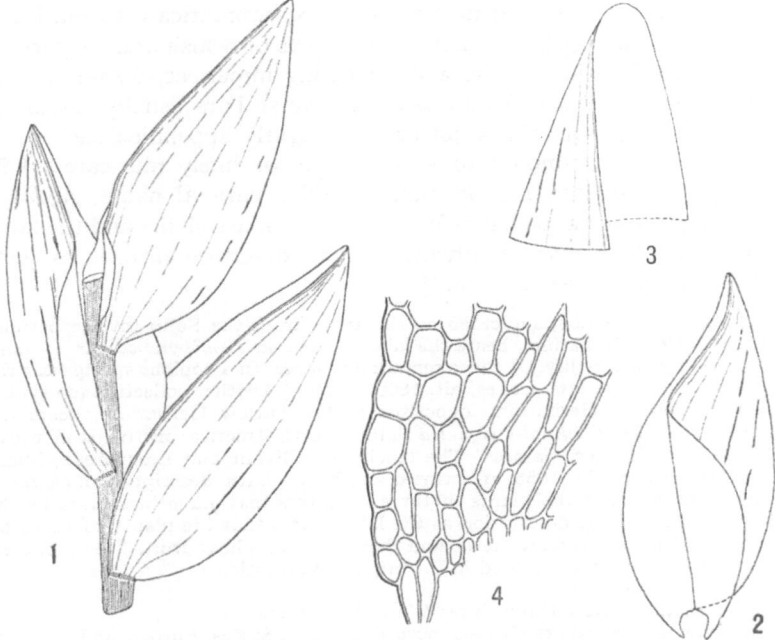

FIG. 3. – *Fontinalis antipyretica* Hedw. var. *gigantea* (Sull.) Sull. – 1.Portion of stem with median cauline leaves. 2.Median cauline leaf. 3.Leaf apex. 4.Alar cells. Drawn from Welch 711, Vermont.

ends attenuate or obtuse, 10–20 μ wide, 6–15 : 1; alar cells enlarged, subrectangular, subquadrate, or subhexagonal, walls yellowish to brown, generally forming distinct auricles, bases usually decurrent, from 0.5–0.75 mm, leaves on same stem varying from not decurrent to decurrent; median branch leaves sometimes similar in shape to median cauline but smaller in size, frequently resembling in appearance the median cauline leaves of *antipyretica*; perichaetial branch 4.5–6 mm in length, perichaetium subcylindrical, 1.5–2 mm in diameter; upper perichaetial leaves oval to suborbicular, obtuse, entire, commonly lacerate with age; calyptra long conical, 1.3–1.5 mm long, 0.5–0.75 mm in diameter; operculum obtuse conical, 0.75–1.5 mm long, 0.75–1 mm in diameter; urn immersed to slightly emergent, upper portion of operculum usually emergent, sometimes all of operculum and upper portion of urn emergent, frequently slightly contracted beneath mouth when dry, urn subcylindrical, suboval, or oval, 2–2.75 mm long, 1–1.9 mm in diameter, 1.3–1.6 : 1; seta 0.25–0.3 mm long; peristome teeth brownish orange, linear-acuminate, often united in pairs at apex, submuricate to muricate, 0.75–1 mm in length, lamellae 13–36; trellis brownish orange, perfect, sometimes broken at base and appearing imperfect, 0.75–1 mm in length, submuricate to muricate, transverse bars usually distinctly appendiculate, sometimes not or only slightly appendiculate; spores yellowish green, smooth to very slightly or finely muricate, 11.9–22.1 μ in diameter, ripe in summer. [The varietal name, *gigantea*, alluding to the giants, has reference to the robust plants and the large median cauline leaves. Sullivant, in 1856, described *gigantea* as "One of the largest mosses known."]

Type.– A collection, indicated as the type, from the Sullivant Herbarium, in the Farlow Herbarium, bears the label, "*Fontinalis antipyretica* var. *gigantea* (olim *F. gigantea* Sulliv.). Leaves remarkably large. In a copious spring near the base of the White Mts. Oakes, alt. 1800'? 1846." Another collection (co-type?) in the Fleischer Herbarium, deposited in the Farlow Herbarium, bears the label, "No. 224. *Fontinalis gigantea* Sulliv. North America: in rivulis montosis Novae Angliae. Leg. James." The labels of Sullivant and Lesquereux, Musci Bor.-Am. Exsic. 224. 1856, are printed and bear a Latin description of *gigantea*. They indicate that the plants of three collectors may have been used as the basis of the original description as the label states that the plants collected by Oakes and Tuckerman are sterile and those by James have fruit. The characteristics of the fruit are included in the original description.

Type collector: Oakes.

Type locality: New Hampshire: White Mountains.

Distribution: Southern Canada, northern United States, Europe, and northern Africa, with the majority of available locality records occurring in areas between approximately 40° and 60° North Latitude.

Additional descriptions: Lindberg, in Not. Sällsk. Fauna et Fl. Fenn. 9: 279. 1868; Schimper, Syn. Musc. 2: 552. 1876; Lesquereux and James, Man. Moss. N. Am., p. 269. 1884; Cardot, Mon. Font., p. 52. 1892; Husnot, Musc. Gall., p. 286. 1892; Limpricht, Laubm. 2: 655 (as vars. *gigantea* and *latifolia*). 1894; Farneti, in Atti Ist. Bot. Univ. Pavia (Ser. 2) 4: 139 (as var. *robusta*). 1897; Grout, M.H.M., p. 396. 1903; Roth, Eur. Laubm. 2: 279 (as vars. *gigantea* and *latifolia*). 1904; Braithwaite, Br. Moss-Fl. 3: 210. 1905; Jennings, Man. Moss. W. Penn., p. 204. 1913; Hj. Möller, in Ark. Bot. 17 (14): 36 (as var. *robusta*). 1922; Brotherus, Laubm. Fenn., p. 395 (as vars. *livonica* and *robusta*). 1923; Jensen, Danm. Moss., p. 193 (as var. *robusta*). 1923; Dixon, Handb. Br. Moss., p. 390. 1924; Mönkemeyer, Laubm. Eur. Erg.-Bd. 4: 657 (as f. *latifolia*), 658 (as f. *gigantea* and as f. *livonica*). 1927; Welch, in Grout, Moss Fl. N. Am. 3: 236. 1934; Jensen, Skand. Bladmossfl., p. 377 (as f. *robusta* and f. *livonica*). 1939; Jennings, Man. Moss. W. Penn. and Adj. Reg., p. 174. 1951.

Illustrations: Sullivant, Icon. Musc., pl. 66. 1864; Grout, M.H.M., pl. 88. 1903; Mönkemeyer, in Pascher, Süsswasserfl. 14: fig. 32 (as f. *gigantea*). 1914; Hj. Möller, in Ark. Bot. 17 (14): fig. 6 and pl. 3⁵ (as var. *livonica*), pl. 3⁶ (as var. *robusta*). 1922; Mönkemeyer, Laubm. Eur. Erg.-Bd. 4: fig. 143b (as f. *gigantea*), fig. 145a (as f. *livonica*). 1927; Welch, in Grout, Moss Fl. N. Am. 3: pl. 74. 1934; Jennings, Man. Moss. W. Penn. and Adj. Reg., pl. 64. 1951. – FIGURE 3.

Specimens examined: AFRICA: ALGERIA: Ouach, May 1888, near Constantine (G-DEL); Trabut, in 1899, Kabylie (Assumed to be the TYPE of *F. antipyretica* var. *rufescens*, PC).

EUROPE: AUSTRIA: Breidler, July 30, 1870, in Steirmark (L).

BELGIUM: Cornet, June 1903, Touillon-Tourneau (BR); Cornet, June 1904, Ruy-de-Chavion (BR); Cornet, July 1921 (BR); Delogne, Nivy (BR); Lochenies, in 1886, Hainaut, Leuze (BR); Naveau, May 1904, Anvers (ILL); Vandenbroeck 462, May 18, 1883, Anvers (BR).

CZECHOSLOVAKIA: E. Bauer, June 9, 1899, Bohemia, alt. 850 m (PC); E. Bauer, June 9, 1899, Bohemia, alt. 900 m, Bryoth. Bohem. 340 (PC, WASH); E. Bauer, Bohemia, Krypt. Exsic. 2090 (K, NY, PC, US); Petrak, Mar. 22, 1913, Bohemia, Ohrensdorf, Fl. Bohem. et Morav. Exsic. 118 (BR, E, FH, S); A. Schmidt, June 1900, Bohemia, near Haida, alt. 400 m, Bauer, Bryoth. Bohem. 246 (PC, US, WASH, YALE); A. Schmidt, July 1911, Bohemia, Petrak, Fl. Bohem. et Morav. Exsic. 117 (BR, E, FH).

DENMARK: Hohenberg, June 2, 1893, Isl. of Möen, Liselund (H); Jensen, June 1900, Ljaelland (S); Hj. Möller, June 1–3, 1893, Isl. of Möen, Liselund (H, S); Hj. Möller, June 2, 1893, Isl. of Möen, Liselund (DPU, DUKE, H, MICH, S); H. G. Simmons, June 2, 1893, Isl. of Möen, Liselund, in still water (Assumed to be the TYPE of *F. antipyretica* var. *monensis*, PC; H, L, S); H. G. Simmons, Isl. of Möen, Liselund, in still water, Cardot, Musci Eur. Exsic. 238 (DUKE, FH, MINN, YALE).

ENGLAND: Barker, Yorkshire, near Malham Cove (BM); Pickard, July 21, 1909, Easington (BM); E. Rogers, Yorkshire, Malham (PC).

FINLAND: H. Lindberg, June 7, 1898 (S).

FRANCE: Bouly de Lesdain, Sept. 30, 1896 (PC); Campbell, Apr. 7, 1938 (BM); Crozals, Hérault, near St. Pons, alt. 900 m (PC); Cussac, May 4, 1857, Billot, Fl. Gall. et Ger. Exsic. 2194 (G-BOIS, G-DEL, MINN, PC); Gasilien 379, June 1885, near Ambert, alt. 1100–1200 m (PC); Guinet, Apr. 7, 1894 (G-DEL); Héribaud, Apr. 1904, Puy-de-Dôme (PC); Husnot, Orne, Mesnil-Hubert, Musci Gall. 673 (CHI, DUKE, FH, G-BOIS, G-DEL, K, PC, WASH, WIS); Husnot, Orne, La Chapelle-Biche, Musci Gall. 931 (CHI, DPU, DUKE, K, PC, WIS); Lachenaud, Mar. 1, 1899, Haute Vienne (PC); Lachenaud, Sept. 1901, Limoges (PC); Loleirol, Corse (PC).

GERMANY: Fuckel, in 1894, Heidelberg (G-BOIS); Goldmann, May 1906, Prussia, Ostenland, near Paderborn (S); Graef, Sept. 20, 1884, Riesengebirge (S); Loeske, July 25, 1910, Bavaria (DUKE); Mönkemeyer, July 1904, Erzgebirge (S); Röll, Apr. 23, 1886, Thüringen (S); Roth, July 1905, Hessen, near Laubach (S); Schultz, in 1887, East Prussia, Königsberg (FH); Toepffer, Apr. 10, 1892, Flora Megapolitana: Schwerin 3003 (PC); Warnstorf, May 1895, Brandenburg (PC); Warnstorf, June 1895 (DUKE).

IRELAND: Clover, May 25, 1907, Tyrone Co., in a well, near Stewartstown (BM); I. Fergusson, Apr. 1877, Lake Muckross (S).

ITALY: Baumgartner, Apr. 16, 1925, Istria, Albona, Bauer, Musci Eur. et Am. Exsic. 1931 (BART, NY); Berger, in 1899, near Tenda (NY).

LATVIA (LIVLAND): Mikutowicz, May 13 and 26, 1907, Kurland, Bryoth. Balt. 294 (B, H, S, US); Von Bock 341, in 1907 (S); Von Bock 212, July 1907, "am Kalksteinen in Sinealiksee bei Fellin in Livland" (S); Von Bock, Aug. 1907, "am Sinealiksee bei Fellin in Livland in 110 m" (Portion of the assumed TYPE [1]

---

[1] The type should be in the Herbarium of Georg Roth, which has not been available. Since the data on these labels are identical with those in the original description, these specimens are considered to be from the collection which the author has assumed to be the type of *F. livonica.*

of *F. livonica*, K, PC); Von Bock, Sept. 1907 (PC); Von Bock, July 1908, "Sinea-liksee bei Fellin, 110 m" (G-BOIS, S); Von Bock, July and August 1908, "Sinea-liksee bei Fellin," Bauer, Musci Eur. Exsic. 1160 (BART, L, WASH); Von Bock, Aug. 1909, "Sinealiksee" (WASH); Von Bock, July 1911, "im Sinealiksee bei Fellin, 110 m" (L).

SARDINIA: Herzog, June 8, 1906, Mt. Limbara (DPU, PC).

SCOTLAND: J. C. Adam, Apr. 15, 1916 (E).

SPAIN: Leroy, Apr. 1, 1933 (BR).

SWEDEN: Åberg 782, Sept. 12, 1915, Jämtland (S); Agelin, July 25, 1930, Uppland (S); Arnell, July 14, 1887, Västergötland (PC, S); Arvén, Apr. 25, 1891, Småland (S); Eriksson, June 11, 1873, Skåne (S); Grotenfelt, June 27, 1875, Gottland (H); Hülphers, Apr. 28, 1918, Västergötland (S); Hülphers, Apr. 15, 1920, Västergötland (S); Hülphers, Apr. 1921, Västergötland (WASH); Lidforss, Jan. 1883, Lund (PC); S. O. Lindberg, June 11, 1865, Gottland (H, K); S. Mede-lius, Aug. 31, 1920, Blekinge (S); Hj. Möller, Jan. 21, 1883, Skåne, Lund (H, S); Hj. Möller, May 1885, Skåne (S); Hj. Möller, Sept. 1890, Skåne (S); Hj. Möller, May 29, 1916, Öland (DPU, MICH); Hj. Möller, June 3, 1917, Öland (DPU, DUKE, H, S); Mosén, Aug. 1864, Stockholm (S); Nordstedt, July 1899, Västergötland (S); Palmgren, Aug. 25, 1917, Stockholm (H); Palmgren, Aug. 25, 1917, Nalka, Södermanland (S); Harry Smith, Aug. 1920, Torne Lappmark (DPU); Stenholm, June 26, 1932 (DUKE, MICH); Sterner, July 11, 1917, Öland (S); K. F. Thedenius, July 31, 1873, Gottland (S); Wallgren, July 1865, Gottland (H); Zetterstedt 451, June 17, 1872, Gottland (S); Zetterstedt, June 29, 1872, Gottland (H).

SWITZERLAND: Bernet (G-BOIS); Bernet, July 1881, Geneva (PC); Bernet, May 20, 1883, Geneva (H, PC); Guinet, Apr. 10, 1904, Geneva (G-DEL).

UNION OF SOVIET SOCIALIST REPUBLICS: Babet, Feb. 1914, Caucasia (H).

NORTH AMERICA: ALASKA: Cooper and Andrews 80, July 4, 1929, Plants of Southeastern Alaska and Adjacent British Columbia: Davidson Glacier (NY).

CANADA: BRITISH COLUMBIA: Brinkman, Nov. 6, 1909, Kamloops Lake District, alt. 840 m, Canadian Mosses 140 (ABS, CM, DPU, E, FH, MICH, US).

CAPE BRETON ISLAND: Macoun, July 16, 1898, Can. Moss. 201 p.p. (CAN, US); Macoun, July 18, 1898, Can. Musci 598a (DUKE, FH, US, WASH); G. E. Nichols, July and Aug. 1915, mountains west of Ingonish (YALE).

LABRADOR: David Potter, July 5, 1937, Great Caribou Isl., Musci of 1937 Macmillan Expedition to Labrador Coast and to Baffin Isl. (DPU); Waghorne, Aug. 13, 1894 (PC).

NEW BRUNSWICK: James Fowler, Kent Co., Bass R., (DPU, FH); James Fow-ler, in 1869, Kent Co. (CHI); James Fowler, June 2, 1879, Fredericton (TRT); Habeeb 174, Sept. 11, 1943, Grand Falls (DPU); Habeeb, Apr. 23, 1944, Frede-ricton, University Forest, Musci Novi Bruns. 28 (DPU); Hay, July 1884, Tobique (CAN); Moser, Aug. 10, 1891, Elmwood, Kings Co. (CAN).

NEWFOUNDLAND: Delamare, Ile Miquelon (FH, NY, PC, S); Delamare, Ile Miquelon, Renauld and Cardot, Musci Am. Sept. Exsic. 71 (CAN, MICH, NY, YALE); Delamare, in 1886, Ile Miquelon (G-DEL); Delamare, 6–9–1894, Ile Mique-lon (S); Long 121, Aug. 16, 1924, Trepassey (BART, MICH, PHIL); Waghorne, July 24, 1890, Blakeston (CAN).

NOVA SCOTIA: Margaret S. Brown 568, Halifax, Melville Park (DPU); Macoun, June 13, 1883, Can. Moss. 201 p.p. (CAN, US); Macoun, July 12, 1883, Tuno (CAN); Macoun, July 11, 1888, Antigonish, near Tracadie Bay (CAN); Prince 6444, Sept. 7, 1925, Guy's Co., near Westcook's Cove (DUKE, MICH, WIS).

ONTARIO: R. F. Cain, Aug. 7, 1939, Lake Timagami, Gull Lake Portage (CHI, DPU, TRT); R. F. Cain, Aug. 23, 1939, Algonquin Park, Opeongo Lake (CHI, DPU, MT, NY, TRT); R. F. Cain, May 21, 1941, Simcoe Co., Sparrow Lake (DPU, TRT); R. F. Cain, May 22, 1941, Simcoe Co., Sparrow Lake (DPU, TRT); R. F. Cain, Aug. 21, 1941, Algonquin Park, near Oxbow Lake (DPU, TRT); R. F. Cain, July 19, 1944, Sudbury District, Lake Penage (DPU, TRT); R. F. Cain, Aug. 16, 1944, Sudbury District, Skead (DPU, TRT); Drexler 133, June 27, 1935, Quetico

Park, Lake Kawnipi, McKenzie Bay (DPU); Drexler 376, Aug. 19, 1935, Quetico Provincial Park, Mack Lake (DPU); James Fowler, June 17, 1902, Plevna (NY, WIS); Fulford, July 11, 1936, Lake Timagami Provincial Forest (CIN, MICH, NY); Hand 466, July 24, 1935, Haliburton District, near Haliburton, "Original specimen 5 feet long" (DPU); Hand 738, Aug. 13, 1941, Timiskiming District, near Latchford (DPU, OS, WASH); Hand 742, Aug. 13, 1941, Timiskiming District, near Latchford (DPU); Macoun, in 1865, Hastings Co. (NY); Macoun 119, July 1865, North Hastings (K, NY); Macoun, July 13, 1865, Hastings Co., Can. Musci 757 p.p. (CAN); Macoun 98, July 1868 (MO); Macoun, in 1870, Hastings Co., Belleville (CAN); Macoun, Aug. 16, 1874, Fl. Can 2222 (NY); Macoun, in 1875, Fl. Can. 259 (K); Macoun, July 8, 1884, Nipigon R., Can. Moss. 201 p.p. (FH); Macoun, July 20, 1884, Nipigon R. (CAN); Macoun, June 12, 1899, Ottawa, Can. Moss. 205 p.p. (CAN, US); Macoun, June 9, 1900, Algonquin Park, Cache Lake (CAN); Macoun 687, June 9, 1900, Algonquin Park (S); Macoun, July 25, 1900, Algonquin Park, Petawana R. (CAN, TRT); Macoun, Aug. 11, 1900, Algonquin Park, Rocky Lake (CAN).

PRINCE EDWARD ISLAND: Macoun, July 11, 1888 (FH).

QUEBEC: Boardman 585, July 21, 1934, Terrebonne Co., near Lake Mercier (CM, MT); Boardman 586, July 25, 1934, Terrebonne Co., Mt. Tremblant (CM, MT); Cléonique-Joseph 8700, Aug. 3, 1935, St. Maurice Co., St. Boniface de Shawinigan (DPU, MT); Collins, Fernald, and Pease, 3788 of Collins, Aug. 22, 1904, Gaspé Co. (PC); Dupret, June 12, 1905, near Montreal (FH, PENN, WASH); Dupret, June 25, 1905, Mt. St. Hilaire, near Montreal (PENN); Dupret, Sept. 8, 1906, Mt. St. Hilaire, near Montreal (CAN, FH); Fabius 1256, June 7, 1947, Knowlton (DPU); Fabius 1263, June 9, 1947, Mt. Shefford (DPU); Fabius 1707, Aug. 19, 1947, Granby (DPU); Fabius 2266, Apr. 26, 1948, Mt. Shefford, alt. 300 m (DPU); Fabius 3126, Nov. 2, 1948, Mt. Shefford (DPU); Fabius 3499, June 9, 1949, Mt. Yamaska, alt. 240 m (DPU); Fabius 3630, Oct. 6, 1949, Mt. Shefford, alt. 300 m (DPU); Gauthier 2146, June 19, 1933, Montcalm Co., Lake Simon, St. Alexis des Monts (DPU, MT); Gauthier 11175, July 13, 1938, Nat. Park of the Laurentides, Camp Mercier, Lake Régis (DPU, MT); Gauthier 11308, July 27, 1938, Nat. Park of the Laurentides (DPU, MT) Gauthier 11332, Aug. 1938, Nat. Park of the Laurentides (DPU, MT; Gauthier 11603, Aug. 7, 1938, Nat. Park of the Laurentides (DPU, MT); Gauthier 11533, Aug. 17, 1938, Nat. Park of the Laurentides (DPU, MT); Kucyniak 42–39, May 9, 1942, Brome Co., Sutton (DPU, MT); Kucyniak 42–162, Aug. 15, 1942, Brome Co., Knowlton (DPU, MT); Kucyniak 44–2, June 18, 1944, Brome Co., Dunkin (DPU, MT); Lepage 1783, June 17, 1940, Rimouski Co., St. Médard (DPU); Lepage 2239, Aug. 7, 1940, Gaspé, Table-Top Mt. (DPU); Lepage 2254, Aug. 7, 1940, Gaspé, Table-Top Mt. (DPU); Lepage 2260, Aug. 7, 1940, Gaspé, Table-Top Mt. (DPU); Lepage 3232, Aug. 16, 1941, Rimouski Co., St. Mathieu Lake (DPU); Marcel, Brûlé, and Lorenzo 35198, July 22, 1935, Gaspé (MT); Marie-Anselme 131, July 13, 1935, Mt. St. Hilaire, near Montreal (DPU, MT); Marie-Anselme 224, Sept. 15, 1935, Laviolette Co., La Tuque (DPU, MT); Marie-Anselme 501, Apr. 4, 1936, near La Tuque (CHI, DPU, MT); Marie-Anselme 838, Aug. 17, 1936, Waterloo (DPU, MT); Marie-Anselme 839, Aug. 20, 1936, Waterloo (CHI, DPU, MT); Marie-Anselme 840, Aug. 20, 1936, Waterloo (CHI, DPU, MT); Marie-Anselme, Oct. 1939, St. Félicien (DUKE); Marie-Anselme 4186, July 1942, Disraeli (DPU, MT); Marie-Anselme 4073, July 17, 1943, Beauceville (DPU, MT); Marie-Jean-Eudes 2986, July 29, 1941, Montcalm Co., Rawdon, alt. 126 m (Herb. of Sisters of St. Anne); Marie-Victorin 4, June 1908, Rouville Co., Mt. Beloeil (DPU, MT, WASH); Marie-Victorin 1749, June 1909, Rouville Co., Mt. St. Hilaire (DPU, MT); Marie-Victorin, July 1913, Temiscouata Co., Lake Pratt (NY, US); Marie-Victorin 9327, July 1913, Temiscouata Co., Lake Pratt (DPU, MT); Marie-Victorin and Rolland-Germain, July 6, 1933, Labelle Co., Bellerive (DPU, MT); Marie-Victorin and Rolland-Germain 45630, July 6, 1933, Labelle Co., Bellerive (DPU, MT); Marie-Victorin, Rolland-Germain, Brunel, and Rousseau 18312, July 13, 1923, Gaspé, St. Marjorique (DPU, MT); Simon 56, Aug. 20, 1931, Rouville Co., Mt. St. Hilaire (MT).

VANCOUVER ISLAND: Macoun, May 30, 1893, Comox, Can. Moss. 201 p.p. (B, E, K, MO, MT, NY, US).

Canada without locality: Macoun, Can. Musci 228 p.p. (FH, K, NY, WIS); Macoun, July 26, 1865, Can. Musci 228 p.p. (MO); Macoun, June 13, 1883, Can. Musci 228 p.p. (MINN).

UNITED STATES: COLORADO: W. B. and M. B. Schofield 7176, Aug. 22, 1956, Routt National Forest, near Rabbit Ear Pass (CAN, DPU).

CONNECTICUT: J. A. Allen 106, Mar. 11, 1876, Mt. Carmel (NY); O. D. Allen, New Haven (NY); Barron (MO); Bishop, Feb. 8, 1880, Plainville (CHI, FH, MICH); Bishop, Mar. 1880 (CHI, DPU, FH, G-DEL, MICH, MINN); Eaton, May 24, 1869, Meriden (US); Eaton, May 19, 1873, Mt. Carmel (CAS); Eaton, Apr. 22, 1874, Bethany (CAS); Eaton, May 24, 1874, Bethany (YALE); Eaton, Dec. 17, 1884, Mt. Carmel (US); Alexander Evans, Apr. 19, 1889, Bethany (YALE); Graves 468, Apr. 7, 1884, New London Co., Waterford (YALE); Haines, in 1877, Bethany (OS); Harger 336, July 7, 1901, Oxford (YALE); Kendall, Goldsborough, and Doolittle, July 13, 1904, Connecticut R. (US); Merriam, Oct. 18, 1875, New Haven (CAS); G. E. Nichols, Cheshire (YALE); G. E. Nichols, Aug. 5, 1905, Southington (YALE); G. E. Nichols, Oct. 12, 1906, Stafford (YALE); G. E. Nichols, Oct. 13, 1906 (YALE); G. E. Nichols, Apr. 16, 1908, Burlington (CIN, YALE); G. E. Nichols, Apr. 17, 1908, Granby (YALE); G. E. Nichols, Apr. 14, 1911, Ledyard (YALE); Setchell, Apr. 12, 1884, Norwich (UC); J. L. Sheldon N143, July 17, 1901, Central Village (NY); J. L. Sheldon, Nov. 5, 1905, Central Village (YALE); J. L. Sheldon, Nov. 8, 1905, Central Village (YALE); J. S. Smith, May 1856, New Haven (YALE); Underwood, July 1887, West Goshen (NY); Underwood, Aug. 1889, West Goshen (NY); Underwood, Aug. 1889, West Goshen, Grout, N. Am. Musci Pl. 137 (ABS, CAN, CIN, CM, DUKE, FH, ILL, MINN, MO, NY, OS, UC, US, WASH, WIS, YALE); Charles Wright, Aug. 5, 1882, Plainville (FH).

DELAWARE: Commons, Sept. 10, 1874, Burris Run, near Centreville (PHIL).

IDAHO: Henderson 3313, July 25, 1895, Wood R. (US).

MAINE: Clara D. Adams, Oct. 17, 1937, Oxford Co., Peru (NY); J. A. Allen, July 1880, Hebron (FH); J. Blake, Apr. 1876, Harrison (OS); Carle, Jan. 11, 1908, Camden (PENN); Chamberlain, June 3, 1907 (H); Chamberlain 3588, July 3, 1907, Otisfield (ABS, FH); Chamberlain and Norton, 3588 of Chamberlain, July 3, 1907, Otisfield (FH); Cheever, Aug. 17, 1904, Jonesport, Indian R. (FH); L. A. Cole, July 14, 1918, Union (US); L. A. Cole 1017 p.p., July 14, 1918, Union (US); Crockett, June 7, 1900, Camden (ABS); Faxon 320, Apr. 11, 1881, Dedham (NY); Faxon, July 23, 1891, Lafayette Co. (NY); F. L. Harvey, in 1890, Orono (NY); E. D. Merrill 68, in 1897, Auburn (DUKE, NY); E. D. Merrill 69 p.p., July 1897, Auburn (NY); E. D. Merrill, Nov. 1898, Chairback Mt. (US); G. K. Merrill 184, Aug. 17, 1919, Union (DUKE, FH); A. H. Norton, July 12, 1923, Kingfield (PSNH); A. H. Norton, July 24, 1940 (DPU); Parlin, Apr. 16, 1938, Oxford Co., Hartford (DPU, PSNH); Parlin, July 16, 1946, Androscoggin Co., Livermore (DPU, PSNH); Patterson 170, July 6, 1929, Mt. Desert Isl., Long Pond (DPU); Redfield, Aug. 8, 1889, Mt. Desert Isl., Turtle Lake (NY); Rowell, July 14, 1938, Androscoggin Co., Livermore (DPU, PSNH); Steyermark 4226, July 18, 1930, Waldo Co., Patrick Mt. (CHI); W. R. Taylor, Aug. 30, 1920, Mt. Desert Isl. (PENN).

MASSACHUSETTS: Ballard, Aug. 9, 1904, Belmont (FH); E. G. Britton, Sept. 1897, Mt. Washington (FH); E. G. Britton, July 28–31, 1901, Stockridge (FH); N. L. Britton, July 28–31, 1901, Stockridge (NY); Alice Carter, Mar. 1888, South Hadley (NY); Cummings, Sept. 29, 1884, Dover (CM, FH, MO, WIS); Darker 6692, May 31, 1936, Westwood (FH, MICH); S. K. Harris, May 7, 1932, Pepperell Springs (FH); Kennedy, Dec. 15, 1894, Randolph (FH); Kennedy, Mar. 18, 1898, Canton (FH); Kingman, Mar. 20, 1908, Woburn (FH); Kingman, May 15, 1909, Norwood (FH); Kingman, Apr. 1, 1912, Andover (FH); A. H. Moore, May 1, 1903, Waltham (FH); E. T. Moul 2490, July 9, 1946, Coonamesset R., near East Falmouth (DPU, PENN); Sanford, Apr. 18, 1910, Swansea (FH); Schuh, in 1891, Vainyard Haven (NY); Seymour, Aug. 1889, Granville (CHI); Seymour, Sept. 4, 1908, Norfolk Co., Blue Hills (DPU); A. M. Small, June 1900, Northburo (ABS); Sterki, May 26, 1911, near South Framingham (CM); W. R. Taylor, July 1923, East Falmouth (PENN).

MICHIGAN: Conard, Aug. 23, 1937, Porcupine Mts., west of Ontonagon, near Lake Superior (CHI, DPU); Gillman, July 12, 1867, Laughing Fish Point, Lake Superior (NY); Ikenberry, July 14, 1936, Ironwood, Black R. Park Road (CHI, DPU); G. E. Nichols, Marquette Co., Lake Huron shore, near Sugar Loaf Mt. (YALE); G. E. Nichols, northern Michigan (UT); G. E. Nichols, Aug. 22, 1934, Marquette Co., near Huron Mt. (DPU, MICH, NY, YALE); G. E. Nichols, June 1937, Marquette Co., Wetmore's Landing, Lake Superior shore, base of Sugar Loaf Mt. (DPU); G. E. Nichols and Steere, Aug. 20–27, 1935, Ontonagon Co., Porcupine Mts. (FH, MICH, YALE); Paul, July 17, 1894, Ontonagon (ILL); Steere, Sept. 1936, Houghton Co., near Laurium (MICH); Thorpe 71, Sept. 1929. Sugar Isl. (MICH).

MINNESOTA: Cheney, July 16, 1891, Cook Co., Mud Lake (WIS); Elftman, June 21, 1897, Cook Co., near Grand Portage (PC); Holzinger and Elftman, June 21, 1897, near Grand Portage (CIN, DUKE, FH, H, MINN, MO, NY, PC, US, WYO); Kretting, July 31, 1937, Cook Co., Pike Lake (MINN); W. G. Moore, July 16, 1936, Lake Co., Little Isabella R. (MINN); Moyle 3645, July 2, 1930, St. Louis Co. (MINN); Moyle 3646, June 25, 1940, Lake Co. (MINN); Moyle 3632, June 27, 1940, Lake Co., Knife R. (MINN); Moyle 3647, June 27, 1940, Lake Co., Knife R. (MINN); Moyle 3626, July 3, 1940, St. Louis Co., near Lester R. (MINN); Moyle 3629, July 4, 1940, Lake Co. (MINN); Moyle 3627, July 8, 1940, St. Louis Co., French R. (MINN); Moyle 3628, July 9, 1940, St. Louis Co. (MINN); Moyle 3613, July 17, 1940, Lake Co., Gooseberry R. (MINN); Moyle 3623, Aug. 1, 1940, Lake Co., Baptism R. (MINN); Rosendahl and Dahlberg, Aug. 1918, St. Louis Co. (MINN).

MONTANA: Holzinger and Blake 42, June 29, 1898, Flathead Co., near Lake McDonald, alt. 1050–1500 m (FH, MINN, MONT, NY, US); Holzinger and Blake, July 14–30, 1898, Flathead Co., near Lake McDonald, alt. 1050–2100 m (CIN); Holzinger and Blake, July 29, 1898, Flathead Co., near Lake McDonald (PC); Maguire and Piranian 5339, July 9, 1934, Glacier Nat. Park (UC); Standley 16031, July 19, 1919, Glacier Nat. Park, Lake McDonald, near Glacier Hotel, alt. 960–1050 m (NY); Standley 18554, Aug. 31, 1919, Glacier Nat. Park, Fish Lake, alt. 1230 m (NY, US); R. S. Williams 391, Aug. 9, 1895, Lake McDonald (CHI, NY).

NEW HAMPSHIRE: Butler, Aug. 1919, Marlow (CHI); Butler, Aug. 1927, Lempster, alt. 420 m (CHI); L. A. Carter, Belmont (ABS); L. A. Carter, Oct. 13, 1898, Gilford (ABS, ILL); L. A. Carter, July 19, 1901, Gilford (ILL); Cleland and Taylor, June 27, 1917, Wing Road (PENN); D. S. and H. B. Correll 11076, Hillsboro Co., near New Boston (DUKE); Croasdale 49, June 24, 1945, Hanover (DPU); Eaton, Oct. 1856, Mt. Wantastiquet (YALE); Farlow, Shelburne (FH); Farlow, Aug. 1882, Shelburne (YALE); E. T. and S. A. Harper, Aug. 4, 1895, Coos Co. (CHI, FH, NY); James, Paco R., White Mts. (ILL); James, 1855, White Mts. (YALE); James, July 1857, Crawford Notch, White Mts. (K, MO, UC); Lampton 617, June 25, 1939, Gregg Lake, alt. 318 m (DPU); Mann, Mt. Hinsdale (CHI); A. H. Moore, Sept. 7, 1901, Coos Co., Randolph Hill, Randolph (CHI); Oakes, White Mts. (OS); Oakes, in 1846, copious spring, White Mts., alt. approximately 540 m (TYPE of *F. antipyretica* var. *gigantea*, FH); C. G. Pringle, June 22, 1880 (CHI).

NEW JERSEY: C. F. Austin, Apr. 1862, Bergen Co. (NY); C. F. Austin, Apr. 1865, near Closter (NY); C. F. Austin, Jan. 1872, Anderson brook (CHI, NY); N. L. Britton, July 23, 1885, Alamuche Mt. (NY).

NEW YORK: Alles, Apr. 8, 1945, White Plains (DPU); Banker, in 1919 (NY); Blanchard, Oct. 28, 1885, Peacham (ILL, NY); E. G. Britton, Sept. 18, 1896, near Lake Placid (NY); E. G. Britton, Sept. 26, 1898, Giant Mt., near Lake Placid (CIN, MICH, NY); E. G. Britton, Sept. 4, 1900, Gooseneck Pond (NY); E. G. Britton, Sept. 5, 1900, Crown Point Gorge (NY); E. G. Britton and A. M. S., June 14, 1901, Essex Co. (US); Bumstead, Orange Co. (DPU, FH); Burnham, Nov. 1, 1896, Washington Co. (NY); Burnham, Dec. 2, 1915 (MO); Clinton, Buffalo (CHI, MICH, NY); Eaton, July 4, 1864, Orange Co. (YALE); Grout, July 5, 1929, Cold Spring Harbor (DUKE); W. P. Harris, Aug. 3, 1900, Essex Co., Chilson Lake (ABS, FH); A. K. Harrison 42, June 24, 1895, Rensselaer Co., Stephentown (US); E. C. Howe, in 1866, Rensselaer Co. (CHI, NY); Knight, Aug. 30, 1887, Adirondack Mts. (NY); Latham 1736, May 1, 1919, Long Isl., Mattituck (DUKE); M. F. Miller, Aug. 1898, Shandaken (ABS, PENN); M. F. Miller, June 5, 1900, Catskill Mts., Shandaken (FH, NY); M. F. Miller, Sept. 20, 1900, Catskill Mts. (MO); M. F. Miller, Sept. 30, 1900, Shandaken (WASH); Muenscher and Isely, 486 of Winne, Sept. 9, 1941, Ulster Co., near Phoenicia, alt. 270 m (DPU, NY); Muenscher and Isely, 509 of Winne, Sept. 9, 1941, Greene Co., near Edgewood, alt. 570 m (DPU, NY); C. F. Parker, in 1864, Sandlake (MICH); Peck, in 1864, Catskill Mts. (NY); Rice, Apr. 1943, Rockland Co., Nyak (DPU); A. H. Smith, Aug. 13, 1934, Adirondack Mts., Catlin Lake (MICH); A. M. Smith, June 14, 1901, Chilson Lake, Eagle Gorge (ABS); Standley and Bollman 12026, Aug. 23, 1915, Dutchess Co. (US); N. Taylor 361, June 3, 1909, Delaware Co., Stamford, alt. 540 m (NY); Underwood, Aug. 1889, West Goshen (WASH); Vail 83, Aug. 29, 1892, Greene Co., Tannersville (NY); Winne 265, July 5, 1941, Washington Co., Pilot Knob Mt., Lake George, alt. 396 m (DPU, NY); Winne 281, July 5, 1941, Washington Co., Pilot Knob Mt. (DPU, NY); Winne 589, Oct. 7, 1941, Greene Co., East Kill, alt. 780 m (DPU, NY); Winne 628, Oct. 8, 1941, Greene Co., near Hunter, alt. 474 m (DPU, NY); Winne 650, Oct. 21, 1941, Greene Co., High Peak Mt., alt. 810 m (DPU, NY); Winne 678, Oct. 21, 1941, Greene Co., near Plaat Clove, alt. 573 m (DPU, NY); Winne 706, Oct. 22, 1941, Greene Co., Kaaterskill Junction (DPU, NY); Winne 714, Oct. 22, 1941, Greene Co., near Stony Clove Notch Lake, alt. 561 m (DPU, NY); Winne, Muenscher, and Isely, 468 of Winne, Sept. 8, 1941, Suffolk Co., near Patchoque, Long Isl., alt. 12 m (DPU, NY); Winne, Muenscher, and Shannon 171, June 10, 1941, Oneida Co., near Point Rock, alt. 270 m (DPU, NY); Wynne 2797, June 1943, Orange Co., Palisades Interstate Park, Bear Mt. Section (DPU).

OHIO: Eckfeldt, in 1870 (PENN).

PENNSYLVANIA: Bartram 238, Feb. 16, 1919, Pike Co. (BART); Best, Sept. 13, 1892, Tobyhanna Mills (NY); Boardman, June 28, 1942, Laurel Hill (CM, MT); E. G. Britton, June 7, 1889, Naomi Pines, Pocono Summit (DUKE, NY); E. G. Britton, July 3, 1899, Pocono (CAS, CHI, CM, DUKE, ILL, MO, NY, UC); Burnett (CIN); Burnett 496, Mar. 17, 1893, McKean Co. (NY); Burnett 496a, July 29, 1894, McKean Co. (NY); Burnett, May 31, 1896, Bradford (ABS, NY, WASH); Burnett 2540, June 27, 1897, McKean Co. (CIN); Burnett and Gates 1704, May 3, 1896, McKean Co. (DUKE); Chamberlain and Bartram, June 14, 1924, Pike Co. (FH); F. S. Chapman, June 1, 1932, Pocono Mts. (CAS); F. S. Chapman, July 3, 1932, Pike Co., Pocono Mts. (DPU, DUKE, WASH, WYO); Conard 46–119, Sept. 20, 1946, Monroe Co., Pocono Manor (DPU); S. K. Eastwood, Sept. 1, 1935, Cameron Co. (CM); Garber, Aug. 1, 1867, Wayne Co. (CHI); Garber, Nov. 1868, Monroe Co. (CHI); Glowenko 941, Dec. 5, 1948, Lackawanna Co., near Elmhurst, alt. 435 m (DPU); T. Green, July 16, 1857, Monroe Co., Pocono Mt. (CHI); T. Green, July 24, 1863, Monroe Co., Polyhanna Creek (CHI); T. Green, Aug. 20, 1863, Monroe Co., Pocono Mt. (PENN); C. Gross, Apr. 18, 1937, Shamokin (DPU); Hepner, Oct. 1932, Westmoreland (CM); Hepner, Oct. 21, 1933, Westmoreland Co. (CM); James, June 1858, Cambria Co. (CHI, PHIL); Kaiser, July 8, 1910, Tobyhanna, alt. 600 m (ABS); Knipe, about 1875, Monroe Co. (CM); Lesquereux, in 1885 (FH); Lewis, July 30, 1927, Monroe Co., Buck Hill Falls (MICH); Lippincott, June 13, 1898, Sullivan Co. (PHIL); E. T. Moul 4457, Sept. 7, 1946, McKean Co., near Corydon (DPU, PENN); E. T. and H. L. Moul 4999, Sept. 26,

1947, Clinton Co., near Ranchtown (DPU, PENN); T. C. Porter, (CHI, PC); T. C. Porter, Aug. 20, 1858, Tobyhanna (PHIL); T. C. Porter, May 1862 (PHIL); T. C. Porter, June 14, 1862, Lancaster Co. (PHIL); Rau, Carbon Co. (NY); Rau, Pike Co. (MICH); Rau 138, Pike Co. (FH); Rau, July 1872, Pike Co. (NY); Rau, July 5, 1877, Pike Co. (OS); Schaeffer, Sept. 13, 1940, Northampton Co., near Slate Valley (DPU, PENN); Schallert, Aug. 25, 1935, Pike Co., Pilford (FH, UC); J. K. Small, June 7-11, 1889, Monroe Co. (US); J. K. Small, Sept. 2-5, 1889, Monroe Co., near Naomi Pines, Pocono Plateau (DUKE, US, WIS); Ward, Carbon Co. (ILL); Wolle, Aug. 1873, Pike Co. (US).

RHODE ISLAND: Bennett, (NY); Bennett, Providence (FH); Mille 10, Sept. 11, 1946, Kingston (DPU); Mille 151, Apr. 7, 1947, Kingston (DPU); Mille 345, Sept. 27, 1947, Kingston (DPU); Olney (NY).

VERMONT: Dutton 965, in 1913, Rochester (CHI, CM, FH, MO, MT); Dutton, July 19, 1923, Salisbury (DUKE); Dutton 2061, July 19, 1923, Salisbury (WASH); Frost, Brattleboro (K); Grout, May 23, 1893, Johnson (FH, NY, UC); Grout, Aug. 17, 1901, N. Am. Musci Pl. 158x (BART, CAS, WASH); Grout, Aug. 2, 1906, New-fane, N. Am. Musci Pl. 281 (CAN, CM, COL, DUKE, FH, ILL, MICH, MINN, MO, NY, OS, UC, US, WASH, WIS, YALE); Grout, Hand-Lens Mosses 100 (FH, MINN, NY); E. T. and H. N. Moldenke 9526, May 9, 1937, Windham Co., Jamaica (ILL, MO, NY); A. J. Sharp, Aug. 14, 1932, near Stratton (DPU); W. R. Taylor, Sept. 15, 1922, Dawlet (PENN); E. W. Thompson, Aug. 1, 1934, Newfane (FH); E. W. Thompson, Aug. 13, 1934, Newfane (FH); Welch 711, Aug. 6, 1932, near Newfane (CHI, DPU, NY).

WASHINGTON: J. A. Allen 82 p.p., Oct. 29, 1898, Cascade Mts., upper Nes-qually Valley (CHI, DUKE, FH, ILL, MICH, MINN, NY, PENN, US, WASH, YALE).

WISCONSIN: E. G. Britton, Aug. 25, 1893, Dells (CIN, MO); Cheney 161, June 15, 1893, Vilas Co., near Lac Vieux Desert Lake (DPU, WIS); Cheney 162, June 15, 1893, Vilas Co., near Lac Vieux Desert Lake (DPU, WIS); Cheney 204, June 16, 1893, Vilas Co., near Lac Vieux Desert Lake (DPU, WIS); Cheney 205, June 16, 1893, Vilas Co., near Lac Vieux Desert Lake (DPU, WIS); Cheney 796, June 26, 1893, Vilas Co., Wisconsin R., near Conover (DPU, WIS); Cheney 966, June 29, 1893, Oneida Co., Wisconsin R., below Eagle R. (DPU, WIS); Cheney 1533, July 10, 1893, Oneida Co., near Newbald (DPU, WIS); Cheney 2058, July 21, 1893, Lincoln Co., Wisconsin R., Nigger Islands, near Tomahawk (DPU, WIS); Cheney 2707, July 30, 1893, Lincoln Co., Wisconsin R., Grandfather Bull Falls (DPU, WIS); Cheney 2916, Aug. 4, 1893, Marathon Co., Wisconsin R., Granite Heights (DPU, WIS); Cheney 3376, June 30, 1894, Portage Co., Wisconsin R., between Knowlton and Stevens Point (CHI, DPU, WIS); Cheney 5943 p.p., Aug. 6, 1896, Bayfield Co., Presque Isle of the Apostle Islands, Lake Superior (DPU, WIS); Cheney 5943a, Aug. 6, 1896, Bayfield Co., Presque Isle of the Apostle Islands, Lake Superior (DPU); Cheney 6313, June 22, 1897, Bayfield Co., Sand Bog, Lake Superior (DPU, WIS); Cheney 6841, July 5, 1897, Bayfield Co., near Herbster (DPU, WIS); Cheney 7346, July 14, 1897, Bayfield Co., near Orienta, Lake Superior (CHI, DPU, WIS); Cheney 9595, May 17, 1898, Sauk Co., Devil's Lake (WIS); Cheney 9798, Feb. 12, 1925, Barron Co., near Barron (DPU, WIS); Cheney 9935, Apr. 9, 1925, Rusk Co. (DPU, WIS); Cheney 9941, Apr. 10, 1925, Barron Co., Doyle Township (DPU, WIS); Cheney 13171, June 25, 1933, Barron Co., Doyle Township (DPU, WIS); Cheney 13227, Barron Co., Doyle (WIS); C. Gross, May 25, 1935, Devil's Lake (DPU); C. Gross, May 30, 1937, Devil's Lake State Park (DPU); C. Gross, Aug. 26, 1937, near Prairie du Sac (CHI, DPU, MICH, NY); T. J. Hale, Baraboo (WIS); Lapham, in 1862, Manitowoc (WIS); Thompson and Jacobson, 5199 of Thompson, July 2, 1943, Douglas Co., Brule R. (WIS); E. H. Walker, May 26, 1928, Devil's Lake (US); L. R. Wilson 2078, July 1926, Douglas Co. (WIS); L. R. Wilson 260, Aug. 23, 1932, Vilas Co., near Nibish Lake (DPU).

WYOMING: A. Nelson 8521 p.p., July 29, 1901, Sheridan Co. (FH, WYO); A. Nelson 9670, Aug. 12, 1912, Albany Co., Medicine Bow Mts., alt. 3000 m (FH, MINN, MO, NY, US, WYO).

United States without locality: Austin, Musci Appal. 243 (CAN, CAS, DUKE, K, NY, PHIL, US, WIS); James, "Con Typus No. 224 of *Fontinalis gigantea*, North America in rivulis montosis Novae Angliae," (From Fleischer Herb.), (FH); Lesquereux, in 1872, New England (PC); Oakes, Tuckerman, and James, Sullivant and Lesquereux, Musci Bor.-Am. 224, in rivulis montosis Novae Angliae (COTYPE, FH; DUKE, K, MICH, MO, NY, WIS); Sullivant and Lesquereux, Musci Bor.-Am. 229 p.p. (TRT); Sullivant and Lesquereux, Musci Bor.– Am. 335 p.p., in rivulis montosis Novae Angliae et Novaeboracensis (CHI, CM, G-BOIS, K, MICH, NY, PHIL, US, WIS, YALE).

The following characteristics usually distinguish mature, complete plants of var. *gigantea* from those of *F. antipyretica:* plants robust, median cauline leaves slightly distant to imbricate, ovate, broadly ovate, suboval, or suborbicular, generally 4–8 mm long by 3–6.5 mm wide, occasionally larger, up to 10 mm in length and 8.5 mm in width, and apices broadly obtuse. The examination of young plants of var. *gigantea* or only branches of mature plants of the same may lead to an incorrect determination, discovered when many packets of the same collection are observed, finding, at least, a few mature, median, cauline leaves with characteristics of var. *gigantea*. If the latter are not found, the determination is very likely to be that of *antipyretica*. If the large size of plants and median cauline leaves were consistent, it would seem logical to consider *gigantea* as a species rather than a variety. But, a great variation in leaf size on plants of *antipyretica* and of var. *gigantea* sometimes results in a mixture of antipyretica-like leaves with gigantea-like leaves on the same plant. These conditions are considered by the writer to be evidence of the close relationship between *antipyretica* and var. *gigantea*. In the instances of difficult recognition between the species and the variety, the author makes the determination on the basis of the length and width ratio in several mature, median, cauline leaves, commonly 1–1.5 : 1 in var. *gigantea* and 1.5–3.5 : 1 in *antipyretica*.

In *F. antipyretica* var. *monensis*, formerly recognized, the habitat factors of calcareous incrustation and still water probably have been causal agents in the spreading position of many of the median cauline leaves and the branches, but, in the opinion of the author, are not diagnostic characteristics to be used in naming the plants. This variety emphasizes the necessity of using the size and the length-width ratio of the median cauline leaves along with leaf shape, cell shape, size, and ratio, the presence of auricles, etc., for determining the specimen. The plants of var. *monensis* illustrate, again, the relationship between *F. antipyretica* and the varieties of this species. If only the leaves on the branches and the younger portions of the stems were to be considered, the plants would be named *F. anti-*

pyretica. The median cauline leaves, however, have the characteristics of var. *gigantea*. The fruit characteristics of var. *monensis* are
comparable to those of var. *gigantea*. The plants of var. *monensis*
have been compared with those of var. *mollis*. There is accord in shape
and length-width ratio in many median cauline leaves, as well as in
other detailed characteristics but not in the prevailing leaf size and
plant robustness. The writer has studied specimens of var. *gigantea*,
collected in the United States and in Canada, which bear strong resemblance to the type plants of var. *monensis*. Thus the author has
concluded that var. *monensis* is a synonym of var. *gigantea*. On July
20, 1938, Dr. Hj. Möller told the writer that H. G. Simmons and he
collected this moss together and that Simmons sent it to Cardot for
determination.

Although the original description of *Fontinalis livonica* gives leaf
dimensions as 4–5 mm long by 4 mm wide, the author has measurements of median cauline leaves 5.5–6 mm long by 5–6 mm wide on the
plants assumed to be portions of the type. The vegetative and fruiting
characteristics of *F. livonica* have been compared with those of var.
gigantea and have been found in agreement. In the original description
of *F. livonica* the habit of the plants is likened to that of *F. antipyretica*
var. *latifolia*, a synonym of *F. antipyretica* var. *gigantea*. In Cardot's
Herbarium in Paris Museum, a notation was observed by the author
on the collection made by Von Bock, Sept. 1907, in Livland, "*Fontinalis livonica* nov. sp. = *Fontinalis antipyretica* var. *monensis*, fide
Cardot." The writer considers both *F. antipyretica* var. *monensis* and
F. livonica as synonyms of *F. antipyretica* var. *gigantea*. The calcareous
deposit on the plants is a habitat factor which may have influenced
the position of the median cauline leaves and the branches to be
frequently spreading but this characteristic, in the opinion of the
author, is not diagnostic for the determination of the plants.

4. **Fontinalis antipyretica** Hedw. var. **mollis** (C. Müll.) Welch, in Grout, Moss Fl. N. Am. 3: 237. 1934.

Fontinalis mollis C. Müll., in Röll, in Bot. Centralbl. 44: 421. 1890.
Fontinalis gigantea Sull. subsp. *mollis* (C. Müll.) Kindb., Sp. Eur. and N. Am.
    Bryin., Part 1: 150. 1896.
Fontinalis utahensis Card. and Thér., in Thériot, in Arch. Bot. 1: 67. 1927.
Fontinalis antipyretica var. *pseudomollis* Card., herb. name, according to Welch,
    in Grout, Moss Fl. N. Am. 3: 237. 1934.
Fontinalis patens Card., herb. name, according to Welch, in Grout, Moss Fl. N.
    Am. 3: 237. 1934.
Fontinalis antipyretica var. *patens* Ren. and Card., herb. name, according to
    Welch, in Grout, Moss Fl. N. Am. 3: 237. 1934.

Plants moderately slender, glossy, green, bright yellowish green, grayish green, brown, or brownish; stems usually subflaccid to flaccid, occasionally slightly rigid to rigid, up to 40 cm long, 0.25–0.35 mm in diameter, foliated to base or slightly denuded at base with age, irregularly pinnately branching; branches usually erect-spreading to spreading, occasionally recurved, varying in length from short to elongate, up to 11 cm long, ends of imbricately foliated stems and branches usually acute, occasionally subobtuse, frequently conspicuously three-angled, pyramidal in outline and triangular in cross

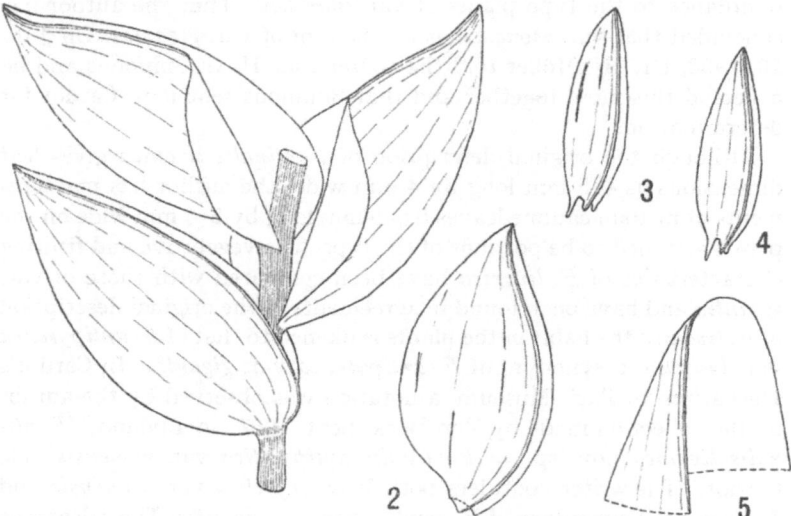

Fig. 4. – *Fontinalis antipyretica* Hedw. var. *mollis* (C. Müll.) Welch – 1. Portion of stem with median cauline leaves. 2.Median cauline leaf. 3 and 4.Median branch leaves. 5.Leaf apex. No. 1 drawn from Ikenberry, Feb. 21, 1935, California, Grout, N. Am. Musci Pl. Suppl. 56; nos. 2–5 from Röll, July 15, 1888, mouth of the Columbia River.

section; median cauline leaves erect-spreading to spreading, slightly distant to distant, bases commonly 0.5–2 mm apart, blades usually flaccid or subflaccid but frequently firm, carinate-conduplicate or deeply concave, keels or median lines generally semicircular or approaching semicircular, sometimes only moderately curved above basal curve, blades usually suboval or suborbicular when unfolded but occasionally subovate, commonly broadest in middle portion, generally decreasing gradually into apical and basal parts, some leaves with one or both margins reflexed near base, blades symmetrical with curvature of the margins of the leaf halves when folded lengthwise usually paralleling the curvature of the keels; apices commonly broadly obtuse, at times narrowly obtuse, sometimes subcucullate or cucullate, leaf tips entire to serrulate; median cauline leaves generally moderate in size, 3.5–5 mm long, occasionally up to 6 mm in length, 2.5–5 mm wide, sometimes up to 6 mm in width, 1–1.6 : 1; median cells of leaves usually linear-rhomboidal or linear with ends attenuate, sometimes subflexuous to flexuous, 10–20 μ wide, 4–15 : 1; alar cells enlarged, rectangular, quadrate, or subhexagonal, greenish, yellowish, brownish, brown, or yellowish brown, with auricles varying from none to slight to distinct, leaf bases not decurrent, briefly decurrent, or decurrent to length of 1 mm; median branch leaves usually similar to median cauline except smaller in size, occasionally some blades resembling those of *antipyretica* in shape but with apices more broadly obtuse; perichaetial branch 4.5–6 mm long, perichaetium suboval to oval, 1.5–2 mm in diameter; upper perichaetial leaves oval to suborbicular, apices broadly obtuse, entire, frequently truncate and lacerate with age; calyptra not seen; operculum obtuse conical, approximately 0.75 mm long; urn immersed to slightly emergent, commonly slightly contracted beneath mouth when dry but occasionally not, oval to subcylindrical, 1.75–3 mm long, 1–1.75 mm in diameter, 1.5–2 : 1; seta 0.25 mm long; peristome teeth brownish orange, linear-acuminate, frequently united in pairs at apex, approximately 1 mm in length, muricate, lamellae 28–40; trellis brownish orange, perfect, approximately 1 mm long, muricate, transverse bars distinctly appendiculate or subappendiculate; spores green, slightly muricate, 22–24 μ in diameter, ripe in summer. [The name, *mollis*, meaning soft or pliant, has reference to the soft or subflaccid to flaccid plants and leaves.]

Type: In the Botanisches Museum, Berlin-Dahlem, Germany. The label reads, "Herbarium C. Mueller-Hal. *Fontinalis mollis* n. sp. Amer. sept. Washington Territorium, Astoria, ad ostium fluminis Columbia: Dr. Julius Roel, lg. 15. Julio 1888." Dr. Mueller's description, hand-written, is mounted on the sheet over the plants, which are in fruit. The author examined this material in the Botanisches Museum in June 1938. In case the type has been destroyed during World War II, the author has selected to substitute for the type, the plants, with fruit, assumed to be a portion of the type collection, in the Herbarium of Cardot, in the Paris Museum, with the label, "*Fontinalis mollis* C. Müll. sp. nova. 15 Juillet 1888. Astoria, Washington. Dr. Julius Röll 292. Bouche du Columbia."

Type collector: Julius Röll 292, July 15, 1888.

Type locality: There is a discrepancy as to the state in which the type collection was made. In the herbaria of the New York Botanical Garden, Botanisches Museum der Universität, Helsingfors, and Naturhistoriska Riksmuseet, Stockholm, portions of the type collection are deposited, with the locality given as Astoria, Oregon. In the herbaria of Royal Botanic Gardens, Kew, Boissier, Geneva, Laboratoire de Cryptogamie, Muséum d'Histoire Naturelle, Paris, and Botanisches Museum, Berlin-Dahlem, as well as in the original description, Washington is the state cited. All labels and citations refer to the Columbia River, and many of them to the mouth of the Columbia River. The original description states, "Washington: ad ostium fluminis Columbia. 15. Julio 1888. (Röll 292)." [1]

Distribution: Southwestern Canada and northern and western United States.

Additional descriptions: Cardot, Mon. Font., p. 90. 1892; Röll, in Hedwigia 32: 298. 1893; Barnes, Gen. and Sp. N. Am. Moss., p. 328. 1896.

Illustrations: Thériot, in Arch. Bot. 1: fig. 1 (as *F. utahensis*). 1927; Welch, in Grout, Moss Fl. N. Am. 3: pl. 74. 1934). – FIGURE 4.

Specimens examined: NORTH AMERICA: CANADA: VANCOUVER ISLAND: Mac-Kenzie 352, Sooke R., among grass plants at edge of brackish water (DPU); Macoun 209, May 26, 1893, Shawnigan Lake (NY); Macoun, May 26, 1893, Shawnigan Lake, Can. Musci 598 p.p. (DUKE, FH, WASH); Macoun, May 26, 1894, Shawnigan Lake (CAN).

UNITED STATES: CALIFORNIA: DuVall, July 20, 1935, Redwood Nat. Forest, near Crescent City (DPU); Ikenberry, Feb. 21, 1935, Auberry (CHI, DPU, MICH); Ikenberry, Feb. 21, 1935, Auberry, Grout, N. Am. Musci Pl. Suppl. 56 (BART, CM, DPU, MINN, NY, OS, UC, UT, YALE); Ikenberry, Feb. 22, 1935, Auberry, alt. 1050 m (DPU); L. and S. Koch 1399, May 5, 1947, Napa Co., near Calistoga (DPU, MICH); L. C. Wheeler 3212, Aug. 23, 1934, Siskiyou Co., Siskiyou Mts., Jaynes Canyon, alt. 1200 m (BART).

IDAHO: DuVall, July 4, 1934, Trude (DPU); T. C. Frye, Sept. 5, 1929, Bear R., west of Cape Horn (DPU, WASH); MacFadden, June 23, 1942, Elmore Co., Boise Nat. Forest, across river from Atlanta (CHI, DPU, MACF, NY, OS); Mac-Fadden, Oct. 8, 1942, Elmore Co., Boise Nat. Forest, Deer Park (CHI, DPU, MACF, NY).

MINNESOTA: Drexler 3781, Aug. 20, 1943, Cook Co., Gunflint Trail, near Grand Morais (DPU, ILL).

MONTANA: Whitham 1494, July 28, 1932, Gallatin Co. Gallatin Nat. Forest, alt. 1740 m (DPU).

OREGON: T. C. Frye, Mar. 25, 1931, Kerby (DPU, WASH); Röll, July 15, 1888, Astoria, mouth of Columbia R. (Portion of TYPE collection, H, NY, S); L. C. Wheeler 2962, July 24, 1934, Jackson Co., Siskiyou Mts., alt. 1420 m (BART).

UTAH TERRITORY: Remy, Sept. 1855, near source of Humboldt R., originally cited as "Utah territory, ruisseau des montagnes vers la source du Humboldt River (California)," (TYPE of *F. utahensis*, PC).

WASHINGTON: Winona Bailey, Mar. 15, 1908, Edmonds, near Hall's Lake (UCLA, WASH); A. S. Foster, July 2, 1904, Cathlamet (PC); A. S. Foster 439b, July 2, 1904, Cathlamet (DUKE); A. S. Foster, July 5, 1914, Clallam Co., Olympic Hot Springs, in forest swamp, alt. 750 m (DPU, WASH); T. C. Frye, Aug. 18, 1914, Ozette, Lake Ozette (DPU, WASH); Piper, Aug. 1892, Hamilton (MICH); Piper 224, Aug. 6, 1895, O'Neill's Camp (PC); Röll 292, July 15, 1888, Astoria, at mouth of Columbia R. (TYPE, B; G-BOIS, K, PC).

The median cauline leaves of var. *mollis* are distinguished from those of species and other varieties in the carinate-conduplicate group by the keels or median lines frequently semicircular or nearly so, blades commonly broadest in the middle portion and suboval to suborbicular in shape, gradually narrowing into broadly obtuse apices, generally 3.5–5 mm long, 2.5–5 mm wide, occasionally up to 6 mm in length and width, 1–1.6 : 1, and frequently flaccid.

---

[1] According to maps, Astoria is in Oregon, on the Columbia River. Perhaps the specimens were collected in Washington, across the Columbia river from Astoria, Oregon.

5. **Fontinalis antipyretica** Hedw. var. **Heldreichii** (C. Müll.) Ruthe, in Geheeb, in Flora 69: 343. 1886.

Fontinalis Heldreichii C. Müll., in Geheeb, in Flora 69: 343. 1886; in De Heldreich, in Akad. Preuss. Wiss. Berlin, Part 1: 158. 1883 (nomen nudum).

Plants medium, yellowish green, golden green, yellowish brown, golden brown, or copper color, very glossy when dry, foliated stems and branches frequently resembling narrow, loose braids; stems subrigid to rigid, up to 40 cm in length and 0.5 mm in diameter, denuded and reddish at base with age, irregularly pinnately branching;

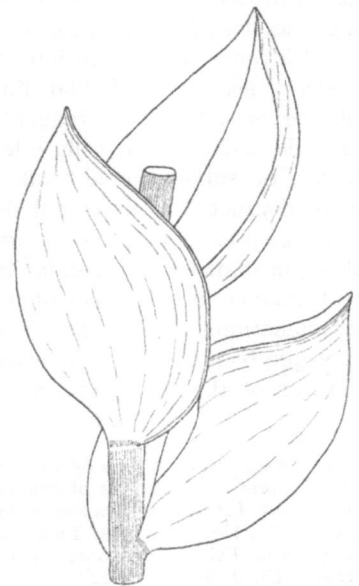

Fig. 5. – *Fontinalis antipyretica* Hedw. var. *Heldreichii* (C. Müll.) Ruthe – Portion of stem with leaves. Drawn from Th. de Heldreich, July 28, 1882, Iter Thessalum 38, Greece.

branches erect, erect-spreading, or spreading, up to 8 cm long, ends of imbricately foliated stems and branches acute or obtuse, club shape, pyramidal or subpyramidal in outline, triangular or subtriangular in cross section; bases of median cauline leaves 1–2 mm apart, occasionally up to 2.5 mm apart, blades firm, erect to erect-spreading, occasionally spreading, carinate-conduplicate, the keel very strongly curved, unfolded blades suborbicular, carinate-conduplicate blades subrhomboidal or rhomboidal, basal margins frequently reflexed on one or both sides, blades unsymmetrical because greatest curvature of keel is approximately at one-half the distance from base to apex and greatest curvature of margins is in basal half, the curvature of the margins of leaf halves when folded lengthwise rarely paralleling the curvature of the keels; apices subobtuse, narrowly obtuse, subacute, acute, or short acuminate, leaf tips serrulate or entire, apical portions often subfalcate, tips frequently extending toward axes because of strongly curved keels, apical region of one leaf half or both leaf halves frequently subcarinate or carinate-conduplicate and producing apices with two or three keels; median cauline leaves generally wider than long or as wide as long, occasionally slightly longer than wide, 4–6 mm wide, 4–5 mm long, 1–1.2 times wider than long to 1.1 times longer than wide; median cells of leaves rhomboidal, linear-rhomboidal, or linear with ends obtuse or attenuate, frequently subflexuous or flexuous, 15.3–25.5 μ wide, 3–8 : 1; alar cells enlarged, subquadrate, subrectangular, suboval, or subhexagonal, walls yellowish brown, aurciles none, slight, or distinct, leaf bases not decurrent, short de-current, or long decurrent, up to 0.75 mm, sometimes ⅜ of the distance to the next leaf below; median leaves of older branches similar to median cauline except smaller, basal leaves of branches and leaves of younger branches sometimes resembling those of *antipyretica*; perichaetium and sporophyte not seen. [The name, *Heldreichii*, has reference to the collector of the type, Th. de Heldreich, Athens, Greece.]

Type: No specimen so indicated has been seen by the writer. The type should be in the Carl Müller Herbarium and portions of the type should be in the herbaria of Th. de Heldreich and of Ruthe. Plants were studied in the Herbarium Delessert, bearing the label, "De Heldreich, Iter Thessalum no. 38. *Fontinalis Heldreichii* K. Müller n. sp. In m. Pelion: reg. super. in rivulis. alt. 4000'. super Zagora. Legit Th. De Heldreich. d. 28 Jul. 1882."

Type collector: Theodor de Heldreich, July 28, 1882.

Type locality: Thessaly: Mount Pelion.

Distribution: Europe: Greece, Thessaly.

Additional descriptions: Kindberg, in Rev. Bryol. 14: 53. 1887; Cardot, Mon. Font., p. 69. 1892; Mönkemeyer, Laubm. Eur. Erg.-Bd. 4: 658. 1927.

Illustrations: FIGURE 5.

Specimens examined: EUROPE: GREECE: De Heldreich, July 28, 1882, Thes-saly, Mt. Pelion, above Zagora, alt. 1200 m, Iter Thessalum 38 (G-DEL); De Heldreich, July 28, 1882 and May 3, 1883, Thessaly, Mt. Pelion, above Zagora, alt. 1200 m (BM, BR, PC, S); De Heldreich, May 3, 1883, Thessaly, Mt. Pelion, above Zagora, alt. 1200 m (PC, S); De Heldreich, Sept. 20, 1887, Thessaly, Mt. Pelion, above Zagora, De Heldreich Herbarium Graecum Normale 1000 (BM, CHI, MINN, PC, S, WASH).

Fontinalis antipyretica var. *Heldreichii* is distinguished from other members of the carinate-conduplicate group by the following combination of characteristics: foliated axes frequently resemble narrow braids, majority of median cauline leaves unsymmetrical, generally wider than long or as wide as long, occasionally slightly longer than wide, rhomboidal or subrhomboidal when carinate-conduplicate, suborbicular when unfolded, keels very strongly curved, greatest curvature of margins in basal half, greatest curvature of keels half distant between base and apex, in blade halves folded length-wise the curves of margins and keels are not approximately parallel, and median cells commonly flexuous or subflexuous.

Although the characteristics of *Heldreichii* seem quite distinct and it might be assumed by some that it should be a species rather than a variety, it resembles, fundamentally, *F. antipyretica* and especially var. *gigantea*. In plants of *F. antipyretica* and var. *gigantea* there has been observed occasional median cauline leaves which approximate the rhomboidal or subrhomboidal shape of those of *Heldreichii* and on specimens of *Heldreichii* occasional leaves which approach the symmetrical form of *gigantea* and *antipyretica*. For these reasons the author has decided to consider *Heldreichii* as a variety of *F. antipyretica*.

6. **Fontinalis antipyretica** Hedw. var. **oreganensis** Ren. and Card.,[1] in Rev. Bryol. 15: 71. 1888.

Fontinalis antipyretica var. *rigens* Ren. and Card., in Röll, in Bot. Centralbl. 44: 421. 1890.[2]
Fontinalis angustifolia Card. sp. nova, herb. name, according to Welch, in Grout, Moss Fl. N. Am. 3: 238. 1934.
Fontinalis Fosteri Card., herb. name, according to Welch, in Grout, Moss Fl. N. Am. 3: 238. 1934.

Plants slender to medium, golden, yellowish green, brownish green, olive green, or copper color, usually glossy in upper portions and dull in lower; stems subflaccid to flaccid and foliated to base when young, slightly rigid to rigid and denuded in basal portions when older, up to 30 cm in length, 0.25–0.4 mm in diameter, irregularly pinnately branching; branches numerous, erect-spreading to spreading, up to 8.5 cm in length, varying from close to distant, ends of subimbricately to imbricately foliated branches and stems frequently attenuate; median cauline leaves firm, distant, with bases up to 1.5 mm apart, blades erect-spreading to spreading, carinate-conduplicate,

---

[1] Although the name of the state is spelled Oregon, Renauld and Cardot were consistent in spelling the name of the variety as *oreganensis* and citing the state as Oregon. Some bryologists have spelled the varietal name as *oregonensis*.
[2] This name was erroneously published as *F. rigens* Ren. and Card. in Welch, in Grout, Moss Fl. N. Am. 3: 237. 1934.

keels above basal curve usually straight, sometimes slightly curved, now and then abruptly curved at apex, occasional leaves with more strongly curved keels and resembling median cauline leaves of *antipyretica* in shape and size, sublanceolate when unfolded, frequently approaching sublinear when conduplicate, the width generally gradually diminishing from base to apex, some leaves with one or both margins reflexed near base; majority of apices subobtuse or obtuse, some acute, leaf tips usually serrulate, occasionally entire; median cauline blades 3.5–6 mm long, occasionally up to 6.5 mm in length, 0.75–2 mm wide, 2–4.5 : 1, sometimes up to 3.5 mm in width in anti-

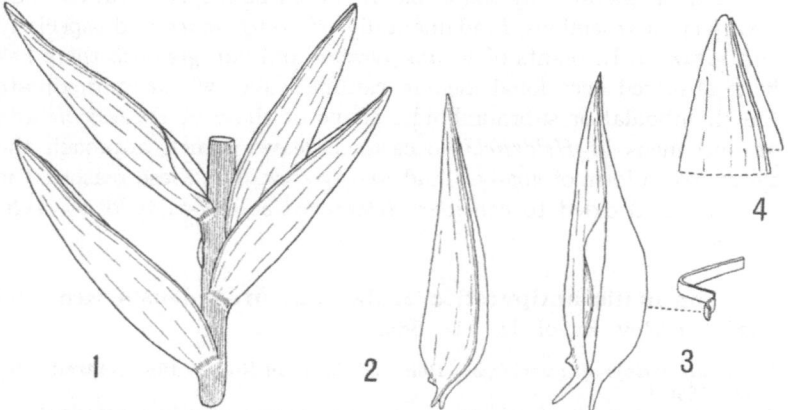

FIG. 6. – *Fontinalis antipyretica* Hedw. var. *oreganensis* Ren. and Card. – 1.Portion of stem with leaves.   2 and 3.Median cauline leaves.   4.Leaf apex. Drawn from Th. Howell 19, Coast Mts., Oregon.

pyretica-like leaves and 1.25–1.7 : 1; median cells of leaves commonly linear with ends attenuate, sometimes linear-rhomboidal, 6.5–13.5 μ wide, 8–17 : 1; alar cells enlarged, very distinct because of contrast in size and shape with basal cells immediately above, frequently in vertical rows, up to 13, also occasionally in horizontal rows, rectangular to subhexagonal, yellowish brown to brown, usually forming conspicuous auricles, leaf bases long decurrent, 0.5–0.75 mm; median branch leaves subcarinate or carinate-conduplicate with keels usually straight but occasionally slightly curved, canaliculate, concave, or plane, lanceolate, 3–4 mm long, sometimes up to 4.5 mm in length, 0.5–0.75 mm wide, occasionally up to 1 mm in width, 4–6 : 1; perichaetial branch 4–5 mm long, perichaetium subcylindrical, 1–1.75 mm in diameter; upper perichaetial leaves suboval to suborbicular, obtuse, entire, usually lacerate with age; calyptra conical, approximately 1.5

mm long and 1 mm in diameter; capsule with urn and base of operculum immersed in perichaetial leaves to urn slightly emergent; seta very short, 0.2–0.25 mm long; operculum conical, 1 mm long, 0.8 mm in diameter; urn ovate-oblong to subcylindrical, 2–3 mm long, 0.8–1.25 mm in diameter, usually contracted beneath mouth when dry; peristome teeth brownish orange, linear-acuminate, often united in pairs at the apex, 0.8–1 mm long, submuricate to muricate, lamellae 25–35, occasionally up to 40; trellis of inner peristome perfect, brownish orange, approximately 1 mm long, submuricate to muricate, transverse bars appendiculate; spores brownish, submuricate, 13.5–17 µ in diameter, ripe in summer. [The varietal name, *oreganensis*, is indicative of Oregon, the state in which Thomas Howell collected the plants which were used as a basis for the original description.]

Type: Not previously designated. In the absence of material so marked, the writer has assumed to be the type the plants in the herbarium of the Laboratoire de Cryptogamie, Muséum d'Histoire Naturelle, Paris, bearing the following label: "19. *Fontinalis antipyretica* L. var. *oreganensis* Renauld and Cardot. Oregon: in swamps on the roots of trees, etc. Top of Coast Mts. Leg. Th. Howell."
Type collector: Thomas Howell.
Type locality: Oregon: top of Coast Mountains.
Distribution: North America: southwestern Canada and western United States.
Additional descriptions: Renauld and Cardot, in Bot. Gaz. 14: 96. 1889; Renauld and Cardot, in Bull. Soc. Bot. Belg. 28: 129. 1889; Cardot, Mon. Font., pp. 54 (as var. *oreganensis*), 55 (as var. *rigens*). 1892; Röll, in Hedwigia 32: 298 (as var. *rigens*). 1893; Barnes, Gen. and Sp. N. Am. Moss., p. 326. 1896; Welch, in Grout, Moss Fl. N. Am. 3: 237. 1934.
Illustrations: Welch, in Grout, Moss Fl. N. Am. 3: pl. 74. 1934. – FIGURE 6.
Specimens examined: NORTH AMERICA: CANADA: BRITISH COLUMBIA: J. W. Bailey, June 1901, Cumberland (DPU, DUKE); J. W. Bailey, June 6, 1901, Cumberland, Grout, N. Am. Musci Pl. 140 (CAN, DPU, DUKE, FH, H, ILL, MINN, NY, OS, PC, UC, US, WASH, WIS, YALE); J. W. Bailey, Aug. 21, 1901, Cumberland (WASH); J. W. Bailey, Aug. 21, 1901, Cumberland, alt. 120 m, Grout, N. Am. Musci Pl. 91 (CAN, DUKE, FH, H, ILL, MICH, MINN, MO, NY, OS, PC, UC, US, WASH, WIS, YALE); T. C. Frye, Aug. 8, 1923, Sooke (DPU); A. J. Hill 101, Mar. 10, 1903, Langley (PC); A. J. Hill, July 1903, New Westminster (DUKE, FH, PENN); A. J. Hill, July 10, 1903, New Westminster (DUKE); A. J. Hill, Aug. 1904, Glenbrook (PENN); A. J. Hill, Aug. 1904, New Westminster (MO); Macoun, Peace R., Can. Musci 227 p.p. (FH, MINN, PC, S); Macoun, July 11, 1890, Kootenay R. (CAN); Macoun, July 20, 1901, Chilliwack R. (CAN); Macoun 186a, July 15, 1909 (K); Macoun, in 1912, Sidney (NY).
QUEEN CHARLOTTE ISLANDS: Spreadborough, July 11, 1910, Skidegate (CAN, CM).
VANCOUVER ISLAND: Conard, Sept. 1947 (DPU); Macoun 155, Esquimault (S); Macoun, May 7, 1875, Can. Musci 757 p.p. (CAN); Macoun, in 1887 (FH); Macoun, May 3, 1887 (H, K); Macoun, May 6, 1893, near Victoria, Can. Moss. 208 p.p. (K, NY); Macoun, May 8, 1893, near Victoria (CAN); Macoun, May 8, 1893, near Victoria, Can. Moss. 208 p.p. (FH, K, MO, NY); Macoun, May 15, 1893, Victoria (K, WASH); Macoun, May 15, 1893, Can. Musci 2 (CAN); Macoun, May 15, 1893, Can. Musci 9 (CAN); Macoun, May 15, 1893, near Victoria, Can. Moss. 208 p.p. (CAN, FH, US); Macoun, May 15, 1893, Can. Musci 598d p.p. (DUKE, FH, MO, NY); Macoun, May 26, 1893, Can. Musci 598e p.p. (PC); Röll 81, May 24, 1888, Victoria (G-BOIS); Röll 683, May 24, 1888, Victoria (NY).

UNITED STATES: IDAHO: T. C. Frye, July 7, 1934, Snake R. (CHI, DPU, WASH); Leiberg 114 = 230 p.p., Kootenai Co. (NY).

NEVADA: C. F. Baker 1464, Aug. 14, 1902, Washoe Co., alt. 2000–2155 m (DUKE, FH, UC, US, WIS, WYO).

OREGON: Daugherty, July 11, 1921, South Beach (DPU, WASH); N. L. Gardner, Apr. 1883, Willamette R. (BART, UC); Henderson, in 1883, Walla Walla R. (YALE); T. Howell 19, Coast Mts. (Assumed to be the TYPE, PC); T. Howell, from Renauld and Cardot herbaria (FH); Lyall, Oregon Boundary Commission, near the 49th parallel of latitude (K); C. G. Pringle 516, Oct. 1881 (NY); C. G. Pringle, Oct. 24, 1881, Winchester Bay (NY).

WASHINGTON: E. Allen, May 26, 1928, Lake Lucern (DPU, WASH); J. A. Allen, Nov. 5, 1898, upper Nesqually R. (NY); J. A. Allen, Feb. 1913, Pierce Co., Grout, N. Am. Musci Pl. 437 (BART, CAS, CM, COL, DUKE, MINN, MO, NY, OS, UC, US, WASH, YALE); J. W. Bailey, June 20, 1904 (DPU); J. W. Bailey, July 15, 1904, Seattle, University Station (WASH); J. W. Bailey, July 15, 1904, Seattle, Grout, N. Am. Musci Pl. 210 (CAN, CM, DPU, DUKE, FH, H, ILL, MINN, MO, NY, OS, PC, UC, US, WASH, WIS, YALE); J. W. Bailey, June 20, 1934, Seattle, Grout, N. Am. Musci Pl. Suppl. 48 (BART, CM, DPU, MINN, NY, OS, UC, UT, YALE); Flett, Oct. 23, 1901, Mt. Rainier (NY, US); Flett, Aug. 24, 1919, Mt. Rainier (NY, US); A. S. Foster 1428 p.p., Nov. 26, 1900, Copalis (NY); A. S. Foster, July 10, 1907, Orcas Isl. (WASH); A. S. Foster, Sept. 20, 1907, Summit (FH, PC); A. S. Foster, in 1908 (PC); A. S. Foster 747, Apr. 1908, Westport (DUKE); A. S. Foster 907c, May 28, 1908, Quinault Indian Agency, Taholah (DPU, DUKE); A. S. Foster, Aug. 18, 1908, near Copalis Crossing (ABS, NY); A. S. Foster 1426, Aug. 18–20, 1908, Copalis (ABS, DPU, FH, PC, WASH); A. S. Foster, Apr. 29, 1909 (DPU, WASH); A. S. Foster, May 1909, Quinault Indian Reservation (PC); A. S. Foster, May 14, 1909, Moclips (DPU, WASH); A. S. Foster 907c, May 14, 1909, Quinault Indian Reservation (DPU, WASH); A. S. Foster 907b, May 15, 1909, near Taholah (DPU, WASH); A. S. Foster 907b, May 24, 1909, Quinault Indian Reservation (DUKE); A. S. Foster, Aug. 1909, Mt. Rainier (PC); A. S. Foster, Aug. 9, 1909, Mt. Rainier Nat. Park, Reflection Lake, base of Pinnacle Peak, alt. 1350 m (WASH); A. S. Foster, Aug. 9, 1909, Mt. Rainier Nat. Park, Reflection Lake, base of Pinnacle Peak, alt. 1500 m (PC); A. S. Foster, Aug. 9, 1909, Mt. Rainier Nat. Park, Reflection Lake, alt. 1800 m (PENN); A. S. Foster 1046, Aug. 9, 1909, Mt. Rainier Nat. Park, Reflection Lake, base of Pinnacle Peak, alt. 1350–1800 m (DUKE, H, WASH); A. S. Foster 1046, Aug. 10, 1909, Mt. Rainier Nat. Park, Reflection Lake (WASH); A. S. Foster, Nov. 26, 1910, Copalis (WASH); A. S. Foster 1426, Nov. 26, 1910, Copalis (NY, YALE); A. S. Foster 1427, Nov. 26, 1910, Copalis (FH, PC, YALE); A. S. Foster 1429, Nov. 27, 1910, Pacific Beach (DPU, PC, WASH, YALE); A. S. Foster, Feb. 26, 1911, Copalis (DPU, WASH); A. S. Foster 1529 (PC); A. S. Foster 1529, Mar. 12, 1911, South Pacific Beach (DPU, NY, PC, YALE); A. S. Foster 1529, Mar. 12, 1911, Copalis (DPU, WASH, YALE); A. S. Foster 1426, Nov. 20, 1912, Copalis (ABS, NY, PC); A. S. Foster 2684, Mar. 20, 1914, Port Angeles (DPU, NY, WASH); A. S. Foster 2792, June 13, 1914, Port Angeles, on crab apple limbs, alt. 540 m (DPU, WASH); T. C. Frye, Apr. 4, 1904, Seattle (WASH); T. C. Frye, July 19, 1904, Orcas Isl. (WASH); T. C. Frye, Aug. 15, 1904, Mt. Rainier (WASH); T. C. Frye, Apr. 12, 1908, Oysterville (WASH); T. C. Frye, Apr. 18, 1908, Westport (DPU, WASH); T. C. Frye, July 10, 1908, San Juan Co. (WASH); T. C. Frye, July 3, 1928, Friday Harbor (DPU, WASH); T. C. Frye, Sept. 10, 1929, Stevens Pass (DPU, WASH); T. C. Frye, Sept. 14, 1932, Stevens Pass (DPU, WASH); T. C. Frye, July 10, 1937, Mt. Rainier Nat. Park (DPU, WASH); N. L. Gardner, Seattle (NY); Grant, in 1916, Sequim (CM, NY); Grant, May 20, 1917, Montesano (NY, WASH); Grant, in 1918, Montesano (WASH); Grant, Aug. 1922, Marysville (DUKE, MICH); G. N. Jones, May 14, 1933, Parkland (DPU, WASH); Karshner and Foster, Aug. 26, 1908, Copalis (DPU, WASH); Piper, Apr. 1892, Seattle (WASH); Piper 133, Apr. 1892, Seattle (NY, UC, YALE); Piper 3a, July 1892, Seattle (WIS); Piper 133, Aug. 1892, Seattle (CIN, FH, NY, WASH);

Röll 665–666 p.p., June 16, 1888, Easton (PC); Röll 665–666 p.p., June 16, 1888, Cascades (NY, PC); Röll 453,[1] Enumclaw (assumed to be the TYPE of *F. antipyretica* var. *rigens*, PC; NY); Röll 453, E. Enumclaw (FH, G-BOIS, PC); Röll 453, July 3, 1888, Enumclaw (G-BOIS); Röll, July 7, 1888, Enumclaw (H); Röll 453, July 7, 1888, Enumclaw (FH, K); Strang, Mar. 15, 1904, Seattle (WASH).
WYOMING: T. C. Frye, July 7, 1934, Yellowstone Nat. Park (DPU, WASH); A. Nelson 2619, Aug. 12, 1896, Lincoln Gulch (FH, MO, NY, WYO).

Fontinalis antipyretica var. *oreganensis* differs from other species and varieties in the carinate-conduplicate group in having the following combination of characteristics: median cauline leaves usually narrow, sublanceolate when unfolded, frequently approaching sublinear when conduplicate, keels above basal curve generally straight or only slightly curved, width of leaves commonly gradually decreasing from basal curve to apex, apices usually subobtuse to obtuse and serrulate, blades 0.75–2 mm wide, 2–4.5 : 1, the occasional antipyretica-like leaves up to 3.5 mm in width, 1.25–1.7 : 1, and median cells of leaves usually linear with ends attenuate, 6.5–13.5 μ wide, 8–17 : 1.

7. **Fontinalis neo-mexicana** Sull. and Lesq., in Musci Bor.-Am. 224b. 1856 [printed description on label]; Sullivant, Icon. Musc. Suppl., p. 76. 1874.

Fontinalis antipyretica Hedw. var. Sullivant and Lesquereux, Musci Bor.-Am. Exsic., p. 56. 1865.
Fontinalis mercediana Lesq., in Mem. Calif. Acad. Sci. 1: 28. 1868.
Fontinalis maritima C. Müll., in Flora 70: 225. 1887.
Fontinalis antipyretica subsp. *neo-mexicana* (Sull. and Lesq.) Card., in Rev. Bryol. 18: 82. 1891.
Fontinalis antipyretica subsp. *columbica* Card., in Rev. Bryol. 18: 82. 1891 (nomen nudum).
Fontinalis columbica (Card.) Card., in Rev. Bryol. 18: 84. 1891 [in descriptive key .
Fontinalis neo-mexicana var. *columbica* (Card.) Card., Mon. Font. p. 61. 1892.

Plants slender to medium, pale green, green, yellowish green, golden brown, or brown, frequently glossy when dry; stems rigid or nearly so, up to 56 cm in length, 0.25–0.5 mm in diameter, denuded and reddish brown to blackish at base with age, irregularly pinnately branching; branches numerous, close, erect to erect-spreading, up to 10 cm in length, ends of imbricately foliated stems and branches attenuate, conspicuously three-angled, elongated pyramidal in outline and triangular in cross section; median cauline leaves commonly

---

[1] Röll 453 was collected as Enumclaw, Washington. Röll 83 was collected at Victoria, Vancouver Island, British Columbia. Both were cited by Renauld and Cardot, no. 453 in first place, with the original description of *F. antipyretica* var. *rigens*. The author regards Röll 83 as *F. patula*.

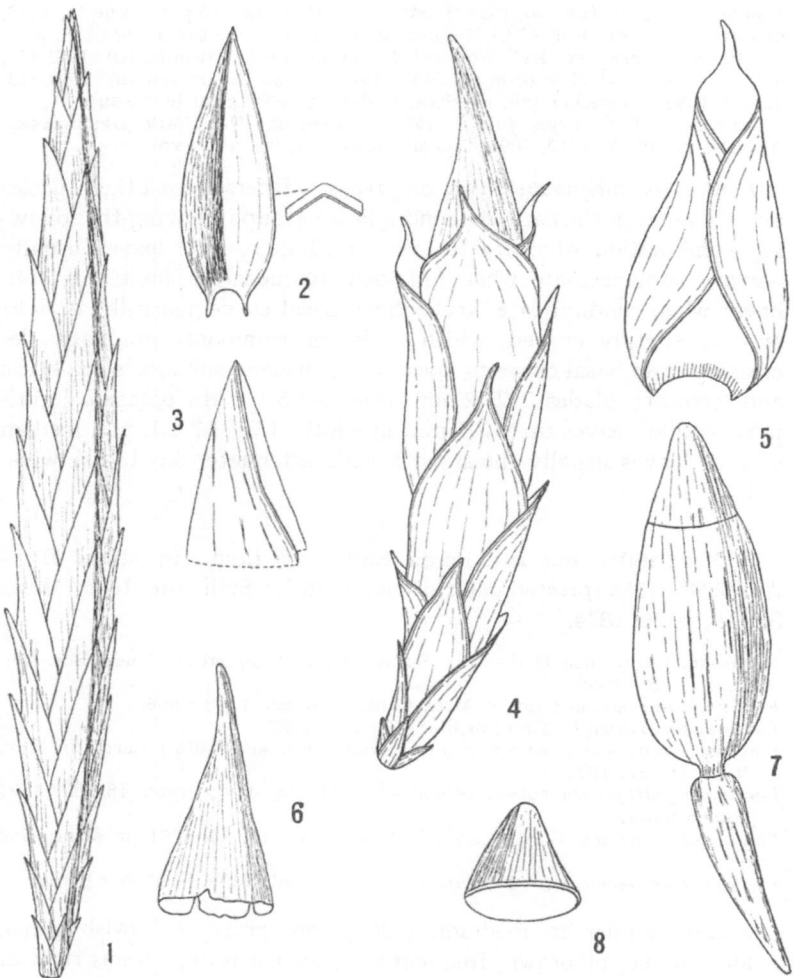

FIG. 7. – *Fontinalis neo-mexicana* Sull. and Lesq. – 1.Apical end of branch showing tristichous leaves and three-angled appearance. 2.Median cauline leaf. 3.Apical portion of median cauline leaf. 4.Perichaetium containing sporophyte with calyptra. 5.Perichaetial leaf. 6.Calyptra. 7.Operculate capsule and seta. 8.Operculum. Nos. 1 and 4–8 drawn after Sullivant, Icon. Musc. Suppl., pl. 57; 2–3 from A. S. Foster 967, Apr. 29, 1909, Washington.

rather close, with bases occasionally up to 1.5 mm apart, blades firm, usually erect-spreading, sometimes spreading, carinate-conduplicate, the keel generally slightly curved but sometimes straight or nearly so and occasionally moderately to strongly curved, often split along keel, oblong-lanceolate if conduplicate and ovate-lanceolate if unfolded; apices commonly acute or subacute, rarely subobtuse or obtuse, leaf tips usually entire but occasionally serrulate; median cauline blades 2.5–5 mm long, sometimes up to 6 mm in length, 1–3 mm wide, 1.5–3 : 1; median cells of leaves linear with ends attenuate, 5–15 μ wide, 10–25 : 1; alar cells slightly enlarged, occasionnally in vertical rows, up to 8, subrectangular to subquadrate, walls hyaline to yellowish in younger leaves, yellowish brown, brown, or reddish brown in older leaves, frequently forming distinct auricles, leaf bases decurrent, 0.25–0.5 mm; median branch leaves comparable to median cauline except smaller in size; perichaetial branch 5–6 mm in length, perichaetium oval to oblong, 1–1.5 mm in diameter; upper perichaetial leaves broadly oval-lanceolate or suboval, abruptly apiculate, lacerate with age; calyptra long conical, approximately 2 mm long, 0.75–1 mm in diameter; capsule usually immersed in perichaetial leaves, occasionally apex of operculum emergent; operculum short conical, apex obtuse, 0.6–0.75 mm in length, 0.5–0.75 mm in diameter; seta very short, approximately 0.3 mm long; urn oval to oblong, 2–2.5 mm long, 1–1.25 mm in diameter, contracted beneath mouth when dry; peristome teeth brownish orange, long acuminate, frequently united in pairs at apex, approximately 0.75 mm long, submuricate to muricate, lamellae 20–25; trellis of inner peristome perfect, brownish orange, approximately 0.75 mm long, submuricate to muricate, many transverse bars appendiculate; spores light green to pale brownish green, smooth to minutely muricate, 15.3–27.2 μ in diameter, ripe in summer. [The specific name, *neo-mexicana*, is indicative of New Mexico, the state in which Charles Wright collected the plants which were used as the basis of the original description.]

Type: Not previously designated. In the absence of material so marked, the writer has assumed to be the type the plants in the Farlow Herbarium, bearing the printed Latin description, in the Exsiccatus of Sullivant and Lesquereux, Musci Boreali-Americani no. 224b. Edition 1. 1856.

Type collector: Charles Wright.

Type locality: New Mexico: in mountain rivulets.

Distribution: North America: Alaska, British Columbia and neighboring islands, western United States from Washington to Montana, New Mexico, and California, in Michigan near Lake Michigan, and, with question, Mexico; [1] South America: Argentina.

---

[1] In the herbaria of the University of California, New York Botanical Garden, Boissier, and Kew, there is a collection from Mexico, without locality and date, made by Lesquereux. In the British Museum of Natural History there is a label which reads, "Lesquereux no. 42, in Mexico." Cardot's note with the specimens in the Boissier Herbarium states when translated, "Probably comes from New Mexico and not Mexico." In Icones Muscorum, Supplement, p. 4, it is stated

Additional descriptions: Austin, Musci Appal., p. 43. 1870; Lesquereux and James, Man. Moss. N. Am., p. 269. 1884; Cardot, Mon. Font. pp. 59 (as *F. neo-mexicana*), 61 (as *F. maritima*). 1892; Barnes, Gen. and Sp. N. Am. Moss., p. 327 (as *F. neo-mexicana* var. *columbica* and as *F. maritima*). 1896; Jones, in Res. Stud. St. Coll. Wash. 1: 183. 1930; Welch, in Grout, Moss Fl. N. Am. 3: 238. 1934.

Illustrations: Sullivant, Icon. Musc. Suppl., pl. 57. 1874; Welch, in Grout, Moss Fl. N. Am. 3: pls. 73 and 74. 1934. – FIGURE 7.

Specimens examined: NORTH AMERICA: ALASKA: Ammann 100, Kiska Isl. (DPU); Bank F-23, May 26, 1945, Adak Isl., alt. 90 m (DPU); Dutilly, Lepage, and O'Neill 22110, June 24, 1947, Glen Road, mile 128 (DPU); S. Fernald 3631, Sitka (DPU); Hardy 68, May 27, 1945, Attu Isl. (DPU, WASH); Hultén 5211, Unalaska Isl. (BART, FH); Kincaid, July 15, 1898, Popof Isl. (DPU); Kincaid, July 1899, Popof Isl. (DPU, MINN); Steenis 4573, Aleutian Islands, Little Sitka Isl. (BART); Steenis 4584, Aleutian Islands, Rat Isl. (BART); Van Schaack 953a, Attu Isl. (DPU, MICH).

CANADA: BRITISH COLUMBIA: J. W. Bailey, June 1901 (PC); J. W. Bailey, Aug. 1901, Cumberland (WASH); J. W. Bailey, Aug. 1901, Cumberland, Grout, N. Am. Musci Pl. 84 (BART, CAN, CAS, CM, DPU, DUKE, FH, ILL, MICH, MINN, MO, NY, OS, PC, UC, US, UT, WASH, YALE); J. W. Bailey, June 1907 (DUKE); Boyd M25, Aug. 19, 1946, Oliver Creek, Lake Cowichan (DPU); Brinkman 68, Nishon-lith Lake, W. Shuswap (DPU, FH); Brinkman, Oct. 9, 1909, Can. Moss. 125, Kamloops District (CM, DPU, E, FH, MICH, US); Haw, in 1891, Salmon Arm (CAN); Lyall, Observatory Inlet (NY); MacFadden (DPU, TRT); MacFadden, Apr. 5, 1926, Bauer, Musc. Eur. and Am. Exsic. 2034 (BART, L, NY, WASH); MacFad-den, Apr. 7, 1926, New Denver, Bonanza Creek (CHI); MacFadden, Apr. 7, 1926, Slocum Lake (DUKE, H); MacFadden, Apr. 7, 1926, Bonanza Creek (NY); Mac-Fadden, Oct. 1928, Bonanza Creek (CIN, MICH, NY, UC, UT, WYO); MacFadden, Oct. 1, 1928, Tranquille Creek near Kamloops (DPU, NY); Macoun, Aug. 26, 1885, Selkirk Range, Can. Musci 103 (CAN); Macoun, Apr. 6, 1889, Hastings (CAN); Macoun, Apr. 6, 1889, Hastings, Can. Musci 599 (DUKE, FH, US, WASH); Macoun 10, Apr. 6, 1889, Burrad Inlet (Assumed to be TYPE of var. *columbica*, PC; S); Macoun, in 1890, Eagle Pass (PC); Macoun, May 30, 1890, Eagle Pass, Can. Moss. 203 p.p. (FH, MT, NY, US); Macoun, May 30, 1890, Eagle Pass, Can. Musci 207 (CAN); Macoun, July 30, 1890, Rogers Pass (CAN); Macoun, July 1, 1893, Comox (CAN); Macoun, June 23, 1905, between Midway and Lake Osoyoos, Ex Herb. Geological Survey of Canada 145 (FH).

QUEEN CHARLOTTE ISLANDS: Spreadborough, July 11, 1910, Skidegate, Geological Survey of Canada 83367 (CAN, CM).

VANCOUVER ISLAND: Gibbs 181 (PC); Hone, in 1903 (PC); Hone 225, July 1903, Port Renfrew (MINN); MacKenzie 341, June 1933, Sooke R. (ABS); Macoun, May 1875, Victoria (NY); Macoun, June 4, 1887 (K, PC); Macoun, June 4, 1887, Can. Musci 129 (CAN); Macoun, 8–6–1887, Can. Musci 229 p.p.[1] (FH, H, K, MINN, MO, NY, PC, UC, US, WIS, YALE); Macoun, Apr. 6, 1889, Nanaimo, Can. Moss. 203 p.p. (CAN, US); Macoun, Apr. 6, 1889, Can. Musci 229 p.p.[1] (FH, NY); Macoun, June 1, 1893, Nanaimo, Can. Moss. 203 p.p. (K, MO); Macoun, July 3, 1908,

---

that Lesquereux collected in the mountainous parts of the Southern States. In the absence of proof that the specimens were collected in Mexico, the author assumes that the collections were made in New Mexico. To date, no collections made in Mexico have been seen unless the ones discussed above were collected in that country.

[1] On labels for Canadian Musci 229, Nova Scotia and Vancouver Island are given as the localities. Since *neo-mexicana* is not otherwise cited from Nova Scotia and since the author has seen no other specimens of *neo-mexicana* collected in Nova Scotia, it has been assumed that the Nova Scotia reference is an error with regard to *F. neo-mexicana*.

Nanaimo, Can. Moss. 205 p.p. (E, K, MO, NY); Waldron 23, July 12, 1901, Port Renfrew (NY).

UNITED STATES: CALIFORNIA: Ames, Auburn (FH, PC); Mrs. R. M. Austin, in 1877, Plumas Co. (CHI, FH); M. S. Baker 309b, Trinity Co. (BART); M. S. Baker 309, Aug. 19, 1926, Trinity Co., Trinity Alps (DPU, MICH); Blacker, in 1880 (G-BOIS); Blacker, in 1882 (FH); Bolander, Merced R. (TYPE of *F. merce-diana*, FH; NY, PC); Bolander 202 (PC, US); Bolander, 1864–1870 (CAS, FH); Bolander 152, 1864–1870 (CHI, FH, NY, PHIL, UC, US, YALE); Cohoe and Jotter, in 1937, Yosemite (DPU); Douglas, in 1835 (K); DuVall, July 20, 1935, Redwood Nat. Forest, near Crescent City (DPU); M. A. Howe, in 1894, Sacramento Valley, Shasta Springs, Renauld and Cardot, Musci Am. Sept. Exsic. 314 (CAN, FH, MICH, NY, PC, YALE); M. A. Howe 103b, July 24 – Aug. 10, 1894, Upper Sacramento (CAS, NY, UC); M. A. Howe 103, Aug. 7, 1894, Shasta Springs (CAS, NY, PC, UC); L. Koch 1363, May 4, 1947, Napa Co., Napa-Monticello Road (DPU, MICH); L. Koch 1819, June 1, 1947, Calaveras Co. (DPU, MICH); L. Koch 2071, June 27, 1947, Santa Cruz Co., State Redwood Park (DPU, MICH); MacFadden, July 13, 1933, Nevada Co., Mt. Oro District (CHI, DUKE, UCLA, WYO); MacFadden, July 1934, Nevada Co., near Nevada City (DPU); MacFadden, Aug. 21, 1934, Placer Co., near Colfax (UCLA); MacFadden, Aug. 21, 1934, Placer Co., Iowa Hill (UCLA, UT, WYO); Pendleton, July 1912, Mt. Shasta, alt. 1020 m (MT); Pendleton, June 4, 1913, Sisson (ABS); Pendleton, June 6, 1913, near Mt. Shasta (FH); Pendleton, June 6, 1913, Mt. Shasta (CM, DPU, WASH); Pringle 26c, Plumas Co. (NY); Setchell, June 8, 1916, Mariposa Co., alt. 1320 m (CAS, CHI, NY, US); Torrey, in 1865, Sierra Nevada (US); Vasey, in 1875, San Francisco (FH, US); Wiggins C54, in 1939, near Yosemite Nat. Park (BART).

COLORADO: D. M. Andrews, in 1898, Boulder (NY); Bethel, July 1904, North Park, alt. 2700 m (NY); Bethel, July 1905, Middle Park (NY); Brandegee (NY); Brandegee, 1874–1878, within 100 miles of Canyon City (MICH, US); Goodding, Aug. 11, 1903, Larimer Co., summit of North Park Range (CM, FH, G-DEL, MO, NY, PC, UC, US, WASH, WYO); E. Hall, Rocky Mts. (H); Kiener 5696, Oct. 13, 1936, Longs Peak valley, Rocky Mt. Nat. Park, alt. 2550 m (CHI, DPU, NEB, NY, UC, UT); McCaskey 200, Aug. 24, 1939, Boulder Co., Science Lodge, alt. 2700 m (DPU); Pammel 256, July 8, 1896, Larimer Co., Little Beaver (MO); Penard 444, Aug. 1891, Caribou, 3000 m (NY, PC); Ramaley 907, Sept. 5, 1901, Allan's Peak near Boulder (WYO); Ramaley, Dodds, and Robbins 2961, July 20, 1907, Duck Lake, Boulder, alt. 3000 m (WYO); Sayre 346, July 1, 1936, Boulder Co., University Camp, alt. 2850 m (DPU); Sayre 697, Aug. 24, 1938, Boulder Co., University Camp, alt. 2850 m (DPU); Vasey, in 1868, Colorado Exploring Expedition (DPU, NY).

IDAHO: Daubenmire 4433, Apr. 30, 1944, Latah Co. (DPU); Daubenmire 4464, July 23, 1944, Boundary Co. (DPU); Elmer, in 1899, Latah Co. (DUKE); Elmer, June 1899, Latah Co., Cedar Mts., (NY, PC); T. C. Frye, Aug. 13, 1925, Trude (DPU, WASH); T. C. Frye, June 25, 1934, Cascade (DPU, WASH); Henderson 2886, Latah Co., Moscow Mts. (NY); M. E. Jones, Sept. 22, 1899, Washington Co. (BM); M. E. Jones 6626, Sept. 22, 1899, Washington Co. (US); Landberg, July 1892 (PC); Leiberg 88 and 114, Kootenai Co. (NY); Leiberg 114, in 1888, Kootenai Co. (MINN, PC, YALE); Leiberg 114, Sept. 1889, Kootenai Co. (CAN, NY, YALE); Leiberg, July 1891, Kootenai Co. (WIS); MacDougal 210, July 27, 1900, between Priest Lake and East Fork (NY); Meyer 80, May 9, 1940, Latah Co. (MO); M. L. Miller, July 20, 1935, Payette Nat. Forest, Cascade (DPU); Pickett 222, Sept. 27, 1914, Latah Co., Cedar Mt., alt. 960 m (BART); Piper 347, Aug. 1901 (WASH); Röll 1125, May 5, 1888 (PC); Rust 1096, Sept. 1917, Coeur d'Alene (NY); J. H. Sandberg, July 1888, Kootenai Co. (MINN); J. H. Sandberg, D. T. MacDougal, and A. A. Heller 1165 p.p., Aug. 1892, Kootenai Co., Hope (US); Umbach, Aug. 25, 1901, Sand Point (CHI, WIS).

MICHIGAN: E. J. Hill 216, 1880, Aug. 23, 1880, Manistee Co., Barr Creek flowing into Bear Lake (CHI, ILL).

MONTANA: H. and V. Bailey 9, June 23, 1939, Flathead Co., between Lake McDonald and Fish Lake, alt. 1200 m. (DPU); Barkley, Oct. 10, 1937, Missoula Co., east of Lolo Mt. Ranger Station, Grout, N. Am. Musci Perf. 332 (BART, CAS, CHI, CM, DPU, MICH, MINN, MT, UC, US, UT, WASH, YALE); Barkley 1640, Oct. 10, 1937, Missoula Co. (DPU, NY); Barkley and Diettert 1767, Sept. 26, 1937, Missoula Co., near Lolo (MO, US, WASH); Barkley and Diettert 1640, Oct. 10, 1937, Missoula Co., near Lolo Mt. Ranger Station (DPU); T. C. Frye, Aug. 20, 1925, Henderson (DPU, WASH); T. C. Frye, Aug. 28, 1929, Glacier Nat. Park (DPU, WASH); T. C. Frye, Aug. 31, 1929, Lolo Hot Springs (DPU, WASH); Röll 1289, Aug. 10, 1888, Ravalli (PC); Rydberg and Bessey, June 28, 1897, Gallatin Co., Spanish Basin, alt. 1950 m (NY); Umbach, in 1903, Lake McDonald area (PC); Umbach 869, Sept. 4, 1903, Lake McDonald area (CHI, FH, NY, US, WIS); Umbach 869, Sept. 4, 1903, Lake McDonald area, Grout, N. Am. Musci Pl. 263 (BART, CAN, CM, DPU, DUKE, FH, ILL, MICH, MINN, MO, NY, OS, UC, US, WASH, WIS, YALE); S. Watson, in 1880, Grasshopper Creek, Northwestern Territories, 10th Census of U.S. Dept. Forestry (FH, NY); Whitham 1495, Sept. 22, 1932, Gallatin Co., alt. 1590 m (DPU); T. G. and E. C. Yuncker 7118, July 17, 1937, Glacier Nat. Park (DPU, NY).

NEW MEXICO: Benedict 2338, Oct. 4, 1926, Santa Fe Canyon (US); Bertaud 207, Aug. 3, 1916, Santa Fe (US); Brouard (Bro. Arsène) 18738, Aug. 8, 1926, Monument Rock, alt. 2420 m (BART, FH, US); Brouard 18621, Oct. 10, 1926, Lake Peak, alt. 3784 m (BART); Ikenberry 13, Aug. 31, 1937, Santa Fe (DPU, MICH); James, Santa Fe (K); Lesquereux, ?Mexico [1] (G-BOIS, K, NY, UC); Lesquereux 42, ?Mexico [1] (BM); Standley, July 30, 1908, Pecos R. Nat. Forest, alt. 2550 m (DUKE, US); Studhalter 1162, Sept. 3, 1933, Las Vegas, Gallinas Canyon (DPU, UT); Wright, in mountains (H, K, MICH, US, WIS); Wright, Sullivant and Lesquereux, Musci Bor.-Am. Exsic. 224b (Assumed to be TYPE in Sullivant's Herb., FH; DUKE, K, MICH, MO, NY, PC, TRT, WIS); Wright (New Mexico, sterile plants) and Hall (Rocky Mts., plants with fruit), Sullivant and Lesquereux, Musci Bor.–Am. Exsic. 334 p.p. (CHI, CM, FH, G-BOIS, MICH, NY, PC, PHIL, WIS, YALE); Wright (New Mexico, sterile plants) and Hall (Rocky Mts., plants with fruit), Austin, Musci Appal. 251b (CAN, CAS, CHI, DUKE, G-BOIS, K, NY, PC, PHIL, US, WIS).

OREGON: Applegate 11842, Aug. 19, 1938, Klamath Co., Crater Lake Nat. Park (DPU); Barrett, Hood R. (US); Blacker, in 1880 (FH, G-BOIS); Bunett, Hood R. (FH, UT); Doty 1221, July 9, 1939, Coos Co., Silver Falls, alt. 150 m (DPU); Drexler 993, Aug. 1936, Cascade Mts., Muir Creek (DPU); DuVall, July 1, 1935, Whitman Nat. Forest, near Halfway, alt. 1500 m (DPU); DuVall, July 13, 1935, Tumalo Creek, near Bend (DPU); A. S. Foster, May 18, 1907 (PC); A. S. Foster, June 17, 1907, Clackamos Co., Salmon R. (YALE); A. S. Foster 539, June 18, 1907, Salmon R., alt. 450 m (ABS, WASH); A. S. Foster, June 19, 1907, Salmon R. (s); A. S. Foster, Apr. and May 1910, Silverton (WASH); A. S. Foster 1213a, June 20, 1910, Silverton (DPU, WASH); A. S. Foster, July 4, 1910, Marion Co., Silver Creek (H); A. S. Foster 1295, July 4, 1910, Marion Co., Silver Creek (DPU, FH, NY, PC, US, WASH, YALE); T. C. Frye, Aug. 17, 1932, Holland (DPU, WASH); T. C. Frye, July 17, 1933, Wapinitia (DPU, WASH); T. C. Frye, July 18, 1933, Mt. Hood (DPU, WASH); N. L. Gardner 1741, Aug. 8, 1888, Deschutes R. (BART, UC); Gilkey 8, June 8, 1940, Linn Co. (DPU, NY); Howell, in 1882, base of Mt. Hood (NY); Howell 34, in 1882, base of Mt. Hood (YALE); Howell, Oct. 1882 (G-DEL); Howell, Sept. 1883, base of Mt. Hood (CM); Howell, Oct. 1883, Cascade Mts. (CHI); Lyall, in 1861, Oregon Boundary Commission, Fort Coville to Rocky Mts. (K, PC, WIS); Mohr, Hood R. (NY); E. P. Sheldon 9028, Sept. 29, 1897, Union Co., Mill Creek, alt. 1500 m (NY); Mrs. R. W. Summers, July 1875, Yaquina (FH); Mrs. R. W. Summers 1380, in 1877 (YALE); Mrs. R. W. Summers, Sept. 1880, Coast Mts. (FH); Van Wert, Oct. 1922, Albany (DUKE, MINN); Van

---

[1] See page 79, note 1.

Wert, May 1923, Alsea (DUKE, MINN); L. C. Wheeler 2963, July 24, 1934, Jackson Co., Siskiyou Mts., alt. 1320 m (BART).
UTAH: D. Hobson, Aug. 2, 1939, Summit Co., Bear R., alt. 2850 m (DPU, UT).
WASHINGTON: J. A. Allen 87, Jan. 8, 1898 (NY, PC, WASH); J. A. Allen, Aug. 1898, Mt. Rainier (WASH); J. A. Allen 80, Aug. 1898, upper Nesqually Valley (CHI, CIN, DPU, DUKE, FH, ILL, MICH, MINN, NY, US, WASH, YALE); J. A. Allen 82 p.p., Oct. 29, 1898, upper Nesqually Valley (NY, WASH); H. and V. Bailey 148, June 15, 1937, Jefferson Co., Elwha R. trail, alt. 576 m (DPU); J. W. Bailey, Sept. 16, 1904, Renton (WASH); J. W. Bailey 1171, Sept. 16, 1904, Renton (WASH); J. W. Bailey, Sept. 1, 1907, Cedar R., Renton (YALE); J. W. Bailey, Sept. 16, 1907, Cedar R., Renton (DPU, OS); J. W. Bailey, July 17, 1930, Mt. Rainier (DPU); J. W. Bailey, Aug. 13, 1931, Big Four (DPU, WASH); Winona Bailey, Aug. 10, 1910, Glacier Park region (DPU, WASH); Bonser, May 6, 1906, near Mica Peak (WASH); Brandegee, in 1880, Yakima Co. (NY); Brandegee, in 1882, Yakima region (NY); Eggers, in 1880, Neah Bay, near Cape Flattery [1] (Assumed to be the TYPE or portion of type of F. maritima, B; PC, S); Eyerdam 874, Nov. 29, 1947, Snohomish Co., Mt. Pilchuck (DPU); Flett 2068, Oct. 19, 1902, Lake Kapowsin (NY); A. S. Foster, Nov. 24, 1904, Hamilton (PC); A. S. Foster 1901, Mar. 15, 1905, Hamilton (WASH); A. S. Foster, Apr. 1, 1905 (WASH); A. S. Foster 439, Oct. 1, 1906, Cathlamet (DUKE); A. S. Foster, Feb. 15, 1907, Cathlamet (DPU, WASH); A. S. Foster, Mar. 5, 1907, Cathlamet (WASH); A. S. Foster, June 29, 1908, Pacific Beach, in salt water (DPU, WASH); A. S. Foster, July 1908, Rainier Nat. Park (DPU, WASH); A. S. Foster 884, Aug. 20, 1908, Chehalis Co. (NY); A. S. Foster, Aug. 27, 1908, Copalis (DPU, WASH); A. S. Foster, Apr. 29, 1909 (PC); A. S. Foster 967, Apr. 29, 1909, Wynooche R., near Montesano (DUKE); A. S. Foster 2396, Aug. 1909, Mt. Rainier, alt. 750 m (S); A. S. Foster, May 28, 1911, Cedar R., Renton (DPU, WASH); A. S. Foster, Aug. 31, 1911, Black R. (DPU, WASH); A. S. Foster 1921, Aug. 31, 1911, Black R., near Gate (DPU, DUKE, NY, WASH); A. S. Foster 1921, Oct. 1911, Black R., Gate (DPU, WASH); A. S. Foster, Oct. 20, 1911, Black R., Gate (DPU, WASH); A. S. Foster 1921a, Mar. 30, 1912, Black R., Gate (DPU, WASH); C. F. Foster, Feb. 1905 (PC); H. R. Foster, Jan. 10, 1896, Snoqualmie Falls (WASH); H. R. Foster, Apr. 20, 1896, Snoqualmie Falls (WASH); T. C. Frye (WASH); T. C. Frye, May 5, 1906, King Co., Cedar Lake (DPU, WASH); T. C. Frye, Aug. 7, 1907, Olympic Mts. (WASH); T. C. Frye, Aug. 7, 1907, Elwha R. Valley, Olympic Mts., alt. 750 m (WASH); T. C. Frye, Jan. 11, 1908, Hall Lake, near Edmunds (WASH); T. C. Frye, Aug. 28, 1909, Pilchuck (DPU, WASH); T. C. Frye, Dec. 26, 1930, North Bend (DPU, WASH); T. C. Frye, Aug. 21, 1931, Clearwater, Clearwater R. (CHI, DPU, WASH); T. C. Frye, July 26, 1934, Lake Josephine, Stevens Pass (DPU, WASH); T. C. Frye, June 1937, Mt. Rainier (DPU, WASH); T. C. Frye 3254, Oct. 1, 1945, Cascade Mts., near Stevens Pass (DPU, UT, WASH); N. L. Gardner, Snoqualmie Falls (NY); N. L. Gardner 28, Snoqualmie (UC); Grant, Apr. 1920, Marysville (MICH, UC, WASH); Grant, Apr. 1926, Snoqualmie Falls (WASH); Grant, Oct. 1926, Hamilton (DPU, WASH); Grant, Oct. 1927, Snohomish Falls (DUKE); Grant, Aug. 1932, Marysville (WASH); C. Harrison, June 14, 1939, Forks (WASH); Henderson, Olympia (FH); Henderson 1741, Olympia (PC); D. C. McArdle, Aug. 26, 1928, Skamania Co., Falls Creek (FH, MICH); C. A. Mosier, June 1892, Seattle (H); C. A. Mosier, Aug. 1892, Seattle (US); C. F. Mosier, Sept. 24, 1892 (PC); Nevius (NY); Nevius, in 1887 (UC); Otis 1403, Mar. 12, 1925, Jefferson Co., Hoh R., alt. 60 m (NY, WYO); A. M. Parker, Aug. 1892, Snoqualmie (NY); Piper 82, Mason Co., Skokomish R. (NY, YALE); Piper, Aug. 15, 1889, Pierce Co. (NY, YALE); Piper, Aug. 1890 (PC); Piper 83, Aug. 1890, Mason Co., Skokomish R. (CHI, DUKE, FH); Piper, Aug. 15, 1890, Pierce Co. (WASH); Piper,

---

[1] In the Herbarium of the Botanisches Museum, Berlin-Dahlem, the original description of F. maritima is pasted beneath the packet of plants from the Herb. C. Müller-Hal. This material is assumed to be the type or a portion of the type of F. maritima C. Müll.

July 1892, Cedar R. (BART, UC); Piper, Aug. 1892, Seattle (WASH); Piper 83, Aug. 1898, Mason Co., upper Skokomish R. (NY); Roberts, Mar. 8, 1925, Cascade Mts., alt. 900 m (BART, CM, DUKE); Röll 492, Cascade Mts., near Weston and Easton (PC); Röll 660, Cascade Mts., near Weston and Easton (NY, PC); Röll 661, Cascade Mts., near Weston and Easton (NY, PC); Röll, June 11, 1888, Cascade Mts., Easton (H); Röll 662, June 11, 1888, Cascade Mts., Easton (B); Röll 663, June 16, 1888, Cascade Mts., near Weston and Easton (PC); Röll 917, June 19, 1888, Cascade Mts., near Weston and Easton (PC); Röll 490–491, June 21, 1888, Cascade Mts., Weston (K); Röll 918, June 21, 1888, Cascade Mts., near Weston and Easton (PC); Röll 490–491, June 29, 1888, Cascade Mts., near Weston and Easton (H, PC); Röll 409, July 7, 1888, Enumclaw (PC); Schallert 718, Sept. 15, 1945, Clallam Co., near mouth of Calawah R. (DPU); Svihla 2132, Aug. 25, 1931, Mt. Rainier, Paradise Valley (DPU, WASH); Svihla, Sept. 6, 1943, Skamania Co., Spirit Lake (DPU, WASH); W. A. Weber 2697, Feb. 4, 1945, Skamania Co., Hemlock Ranger Station (DPU); Willis, Apr. 26, 1883, Boisé Creek, near Wilkeson, alt. 750 m (NY).

WYOMING: Compton, in 1883, Yellowstone Nat. Park (NY); Conard 48–271, Sept. 4, 1948, Yellowstone Nat. Park, Firehole R. (DPU); Drexler 847, Sept. 1936, Yellowstone Nat. Park (DPU); T. C. Frye, July 6, 1934, Yellowstone Nat. Park (DPU, WASH); Hapeman, Sept. 1892, Big Horn Mts. (FH, US); A. Nelson 2371 p.p., July 16, 1896, Little Goose Creek (CHI, WYO); A. Nelson 2506, July 25, 1896, Buffalo (E, FH, MO, NY, PC, WYO); A. Nelson 6293, Aug. 3, 1899, Yellowstone Nat. Park (WYO); A. Nelson 7540, July 10, 1900, Albany Co., Laramie Peak (CHI, CM, COLO, FH, K, MINN, MO, NY, PC, US, WYO); A. and E. Nelson 6293, Aug. 3, 1899, Yellowstone Nat. Park, Spring Creek (CHI, CM, FH, ILL, K, MINN, MO, NY, PC, US, WYO); A. and E. Nelson 6320, Aug. 6, 1899, Yellowstone Nat. Park, Yellowstone Lake (CHI, CM, FH, G-DEL, ILL, K, MINN, MO, NY, PC, US, WYO); Newberry, Aug. 1883, Yellowstone Nat. Park, Firehole R. (NY); C. L. Porter 883, June 27, 1931, Carbon Co., Brush Creek (WYO); Rydberg and Bessey, Aug. 10, 1897, Yellowstone Nat. Park, East de Lacy's Creek, alt. 2250 m (NY); Studhalter 2532, Yellowstone Nat. Park (DPU, DUKE, UT); Welch 11908, Aug. 29, 1952, Yellowstone Nat. Park, in Gibbon R., above Virginia Cascades (DPU).

Locality indefinite: Christ 47, in 1932 (WASH); Hall, Rocky Mts. (NY); Hall, in 1862, Rocky Mts., Bear Creek (CHI); Hall and Harbour, in 1862, Rocky Mts. (YALE); Lemmon, Sierra Nevada Mts. (FH); Loew (US); Vasey, Rocky Mts. (US).

SOUTH AMERICA: ARGENTINA: E. Palmer, in 1855, Corrientes (US).

Fontinalis neo-mexicana differs from other species and varieties in the carinate-conduplicate group in having the following combination of characteristics: foliated branches frequently resemble miniature feathers, ends of foliated branches and stems attenuate, conspicuously three-angled, elongated pyramidal in outline and triangular in cross section, leaf apices usually acute and entire, keels commonly only slightly curved above basal curve, median leaf cells linear-attenuate, and perichaetial leaves abruptly apiculate.

8. Fontinalis patula Card., in Rev. Bryol. 23: 67. 1896.

Fontinalis antipyretica Hedw. var. *patula* (Card.) Welch, in Grout, Moss Fl. N. Am. 3: 238. 1934.

Plants slightly robust to robust, green, yellowish green, or brownish, slightly glossy to glossy; stems rigid or nearly so, up to 63 cm in length, 0.3—0.5 mm in diameter, foliate to base or slightly denuded, irregularly pinnately branching; branches erect-spreading to spreading, varying in length, frequently elongate, up to 12 cm long, ends of sub-imbricately or imbricately foliated stems and branches acute to acuminate; median cauline leaves erect-spreading to spreading, distant, bases up to 1.5 mm apart, blades firm, keeled-conduplicate, or carinate and open, or concave, keels or median lines usually straight from above basal curve to apical region, occasionally slightly curved, sometimes distinctly curved as in *antipyretica*, frequently

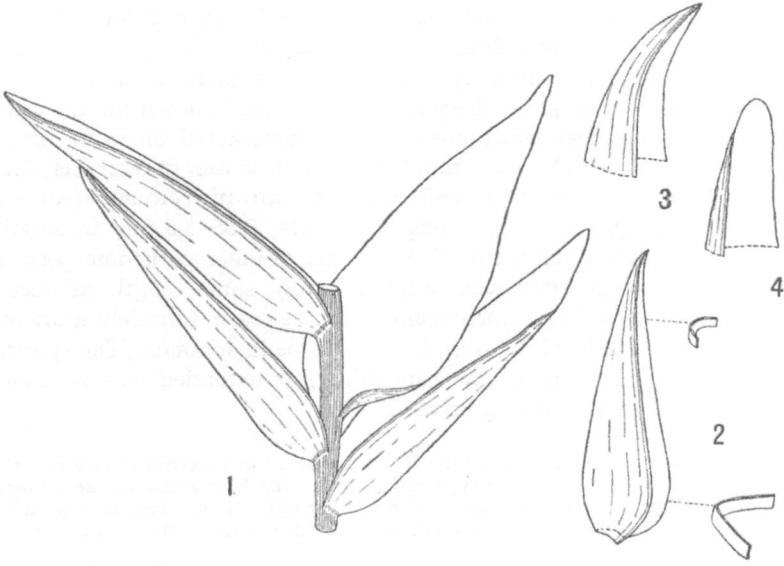

Fig. 8. – *Fontinalis patula* Card. – 1.Portion of stem with median cauline leaves. 2.Median cauline leaf. 3 and 4.Leaf apices. No. 1 drawn from T. C. Frye, Aug. 13, 1925, Idaho; nos. 2–4 from Macoun, May 25, 1893, Vancouver Island.

briefly and abruptly curved at apex, sometimes split along keel, blades ovate-lanceolate to broadly ovate-lanceolate when unfolded, broadest in upper portion of basal fourth or lower part of next fourth above and gradually tapering to upper ⅓ or ¼ which is commonly conspicuously narrowed into a briefly and broadly acuminate apex; majority of apices plane or nearly so, some concave or subconcave, others conduplicate, leaf tips acute, subobtuse, or obtuse, entire to serrulate; median cauline leaves 4–6 mm long, occasionally up to 7.5 mm in length, 1.5–2.5 mm wide, sometimes up to 5 mm in width, 1.5–3 : 1; median cells of leaves usually linear with ends attenuate, oc-

casionally linear-rhomboidal, commonly 8.5–15 μ wide, sometimes
up to 24 μ in width, 10–20 : 1; alar cells enlarged, subrectangular,
rectangular, or subhexagonal, brownish or yellowish brown, commonly
forming distinct auricles, leaf bases not decurrent to decurrent up to
1 mm, occasionally base decurrent on one side and not decurrent on
the other; median branch leaves similar to median cauline except
usually reduced in size; perichaetial branch 4–5 mm long, perichaetium oval to oblong, 1.25–1.5 mm in diameter; upper perichaetial
leaves broadly ovate to suborbicular, rounded apical portion entire,
lacerate with age; calyptra long conical, 1.5–1.75 mm long, 0.5–0.7
mm in diameter; operculum obtuse conical, 1 mm long, 0.75–0.8 mm
in diameter; urn commonly immersed but sometimes immersed to
emergent on same plant, frequently contracted beneath mouth when
dry but sometimes contracted and not contracted on same plant,
oblong to subcylindric, 2–3 mm long, 0.75–1.25 mm in diameter, 2.5–
2.6 : 1; seta 0.25 mm long; peristome teeth brownish orange, frequently united in pairs at apex, long acuminate, 0.85–1.5 mm in length,
slightly muricate, lamellae 25–50; trellis of inner peristome perfect,
brownish orange, muricate, approximately same length as teeth,
lower transverse bars appendiculate; spores green, minutely muricate
to almost smooth, 12–17 μ in diameter, ripe in summer. [The specific
name, *pàtula*, meaning open, spreading, or extended, has reference
to the open and spreading leaves.]

Type: Not previously designated. In the absence of material so marked, the
writer has assumed to be the type the plants in the Herbarium of the Labora-
toire de Cryptogamie, Muséum d'Histoire Naturelle, Paris, bearing the label,
"*Fontinalis patula* Card. sp. nova. On stones in the Colquity River near Victoria.
25. 5. 93."
Type collector: John Macoun, May 25, 1893.
Type locality: British Columbia: Vancouver Island, near Victoria, in Colquity
River.
Distribution: North America: Alaska, western Canada and United States,
Michigan, Connecticut, and Cape Breton Island.
Illustrations: Welch, in Grout, Moss Fl. N. Am. 3: pl. 74 (as *F. antipyretica*
var. *patula*). 1934. – FIGURE 8.
Specimens examined: NORTH AMERICA: ALASKA: J. P. Anderson 639, Sept. 9,
1917, Mendenhall (NY); Eyerdam 580, May 20 and June 25, 1932, Unalaska Isl.
(BART, CHI, FH, G-DEL, ILL, UCCL); Eyerdam 581, July 11, 1932, Atka Isl. (BART,
CHI, FH, G-DEL); Lepage 22652, Aug. 27, 1948, Alaska Peninsula, Naknek (DPU);
Rudd, July.1935, Akutan (DPU); Stair 4888, July 14, 1945, Yakutat, bottom
of Ankau Creek in four feet of water (DPU).
CANADA: BRITISH COLUMBIA: J. W. Bailey, Aug. 23, 1901, Courtenay R.
(PC, PENN); J. W. Bailey, Aug. 23, 1901, Courtenay R., alt. 180 m, Grout, N.
Am. Musci Pl. 158 (CAN, CM, DUKE, FH, H, ILL, MICH, MINN, MO, OS, NY, PC, UC,
US, WASH, WIS, YALE); Brinkman 444, Aug. 8, 1911, Tetachuk Lake, alt. 1080 m
(ABS, FH, MICH); A. J. Hill, Mar. 29, 1901, Ruskin, Fraser R. (NY); Macoun, Can.
Musci 227 p.p. (NY, PC); Macoun, July 31, 1889, Eagle R. at Griffin Lake, Can.
Musci 368 (CAN); Macoun, in 1912, Sidney (NY).

CAPE BRETON ISLAND: G. E. Nichols, July to Aug. 1914, South Ingonish (YALE); G. E. Nichols, July to Aug. 1915, near mouth of Barrasois (YALE).

VANCOUVER ISLAND: Macoun, May 10, 1875, near Victoria (CAN); Macoun, Apr. 21, 1887, Lost Lake, Can. Musci 104 (CAN); Macoun, May 8, 1893, near Victoria, Can. Musci 598d, p.p. (K); Macoun, May 15, 1893 (CAN); Macoun, May 15, 1893, Victoria, Can. Musci 598d, p.p. (PC); Macoun, May 17, 1893, near Victoria (CAN); Macoun, May 25, 1893, Colquity R. near Victoria (Assumed to be the TYPE, PC); Röll 83,[1] May 23, 1888, Victoria (PC); Röll 87, May 23, 1888, Victoria (PC); Röll, June 24, 1888, Victoria (K); Röll 89 p.p., June 24, 1888, Victoria (G-BOIS); Svihla 2177, Aug. 9, 1931, Salt Spring Isl. (DPU, WASH); Waldron, July 9, 1901, Port Renfrew (NY).

UNITED STATES: CALIFORNIA: Bioletti, in 1900, Yosemite region (CAS, NY); Bolander, 1864–1870 (NY, US); Bolander 68 and 79, 1864–1870 (NY, US); Bolander 79 p.p., 1864–1870 (CHI, PHIL, UC, US, YALE); H. E. Brown, June 25, 1897, Mt. Shasta (US); Lehnert, Dec. 1884 (US); S. B. and W. F. Parish 1685, Aug. 1884, San Bernardino Mts. (FH); Whitehouse 22139, Aug. 21, 1949, Tuolumne Co., Yosemite Nat. Park, near head of north fork of Tuolumne R. (DPU, SMU).

CONNECTICUT: Pease, Jan. 4, 1875, Somers (YALE).

IDAHO: Daubenmire 4243, Aug. 2, 1942, Nez Perce Nat. Forest, bottom of Bear Lake (DPU, TENN); DuVall, July 4, 1934, Trude (DPU); T. C. Frye, Aug. 13, 1925, Trude (DPU, WASH); T. C. Frye, Sept. 6, 1929, Deadwood R., east of Cascade (DPU, WASH); G. N. Jones, June 28, 1930, Kootenai Co., Spirit Lake and Lake Tessemini (DPU); Leiberg 85, in 1888, Kootenai Co, North Fork Basin, Lake Coeur d'Alene (CHI, NY, PC, YALE); MacFadden, Sept. 20, 1942, Elmore Co., Boise Nat. Forest, China Basin, Atlanta District (CHI, DPU, NY); Röll, Aug. 6, 1888, Coeur-d'Alene (H); Röll 1196, Aug. 6, 1888, Coeur d'Alene (G-BOIS, PC); Röll 1197, Aug. 6, 1888, Coeur d'Alene (PC, US); Röll 1200, Aug. 6, 1888, Coeur d'Alene (PC); Sandberg, MacDougall, and Heller 1165 p.p., Aug. 1892, Kootenai Co., Hope (US).

MICHIGAN: G. E. Nichols, June 1937, Marquette Co., Huron Mt., Rush Lake (DPU, MICH, YALE); Steere, Sept. 1935, Keweenaw Co., Horseshoe Harbor (DPU, MICH).

MONTANA: T. C. Frye, Aug. 20, 1925, Henderson (DPU, WASH); F. H. Rose and M. Forbes 4274, Aug. 11, 1941, Missoula Co., east of Missoula, alt. 963 m (CHI, DPU); Standley 17681, Aug. 16, 1919, Glacier Nat. Park, alt. 1440–1530 m (NY, US); Standley 17964, Aug. 23, 1919, near Snyder Lake (NY, US); Umbach, in 1903, near Lake McDonald (PC); Umbach 777, Aug. 28, 1903, near Lake McDonald (CHI, FH, ILL, NY, US, WIS); R. S. Williams 362, Sept. 5, 1898, Columbia Falls (CAS, K, MINN, MO, NY); T. G. and E. C. Yuncker 7106, July 25, 1937, Glacier Nat. Park, Camas Creek, alt. 1050 m (CHI, DPU, MICH, NY); T. G. and E. C. Yuncker 7119, Aug. 3, 1937, Glacier Nat. Park, Snyder Lake, alt. 1650 m (CHI, DPU, NY).

OREGON: A. S. Foster, June 17, 1907, Clackamas Co., in Salmon R. (DPU, WASH); A. S. Foster 1213a p.p., Sept. 12, 1910, Silverton, Powers Creek (DPU, WASH); Henderson, near Portland, Willamette R. (FH); Henderson 1742, near Portland, Willamette R. (NY); Howell, Oct. 1885, Pacific Coast, Sauvie's Isl. (BM, PC); Howell, Feb. 1886, Sauvie's Isl. (BR, PC); Howell, Apr. 7, 1886, Sauvie's Isl. (FH); L. Porter, Aug. 1922, Cascadia (MINN).

WASHINGTON: J. A. Allen, in 1898, upper Nesqually R. (NY); J. A. Allen, May 11, 1898, upper Nesqually R. (NY); J. W. Bailey, Seattle, Lake Washington, Union Bay (WASH); J. W. Bailey, Oct. 24, 1904, Seattle, Lake Washington, Union Bay (WASH); J. W. Bailey 144, Oct. 5, 1907, Seattle, Lake Union (DUKE); Bardell, Nov. 19, 1910, Seattle (MO); Brandegee, in 1882, Yakima region (NY); Cowles 782, July 20, 1907, Mt. Rainier, Reflection Lake (CHI, ILL); DuVall, Aug. 18, 1934, Mt. Rainier Nat. Park (DPU); Flett, in 1903, Legu Lake (PC);

---

[1] See page 77, note 1.

A. S. Foster, (PC); A. S. Foster, Mar. 30, 1907, Cathlamet (NY); A. S. Foster, Mar. 3, 1912, Gate (DPU, WASH); A. S. Foster, Apr. 1920, Snoqualmie Falls (UC); H. R. Foster, Apr. 20, 1896, Snoqualmie Falls (WASH); T. C. Frye, Aug. 1, 1904, Seattle (WASH); T. C. Frye, Apr. 13, 1907, Cathlamet (DPU, WASH); T. C. Frye, July 27, 1911, Ridgefield (WASH); T. C. Frye, Apr. 10, 1932, near Brothell (CHI, DPU, OS, WASH); N. L. Gardner, Snoqualmie Falls (NY); N. L. Gardner 27, Snoqualmie Falls (UC); Grant, Quinault Lake (US); Grant, Apr. 1920, Marysville (MICH, WASH, WYO); Grant, Oct. 1920, Marysville (DUKE); Grant 6510, in 1926, Marysville (BART); Grant, Mar. 1926, Marysville (WASH); Grant, Jan. 1927, Marysville (WASH); Grant, July 1927, Marysville (DPU); Grout, Sept. 29, 1902, Seattle (WASH); Jao 281, Sept. 29, 1935, San Juan Isl. (MICH); Lyall, 1858–1859, near 49th parallel of Latitude, Oregon Boundary Commission (K); R. E. McArdle, July 7, 1929, Skamania Co., Spirit Lake (FH); R. E. McArdle, Aug. 7, 1929, Skamania Co., Spirit Lake, alt. 960 m, from bottom of lake, 1.8 m under water (MICH); Piper 37, Apr. 13, 1890, Seattle (DUKE); Piper, 37. May 27, 1890, Seattle (NY, YALE); Piper 1099, May 27, 1890, Seattle, Lake Washington (WASH); Piper 37, May 1891, Seattle, Mud Lake (CHI, FH, NY); Savage, July 12, 1898, Index (NY); Savage, Cameron, and Lenocker, July 1898, Snohomish Co., Index (CHI, IOWA); E. G. Simmons 1756, Aug. 2, 1948, Pierce Co., Mt. Rainier Nat. Park, alt. 1050 m (DPU); Suksdorf, July 12, 1890, Whatcom Co., Padden Lake, Musci Occidentali- Americani 30 (NY, YALE); Suksdorf, Aug. 29, 1890, Skamania Co. (NY); Svihla 778, Aug. 31, 1930, Olympic Peninsula, Boulder Lake (MICH); Svihla 2176, Aug. 29, 1931, Mt. Rainier region, Reflection Lake (DPU, WASH); Svihla, Sept. 6, 1943, Skamania Co., Spirit Lake (DPU, WASH).

WYOMING: Conard 48–352, Aug. 30, 1948, near Beartooth Lake, alt. 3000–3300 m (DPU); Welch 15074, Aug. 20, 1953, near Beartooth Lake (DPU); Welch 15081, 15082, 15084, 15091, and 16060, Aug. 22, 1953, Park Co., in Clark Fork of Yellowstone R., Shoshone National Forest (DPU); Welch 15083, Aug. 19, 1953, in creek flowing into Beartooth Lake (DPU); Welch 16057, 16058, and 16059, Aug. 20, 1953, near Beartooth Lake (DPU); Whitehouse 27488, Aug. 22, 1953, Park Co., near Clark Fork of Yellowstone R. (DPU, SMU).

Although there are occasional leaves which resemble *F. antipyretica*, the great majority of the median cauline leaves of *F. patula* may be readily distinguished from *antipyretica* by the keels or median lines straight above the basal curve, apices briefly and broadly acuminate and plane or subconcave, and median cells usually linear with ends attenuate. Because of an approximation in leaf size between *F. antipyretica* var. *gigantea* and robust plants of *F. patula*, difficulty in determination might occur if the keels of *patula* were slightly curved above the basal curve and if the keels of *gigantea* were not as strongly curved as usual. In these conditions, if the majority of keels of median cauline leaves are straight, the leaves concave or carinate and open, and apices open or nearly so and frequently briefly and abruptly curved, the plant is considered to be *patula*. In *F. Howellii* and in *F. patula* the keels are usually straight above basal curve. In *Howellii* the majority of median cauline leaves are definitely carinate-conduplicate, the keels frequently slightly concave, and the majority of apices long and narrowly acuminate, in contrast with median cauline leaves keeled-conduplicate, or carinate and open, or concave, and the majority of apices briefly and broadly acuminate. *F. antipyretica* var. *oreganensis* has keels straight, also, but usually the narrow, sublanceolate median cauline leaves of *oreganensis*, commonly 0.75–2 mm wide, and the width gradually diminishing from base to subobtuse or obtuse apex will contrast distinctly with the broad, ovate-lanceolate median cauline leaves of *patula*, commonly 1.5–2.5 mm wide, with apices briefly and broadly acuminate.

Cardot published *patula* as a specific name although on the label of the assumed type he had also indicated the varietal relationship with *antipyretica*. In 1934, the author published *patula* as a variety of *antipyretica* because of the occasional occurrence of leaves which resemble *antipyretica* among the leaves with distinct *patula* characteristics. At present it seems that the following differences are sufficient to consider *patula* as having specific rank: majority of keels or median lines straight above the basal curve, apices commonly open or nearly so, median cells usually linear with ends attenuate, and teeth of peristome up to 1.5 mm long, with lamellae up to 50 in number, in contrast with *antipyretica* in which majority of keels of the median cauline leaves are distinctly curved above the basal curve, apices commonly carinate-conduplicate, median cells usually linear-rhomboidal, and teeth of peristome up to 1 mm long, with lamellae up to 35 in number.

9. **Fontinalis Howellii** Ren. and Card., in Bot. Gaz. 13: 200. 1888.

Fontinalis Kindbergii Ren. and Card., in Bot. Gaz. 15: 58. 1890.
Fontinalis antipyretica Hedw. var. *ambigua* Card., in Rev. Bryol. 18: 82. 1891; syn. of *F. Kindbergii*, according to Cardot, Mon. Font., p. 63. 1892 (nomen nudum).
Fontinalis subbiformis Ren. and Card. in litt., as a syn. of *F. antipyretica* var. *ambigua*, according to Cardot, in Rev. Bryol. 18: 82. 1891; syn. of *F. Kindbergii*, according to Cardot, Mon. Font., p. 63. 1892.
Fontinalis antipyretica subsp. *Kindbergii* (Ren. and Card.) Card., in Rev. Bryol. 18: 82. 1891.
Fontinalis antipyretica var. *cuspidata* C. Müll., in MS., according to Cardot, Mon. Font., p. 63. 1892.
Fontinalis antipyretica var. *purpurascens* C. Müll., in MS., according to Cardot, Mon. Font., p. 63. 1892.
Fontinalis neo-mexicana Sull. and Lesq. var. *robusta* C. Müll., in MS., according to Cardot, Mon. Font., p. 63. 1892.
Fontinalis Kindbergii f. *robustior* Card., Mon. Font., p. 64. 1892.
Fontinalis Kindbergii f. *gracilior* Card., Mon. Font., pp. 64 and 65. 1892.
Fontinalis Kindbergii f. *robusta* Card., Mon. Font., p. 65. 1892.
Fontinalis Kindbergii var. *Howellii* (Ren. and Card.) Barnes, Gen. and Sp. N. Am. Moss., p. 327. 1896.
Fontinalis Kindbergii var. *gracilis* Card., herb. name, according to Welch, in Grout, Moss Fl. N. Am. 3: 239. 1934.

Fig. 9. *Fontinalis Howellii* Ren. and Card. – 1.Plant portion, dendroid, branches erect-spreading to spreading, short to long, straight to recurved and subarcuate. 2.Portion of stem with median cauline leaves. 3.Median portion of branch with leaves. 4 and 5.Median cauline leaves, opened, showing broad blades with narrow apices, the latter beginning to split along the keel. 6.Median branch leaf. No. 1 drawn from Thomas Howell 44/No. 20, Oregon; nos. 2–3 from Mehner, July 1913, Washington; nos. 4–6 after Renauld and Cardot, in Bot. Gaz. 13: pl. 18. 1888.

Fontinalis Kindbergii subsp. *eu-Kindbergii* Giacom., in Atti Ist. Bot. Univ.,
  Pavia, Lab. Crittog. (Ser. 5) 4 (2): 250. 1947.
Fontinalis antipyretica subsp. *Kindbergii* (Ren. and Card.) Card. f. *gracilior*
  Card., in Podpĕra, Consp. Musc. Eur., p. 507. 1954.
Fontinalis antipyretica subsp. *Kindbergii* (Ren. and Card.) Card. f. *robustior*
  Card., in Podpĕra, Consp. Musc. Eur., p. 507. 1954.

Plants medium to robust, ferruginous, copper color, reddish,
golden, golden brown, brownish green, yellowish green, or pale green,
slightly glossy to very glossy when dry; stems rigid, up to 32 cm in
length, 0.5–0.75 mm in diameter, usually denuded and blackish at
base with age, commonly very glossy as though varnished, irregularly
pinnately branching; branches numerous, frequently short, 0.5–4 cm,
but occasionally up to 15 cm in length, close, erect-spreading to
spreading, occasionally branches recurved-spreading at base and
erect-spreading in upper portions, the shorter ones often spreading,
subarcuate with ends extending toward base of stem, occasionally
shorter ones erect-spreading; ends of closely to loosely imbricately
foliated stems and branches sometimes subattenuate, frequently obtuse; median cauline leaves erect to erect-spreading, distant, bases
up to 1.5 mm apart, occasionally up to 3 mm, blades firm, usually
carinate-conduplicate or subcarinate-conduplicate, sometimes deeply
concave, keels or median lines usually straight above the basal curve,
occasionally gradually and slightly curving towards margins or
slightly concave in middle portion, sometimes abruptly curved at the
apex, ovate-lanceolate when unfolded, blades broadest in upper part
of basal fourth and narrowing gradually to middle portion, some
rapidly decreasing in width in upper half and others in upper fourth,
majority long acuminate in upper fourth of blade, some short acuminate; leaf tips acute or narrowly obtuse, entire or serrulate, occasionally margins of apices slightly involute; median cauline leaves
5–7 mm long, sometimes 9 mm in length, 1.5–3.5 mm wide, occasionally up to 5 mm in width, 1.5–3 : 1; median cells of leaves linear-
rhomboidal to linear with ends attenuate, 8.5–20 μ wide, 10–25 : 1;
alar cells enlarged, subrectangular, subquadrate, to subhexagonal,
yellowish, brownish, or ferruginous, auricles none to slight, leaf bases
not decurrent to decurrent on same axis, decurrent up to 0.75 mm;
bases of median branch leaves up to 1 mm apart, blades erect-
spreading, carinate-conduplicate, subcarinate-conduplicate, or deeply
concave, keels or median lines generally straight above basal curve,
occasionally abruptly curved at apex, sometimes recurved, rarely
subfalcate secund, narrowly ovate-lanceolate when unfolded, blades
gradually narrowing from upper part of basal fourth into long acuminate apex, margins sometimes subinvolute, apical portions of

younger ones frequently canaliculate, leaves 3–6 mm long, 1–2 mm wide, 2–4 : 1; perichaetial branch 4–6 mm in length, perichaetium oblong, 1.25 mm in diameter; upper perichaetial leaves oval to suborbicular, broadly obtuse or rounded apical portion, entire, usually lacerate with age; calyptra long conical, 2 mm long 0.5–0.75 mm in diameter; operculum short conical, 0.65–0.75 mm long, 0.75–0.8 mm in diameter; urn immersed to slightly emergent, frequently contracted beneath mouth when dry, oblong to subcylindric, 2–2.5 mm long, 0.5–1.25 mm in diameter, 2.6 : 1; peristome teeth brownish orange, frequently united in pairs at apex, long acuminate, 0.75–1 mm in length, minutely muricate, lamellae 20–38; trellis of inner peristome perfect, brownish orange, 0.75–1 mm in length, muricate, transverse bars appendiculate; spores light green to yellowish green, submuricate to muricate, occasionally smooth, 12–18 µ in diameter, ripe in spring and summer. [The specific name, *Howellii*, honors Thomas Howell, who collected the plants which were used as a basis of the original description.]

Type: Not previously designated. In the absence of material so marked the writer has assumed to be the type the plants in the Herbarium of the Laboratoire de Cryptogamie, Muséum d'Histoire Naturelle, Paris, bearing the label, "*Fontinalis Howellii* Ren. and Card. sp. nova. 44/No. 20, Oregon: in swamps on old logs, etc. Leg. Th. Howell."

Type collector: Thomas Howell, in 1887, according to Renauld and Cardot, in Bull. Soc. Bot. Belg. 27: 134. 1888.

Type locality: Oregon: in swamp.

Distribution: North America: northwestern United States and southwestern Canada, and northeastern United States and southeastern Canada; South America: Peru.

Additional descriptions: Renauld and Cardot, in Bull. Soc. Bot. Belg. 27: 133. 1888; Renauld and Cardot, in Bull. Soc. Bot. Belg. 29: 155 (as *F. Kindbergii*). 1890; Renauld and Cardot, in Beihefte Bot. Centralbl. 1: 103 (as *F. Kindbergii*). 1891; Cardot, Mon. Font., pp. 62 (as *F. Kindbergii*), 66 (as *F. Howellii*). 1892; Limpricht, Laubm. 2: 660 (as *F. Kindbergii*). 1894; Barnes, Gen. and Sp. N. Am. Moss., p. 327 (as *F. Kindbergii*). 1896; Mönkemeyer, in Pascher, Süsswasserfl. 14: 105 (as *F. Kindbergii*), 106 (as f. *gracilior* and as f. *robustior*). 1914; Hj. Möller, in Ark. Bot. 17 (14): 48 (as *F. Kindbergii*). 1922; Brotherus, Laubm. Fenn., p. 396 (as *F. Kindbergii*). 1923; Jensen, Danm. Moss., p. 194 (as *F. Kindbergii*). 1923; Mönkemeyer, Laubm. Eur. Erg.-Bd. 4: 660 (as *F. Kindbergii*). 1927; Welch, in Grout, Moss Fl. N. Am. 3: 239 (as *F. Kindbergii*), 240 (as *F. Howellii*). 1934; Jensen, Skand. Bladmossfl., pp. 378 (as *F. Kindbergii*), 379 (as f. *robustior*). 1939.

Illustrations: Renauld and Cardot, in Bot. Gaz. 13: pl. 18. 1888; Renauld and Cardot, in Bull. Soc. Bot. Belg. 27: pl. 8. 1888; Renauld and Cardot, in Bot. Gaz. 15: pl. 9a (as *F. Kindbergii*). 1890; Renauld and Cardot, in Bull. Soc. Bot. Belg. 29: pl. 6a (as *F. Kindbergii*). 1890; Hj. Möller, in Ark. Bot. 17 (14); figs. 15 and 16, and pl. 6[11] (as *F. Kindbergii*). 1922; Welch, in Grout, Moss Fl. N. Am. 3: pl. 73 (as *F. Kindbergii* and as *F. Howellii*). 1934. – FIGURE 9.

Specimens examined: NORTH AMERICA: CANADA: BRITISH COLUMBIA: J. W. Bailey 4, Aug. 23, 1901 (PC); Brinkman, Sept. 26, 1908, Seymour Arm, Shuswap Lake, Can. Moss. 614 (E, K); Brinkman 246, July 12, 1910, Skuhun Creek,

Nicola R., alt. 1350 m (ABS); Dr. Dawson, Aug. 27, 1888, alt. 1500 m (CAN); MacFadden, Sept. 20, 1926 (DPU); MacFadden, Nov. 20, 1926, New Denver (CIN, DPU, H, MICH, NY, UC, WYO); MacFadden, June 6, 1927, Selkirk and Rocky Mts., near South Slocan (DPU, NY); Macoun, July 1, 1905, Lake House, Skagit R. (CAN, FH); Macoun, July 6, 1911, Beaver Lake (V); Macoun, in 1912, Sidney (NY); G. A. Potter, Howe Sound (NY).

CAPE BRETON ISLAND: Macoun, July 14, 1898, Can. Musci 598 p.p. (DUKE, FH, PC, US, WASH).

ONTARIO: Macoun, June 15, 1891, Ottawa, Can. Moss. 205 p.p. (CAN, FH, MT, US).

VANCOUVER ISLAND: J. R. Anderson, May 1897, Lake Shawnigan (MO); Mackenzie 102, near Victoria (ABS); Mackenzie, July 1930, near Victoria, alt. 450 m (CHI, DPU, MICH, NY); Mackenzie 154, Sept. 1934, near Duncan, north of Victoria (ABS, DUKE); Macoun, near Esquimalt, Can. Moss. 202 (K); Macoun, July 1875, Fl. Can. 751 (CAN); Macoun, in 1887 (FH, H); Macoun, June 8, 1887, Mt. Benson, Can. Musci 133 (CAN); Macoun, June 8, 1887, Mt. Benson, Can. Musci 227 p.p. (FH, MO); Macoun, June 23, 1887, Victoria, Can. Musci 598b (DUKE, FH, WASH); Macoun, June 25, 1887, near Esquimalt, Can. Moss. 202 (FH, K, MO, MT, NY, US); Macoun, June 25, 1887 (Assumed to be TYPE of *F. Kindbergii*, "Canadian Musci 233. *F. Kindbergii* Cardot (MS.)," PC; DUKE, FH, K, MINN, MO, NY, UC, US, WIS, YALE); Macoun, June 25, 1887, Victoria, Can. Musci 598b (US); Macoun, June 28, 1887, Cedar Hill, near Victoria, Can. Musci 251 (CAN); Macoun 163, June 18, 1908, Shawnigan Lake (CAN, S); Röll, May 23, 1888 (H); Röll 84, May 23, 1888, Victoria (PC); Röll 82, May 23 and 24, 1888, Victoria (PC); Röll 89 p.p., May 23 and 24, 1888, Victoria (PC); Röll, May 25, 1888, Victoria (H); Röll 86, May 28, 1888, Victoria (PC).

UNITED STATES: CALIFORNIA: M. A. Howe, July 24 – Aug. 10, 1894, upper Sacramento, Muir's Peak, north of Sisson (NY).

IDAHO: J. H. Sandberg, in 1892 (US).

MAINE: R. Lowe, Oct. 6, 1935, West Gray (DPU, DUKE).

MASSACHUSETTS: Kingman, Sept. 4, 1908, Middlesex Co., Wakefield (FH).

MONTANA: Standley 16979, Aug. 5, 1919, near Lake McDermott, alt. 1450–1740 m (NY, US); R. S. Williams 361, Aug. 9, 1895, Lake McDonald (CAS, K, MINN, MO, NY).

OREGON: Doty 2947, June 26, 1939, Coos Co., Snag Lake (DPU); DuVall, July 13, 1935, Wizzard Isl. in Crater Lake (DPU); A. S. Foster 1430 p.p., July 20, 1910, near Silverton (PC, YALE); A. S. Foster, Aug. 4, 1910, near Silverton (DPU, WASH); A. S. Foster 1431, Aug. 4, 1910, near Silverton (WASH); Henderson, Lost Lake, Cascade Mts. (FH); Henderson 1282, Lost Lake, Cascade Mts. (PC); Henderson 1744, Willamette R., near Portland (PC); Th. Howell (PC); Th. Howell 44/no. 20 (Assumed to be the TYPE, PC); Th. Howell, "*F. Howellii* Ren. and Card. sp. nova. Herb. J. Cardot, Mousses de l'Amérique du Nord" (FH, WIS); Th. Howell, "Am. bor. Oregon. Hb. J. Cardot" (K, NY, US); Th .Howell 33, Sept. 1882, St. Helen (YALE); Th. Howell, Sept. 1884 (CAN, FH, K, NY, US, WIS, YALE).

WASHINGTON: Alcorn, Jan. 31, 1934, South Tacoma (DPU, WASH); J. A. Allen, Feb. 11, 1898, Bear Prairie, near Mt. Rainier (NY); J. A. Allen, Sept. 19, 1898, Mt. Rainier region, Goat Mts., alt. 1500 m (WASH); J. A. Allen 81a, Sept. 19, 1898, Goat Mts., alt. 1500 m (CHI, DUKE, FH, ILL, MICH, MINN, NY, US, WASH, YALE); J. A. Allen 98, Sept. 24, 1898, Goat Mts., alt. 1500 m (PC); J. A. Allen, Nov. 5, 1898, upper Nesqually Valley (PC); J. A. Allen 81b, Nov. 5, 1898, upper valley of Nesqually R. (CHI, DUKE, FH, ILL, MICH, MINN, NY, US, WASH, YALE); O. D. Allen, Mar. 16, 1901, Pierce Co., Roy (NY); J. W. Bailey, June 1904, Seattle (PC); Flett, July 26, 1914, Mt. Rainier, alt. 690 m (NY, US); Flett, Oct. 5, 1933, Tacoma (PC); A. S. Foster, Apr. 5, 1905, Hamilton (WASH); A. S. Foster 1045, Aug. 9, 1909, alt. 1200 m (DUKE); A. S. Foster 1428 p.p., Nov. 26, 1910, Pacific Beach (YALE); T. C. Frye, Aug. 19, 1909, Lynden (WASH); T. C. Frye,

Nov. 15, 1915, Seattle (DPU, WASH); T. C. Frye, Sept. 10, 1929, Stevens Pass (DPU, WASH); T. C. Frye, July 10, 1937, Mt. Rainier Nat. Park, near Silver Falls (DPU, WASH); T. C. Frye 3200, Sept. 30, 1943, Snoqualmie Pass, alt. 1050 m (CHI, DPU, WASH); C. Harrison, July 11, 1939, Whidby Isl. (WASH); Kienholz, July 5, 1929, Cascade Mts., Columbia Nat. Forest, Goose Lake, alt. 900 m, "One whole end of lake has about 4 ft. of this moss growing on (and in, I suppose) a deep layer of alluvial mud. It grows submerged a foot or two, depending on the depth." (BART, NY); Mehner, July 1913, Centralia (DPU, WASH); Millican, July 16, 1925, near East Sound, alt. 360 m (DPU, WASH); Roberts, Apr. 5, 1925, Whidby Isl. (BART, CM, DUKE, FH, MICH, PENN, WYO); Röll, June 1888, Easton (H); Röll 821–822, June 5, 1888, Kachess Lake, near Easton (K, PC); Röll 668a, June 11, 1888, Easton (PC); Röll 823, June 12, 1888, Kachess Lake, near Easton (PC); Röll 668, June 15, 1888, Roslyn (PC); Röll 665 and 666 p.p., June 16, 1888, Easton (NY); Röll 667, June 16, 1888, Easton (PC); Röll 207, July 1, 1888, Tacoma (PC); Röll, July 3, 1888, Tacoma (G-BOIS, H); C. Smith, July 18, 1923, Orcas Isl., Twin Lakes (DPU, WASH); Suksdorf, Aug. 26, 1890, Skamania Co. (NY); Suksdorf, Aug. 27, 1890, Skamania Co. (NY, YALE); Svihla 2206, Sept. 3, 1931, Cascade Mts., Lake Kachess (DPU, WASH); Wentworth, July 22, 1923, Orcas Isl. (DPU, WASH).

SOUTH AMERICA: PERU: Ruiz and Pavón (BM, K, NY, PC).

Fontinalis Howellii differs from other species and varieties in the carinate-conduplicate group in having the following combination of characteristics: the long branches commonly erect-spreading and possessing short branches erect-spreading to spreading, straight to recurved and subarcuate, median cauline leaves large, 5–9 mm long, 1.5–5 mm wide, usually carinate-conduplicate with keels or median lines straight above basal curve, ovate-lanceolate when unfolded, and blade rapidly decreasing in width from lower two-thirds or three-fourths into long acuminate apex.

The types of *F. Howellii* and *F. Kindbergii* have been studied, and plants determined by Cardot as belonging to these two species have been examined. The author has not found constant characteristics upon which to separate these two species. Apparently Cardot had difficulty in distinguishing *Howellii* and *Kindbergii* because he states in his discussion of these species in his monograph that perhaps *Howellii* is a habitat variety or form of *Kindbergii*. Notations on his herbarium sheets in the Paris Museum indicate changes in his determinations from *Kindbergii* to *Howellii*. Barnes followed Cardot's suggestion in the monograph and described *F. Kindbergii* var. *Howellii*.

The original description of *F. Howellii* was published in 1888 and that of *F. Kindbergii* in 1890. According to the International Rules of Nomenclature, *F. Howellii* is the name with priority and not *F. Kindbergii*. The collections determined as *Kindbergii* are much more numerous than those of *Howellii*. Perhaps this was a factor in Cardot's suggestion of the priority of *Kindbergii*.

Cardot mentions in his descriptions the spreading and frequently recurved branches of *Howellii* and in his discussions the rigid, almost dendroid habit of this species. In the author's monograph of the Fontinalaceae in Grout's Moss Flora of North America, the separation of these two species was made largely upon this characteristic. Since that time, having seen many additional collections of *Fontinalis* which belong to one or the other of these two species, the problem of separating the two has increased. If one could follow the habit of the same plant from early stages to extreme age and also note the effect of growth in water as well as through and following a period of drought, one might better be able to make the correct decision. The writer hopes for this opportunity, and in the meantime, is relying upon herbarium materials as a basis of study.

In the habit of plants fitting the description of *F. Kindbergii*, the branches are long, straight, and extend from the stem at an angle of approximately 45°. The branches of these branches are usually short, spreading, and arcuate, very frequently recurved. Specimens have been studied in which a few of these short recurved branches have greatly elongated, showing at the base the arcuate condition and then growing at an approximate angle of 45° with the branch from which it originated. Its branches, also, are usually short and recurved. In plants of very robust habit these rigid axes with short, spreading, recurved branches suggest a dendroid habit. If one studies a robust, dendroid branch, one might apply the name of *F. Howellii*. If one examines plants with long, straight branches only, growing at an angle of approximately 45°, one could determine these as *F. Kindbergii*. Having seen both conditions on the same plant and not being able to discover constant morphological differences in stems, leaves, or fruits, the writer now considers that the plants should not be classified as two species upon the basis of the rigid, dendroid habit. Because of priority, the name to be applied is *F. Howellii*, although the collections of plants not showing the dendroid characteristic are much more numerous.

10. **Fontinalis chrysophylla** Card., in Rev. Bryol. 18 : 84. 1891 [in descriptive key]; Cardot, Mon. Font., p. 67. 1892; Cardot, in Rev. Bryol. 18 : 82. 1891 (nomen nudum).

Plants slender, ferruginous, reddish yellow, brownish yellow, golden yellow, sometimes brownish in lower portions, glossy when dry; stems rigid or nearly so, up to 20 cm in length, 0.2–0.3 mm in diameter, denuded and reddish brown to blackish at base with age, irregularly pinnately branching; branches plumose, commonly spreading, occasionally erect-spreading, varying from close to distant, ends of closely to loosely imbricately foliated stems and branches sometimes subattenuate, frequently obtuse; median cauline leaves with bases up to 0.5 mm apart, firm, commonly erect-spreading, occasionally spreading, generally carinate-conduplicate, sometimes carinate, subcarinate, or concave, keel or median line straight, blades narrowly lanceolate, width gradually decreasing from base

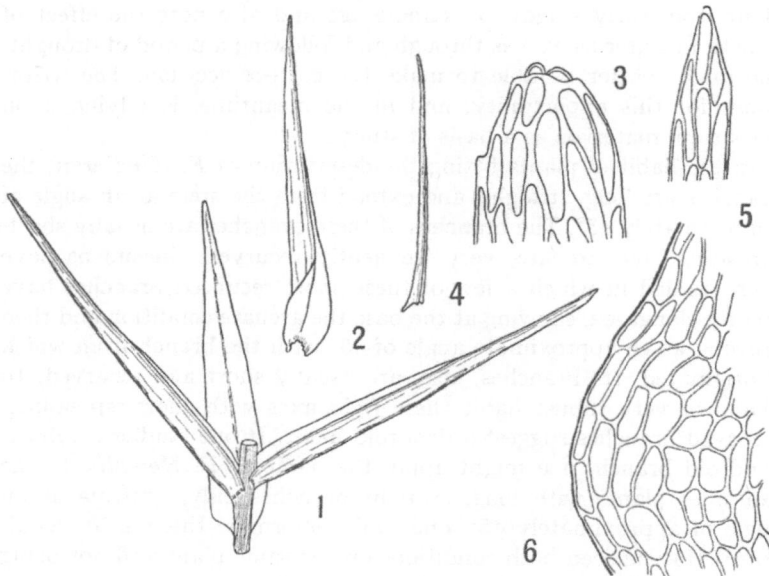

Fig. 10. – *Fontinalis chrysophylla* Card. – 1.Portion of stem with leaves. 2.Median cauline leaf.   3.Apex of cauline leaf.   4.Median branch leaf.   5.Apex of branch leaf.   6.Alar cells of median cauline leaf. Nos. 1–6 drawn from Henderson 1907, Washington.

into long acuminate apex; leaf tips acute, subacute, or narrowly obtuse, serrulate, subserrulate, or entire; median cauline leaves 4–5 mm long, 0.5–1 mm wide, 6–8 : 1; median cells of leaves linear, ends attenuate, 6.8–8.5 μ wide, 15–30 : 1; alar cells enlarged, very conspicuous, quadrate, subrectangular, subhexagonal, sometimes in vertical rows, 6–8, brown or brownish yellow, forming distinct auricles, bases decurrent to long decurrent, sometimes extending nearly to adjacent leaf base; median branch leaves comparable to median cauline in shape, usually narrower in width, some almost awl-shaped, apices carinate-conduplicate, carinate, tubular, concave, or subconcave, 3–5 mm long, 0.2–0.6 mm wide, 8–18 : 1; sporophyte not seen. Cardot received one old capsule from Henderson and described it as follows: perichaetium oblong, upper perichaetial leaves broadly oval, obtuse, with age lacerate at apex; capsule completely immersed, oblong to subcylindric. [The specific name, *chrysophylla*, meaning golden leaves, is indicative of the golden color of the leaves.]

Type: Not previously designated. In the absence of material so marked, the writer has assumed to be the type the plants in the Herbarium of the Laboratoire de Cryptogamie, Muséum d'Histoire Naturelle, Paris, bearing the label, "*Fontinalis chrysophylla* Cardot sp. nova. Washington. Olympic Mountains. Leg. L. F. Henderson. No. 1907."

Type collector: L. F. Henderson, No. 1907.

Type locality: Washington: Olympic Mountains.

Distribution: North America: Washington.

Additional descriptions: Barnes, Gen. and Sp. N. Am. Moss., p. 327. 1896; Welch, in Grout, Moss Fl. N. Am. 3: 240. 1934.

Illustrations: Welch, in Grout, Moss Fl. N. Am. 3: pl. 74. 1934. – FIGURE 10.

Specimens examined: NORTH AMERICA: UNITED STATES: WASHINGTON: Henderson 1907 (Assumed to be the TYPE, PC; FH, S); Henderson, July 22, 1890, swamp near the upper Skokomish R., Camp 6, Olympic Mts. (NY).

Fontinalis chrysophylla differs from other species and varieties in the carinate-conduplicate group in having the following combination of characteristics: plants slender, median cauline leaves generally carinate-conduplicate, narrowly lanceolate when unfolded, approaching linear when conduplicate, long acuminate, width gradually decreasing from base to apex, keel straight from the slight basal curve to the apex, blade 0.5–1 mm wide, 6–8 : 1, median cells of leaves 6.8–8.5 μ wide, 15–30 : 1.

Cardot and Welch in monographs previously published have considered the branch and cauline leaves of *F. chrysophylla* as being slightly dimorphic. The author now considers this to be an error because median cauline leaves and median leaves of the oldest branches seem to differ only in size and that to a degree to be expected because of the difference in age.

GROUP 2: LEAVES CONCAVE

Leaves generally concave throughout, but intermixed with these blades may be those which are plane, subconcave or concave at base and subconcave to plane above, deeply concave to canaliculate, or deeply concave to convolute tubulose.

11. **Fontinalis squamosa** Hedw., Sp. Musc., p. 299. 1801.

Fontinalis squamosa var. *angustifoliam* Hampe, in Linnaea 13: 45. 1839 (nomen nudum).
Fontinalis squamosa var. *subuliformem* Hampe, in Linnaea 13: 45. 1839 (nomen nudum).
Fontinalis squamosa var. *trichodem* Hampe, in Linnaea 13: 45. 1839 (nomen nudum.)
Pilotrichum squamosum C. Müll., Syn. Musc. 2: 149. 1850.
Fontinalis arduennensis Grav. in Piré, in Bull. Soc. Bot. Belg. 10: 105. 1871. [On basis of description; type not seen.]
Fontinalis squamosa f. *latifolia* Grav., in Cardot, Mon. Font., p. 83. 1892.
Fontinalis squamosa f. *julacea* Card., Mon. Font., p. 83. 1892.
Fontinalis Dixoni Card., in Rev. Bryol. 23: 70. 1896.
Fontinalis squamosa subsp. *Dixoni* (Card.) Dixon, Stud. Handb., p. 394. 1904.
Fontinalis Prageri Warnst., in Prager, Allg. Bot. Zeit. 13: 124. 1907.
Fontinalis squamosa var. *capillaris* Luis., in Ann. Sci. Acad. Polyt. Porto 2: 240. 1907.
Fontinalis squamosa var. *angustifolia* Luis., Musci Salmant., p. 199. 1924. [On basis of description; type not seen.]
Fontinalis squamosa var. *intorta* Luis., Musci Salmant., p. 199. 1924. [On basis of description; type not seen.]
Fontinalis squamosa var. *latifolia* (Grav.) Luis., Musci Salmant., p. 199. 1924. [On basis of description; type not seen.]
Fontinalis squamosa f. *tumida* Loeske, in Verdoorn, Bry. Arduen. Exsic. 8. 1927 [printed Latin description on label].
Fontinalis squamosa f. *angustifolia* (Luis.) Podp., Consp. Musc. Eur., p. 508. 1954.
Fontinalis squamosa f. *intorta* (Luis.) Podp., Consp. Musc. Eur., p. 508. 1954.
Fontinalis dalecarlica subsp. *Prageri* (Warnst.) Podp., Consp. Musc. Eur., p. 509. 1954.

Plants commonly medium in size but varying from slender to slightly robust, usually blackish green or brownish green, occasionally yellowish green or copper colored; frequently very glossy; stems somewhat rigid, up to 40 cm in length, 0.25–0.5 mm in diameter, generally denuded and dark at base with age, irregularly pinnately branching; branches numerous, erect to erect-spreading, close to slightly distant, occasionally appearing subfasciculate, up to 9 cm in length, ends of foliated stems and branches attenuate; median cauline leaves subimbricate to imbricate, close, bases usually up to 0.5 mm apart, occasionally up to 1 mm apart, blades firm, erect to

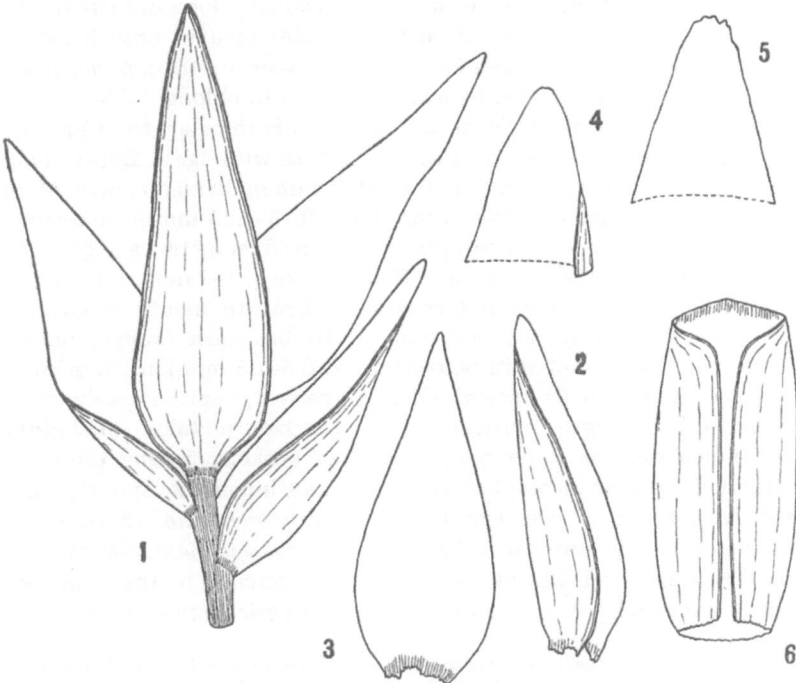

FIG. 11. – *Fontinalis squamosa* Hedw. – 1.Portion of stem with median cauline leaves. 2 and 3.Median cauline leaves. 4 and 5.Apices of median cauline leaves. 6.Perichaetial leaf. Nos. 1 and 4–6 drawn from W. Wilson, Sept. 1828, Wales; nos. 2 and 3 from J. Cardot, Aug. 6, 1884, Belgium.

erect-spreading, concave, usually ovate-lanceolate, occasionally lanceolate or ovate, margins sometimes slightly involute in apical portions of blades; apices acuminate, acute, subacute, or obtuse, usually entire, sometimes serrulate, occasionally twisted one-half turn, blades 2.5–4.5 mm long, occasionally up to 5.5 mm in length, 0.75–2 mm wide, occasionally up to 3 mm in width, 1.8–3.2 : 1; median cells of leaves linear with ends attenuate to subobtuse, subflexuous to flexuous, 5.1–14 μ wide, 7.5–20 : 1; one or two rows, occasionally three to five, of the marginal cells in some portion of the mature blades frequently abruptly narrower and with walls thicker and darker than adjacent cells, forming an indistinct to distinct border; alar cells enlarged, quadrate, rectangular, or subhexagonal, walls yellowish, brownish, or subhyaline, commonly forming distinct, auricles, occasionally auricles slight or none, leaf bases usually short decurrent, sometimes not decurrent, occasionally long decurrent and extending almost to the adjacent leaf base; median branch leaves comparable to median cauline except smaller in size; perichaetial branch 5–6 mm in length, perichaetium oval to oblong, 1.25–2.5 mm in diameter; upper perichaetial leaves suborbicular to orbicular, broadly obtuse to truncate at apex, lacerate with age; calyptra long conical, 1.75–2.5 mm in length, 0.65–0.85 mm in diameter; operculum short conical, obtuse, 0.75–1.5 mm long, 0.75–1.65 mm in diameter; seta 0.25–0.4 mm long; urn usually immersed, sometimes slightly to half emergent, suboval to oval, 1.75–3 mm long, 1–2 mm in diameter, 1.37–1.66 : 1, commonly not contracted beneath mouth when dry, occasionally slightly so; peristome teeth brownish orange, linear-acuminate, often united in pairs at apex, 0.5–1.5 mm long, muricate, lamellae 22–40; trellis brownish orange, perfect, approximately equal with teeth in length, muricate, transverse bars nodulose to slightly appendiculate, the lower ones sometimes broken, causing the cross strands to appear incomplete and the lattice imperfect; spores green, yellowish green, or yellowish brown, slightly muricate, 15.3–30 μ in diameter, ripe in summer. [The specific name, *squamosa*, was applied to this moss by Dillenius in 1741, with reference to the scale-like appearance and position of the cauline and perichaetial leaves.]

Type: Not previously designated. The specimens in the J. J. Dillenius Collection in the Oxford University Herbarium, in Oxford, England, first described and illustrated by Dillenius as *Fontinalis squamosa tenuis sericea atro-virens,* and later known as *Fontinalis squamosa* Hedw., have been assumed to be the type.
Type collector: Not known.
Type locality: "In a rivulet near Llanberris, Aug. and Sept.," according to Dillenius, Hist. Musc., p. 6. 1763. The author assumes this locality to be Llanberis, Carnarvonshire, North Wales.
Distribution: Fontinalis squamosa occurs in northern Africa and western Europe.
Additional descriptions: Dillenius, Hist. Musc., p. 258. 1741; Linnaeus, Sp. Pl. 2: 1108. 1753; Linnaeus, Sp. Pl. 2: 1571. 1763; Hedwig, Musc. Frond. 3: 32. 1792; Bridel, Musc. Recent. 2 (3): 160. 1803; Hedwig, Schwaegrichen, Sp. Musc., Suppl. 1: 307. 1816; Bridel, Musc. Recent., Suppl. 3: 108. 1817; Bridel, Bry. Univ. 2: 657. 1827; Bruch and Schimper, Bry. Eur., Fasc. 16, 5: 5. 1842; Hartman, Handb. Skand. Fl., p. 434. 1843; Bruch and Schimper, Bry. Eur., Fasc. 31, 5: 6. 1846; Schimper, Syn. Musc., p. 456. 1860; Milde, Bry. Siles., p. 276. 1869; Schimper, Syn. Musc. 2: 554. 1876; Cardot, Mon. Font.,

p. 80. 1892; Husnot, Musc. Gall., p. 286. 1892; Limpricht, Laubm. 2: 666. 1894; Limpricht, Laubm. 3: 802 (as *F. Dixoni*). 1903; Roth, Eur. Laubm. 2: 289 (as *F. squamosa*), 290 (as *F. Dixoni*). 1904; Braithwaite, Br. Moss-Fl. 3: 212 (as *F. squamosa*), 214 (as *F. Dixoni*). 1905; Mönkemeyer, in Pascher, Süsswasserfl. 14: 106. 1914; Hj. Möller, in Ark. Bot. 17 (14); 51. 1922; Brotherus, Laubm. Fenn., p. 397. 1923; Dixon, Handb. Br. Moss., pp. 392 (as *F. squamosa*), 393 (as *F. Dixoni*). 1924; Mönkemeyer, Laubm. Eur. Erg-Bd. 4: 661 (as *F. squamosa*), 662 (as var. *latifolia* and as *F. Dixoni*). 1927; Verdoorn, Bry. Arduen. Exsic. 7 (as f. *julacea*) [printed Latin description on label]. 1927; Jensen, Skand. Blad-mossfl., pp. 380 (as *F. squamosa*), 381 (as *F. Dixoni*). 1939.

Illustrations: Dillenius, Hist. Musc., pl. 33, fig. 3. 1741; Dillenius, Hist. Musc., pl. 33, fig. 3. 1763; Hedwig, Musc. Frond. 3: pl. 12. 1792; Bruch and Schimper, Bry. Eur., Fasc. 16, 5: pl. 430. 1842; Renauld and Cardot, in Bot. Gaz. 14: pl. 14. 1889; Husnot, Musc. Gall., pl. 80. 1892; Roth, Eur. Laubm. 2: pl. 30 (as *F. squamosa*), pl. 32 (as *F. Dixoni*). 1904; Brotherus, in Engler and Prantl, Nat. Pflanz.: Bryophyta 2: fig. 546. 1905; Braithwaite, Br. Moss-Fl. 3: pl. 122 (as *F. Dixoni*), pl. 123 (as *F. squamosa*). 1905; Prager, in Allg. Bot. Zeit. 13: 124, figs. a and sp (as *F. Prageri*). 1907; Mönkemeyer, in Pascher, Süsswasserfl. 14: fig. 32. 1914; Hj. Möller, in Ark. Bot. 17 (14): figs. 17–18 and pl. 6^{12}. 1922; Dixon, Handb. Br. Moss., pl. 48 (as *F. squamosa* and as *F. Dixoni*). 1924; Brotherus, in Engler and Prantl, Nat. Pflanz., Musci 11 (2): fig. 476. 1925; Mönkemeyer, Laubm. Eur. Erg-Bd. 4: fig. 143d. 1927. – FIGURE 11.

Specimens examined: AFRICA: ALGERIA: Durieu (BM, PC).
EUROPE: BELGIUM: Cardot, Mar. 7, 1884, Luxembourg, Poix, near St. Hubert (TYPE of f. *julacea*, PC); Cardot, Mar. 7 and 8, 1884 (PC); Cardot, Mar. 8, 1884 (PC); Cardot, Aug. 6, 1884 (PC); Cardot, Apr. 12, 1889 (US); Cornet 96, Oct. 1900 (BR); Cornet, June 1904 (BR); Cornet, Sept. 1904, Vesdre R. (BR); Cornet, Aug. 1921, in the Marche (BR); Delonge, Sept. 1868, Luxembourg, Semoy R., Gravet, Bryoth. Belg. 334 (K, NY); Delonge 30, Sept. 1868, Luxembourg, in Semoy R., in Ardennes Mts., alt. 200 m (BR, L); Gravet, Nov. 1869, Namur (PC); Gravet, in 1871, Namur (G-DEL); Gravet, in Nov. 1871, Namur, Bryoth. Belg. 231 (Assumed to be a portion of the TYPE of f. *latifolia* Grav., K, NY); Gravet, Apr. 20, 1881, Namur (BR, H, PC); Gravet, Mar. 1886, Namur, Gedinne, alt. 400 m (R); Gravet, May 1886, Namur, Gedinne, Husnot, Musci Gall. 775 (CHI, FH, K, PC, WASH, WIS); Mansion 455, Amblève R. (BR); Mansion 200a, Sept. 12, 1900, Amblève R. (BR); Mansion 53, July 15, 1904, Bovigny (BR); Mitten, in 1877, Amblève R. (NY); Verdoorn, Aug. 1927, Amblève R., alt. 400 m., Bry. Arduen. Exsic. 7 (CHI, DUKE, K, L, NY, UC, YALE); Verdoorn, Aug. 1927, Amblève R., alt. 395 m, Verdoorn, Bry. Arduen. Exsic. 8 (Assumed to be portion of TYPE of var. *tumida*, CHI, DUKE, K, L, NY, UC, YALE).
CORSICA: Philibert, July 1875 (PC).
CZECHOSLOVAKIA: E. Bauer, Bohemia, Böhmer Wald Mts., alt. 700 m, Krypt. Exsic. 594 (H, K, NY, US); E. Bauer, July 25, 1897, Bohemia, Böhmer Wald Mts., alt. 1000 m, Bryoth. Bohem. 43 (PC, WASH); E. Bauer, Oct. 21, 1897, Bohemia, Böhmer Wald Mts., alt. 700 m, Bryoth. Bohem. 44 (PC, WASH); E. Bauer, June 14, 1903, Bohemia, Riesengebirge, alt. 1400 m, Bryoth. Bohem. 344 (FH, PC, WASH); Blumrich, Aug. 14, 1911, Bohemia, Isergebirge, alt. 550–600 m, E. Bauer, Musci Eur. Exsic. 1165 (BART, L, WASH); Matouschek, Bohemia, Isergebirge, alt. 700 m, Krypt. Exsic. 594b (NY, US); Matouschek, Sept. 1902, Bohemia, Isergebirge, alt. 600 m, Migula, Crypt. Ger., Aust., and Helv. Exsic. 69 (ILL, MICH, NY); Milde, Sudeten (H); Pilous, July 1947, Bohemia, Brdy, Třítrubečný R., near Strašice, alt. 630 m (DPU); Schiffner, July 29, 1887, Bohemia (MICH); Schiffner, July 30, 1887, Bohemia (US); Schmidt, Sept. 1885, Bohemia, Isergebirge (H); Von Bock, Oct. 1909, Bohemia, Eichwald, alt. 420 m (G-DEL, S).
ENGLAND: Barker, Aug. 1870, Yorkshire (UC); Barker, July 16, 1899, Derby-shire, near Whaley Bridge (BM); Barker, July or Aug. 1900, Derbyshire, near

Whaley Bridge, R. Wye (K); Binstead, Oct. 1914, R. Wye, near Hereford (E); Binstead, Oct. 10, 1914, Breinton, near Hereford, R. Wye (BM); Binstead, Aug. 29, 1925, in R. Wye (BM); Binstead, Sept. 12, 1925 (H); Binstead, July 1929, Herefordshire, Clifford, R. Wye (BM); Binstead, Aug. 1933, R. Wye (DPU, DUKE); Binstead and Rhodes, Aug. 17, 1925, Herefordshire, R. Wye (BM); Binstead and Rhodes, Aug. 19, 1925, Herefordshire, R. Wye (BM); Binstead and Rhodes, Aug. 29, 1925, Herefordshire, R. Wye (BM); Boner, in 1847, Dartmoor, East Dart (NY); Boswell, Sept. 1878, Yorkshire (FH); Catcheside, Aug. 28, 1940, Westmorland Co., near Grasmere, R. Rothay (CGE, DPU); Crossland and Needham, Nov. 13, 1904, Yorkshire, Hebden Bridge, Hebden R. (E); Curnow, July 30, 1865, Cornwall, near Penzance, W. Wilson, Musci Britannici 443 (FH); H. N. Dixon, July 19, 1897 (PC); J. B. Duncan, Aug. 1933, Northumberland Co. (BART); Hunt, Yorkshire, Hebden Bridge (H); Hunt, June 1863, Yorkshire, Hebden Valley (FH); Hunt, June 1865, Yorkshire, Hebden Bridge (H, K, NY); Hunt, July 1865, Yorkshire (NY); Hunt, in 1867, Yorkshire, Hebden Valley (BR); Lamb, July 19, 1946, Cornwall, near Boscastle, attached to schistose rocks, alt. approximately 60 m (DPU); Needham, in 1897, Yorkshire, Hebden Bridge (BM); Schimper, Yorkshire, Todmorden (NY); Schimper, June 1865, Yorkshire, Hebden Bridge, near Todmorden (DPU); Schimper, June 10, 1865, Yorkshire, Hebden Bridge, near Todmorden (DPU, G-BOIS, MINN, NY, PC, WIS); W. West, Aug. 1885, Yorkshire, Hebden Bridge (K); W. Wilson, July 14, 1843, Yorkshire (S); W. Wilson, June 1865, Yorkshire, Hebden Bridge (DPU, DUKE, S); W. Wilson, June 1865, Yorkshire, Hebden Wood, Musci Britannici 443 (FH, CIN, MO, NY, YALE); Wood, May 1816, Deben R. valley (K); J. B. Wood, July 6, 1878, Yorkshire, near Hebden Bridge (S).

FRANCE: P. Allorge, Aug. 5, 1924, Ardennes, between Rocroi and Revin (PC); Brevière (PC); Brun, Dec. 1910, Allier, Le Breuil (PC); Camus, Oct. 4, 1878, Finistère (PC); Camus, May 30, 1881, Finistère (PC); Camus, July 4, 1895 (PC); Camus, Sept. 20, 1896 (CHI); Camus, Aug. 21, 1898, Loire-Inférieure and Vendée (PC); Camus, July 3, 1900 (PC); Camus, July 29, 1900, Finistère, Dismier, Bryoth. Gall. 164 (BART, CM, WASH, YALE); Camus, Aug. 30, 1901, Finistère (PC); Camus, Sept. 9, 1901, Finistère (MINN); Camus, Sept. 19, 1910 (CHI, PC); Cardot, May 23, 1883, Arvernia (PC); Charrier, Sept. 24, 1935, Loire-Inférieure (BART, CAS, DUKE, MICH, WASH); Crozals, Hérault. alt. 120 m (PC); De Brébisson, Normandy (K, US); De Brébisson 52, in 1829, Calvados, Vire (BR, G-DEL); Desmazières, Plantes Cryptogames de France 533 (BM); Desmazières, Plantes Cryptogames de France 1133 (K, NY); Dismier, July 1896, Aven R. (CHI); Dismier, July 6, 1896, Aven R. (PC); Dismier, Aug. 11, 1909, Pyrénées Mts. (PC); Dismier, July 20, 1930, Dourbie R., Bryoth. Gall. 363 (BART, CM, WASH, YALE); Fleischer, May 10, 1908, Pyrénées Mts., Loire-Inférieure, St. Etienne, alt. 500 m, Fleischer and Warnstorf, Bryoth. Eur. Merid. 366 (H); Husnot and Ménager, Sept. 17, 1891, Orne, attached to granite rock, Société Rochelaise 3199 (PC); Jeanpert, Aug. 2, 1901 (CHI); Jeanpert, Aug. 4, 1901, Finistère (CHI); Lachenaud, June 29, 1898 (PC); Lachenaud, Feb. 16, 1899, Haute-Vienne, Vienne R. at Limoges (PENN, S); Lachenaud, Feb. 16, 1899, Haute-Vienne, Vienne R. at Isle (PC); Lachenaud, Aug. 10, 1900, central France (DPU, PC); Lachenaud, Feb. 26, 1901, Haute-Vienne (K); Lenormand, Vire (PC); Lenormand, Mar. 1841, Vire (CHI, G-BOIS); Lenormand, Mar. 1841, Vire, Fl. Gall. and Ger. Exsic. 587 (CHI, K, MINN, PC); Lenormand, July 1852, Calvados, near Vire, Schultz and Winter Herbarium Normale Cryptogamia 25 (PC); Mougeot and Nestler, Vosges, Stirpes Cryptogamae Vogeso-Rhenanae 430 (CHI, E, FH, K, NY); Pelvet, Normandy, Vire (H, K, MINN); Pelvet, Calvados, Vire, Husnot, Musci Gall. 88 (CHI, FH, K, NY, PC, WASH, WIS); Schimper, Normandy (H, MINN); Schimper, Calvados (US); Schimper, near Vire (L); Schimper, in 1840, Musci Europaei Stirpes Normales, Normandy (NY); Schimper, in 1845, Normandy (DPU, PC, S).

GERMANY: E. Bauer, June 1904, Zwiesel, Böhmer Wald Mts., Musci Eur. Exsic. 500 (BART, PC); W. Baur, in 1890, Baden (PC); Bertram, Flora Hercynica

(FH, MICH, NY, S); Bertram, Harz, Rabenhorst, Bryoth. Eur. 1314 (FH, NY, WASH); Bertram, Aug. 1874, Flora Westfalen (NY); Blandow 200, New Brandenburg (UC); Bornmüller, Oct. 1928, Thuringia (DUKE); Brann, in May 1877, Flora des Harzes, Warnstorf, Deut. Laubm. (BR); Brann, June 1878, Flora des Harzes, Warnstorf, Deut. Laubm. (G-BOIS); A. Braun (K); A. Braun, in 1824, Black Forest Mts. (US); A. Braun, July 1833, Württemberg, Wildbad (K); Breidler, July 30, 1870, Steiermark (H); Breutel, Harz Mts. (H); Breutel and Spohrleder, Heindorfer Waterfalls and Harz Mts., respectively, Breutel, Musci Frond. Exsic. 185 (NY, PC); Everken, in 1869, Isergebirge, Flora of Silesia (K); Familler, Aug. 1907, Bavaria, Fl. Exsic. Bav.: Bry. 678 (PC); Flössner, July 14, 1929, Saxony, in the Erzgebirge near Neuhausen, Kopsch, Bryoth. Sax. 342 (MICH); Freiberg, Nov. 1915 and Apr. 1916, Rheinprovinz, alt. 420 m., Bauer, Musci Eur. Exsic. 1396 (BART, L, NY, WASH); Freiberg, Sept. 30, 1933, Rhena, alt. 450 m, Wirtgen, Herbar. Plant. Critic., Select. Hybrid. Florae Rhenanae 1135 (BR); Funck, Bavaria, Fichtelgebirge, Gefrees (K); Funck, Bavaria, Fichtelgebirge, Cryptogamische Gewächse des Fichtelgebirges 77 (L, NY); Funck, Bavaria, Fichtelgebirge, Cryptogamische Gewächse, besonders des Fichtelgebirges 117 (CHI, H, UC); Hampe, Hercynia (H); Hepp, July 1816, Rheinpfalz, near Trippstadt, Rabenhorst, Bryoth. Eur. 631 (G-DEL, NY, WASH); Jaeger, in 1864, Baden (NY); Kern, July 12, 1886, Silesia, Riesengebirge (CAN, PC); Koch, Rheinpfalz, near Trippstadt (K); Kolb, June 1875, Württemberg, in the Enz near Wildbad (CHI, ILL, NY, WASH); Kolb, June 1876, Württemberg, Enz R., Wildbad, alt. 420 m (R); Kolbynz, June 1877, Württemberg, on granite, alt. 420 m, Warnstorf, Deut. Laubm. (BR, G-BOIS); Kopsch, Sept. 1929, Saxony, Erzgebirge, near Wiesenbad, alt. 410 m (CAS); Krieger, Apr. – Aug. 1904, Sachsen, Königsbronn, alt. 140 m, Bauer, Musci Eur. Exsic. 559 (BART, PC); H. Lange, Nov. 20, 1925, Saxony, Erzgebirge, near Wiesenbad, Kopsch, Bryoth. Sax. 341 (MICH); Lehnert, Württemberg, Wildbad (US); Lickenberger 246, Aug. 1865, Oberbaden (DUKE, NY); Limpricht, in 1864, Silesia, Bryoth. Siles. 21 (PC); Limpricht, July 23, 1866, Silesia, Riesengebirge, Bryoth. Siles. 33 (NY, PC); Limpricht, in 1867, Silesia, Breslau (PC); Limpricht, Aug. 1867, Silesia, Bunzlau, alt. 500 m, Bryoth. Siles. 33b (NY); Loeske, May 15, 1892, Harz, on granite, alt. 400 m (H); Loeske, July 20, 1901, Harz Mts. (DUKE); Milde, Silesia, Riesengebirge (FH, MINN); Milde, Aug. 1860, Silesia, Riesengebirge, Rabenhorst, Bryoth. Eur. 630 (FH, NY); Prager, July 11, 1904, Prussia, Silesia, "Im Wasser des Eulengrundes, lang an Steinen flutend," in Riesengebirge, alt. 750 m. (Assumed to be a portion of the TYPE of *F. Prageri*, PC; S);[1] Prager, Aug. 1, 1904, Prussia, Riesengebirge (PC); Prager, July 20, 1908, Prussia, Silesia, Riesengebirge, Eulengrund, alt. 900 m (PC); Prager, July 20, 1908, Prussia, Silesia, Riesengebirge, Eulengrund, alt. 900 m, Bauer, Musci Eur. Exsic. 558 (BART, PC); Prager, July 22, 1908, Prussian Silesia, Riesengebirge, near Ober-Giersdorf, Bauer, Musci Eur. Exsic. 499 (BART, PC); Progel, Aug. 30, 1887, Bavaria, Bayrischer Wald, alt. 800 m (G-BOIS); Rabenhorst, Saxony, in the Biela, near Neidberg, Bryoth. Eur. 432 (NY); Rainer, Aug. 1874, Rheinprovinz (CAS); Reineck, Aug. 1914, Thuringia (G-DEL); Reinsch, Rheinpreussen, in the Laar (FH, K); Riehmer, Sept. 11, 1921, Saxony, Erzgebirge, near Geising, Kopsch, Bryoth. Sax. 259 (BM, MICH); Roemer, in 1876, Rheinprovinz (NY); Roth, June 1884, Hessen, Odenwald (CHI, DPU, US); Schiffner, July 29, 1889 (CHI); Schwab, Aug. 15, 1900, Bavaria, Fichtelgebirge (DPU); Schwab, Aug. 15, 1900, Bavaria, Fichtelgebirge, Familler, Fl. Exsic. Bav.: Bry. 27 (PC); Sendtner, July 15, 1838 (UC); Sickenberger 246, Aug. 1865, Oberbaden (NY); Spindler, Nov. 14, 1906, Vogtland (S); Spindler, Nov. 1907, Vogtland (MT); Vigener, in 1874, Westfalen

[1] The plants in Cardot's Herbarium in Laboratoire de Cryptogamie, Muséum d'Histoire Naturelle, Paris, were given to Cardot by C. Warnstorf. The specimens in the Herbarium of the Paris Museum were from the Herbarium of Loeske. The material in the Herbarium of Hj. Möller in Stockholm had been sent to Möller by Roth.

(NY); Wälde, in 1900, Württemberg (MINN); Wälde, in 1902 (CAS); C. Warnstorf, Aug. 1875, Deut. Laubm. (BR); C. Warnstorf 218, Aug. 1875, Rheinprovinz, Deut. Laubm. (G-BOIS); J. Warnstorf, July 14, 1908, Riesengebirge (DUKE, MT); Winter, July 1865, Rheinprovinz, in the Idarbach between Oberstein and Idar, Rabenhorst, Bryoth. Eur. 927 (NY, WASH); Winter, Aug. 27, 1865, Rheinprovinz, in the Idarbach between Oberstein and Idar (K); Winter, July 1869, Rheinprovinz, between Oberstein and Idar (PC).

IRELAND: I. Carroll, May 1851, Tipperary (YALE); Conard, Aug. 8, 1935, Kerry Co., Great Coombe of Mt. Brandon, Cloghane (DPU, FH); Jones, Owen, and Duncan, Aug. 1900, Killarney (BM); D. McArdle, May 1901 (BM).

ITALY: Coll.? Date? (G-BOIS).

NORWAY: Bryhn, July 1889, Stavanger Amt (PC, s); Bryhn, Aug. 1889, Søndre Bergenhus Amt, Bergen (DPU, H, s); Bryhn, Aug. 4, 1889, Bratsberg Amt (s); Bryhn, July 1902, Buskeruds Amt, near Ekersund, alt. 5 m (H, PC, s); Bryhn, July 1902, Buskeruds Amt, near Ekersund, alt. 2 m (H, PC, s); Bryhn, Nov. 1907, Søndre Bergenhus Amt, Bergen (H); Kaalaas, Aug. 3, 1886, Stavenger Amt (H, s); Kaalaas, July 29, 1889, Søndre Bergenhus Amt, Rosendal in Hardanger (s); Kaalaas, Aug. 3, 1892, Romsdals Amt (s); Kaalaas, July 17, 1900, Lister und Mandals Amt (s); Kaalaas, Aug. 7, 1900, Sogn, Sogndal (H); Kaalaas, Aug. 7, 1900, Søndre Bergenhus Amt (s); Kaalaas, July 18, 1910, Nordre Bergenhus Amt (s); Nyman, July 7, 1893, Stavanger Amt (s); Nyman, July 8, 1893, Lyse fjord (DPU, s); H. Persson, July 17, 1936, Stavanger Amt (s); Wulfsberg, July 15, 1867, Søndre Bergenhus Amt (K, s).

PORTUGAL: Birger, in 1921 (s); Luisier, Aug. 1907, Felgueiras, Jogueiros R. (Assumed to be portion of TYPE of var. *capillaris*, PC, s); Luisier, Dec. 5, 1907, Felgueiras, Jogueiros R., Bryoth. Lus. 10 (K); Welwitsch 260, Aug. 1848, Estrella Mts. (K).

RUMANIA: Péterfi, July 27, 1915, Transylvania, Cluj, alt. 620 m, Fl. Rom. Exsic. 715 (BR, US).

SCOTLAND: J. C. Adam, Aug. 24, 1916, Perthshire, Killin (E); Catcheside, Mar. 9, 1929, Dumbartonshire, near Milngavie (CGE, DPU, WASH); Croall, July 1, 1849, North Esk R. (BM); J. Dikcson 25 p.p., mountain streams (BM); Greville (NY); McAndrew, June 1901, Cairn Edward Burn, New Galloway (E).

SPAIN: P. Allorge, Sept. 21, 1933, Orense, Viana del Bollo, Bibey R. (DPU, PC); P. Allorge, Sept. 23, 1933, Orense, Verín, Tamega R. (DPU, PC); P. Allorge, Sept. 24, 1933, Orense, Minho R. (DPU, PC); Casares-Gil, July 1909, Valley of Arán (H); Durieu, in 1835, Austrias (PC); Durieu, June 9, 1835, Plant. Select. Hispano-Lusit. Sect. 1. Asturicae 134 (BR); Durieu, July 23, 1835, Plant. Select. Hispano-Lusit. Sect. 1. Asturicae 144 (MINN, PC, US); Luisier, Apr. 13, 1912, Pontevedra (US).

SWEDEN: Åberg 140, Aug. 12, 1912, Jämtland (s); Åberg 468, Aug. 4, 1913, Jämtland (s); Åberg 844, Sept. 7, 1916, Jämtland (s); Åberg, Sept. 2, 1920, Jämtland (DPU); Åberg 1600, Sept. 2, 1920, Jämtland (FH, s); Alm, June 3, 1917, Bohüslan (s); Arnell, Aug. 15, 1894, Medelpad (s); Behm, July 1865, Jämtland (s); Hamnström, July 12, 1873, Västmanland (s); Hässler, Sept. 18, 1921, Västergötland (s); Hellsing, Aug. 4, 1918, Närke, Musci Scandinavici 105 (s); Hellsing 211, Sept. 13, 1918, Lule Lappmark (s); Holmgren, Aug. 1, 1867, Lule Lappmark (s); Hj. Möller (DPU); Olossohn, Aug. 1912 (MT); Östman, Nov. 9, 1895, Gotland, Othem (DPU); Tärnlund, June 16, 1905, Västergötland (DPU); H. Thedenius, Västergötland (DPU, DUKE, s); Wahlstedt, in 1866, Västergötland (CHI).

SWITZERLAND: Bernet, Apr. 8, 1883, Geneva (G-BOIS); Bernet, Apr. 18, 1886, near Geneva, Rhône R. (BR); Bernet, Apr. 1890, near Geneva, Rhône R. (PC); Guinet, Mar. 16, 1890, Geneva, Rhône R. (G-DEL); Guinet, Mar. 23, 1890, Geneva, Rhône R. (G-DEL); Seringe (NY).

WALES: Binstead, Aug. 1933, R. Wye (CGE, DPU); Catcheside, Aug. 30, 1924, Carnarvonshire, Llanberis (CGE, DPU); H. N. Dixon, Aug. 1888, R. Colwyn,

Beddgelert, "in aqua fluente," (TYPE of *F. Dixoni* Card. "sp. nova," PC; portion of Type, "orig. specimen," Aug. 1, 1888, BM); H. N. Dixon, July 10, 1901, Llyn Caer, Cader Idris Mt. (BART); Hamilton, May 30, 1888, Merioneth Co., Barmouth (K); Hamilton, June 30, 1888 (K); Hamilton, Apr. 1895, R. Dee, Llangollen (K); D. A. Jones, May 1902, Merioneth Co., near Harlech (BM); D. A. Jones, Dec. 1902, Merioneth Co., near Harlech (E); D. A. Jones, June 1907, Merioneth Co., near Harlech (BM); D. A. Jones 450, June 1907, Merioneth Co., near Harlech (DPU, H); D. A. Jones, July 1912, Merioneth Co., near Harlech (WASH); D. A. Jones, Oct. 1913, Merioneth Co., near Harlech (BM); D. A. Jones and S. J. Owen, Feb. 1904 (E); D. A. Jones and J. Rhodes, Sept. 1911, Merioneth Co., near Harlech (CAS); Mitten, Aug. 1867, Berwyn Mt. (NY); S. J. Owen and D. A. Jones, Sept. 1905 (E); Painter, July 1895, Carnarvonshire, Bettws-Y-Coed (K); J. Rhodes, Sept. 1911 (WASH); P. G. M. Rhodes, June 1910, Merioneth Co., near Harlech, alt. 150 m (FH); Richards, Aug. 1947, Montgomeryshire, Machynlleth, R. Dovey (CGE, DPU); Simmonds, Apr. 14, 1943, Merioneth Co., Dovey Junction, R. Dovey (CGE, DPU); Welch 5696, Aug. 29, 1938, Pistyll Rhaiadr, near Llangollen (CHI, DPU, MICH, NY); Welch 5816, Aug. 30, 1938, World's End, near Llangollen (DPU); Welch 5662, Aug. 31, 1938, Ceiriog (DPU); W. Wilson, Sept. 1828, Carnarvonshire, near Aber (BR, DPU, MICH); W. Wilson, Sept. 5, 1828, Carnarvonshire, Aber (K).

Fontinalis squamosa may be separated from the other members of the concave group by the following combination of characteristics: majority of the mature median cauline leaves dark, very glossy, close, up to 0.5 mm apart, subimbricate to imbricate, ovate-lanceolate with apices acuminate to obtuse, entire, one to two rows of marginal cells in some portion of the blade abruptly narrower than adjacent cells and with thicker walls, thus forming an indistinct to distinct border, and apices of perichaetial leaves broadly obtuse or truncate.

The author has placed the *latifolia* varieties and forms in the synonymy of the species because the majority of the leaves on these plants are very comparable to those of *squamosa*, although occasional blades may be broader and sometimes longer than the usual dimensions of the median cauline leaves of *F. squamosa*.

12. **Fontinalis squamosa** Hedw. var. **Curnowii** Card., Mon. Font., p. 84. 1892.

Apices of upper perichaetial leaves abruptly apiculate, or subacute to acute; otherwise, plants similar to those of *F. squamosa*. [The varietal name, *Curnowii*, has reference to the collector of the type specimen, W. Curnow.]

Type: Not previously designated. The author has assumed to be the type, Rabenhorst, Bryoth. Eur. 926, in the herbarium of the Laboratoire de Cryptogamie, Muséum d'Histoire Naturelle, Paris, distributed as *F. squamosa*, and upon the sheet of which is written, "*Fontinalis squamosa* L. var. *Curnowii* Card."

Type collector: W. Curnow, June 1865.
Type locality: England: Cornwall, near Penzance.
Distribution: England, France, and Norway.

Additional descriptions: Braithwaite, Br. Moss-Fl. 3: 213. 1905; Dixon, Handb. Br. Moss., p. 392. 1924.

Illustration: FIGURE 12.

Specimens examined: EUROPE: ENGLAND: Curnow, June 1865, Cornwall, near Penzance, (E, L); Curnow, June 1865, Cornwall, near Penzance, Rabenhorst, Bryoth. Eur. 926 (Assumed to be TYPE of *F. squamosa* var. *Curnowii*, PC; FH, G-BOIS, NY, S); Curnow, June 30, 1865, Cornwall, near Penzance (BR, FH); Rogers, in 1878, Taxal (PC); W. Wilson, Sept. 2, 1865, Cornwall, Penzance (BM, DPU, MICH).

FRANCE: Lenormand, Calvados, Vire (PC).

NORWAY: Kaalaas, Aug. 2, 1898, Sogn (H).

Although var. *Curnowii* is separated from *F. squamosa* by perichaetial leaves apiculate or subacute to acute, the condition is sometimes difficult to determine if the leaves are lacerated or if the apicula is very slight. Some perichaetial leaves on the plants of Rabenhorst,

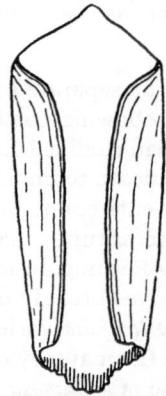

FIG. 12. – *Fontinalis squamosa* Hedw. var. *Curnowii* Card. Perichaetial leaf. Drawn from W. Wilson, Sept. 2, 1865, England.

Bryoth. Eur. 926 are round obtuse although the majority are subapiculate to apiculate, showing degrees of transition. The author has determined the plants to be var. *Curnowii* in instances of the majority of perichaetial leaves apiculate or acute.

13. **Fontinalis dalecarlica** Br. and Schimp., Bry. Eur., Fasc. 31, 5: 7. 1846; Schimper, in Fl. 28: 146, 1845 (nomen nudum).

Fontinalis squamosa Hedw. var. Sullivant, Musci Allegh., p. 46. 1846.
Fontinalis squamosa var. *dalecarlica* (Br. and Schimp.) C. J. Hartm., Handb. Skand. Fl., p. 341. 1849.
Pilotrichum dalecarlicum C. Müll., Syn. Musc. 2: 149. 1850.
Fontinalis dichelymoides Lindb., in Öfvers. Finska Vet.-Soc. Förh. 12: 76. 1870.
Fontinalis dalecarlica var. *curvata* Lindb., in Acta Soc. Sci. Fenn. 10: 94. 1871.

Fontinalis seriata Lindb., in Bot. Not. 1882, p. 26. 1882.
Fontinalis subconnivens Lindb., in Kindberg, in Bot. Not. 1882, p. 146. 1882 (nomen nudum).
Fontinalis dalecarlica var. *seriata* (Lindb.) Kindb., in Bih. K. Svenska Vet. - Akad. Handl. 7 (9): 51. 1883.
Fontinalis dalecarlica var. *baltica* Limpr. in litt. to Lutzow, according to Von Klinggraeff, in Schrift. Naturf. Gesell. Danzig (N.F.) 6: 24. 1883.
Fontinalis squamosa var. *elongata* Jens., in Lange, Consp. Fl. Groenl. 1 (2): 342. 1887. [on basis of description; type not seen.]
Fontinalis squamosa subsp. *dalecarlica* (Br. and Schimp.)Card., in Rev. Bryol. 18: 83. 1891.
Fontinalis dalecarlica var. *gracilescens* Warnst., in Cardot, Mon. Font., p. 88. 1892; var. *gracilescens* Warnst. in litt., according to Cardot, Mon. Font., p. 88. 1892.
Fontinalis Lescurii Sull. var. ? Herb. Barnes, according to Cardot, Mon. Font., p. 122. 1892. [Plants regarded by Cardot as *F. dichelymoides* Lindb.]
Fontinalis baltica (Limpr.) Von Klinggr., Leber- u. Laubm. West- u. Ostpreuss., p. 230. 1893.
Fontinalis Cavaraeana Farn., in Atti Ist. Bot. Univ., Pavia (Nuov. Ser.) 3: 70. 1893. [Based on description; type not seen.]
Fontinalis dalecarlica var. *atra* Limpr., Laubm. 2: 671. 1894.
Fontinalis microphylla Schimp., in Limpricht, Laubm. 2: 671. 1894; Schimper in litt. to Caspary, according to V. Klinggraeff, Topogr. Fl. Westpreuss., p. 112. 1880.
Fontinalis dichelymoides subsp. *microphylla* (Schimp.) Kindb., Sp. Eur. and N. Am. Bryin., Part 1: 147. 1896.
Fontinalis Berneti Card., in Roth, in Hedwigia 49: 221. 1910.
Fontinalis microphylla f. *subfalcata* Dietz. and Freib., in Bauer, Musci Eur. Exsic. 1164. 1915, [printed description on label].
Fontinalis dalecarlica f. *filiformis* Hj. Möll., in Ark. Bot. 17 (14): 57. 1922.
Fontinalis dalecarlica f. *laxa* Hj. Möll., in Ark. Bot. 17 (14): 57. 1922.
Fontinalis dalecarlica var. *microphylla* (Schimp.) Limpr., in Hj. Möller, in Ark. Bot. 17 (14): 62. 1922; var. *microphylla* Limpr. in litt. as syn. of *F. microphylla*, according to Limpricht, Laubm. 2: 671. 1894.
Fontinalis dalecarlica f. *curvata* (Lindb.) Jens., Skand. Bladmossfl., p. 381. 1939.
Fontinalis dalecarlica f. *microphylla* (Schimp.) Perss., in Bot. Not. 1942, p. 323. 1942.
Fontinalis dalecarlica f. *atra* (Limpr.) Podp., Consp. Musc. Eur., p. 509. 1954.
Fontinalis dalecarlica subsp. *baltica* (Limpr.) Podp., Consp. Musc. Eur., p. 509. 1954.

Plants slender, yellowish green, olive green, or dark green, younger portions of plants glossy, older parts frequently dull; stems rigid or nearly so, up to 90 cm in length, 0.3–0.4 mm in diameter, denuded and blackish at base with age, irregularly pinnately branching; branches numerous, slender, elongate, erect to slightly erect-spreading, close, frequently appearing fasciculate, up to 9 cm in length, ends of foliated stems and branches attenuate; median cauline leaves subimbricate to imbricate, bases up to 1 mm apart, blades firm, erect to slightly erect-spreading, generally concave or subconcave throughout, sometimes concave below and plane above, ovate-lanceolate, narrowly ovate-lanceolate, or narrowly lanceolate, margins sometimes slightly involute in apical portions of blades; apices usually acute or acumi-

nate, sometimes subacute to very narrowly obtuse, leaf tips commonly
entire but frequently serrulate, now and then twisted one-half turn;
blades 2–4 mm long, occasionally up to 4.75 mm in length, 0.5–1.25
mm wide, occasionally up to 1.5 mm in width, usually 3–3.5 : 1,
sometimes up to 4.5 : 1; median cells of leaves linear with ends attenuate, frequently flexuous or subflexuous, 5–15 μ wide, commonly

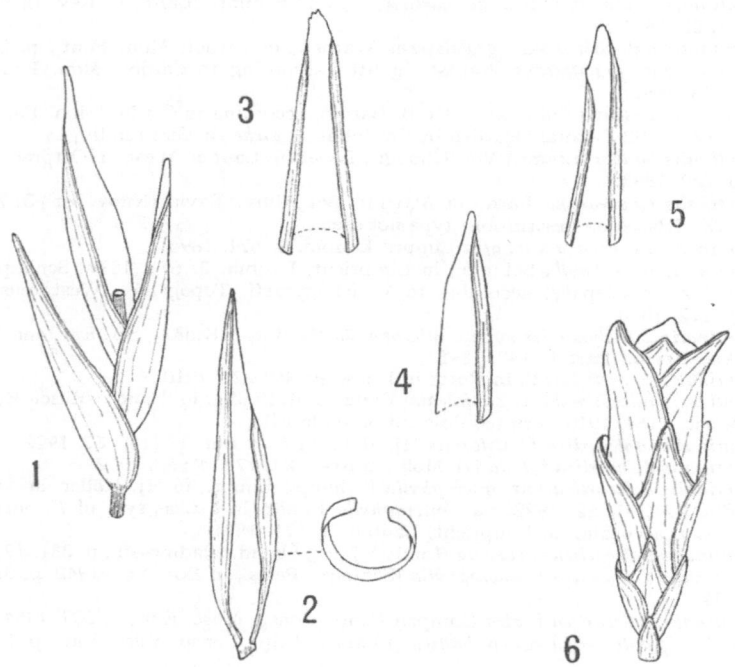

Fig. 13. – *Fontinalis dalecarlica* Br. and Schimp. – 1. Portion of stem
with median cauline leaves.   2.Median cauline leaf. 3, 4, and 5.Apices
of median cauline leaves.   6.Perichaetium containing sporophyte. Nos.
1, 2, and 6 drawn after Bruch and Schimper, Bry. Eur., Fasc. 31,
5: pl. 431; nos. 3–5 from Welch 713, Vermont.

10–15 : 1, sometimes up to 30 : 1, cells gradually and slightly narrowing toward leaf margins; alar cells enlarged, quadrate, rectangular,
or subhexagonal, walls yellowish, brownish, or subhyaline, commonly
forming distinct auricles, sometimes auricles slight to none, leaf bases
subclasping to clasping, usually short decurrent, 0.25–0.5 mm, occasionally almost to adjacent leaf base; median branch leaves comparable to median cauline except smaller in size; perichaetial branch

4.5–6 mm in length, perichaetium oval to subcylindrical, 1.25–1.5 mm in diameter; upper perichaetial leaves suboval, commonly narrowed at apex, acute or apiculate, entire, lacerate with age; calyptra long conical, 1.25 mm in length, 0.5–0.7 mm in diameter; capsule immersed; operculum very short conical, apex obtuse, usually 0.4–0.6 mm long, occasionally 1 mm in length, 0.5–0.75 mm in diameter; seta 0.75–1 mm long; urn oval, oblong, or subcylindric, 1.25–2.4 mm long, 0.5–1.4 mm in diameter, 1.75–2.6 : 1, occasionally contracted beneath mouth when dry; peristome teeth brownish orange, linear, long acuminate, frequently united in pairs at apex, 0.3–0.75 mm in length, submuricate to muricate, lamellae 12–22; trellis of inner peristome imperfect, brownish orange, sometimes yellowish orange, 0.3–0.75 mm long, slightly muricate, transverse bars of the lower half incomplete, not appendiculate, sometimes almost smooth; spores commonly light green, occasionally yellowish green, usually muricate, sometimes almost smooth, 25–35 μ in diameter, ripe in summer. [According to Bryologia Europaea, the specific name, *dalecarlica*, has reference to Dalecarlia, a west midland region of Sweden.]

Type: Not previously designated. In the absence of material so marked the writer has assumed to be the type or a portion of the type the plants in the herbarium of the Laboratoire de Cryptogamie, Muséum d'Histoire Naturelle, Paris, bearing the label, "*Fontinalis dalecarlica* mihi. Spec. nova! In Dalecarliae rivulis. August, 1844. Schimper leg."

Type collector: W. P. Schimper, in 1844.

Type locality: Sweden: Dalecarlia.

Distribution: Eastern half of the United States and Canada, in Greenland, and in western Europe.

Additional descriptions: Schimper, Nya Moss., p. 168. 1848; Sullivant, Musci and Hep. U.S., p. 655 (55). 1856; Schimper, Syn. Musc., p. 457. 1860; Sullivant, Icon. Musc., p. 102. 1864; Lange, Fl. Dan., Fasc. 46, p. 17. 1867; Schimper, Syn. Musc., pp. 554 (as *F. dalecarlica*), 559 (as *F. dichelymoides*). 1876; Lesquereux and James, Man. Moss. N. Am., p. 270. 1884; Cardot, Mon. Font., pp. 86 (as *F. dalecarlica*), 107 (as *F. seriata*), 122 (as *F. dichelymoides*). 1892; Limpricht, Laubm. 2: 667 (as *F. baltica*), 669 (as *F. seriata* and *F. dalecarlica*), 671 (as *F. dichelymoides*), 673 (as *F. Cavaraeana*). 1894; Barnes, Gen. and Sp. N. Am Moss., p. 331 (as *F. dichelymoides*). 1896; Grout, M. H. M., p. 398. 1903; Roth, Eur. Laubm. 2: 287 (as *F. seriata*), 290 (as *F. baltica*), 291 (as *F. dalecarlica*), 292 (as var. *atra* and as *F. microphylla*), 293 (as *F. dichelymoides*). 1904; Braithwaite, Br. Moss-Fl. 3: 213. 1905; Jennings, Man. Moss. W. Penn., p. 206. 1913; Mönkemeyer, in Pascher, Süsswasserfl. 14: 106. 1914; Hj. Möller, in Ark. Bot. 17 (14): 54 (as *F. dalecarlica*), 57 (as var. *curvata*), 63 (as *F. seriata*), 66 (as *F. dichelymoides*). 1922; Brotherus, Laubm. Fenn., pp. 397 (as *F. dalecarlica*), 398 (as var. *microphylla* and as *F. seriata*), 399 (as *F. dichelymoides*). 1923; Jensen, Danm. Moss., p. 191 (as *F. dalecarlica* and as var. *microphylla*). 1923; Dixon, Handb. Br. Moss., p. 392. 1924; Mönkemeyer, Laubm. Eur. Erg.-Bd. 4: 662 (as *F. dalecarlica*), 664 (as vars. *microphylla* and *baltica*); 665 (as *F. seriata* and *F. dichelymoides*). 1927; Welch, in Grout, Moss Fl. N. Am. 3: 245 (as *F. dalecarlica*), 256 (as *F. dichelymoides*). 1934; Jensen, Skand. Bladmossfl., pp. 381 (as *F. dalecarlica*, f. *filiformis*, f. *laxa*, var. *microphylla*, as *F. seriata*), p. 382

(as *F. dichelymoides*). 1939; Jennings, Man. Moss. W. Penn. and Adj. Reg., p. 175. 1951.

Illustrations: Bruch and Schimper, Bry. Eur., Fasc. 31, 5: pl. 431. 1846; Schimper, Nya Moss., pl. 16. 1848; Lange, Fl. Dan., Fasc. 46, 16: pl. 2751. 1867; Husnot, Musc. Gall., pl. 80. 1892; Farneti, in Atti Ist. Bot. Univ., Pavia (Nuov. Ser.) 3: pl. 24 (as *F. Cavaraeana*). 1893; Grout, M. H. M., fig. 218 and pl. 88. 1903; Roth, Eur. Laubm. 2: pl. 30 (as *F. dalecarlica*), pl. 31 (as *F. baltica, F. microphylla, F. dichelymoides, F. Berneti, F. seriata*). 1904; Brotherus, in Engler and Prantl, Nat. Pflanz.: Bryophyta 2: fig. 546 (*F. dichelymoides*). 1905; Braithwaite, Br. Moss-Fl. 3: pl. 123. 1905; Jennings, Man. Moss. W. Penn., pl. 30. 1913; Hj. Möller, in Ark. Bot. 17 (14): figs. 19–21 (as *F. dalecarlica*), fig. 22 and pl. 7[13] (as var. *microphylla*), figs. 23–25 and pl. 7[14] (as *F. seriata*), figs. 26–28 and pl. 8[15] (as *F. dichelymoides*). 1922; Brotherus, Laubm. Fenn., fig. 68. 1923; Jensen, Danm. Moss., pl. 8 (as *F. dalecarlica* and as var. *microphylla*). 1923; Dixon, Handb. Br. Moss., pl. 48. 1924; Brotherus, in Engler and Prantl, Nat. Pflanz., Musci 11 (2): fig. 476 (as *F. dichelymoides*). 1925; Mönkemeyer, Laubm. Eur. Erg.-Bd. 4: fig. 145 (as *F. dichelymoides* and *F. dalecarlica*). 1927; Welch, in Grout, Moss Fl. N. Am. 3: pl. 76 (as *F. dalecarlica*), pl. 79 (as *F. dichelymoides*). 1934; Jennings, Man. Moss. W. Penn. and Adj. Reg., pl. 34. 1951. – FIGURE 13.

Specimens examined: EUROPE: BELGIUM: Halin, Sept. 18, 1898, Vesdre R. (BR); Halin, Aug. 1899 (PC); Halin, Sept. 5, 1904, Vesdre R. (BR).

DENMARK: Wesenberg-Lund and Jensen, Sept. and Oct. 1908, Seeland, Bauer, Musci Eur. Exsic. 565 (BART, PC, S); Wesenberg-Lund and Jensen, Oct. 17, 1908, Seeland (DPU, S).

ENGLAND: H. N. Dixon, July 1894, Princeton, Dartmoor (BM, PC); J. Fowler 89, June 1865, Yorkshire (PC); Sherrin, July 2, 1918, Lake Devon, Port Bridge (S); A. Wilson, June 1917, R. Wenning, West Lancashire (BM, E).

FINLAND: V. F. Brotherus, July 7, 1869, Tavastia, Viitasaari (NY); V. F. Brotherus, Aug. 1869, Tavastia borealis, near Viitasaari, Lake Piojärvi, Musci Fenn. Exsic. 457 (L); V. F. Brotherus 442, Aug. 1869, Tavastia borealis, near Viitasaari, Lake Piojärvi (S); V. F. Brotherus, Aug. 7, 1869, Tavastia borealis, near Viitasaari, Lake Piojärvi (Assumed to be the TYPE of *F. dichelymoides*, H; CAN, DPU, K, MINN, NY, PC, S, US); V. F. Brotherus, July 25, 1870, Tavastia borealis, near Viitasaari, Lake Piojärvi (NY, WASH); V. F. Brotherus, July 25, 1870, Tavastia borealis, near Viitasaari, Lake Piojärvi, Musci Fenn. Exsic. 24 (G-BOIS, K, NY); V. F. Brotherus, Aug. 1873, Ostrobothnia, Musci Fenn. Exsic. 131a (K, PC); V. F. Brotherus, July 1887, Lapponia murmanica (G-BOIS); V. F. Brotherus, July 6, 1887, Lapponia murmanica (H); V. F. Brotherus, June 6, 1910, Bryoth. Fenn. 359 (CHI, DPU, L, YALE); Hjelt and Hult, Aug. 7, 1877, Lapponia kemensis (H, S); Hult and Kihlman 230, Aug. 14, 1880, Lapponia inarensis (H); Kihlman, June 21, 1892, Lapponia imandrensis (S); Kihlman 520, June 21, 1892, Lapponia imandrensis (S); Kihlman 521, June 21, 1892, Lapponia imandrensis (H); Kihlman 521, June 30, 1892, Lapponia imandrensis (H, S); Lackström (NY); Lackström, in 1867 (UC); Lackström, Aug. 1868 (WASH); Lackström, July 9, 1873, Brotherus, Musci Fenn. Exsic. 131b (BM, K, L); H. B. Lindberg, Oct. 11, 1884, Helsingfors, Tali-å (DPU, NY); H. B. Lindberg, Nov. 5, 1900, Helsingfors, Tali-å, Bauer, Musci Eur. Exsic. 492 (BART, PC, S); S. O. Lindberg, June 18, 1867, Hogland (H); S. O. Lindberg, in 1868, Helsingfors (DPU); S. O. Lindberg, Oct. 11, 1868, Helsingfors, Tali-å (BM, CAN, DPU, H, K, L, MINN, MO, NY, PC, S, US, WIS); S. O. Lindberg, Oct. 11, 1868, Helsingfors, Tali-å (Assumed to be the TYPE of var. *curvata*, H; S, L); S. O. Lindberg, Oct. 11, 1868, Helsingfors, Talia-å (Assumed to be a portion of the TYPE of var. *atra*, H; DPU, S); S. O. Lindberg, in 1873, Tavastia borealis, near Viitasaari, Lake Piojärvi, R. Hartman, Bry. Scand. Exsic. 412 (K); Lundström and Palmén, Aug. 14, 1865, Karelia borealis (UC); Lundström and Palus, Aug. 14, 1865 (MINN, S); Norrlin, Oct. 1868 (K); F. Nylander, Ostrobothnia borealis (H); W. Nylander, eastern Finland (TRT); Palmén, Aug. 1865 (US).

GERMANY: Caspary, in shallow water of Lake Czarny near Kowalle (Assumed to be the plants of *F. microphylla* about which Schimper wrote to Caspary, according to Von Klinggraeff, ex Herb. Schimper, K); Caspary, Aug. 26, 1876 and 6, 1877, West Prussia, Castens, See Czarny, "Original-Exemplar" (Assumed to be a portion of the collection of *F. microphylla* about which Schimper wrote to Caspary, according to v. Klinggraeff, H); Caspary, Aug. 28, 1880, East Prussia, Allenstein, Torfsee Tielk (Assumed to be a portion of the TYPE or topotype of *F. microphylla*, according to Limpricht, H); Caspary, Aug. 1, 1884, West Prussia, Neustadt (H); Caspary, Aug. 7, 1884, West Prussia, Neustadt (H); Caspary, Aug. 11, 1884, West Prussia, Neustadt (H); Caspary, Aug. 15, 1884, West Prussia, Neustadt (H); Caspary, Aug. 16, 1884, West Prussia, Neustadt (H); Freiberg, Sept. 3, 1909, near Jellenschehütte, alt. 154 m (S); Freiberg, Sept. 4, 1909, Torfsee Wook, near Wahlendorf, alt. 154 m (S); Freiberg, in 1910, West Prussia, Torfsee Wook, near Wahlendorf (PC); Freiberg, in 1910, West Prussia, near Waltersdorf, at Schlochau (PC); Freiberg, May 29, 1910, West Prussia, near Waltersdorf (H); Freiberg, Aug. 9, 1910, West Prussia, Neustadt, Torfsee Wook, near Wahlendorf, alt. 145 m, Bauer, Musci Eur. Exsic. 1163 (BART, L, S, WASH); Freiberg, Aug. 17, 1910, West Prussia, Neustadt, Torfsee Wook, near Wahlendorf (H, S); Freiberg, Nov. 22, 1911, East Prussia, Allenstein, Tilksee, near Ganglau, alt. 125 m. (H); Freiberg, Nov. 22, 1911, East Prussia, Allenstein, Torfsee Tielk, near Ganglau, alt. 120 m (BART, L, WASH); Freiberg, Oct. 27, 1912, East Prussia, Allenstein, Torfsee Tielk (MT), Freiberg, Oct. 28, 1912, East Prussia, Allenstein, Bszeneksee, near Ganglau, Bauer, Musci Eur. Exsic. 1164 (Assumed to be portion of TYPE of *F. microphylla* f. *subfalcata*, BART, L, WASH); Freiberg, Oct. 1913, East Prussia, Allenstein, Bszeneksee, near Ganglau (S); Freiberg, May 2, 1914, East Prussia (DUKE); H. Gross, June 15, 1921, East Prussia, Allenstein (DUKE, MT); Hintze, July 1911, Pommerania, Bublitz (PC); Hintze, Aug. 30, 1912, Pommerania, in Wardelsee (BART, L, WASH); J. Lange, June 26, 1884, West Prussia, Neustadt, Wygodda (H); Lützow, July 1880, West Prussia, in Lake Karpionki, near Wahlendorf (Assumed to be the TYPE of *F. dalecarlica* var. *gracilescens*, PC); Lützow, Aug. 3, 1880, West Prussia, Neustadt, near Wahlendorf, in Karpionki-See (B); Lützow, in 1881, West Prussia (PC); Lützow, July 1881, West Prussia, Neustadt, Steinkruger See (Assumed to be a portion of TYPE or topotype of *F. baltica*, BR, G-BOIS, PC, S); Lützow, July 1882, West Prussia, Neustadt (R); Lützow, Aug. 1882, West Prussia, Neustadt (H, PC); Lützow, in 1884, West Prussia, Neustadt, Steinkruger See (S); Lützow, Oct. 1884, West Prussia, Neustadt, Steinkruger See (PC); Lützow, July 1885, West Prussia, Neustadt (S); Lützow, Oct. 1885, West Prussia, Neustadt (S); Lützow, Oct. 5, 1885, West Prussia, Morsnitza-See, near Wygodda (S); Roemer, Rheinprovinz (NY); Steffen, June 15, 1921, East Prussia, Allenstein (DPU).

IRELAND: M. L. Anderson, in 1877, O'Sullivan's Cascade (H); Conard, Aug. 8, 1935, Kerry Co., Mt. Brandon (DPU).

ITALY: Sbarbaro (WASH).

LATVIA (LIVLAND): Werner, July 28, 1909, Wolmar, depth 1 m, Mikutowicz, Bryoth. Balt. 398 (H, US).

NORWAY: Bryhn, Aug. 4, 1885, Hedemarkens Amt, Elverum, Glommen R. (H); Bryhn, Aug. 5, 1885, Hedemarkens Amt, Elverum, Glommen R. (H, PC, S); Bryhn, Aug. 1892, Hedemarkens Amt, Elverum, Glommen R. (S); Bryhn, July 1900, Buskeruds Amt (MINN, S); Bryhn, July 1902, Ekersund (DPU, S); Conradi, July 14, 1899, Hamar (DPU, S); Fries, Sept. 25, 1863, Skarsvaag in Mageröy (H); Fries and Hensehen, July 26, 1863, Mageröy (S); Greve, Nov. 25, 1900, Søndre Bergenhus Amt (S); Hagen, July 30, 1891, Nedenes Amt (CAN, PC, S); Hagen, Aug. 22, 1896, Hedemarkens Amt, Elverum, Glommen R. (PC); Jörgensen, July 30, 1895, Søndre Bergenhus Amt (S); Jörgensen, Aug. 5, 1902, Söndfjord, alt. 250 m, Bauer, Musci Eur. Exsic. 498 (BART, PC, S); Kaurin, Aug. 1894, Finnmarkens Amt (H); Kiaer, July 26, 1884, Larviks (S); Lindblom, June 1826 (K);

Nyman, July 1893, Lister und Mandals Amt (DPU, s); Sörensen, June 25, 1911 (WASH); Tullgren, Aug. 15, 1891, Kristians Amt (CHI, s, US); Tullgren, Aug. 5, 1904, Kristians Amt (MT, s); Wulfsberg, July 19, 1867, Nordre Bergenhus Amt (s); Zetterstedt, Aug. 10, 1878, Gutbrandsdalen (s); Zetterstedt and Wickbom, Aug. 1870, Gutbrandsdalen, Migula, Krypt. Ger., Aust., and Helv. Exsic., Fasc. 31 and 32, 316 (ILL, MICH, MINN, NY, PC); Zetterstedt and Wickbom 175a, Aug. 16, 1870 (FH, PC).

SCOTLAND: McAndrew, Apr. 1892, New Galloway (E); McAndrew, Jan. 1898, New Galloway (E); C. McIntosh, Feb. 1902, Perthshire (PC).

SWEDEN: Åberg 1599, July 13, 1920, Jämtland (s); Åberg 1911, Aug. 19, 1921, Värmland (s); Åberg 1912, Sept. 1, 1921, Värmland (s); Åberg 1913, Sept. 19, 1921, Västergötland (s); Åberg 3047, Sept. 19, 1925, Västergötland (s); Åberg 3331, July 27, 1926, Värmland (s); Åberg 4696, Aug. 5, 1931, Dalarna (s); Adlerz, Aug. 8, 1904, Nericia (s); Andersson, Aug. 3, 1839, Värmland (s); Ångström, July 1868, Arnäs, Rabenhorst, Bryoth. Eur. 1179 (NY, UC); Arnell, Aug. 1881, Hernosand, Husnot, Musci Gall. 674 (CHI, DPU, FH, K, PC, WASH, WIS); Arnell, Aug. 9, 1881, Ångermanland (DPU, FH, G-BOIS, S, UC, US, WIS); Arnell, Aug. 10, 1881, Ångermanland (H); Arnell, Aug. 9, 1887, Noraström (DUKE); Arnell and Jensen, July 13, 1909, Hälsingland, Bauer, Musci Eur. Exsic. 561 (BART, PC); Arvén, Aug. 14, 1903 (FH); Arvén, July 14, 1914, Ångermanland (CHI); Behm, in 1886, Jämtland (DPU, s); Berggren, in 1859, Västergötland, Hunneberg (s); Clason, Dalecarlia (K); Clason, Aug. 1831, Dalecarlia (s); Grape, Mar. 1880, Blekinge (s); Hartman, in 1891, Gästrikland (G-BOIS); C. Hartman, Gästrikland (TRT); R. Hartman, in 1858, Gästrikland, Bry. Scand. Exsic. 84 (K, s); Hasslow, Aug. 29, 1921, Skåne (FH); Hellsing, Aug. 1, 1893, Dalarna (US); Hellsing, Aug. 7, 1898, Västmanland (CHI); Hellsing, Oct. 2, 1914, Bohuslän, Musci Scandinavici 279 (s); Hellsing, Nov. 5, 1914, Västergötland, Musci Scandinavici 303 (s); Hellsing, Nov. 12, 1914, Västergötland, Musci Scandinavici 306 (s); Hellsing, June 22, 1915, Bohuslän, Musci Scandinavici 398 (s); Hellsing, July 6, 1918, Hälsingland (s); Hellsing, Aug. 18, 1919, Västergötland, Musci Scandinavici 278 (s); Hellsing, Sept. 16, 1919, Värmland, Musci Scandinavici 298 (s); Hellsing, Oct. 16, 1920, Närke, Musci Scandinavici 370 (s); Hülphers, Aug. 1923, Västergötland (WASH); Indebetou, May 1875 (WASH); Indebetou, in 1880, Dalarna, Avesta, Lindsnäs (H, PC); Indebetou, July 1880, Dalarna, Avesta, Lindsnäs (K); Indebetou, Aug. 1880, Dalarna, Avesta, Lindsnäs (s); Indebetou, Sept. 1880, Dalarna, near Avesta, Lindsnäs (Assumed to be the TYPE of *F. seriata*, H; BM, FH, K, L, MO, NY, s); Indebetou, Sept. 1880, Dalarna, Asköforsen (s); Indebetou, Oct. 1882, Dalarna, Avesta, Lindsnäs (CHI, DPU, DUKE, H, MICH, MINN, MT, NY, PC, s, WASH); Jäderholm, Aug. 1, 1893, Dalecarlia (DPU); Jäderholm, July 29, 1910, Lapponia Tornensis (DPU); Jäderholm, July 13, 1911, Lapponia Tornensis, in Lake Tornejaur, at Vuoskaluokta (s); Johanson, July 1877, Bohuslän (WASH); Korssell, July 1880, Västergötland (G-BOIS); Larsson, Aug. 1, 1921, Dalsland, Lake Tydjesjön (DPU, s, WASH); Lillieroth, July 30, 1934, Småland (UC); S. O. Lindberg, Dalarna (K, PC); S. O. Lindberg, June 1854, Dalarna (s); S. O. Lindberg, July 1854, Dalarna (K, YALE); S. O. Lindberg, Aug. 1854 (CAN); S. O. Lindberg, June 1859, Hunneberg (K, US); Lönnkvist, July 1880, Jämtland (DPU); S. Medelius, July 30, 1920, Blekinge (s); Hj. Möller, June 29, 1909, Dalarna (H); Hj. Möller, July 1919, Lule Lappmark, alt. 300 m (s); Hj. Möller, July 5, 1921, Lule Lappmark, alt. 350 m (DPU, MICH, s); Nordström, Aug. 19, 1915, Härjedalen (s); Nyman, Aug. 13, 1888, Östergötland (NY); Oldberg, Aug. 1868, Gästrikland (US); Olsson, in 1856 (PC); Olsson, Aug. 1878, Östergötland (DPU, s, UC); Rubenson, May 1902, Västergötland (CHI); Sandberg, July 1921, Halland (s); Scheutz, Småland (CHI); Scheutz, in 1865, Småland (DPU); Scheutz, July 15, 1871, Blekinge (BM, BR, PC, US); Schimper, Dalecarlia (K, NY); Schimper, in 1844, Dalecarlia (PC); Schimper, Aug. 1844 (Assumed to be the TYPE or a portion of the Type of *F. dalecarlica*, PC); Schimper, in 1845, Dalecarlia (s); Sillén 50 p.p., Aug. 1875, Gästrikland (s); Harry Smith, Aug. 19,

1920, Torne Lappmark (DPU); Stenholm, Aug. 1, 1919, Västergötland (s); Stenholm, Sept. 4, 1920, Göteborg (DPU, DUKE); Stenholm, Sept. 22, 1920, Bohuslän (DPU, s); Stenholm, Sept. 25, 1920, Halland (s); Stenholm, Oct. 2, 1920, Halland (DPU, DUKE, s); Stenholm, Aug. 2, 1923, Lappmark (DUKE); Stenholm, Sept. 30, 1931, Västergötland, Göteborg (DUKE, WYO); Stenholm, July 20, 1932, Halland (MICH); Stenholm, July 26, 1932, Halland (DUKE); Svensson, Aug. 1934, Dalsland (UC); Swartz, in 1807, Dalarna (s); Tärnlund, June 16, 1934, Västmanland (DPU); H. Thedenius, in 1863, Gästrikland (s); K. F. Thedenius, Bohuslän (FH, s); K. F. Thedenius, Sept. 1834, Gästrikland (s); K. F. Thedenius, in 1838, Gästrikland (s); Wahlstedt, July 1867, Västergötland, (CHI, DPU, DUKE, NY, PC, US); Waldheim, July 27, 1937, Närke, Lake St. Gålsjön (DPU, s); Westling, Aug. 1863, Gästrikland (CHI); Zetterstedt, July 29, 1865, Småland (FH); Zetterstedt 175b, July 29, 1865, Småland (PC).

SWITZERLAND: Bernet, in 1883, near Geneva, "sur des blocs de nagelflue," in Rhône R. (Assumed to be a portion of the TYPE of *F. Berneti*, K, NY, S); Bernet, Apr. 8, 1883, Berne, in Rhône R. (G-BOIS, G-DEL); Bernet, June 8, 1883, Geneva, "sur des blocs de nagelflue," in Rhône R. (Assumed to be the TYPE of *F. Berneti*, PC); Bernet, Apr. 12, 1886, Geneva, Rhône R. (BR, FH, G-BOIS).

UNION OF SOVIET SOCIALIST REPUBLICS: V. F. Brotherus, Aug. 1872, Russia, Orlov (L, MO); V. F. Brotherus 78, Aug. 1872, Russia, Lapponia Orel (Orlov), (NY); V. F. Brotherus 181, Aug. 1872, Russia, Lapponia Orel (Orlov), (H); Sahlberg, July 28, 1869, Russia, Lapponia (H); Sahlberg, Aug. 7, 1870, Russia, Lapponia (H).

NORTH AMERICA: CANADA: ANTICOSTI ISLAND: Macoun, Aug. 16, 1883 (CAN); Macoun, Aug. 26, 1883, Can. Moss. 205 p.p. (CAN, US).

CAPE BRETON ISLAND: Macoun, Aug. 4, 1898, Can. Moss. 204 (CAN, FH, K, MO, MT, NY, US); Macoun, Aug. 4, 1898, Can. Musci 600 (DUKE); Macoun, Aug. 5, 1898 (CAN); Macoun, Aug. 5, 1898, Can. Musci 600a (CAN, DUKE, FH, K, MO, NY, PC, US, WASH); Macoun, July 4, 1899, Can. Moss. 204 (CAN, US); G. E. Nichols, July and Aug. 1909 (YALE); G. E. Nichols, Aug. 12, 1909 (PC); G. E. Nichols, July and Aug. 1914 (NY, YALE).

LABRADOR: J. A. Allen, July 29, 1882, Bonne Espérance (NY); Low 53, Aug. 30, 1896 (CAN); D. Potter 11, July 5, 1937 (DPU); Schofield, Aug. 10, 1949, Traverspine R., near Goose Bay (DPU, MICH); Waghorne, Apr. 8, 1894 (PC); Waghorne, July 28, 1894, Forteau (BM, G-BOIS, MO); Waghorne 3, July 28, 1894, Forteau (MINN, MO, US); Wickes, Aug. 4, 1938 (DUKE); Wickes 45, Aug. 4, 1938 (FH, US); Wickes 25, July 18, 1940, Cartwright (DPU).

NEW BRUNSWICK: Davidson 9, July 22, 1944, North Head, Grand Manan (DPU); James Fowler, in 1868, Bass R. (H, NY); James Fowler, June 10, 1869, Kent Co. (CHI); James Fowler, July 17, 1873, Bass R. (US); James Fowler, July 30, 1873, Bass R., Bathurst (MO); James Fowler, July 3, 1892 (TRT); Habeeb 275, Apr. 23, 1944, Fredericton, (DPU); Habeeb and Davidson, Aug. 6, 1944, Fredericton, Habeeb, Musci Novi Bruns. 52 (DPU); Macoun, July 3, 1899, Woodstock, Can. Moss. 204 (CAN, US); W. R. Taylor, Sept. 8, 1922, Grand Manan (MICH, PENN, YALE).

NEWFOUNDLAND: Macoun, Oct. 1, 1891 (CAN); Waghorne, Willow Bay (CAN); Waghorne, July 13, 1890, Green Harbor (CAN); Waghorne, in 1891, White Bay (BM, FH, G-BOIS, MINN, MO, NY, WIS); Waghorne, July 1, 1893 (PC).

NOVA SCOTIA: M. S. Brown, Aug. 18, 1923, Port Mouton (DPU); M. S. Brown 569, Aug. 1939, Sawlor Lake, Hubbard (DPU); Macoun, June 13, 1883, Bruno, Can. Musci 600 (FH, US, WASH); Macoun, June 13, 1883, Can. Musci 230 (UC, US); Macoun, July 9, 1883 (CAN); Macoun, July 17, 1883, Yarmouth (CAN); Macoun, July 23, 1883, Can. Musci 230 (FH, K, MO, NY, WIS, YALE); Macoun, June 13, 1885, Can. Musci 230 (MINN); Macoun, in 1889, Can. Musci 230 (PC); Prince 6458, Sept. 4, 1925 (MICH, WIS); Prince 6187, Sept. 3, 1928 (DPU, DUKE, MICH, WIS).

NOVA SCOTIA and ONTARIO: Macoun, Can. Musci 231 p.p. (FH, MO, NY, US); Macoun, in 1889, Can. Musci 231 p.p. (PC).

ONTARIO: H. H. Brown 117, Sept. 6, 1940, Sydney, Muskoka (DPU, TRT); R. F. Cain, July 27, 1939, Lake Timagami (DPU, TRT); R. F. Cain, Aug. 23, 1939, Algonquin Park (DPU, TRT); R. F. Cain, Aug. 30, 1939, Algonquin Park (DPU, DUKE, NY, TRT); R. F. Cain, Aug. 3, 1940, Brant Co. (DPU, TRT); R. F. Cain, Aug. 9, 1940, Whitney (DPU, TRT); R. F. Cain, Aug. 10, 1940, Algonquin Park (DPU, TRT); R. F. Cain, Aug. 26, 1940, Algonquin Park (DPU, TRT); R. F. Cain, Aug. 27, 1940, Whitney (DPU, NY, TRT); R. F. Cain, June 22, 1941, Bruce Co. (DPU, DUKE, MT, NY, TRT); R. F. Cain, Aug. 20, 1941, Algonquin Park (DPU, TRT); R. F. Cain, Aug. 21, 1941, Algonquin Park (DPU, TRT); R. F. Cain, July 22, 1944, Sudbury Dist (DPU, TRT); R. F. Cain, Aug. 2, 1944, Sudbury Dist. (DPU, TRT); R. F. Cain, Aug. 15, 1944, Sudbury Dist. (DPU, TRT); Drummond, Holland Landing, Musci Am. 233 (CAN, K, MICH, NY, PHIL); Elliott, May 28, 1889, Plover Mills (CAN); Elliott, June 1, 1889, near London (CAN); Hand 617, June 1938, Bruce Co., Sable Falls (DPU); Hand 649, July 5, 1940, Kent Co. (DPU); Hand 756, Aug. 13, 1941, Nysissing Dist., Duchesnay R. (DPU, OS); Hand 754, Aug. 16, 1941, Parry Sound Dist. (DPU); Hand 755, Aug. 16, 1941, Parry Sound Dist. (DPU); Hand 753, Aug. 17, 1941, Muskoka Dist. (DPU); Macoun 122, Hastings Co. (K); Macoun 191 ? = 190b, Hastings Co. (MO, NY); Macoun, in 1865, North Hastings (NY); Macoun 101, July 17, 1867, North Hastings (MO); Macoun, July 27, 1868, Peterboro Co. (CAN); Macoun, in 1874, Fl. Can. 255 (K); Macoun, Aug. 16, 1874, Lake Region, Fl. Can. 2224 (NY); Macoun, July 10, 1875, Fl. Can. 2223 (NY); Macoun, July 25, 1900, Algonquin Park, Long Portage, Petawana R. (TRT); Moxley 7096, June 22, 1941, Bruce Co. (DPU); Moxley 1213, June 29, 1942, Bruce Co. (DPU, TRT); Sprules, Aug. 1938, Algonquin Park (DPU, TRT); Sprules, Aug. 20, 1938, Algonquin Park (DPU, TRT); Sprules, Sept. 9, 1939, Algonquin Park, near Whitney (DPU, TRT).

QUEBEC: Catcheside 47186, Aug. 29, 1947, Pontiac Co., Indian Brook (CGE, DPU); Cléonique-Joseph 10660, Aug. 6, 1938, St. Maurice Co., Point du Lac (DPU, MT); Collins 4310, Aug. 2, 1906, Gaspé Co. (PC); Dupret, Aug. 31, 1905, St. Jérôme (US); Dupret, Aug. 17, 1910, near Quebec (ABS, PC, US); Dupret, Aug. 18, 1910, near Quebec (PC); Dupret, Aug. 31, 1915, Montfort (DPU, DUKE); Dupret 449, Aug. 31, 1915, Laurentian Hills, alt. 600 m (DPU); Dupret 10470, Aug. 31, 1915, Terrebonne Co., Montfort (DPU, MT); Dupret, Sept. 1915, near Montreal (ABS, US); Dupret, Aug. 22, 1925, St. Jérôme (WASH); Dupret 57, Aug. 27, 1927, Argenteuil Co. (MT); Fabius 1516, July 25, 1947, Waterloo (DPU); Fabius 1582, Aug. 7, 1947, Mt. Shefford (DPU); Fabius 1591, Aug. 7, 1947, Mt. Shefford (DPU); Fabius 2334, May 13, 1948, Mt. Shefford (DPU); Fabius 2601, July 27, 1948, Mt. Yamaska (DPU); Fabius 2612, July 27, 1948, Mt. Yamaska (DPU); Fabius 3477, June 9, 1949, Mt. Yamaska, alt. 300 m (DPU); Gauthier 416, Aug. 29, 1933, Lac Fernald (DPU, MT); Gauthier 2162, July 21, 1934, Berthier Co. (DPU, MT); Gauthier 11204, July 18, 1938, Laurentides Nat. Park (DPU, MICH); Gauthier 11241, July 21, 1938, Laurentides Nat. Park (DPU, MICH); Gauthier 11304, July 29, 1938, Laurentides Nat. Park (DPU, MICH); Gauthier 11412, Aug. 10, 1938, Laurentides Nat. Park (DPU, MICH); Hedley, Raymond, and Kucyniak 45–82, Aug. 1, 1945, Charlevoix Co. (DPU, MT); Lepage 2251, Aug. 7, 1940, Gaspé Co. (DPU); Lepage 2253, Aug. 7, 1940, Gaspé Co. (DPU, MT); Lepage 3008, July 15, 1941, near Rimouski (DPU); Lepage 9870, July 23, 1945, Ungava (DPU); Lepage and Dutilly 4379, July 31, 1943, Marten R. (DPU); Lepage and Dutilly 4553, July 31, 1943, Marten R. (DPU); Lepage and Dutilly 4312, Aug. 4, 1943, Marten R. (DPU); Macoun, May 16, 1896, near Hull (TRT); Marie-Anselme 63, in 1934, Tukesbury (TRT); Marie-Anselme 211, Apr. 7, 1935, La Tuque (DPU); Marie-Anselme 427, Apr. 8, 1935, Tukesbury (DPU); Marie-Anselme, Sept. 2, 1935, La Tuque (WASH); Marie-Anselme 194, Sept. 2, 1935, Lake Wayagamack (DPU); Marie-Anselme 1, Sept. 7, 1935, Laviolette Co. (DPU, MT); Marie-Anselme 210, Sept. 7, 1935, near La Tuque (DPU); Marie-Anselme 197, Nov. 2, 1935, La Tuque (DPU); Marie-Anselme, Apr. 4, 1936, La Tuque (DPU); Marie-Anselme 503, Apr. 4, 1936, near Dufendu Lake (DPU);

Marie-Anselme 508, Apr. 4, 1936, La Tuque (DPU); Marie-Anselme, Apr. 10, 1936, La Tuque (DPU); Marie-Anselme, Apr. 11, 1936, La Tuque (DPU); Marie-Anselme, Apr. 18, 1936, La Tuque (DPU); Marie-Anselme, May 1936, La Tuque (BART, DPU, TRT); Marie-Anselme, May 29, 1936, Lake Houle (DPU, OS, TRT); Marie-Anselme 692, June 16, 1936, Emmanuel Lake (DPU); Marie-Anselme 705, June 26, 1936, near La Tuque (DPU); Marie-Anselme 725, July 2, 1936, Lake Bourgeois, near La Tuque (DPU); Marie-Anselme 726, July 2, 1936, Lake La Cache, near La Tuque (DPU); Marie-Anselme 762, July 16, 1936, Bouleau Creek, near La Tuque (DPU); Marie-Anselme 764, July 16, 1936, Lake Moose, near La Tuque (DPU); Marie-Anselme 841, Aug. 12, 1936, Waterloo (DPU); Marie-Anselme 204, Sept. 2, 1936, La Tuque (DPU); Marie-Anselme 3228, Sept. 23, 1939, St. Félicien (DUKE); Marie-Anselme, Sept. 1940, St. Félicien (TRT); Marie-Anselme 4075, July 17, 1943, Beauceville (DPU); Marie-Victorin 2, in 1915, Laval Co. (DPU, MT); Marie-Victorin 26, June 30, 1919, Portneuf Co. (DPU, FH, MT, NY, PC, US); Marie-Victorin and Rolland-Germain 49574, July 20, 1925, Romaine R. (DPU, MT); Marie-Victorin and Rolland-Germain 1948, July 8, 1941, Lake St. Joseph (DPU, MT); Marie-Victorin and Rolland-Germain, Aug. 8, 1943, St. Adolphe (DPU, MT); Vinette 41, July 27, 1940, Argenteuil-Laurentians (DPU, TRT).

SAINT-PIERRE and MIQUELON: Le Gallo 30, Aug. 19, 1943, St. Pierre Isl., alt. 50 m (DPU, MT).

SASKATCHEWAN: Macoun, Sept. 24, 1872, Athabaska Plains (CAN); Macoun, July 4, 1895, Cypress Hills (E, K).

Canada without locality: Macoun, Moss. Can. 190 (= 190a), (NY).

GREENLAND: Hornemann (BM); Vahl, Sept. 1829, Fredericksdal (H).

UNITED STATES: CONNECTICUT: J. A. Allen, Apr. 19, 1880, East Haven (PC); J. A. Allen, May 1880, New Haven (FH); O. D. Allen, Feb. 27, 1878, near New Haven (NY, PC); Chatterton, Nov. 16, 1890, New Haven (NY); B. H. Clark, May 17, 1931, Barkhamsted (DPU, NY); Eaton, Nov. 1866, New Haven (YALE); Eaton, Sept. 23, 1872, Wintergreen Falls (CAS); Eaton, Oct. 1877, Wintergreen Falls, Hamden, "Station watched since 1856, but fruit never found" (US); Eaton, Apr. 18, 1890, Hamden (YALE); A. W. Evans, June 1891, Hamden (US); A. W. Evans, July 1891, Hamden (H); Kendall, Goldsborough, and Doolittle, July 17, 1904, Connecticut R. (US); G. E. Nichols, Salisbury (CIN, YALE); G. E. Nichols, Burlington (YALE); G. E. Nichols, Aug. 15, 1903, Southington (YALE); G. E. Nichols, Apr. 1905, Salisbury (YALE); G. E. Nichols, Oct. 13, 1906, Vernon (YALE); G.E. Nichols, Mar. 29, 1907, Chester (YALE); G. E. Nichols, May 30, 1907, Beacon Falls (YALE); G. E. Nichols, Dec. 27, 1907, Hamden (ABS, WASH); G. E. Nichols, Apr. 17, 1908, Burlington (YALE); G. E. Nichols, Dec. 11, 1909, Naugatuck (CIN, YALE); G. E. Nichols, Nov. 14, 1911, Putnam (YALE); Setchell, in 1885, New Haven (NY); Setchell, May 30, 1885, Meriden Hills (UC); Wareham, July 5, 1946, near Norwich (DPU).

DELAWARE: H. C. Wood, July 1860, Monroe Co. (PHIL).

FLORIDA: Collector ? May 1891 (DPU, DS).

GEORGIA: G. Carroll, Apr. 15, 1941, Lookout Mt., Dade Co. (CIN, DPU); S. S. Clarke, Dec. 8, 1895, Thomasville (WASH); W. H. Duncan 3301, May 25, 1941, Rabun Co. (DPU); R. M. Harper 1808a, Nov. 12, 1902, Bibb Co., in Ocmulgee R., alt. 30 m (BM, NY, PC, US); J. K. Small, Aug. 6, 1895, Ringgold, alt. 450 m, Mosses of the Southern United States 32 (BART, BM, CAN, CM, DPU, FH, MINN, MO, NY, OS, UC, US, WASH, WIS).

INDIANA: Deam 29527, Aug. 24, 1919, White Co., near Monticello (BUT, DPU); Kriebel 172, Aug. 13, 1933, Lawrence Co., near Bedford (DPU).

MAINE: C. D. Adams, in 1937, Oxford Co. (NY); C. D. Adams, June 5, 1937, Oxford Co., Hartford (DPU, PSNH); C. D. Adams, July 21, 1937, Penobscot Co., Carroll (DPU, NY, PSNH); J. A. Allen, Aug. 24, 1877, Oxford Co. (NY); O. D. Allen, July 20, 1876, Buckfield (NY); O. D. Allen, Aug. 25, 1877, Oxford Co. (YALE); Bacon 254, May 29, 1932, Woodstock (NY); Bacon, June 17, 1932, Norway (DPU, NY, PSNH); Burrage, Aug. 1, 1883, Mount Desert Isl., Sargent's Mt., (NY); R. M. Carle 3584, July 10, 1906, Camden (PC); R. M. Carle 3585, July 10, 1906 (PC); Chamberlain 3318, July 5, 1906, Somerset Co., alt. 300 m (ABS, PC); Collins 2368, July 16, 1900 (PC); Collins 3163, June 30, 1904, Pleasant Ridge (PC); Crockett, May 16, 1900, Camden (NY); Crockett, May 18, 1900, Camden (ABS, NY); Faxon, July 23, 1891, near Franconia (NY); R. L. Lowe, July 15, 1939, Washington Co. (DPU); R. L. Lowe, Oct. 17, 1939, Stow (DPU); E. D. Merrill 67, Apr. 1897, Orland (DUKE); E. D. Merrill, Apr. 25, 1897, Orland (US); E. D. Merrill 67, Apr. 1898, Orland (NY); G. K. Merrill 170, June 29, 1919, West Rockport (FH); G. K. Merrill 170, July 15, 1919, West Rockport (FH); G. K. Merrill 186, Aug. 17, 1919, Union (DUKE, FH); A. H. Norton, July 24, 1940, Burnham (DPU, PSNH); Norton and Fanning, Mar. 25, 1938, Isle au Haut, North End (DPU, PSNH); Norton and Haven, July 11, 1927, Cherryfield (DPU, PSNH); Parlin, July 5, 1934, Oxford Co., Buckfield, Streaked Mt. (DPU, NY, PSNH); Parlin, July 18, 1939, Oxford Co., Buckfield (DPU, PSNH); Parlin, June 25, 1946, Oxford Co., Waterford (DPU, PSNH); Patterson 28, June 21, 1929, Mount Desert Isl. (DPU); Rand, July 14, 1898, Mount Desert Isl. (FH); Redfield, July 30, 1890, Mount Desert Isl. (NY); Stebbins C680, Sept. 2, 1929, Mount Desert Isl. (NY); M. L. Stevens, July 1906, Somerset Co. (FH); W. Williams, Apr. 18, 1946, Franklin Co., Rangeley (DPU, PSNH, US).

MARYLAND: C. Gross, Mar. 20, 1937, near Frederick (DPU); C. Gross, May 21, 1937, Yellow Springs, near Frederick (CHI, DPU, NY); Owens, Feb. 22, 1938, Frederick Co., Catoctin Mts., alt. 360 m (DPU, MD).

MASSACHUSETTS: E. G. Britton, Sept. 22, 1897, Mt. Everett, Mt. Washington (FH); A. W. Chapman (NY); C. H. Clarke, Aug. 18, 1900, Magnolia (ABS); C. H. Clarke, Aug. 18, 1905, near Magnolia (FH); C. H. Clarke, July 1907, Magnolia (FH); E. M. Dunham, June 17, 1905, Worcester Co., Holden (NY); E. M. Dunham, Aug. 1906, Worcester Co., Holden (FH); Faxon 325, Oct. 21, 1881, Braintree (NY); James, Salem R. (PC); Setchell, Oct. 14, 1894, Mt. Washington (UC); Seymour, Aug. 1889, Granville (CHI); Wareham, Sept. 22, 1946, at Old North Bridge (DPU); Winslow, July 1912, Cheshire (FH).

MINNESOTA: Arthur, July 21, 1886, Lake Vermilion, (NY, PC); Arthur 30, July 21, 1886, Lake Vermilion (MINN, WIS); L. H. Bailey, July 19, 1886, Lake Vermilion (PC); L. H. Bailey 82, July 19, 1886, Lake Vermilion (MINN, NY, WIS); MacMullan, near International Boundary (PC).

NEW HAMPSHIRE: J. A. Allen, July 9, 1880, Jackson (MICH); J. A. Allen, in 1883, Ellis R. (FH); O. D. Allen, July 25, 1876, Shelburne (NY); O. D. Allen, July 9, 1880, Jackson (NY); O. D. Allen, July 15, 1880, White Mts. (NY); C. F. Austin, in 1872, White Mts. (H); W. B. Bailey, in 1882, Franconia (PC); W. B. Bailey, Aug. 1882, Franconia (WIS); Barber, Plymouth (FH); Bullard, July 15, 1938, Hill (FH); L. A. Carter, in 1901, Laconia (DUKE); L. A. Carter, July 19, 1901, Gilford (ABS, ILL); L. A. Carter, Sept. 5, 1904, Gilford (MO); L. A. Carter, Sept. 5, 1904, Belknap Co. (WASH); L. A. and W. A. Carter, May 7, 1901, Gilford (ABS); W. A. Carter, May 7, 1901, Gilford (PENN); Cummings, July 20, 1878, Plymouth (CHI); Cummings, July 20, 1879, Plymouth (FH); Cummings, Aug. 1884, North Woodstock (CM); A. W. Evans, Aug. 1890, Jackson (YALE); Farlow, Aug. 1882, Shelburne (FH, YALE); Farlow, Aug. 1904, Bolles (FH); Faxon 323, June 6, 1881, Silver Cascades, White Mt. Notch (NY); Faxon 321, June 7, 1883, Crawford Notch (NY); Faxon 322, June 18, 1883, White Mts., Gate of Notch (NY); Faxon 554, June 1, 1890, Easton (NY); Grout 99, July 13, 1900, Plymouth (ABS, MINN, NY, WASH); E. T. and S. A. Harper, Aug. 1895, Coos Co., Twin Mt. (CHI, FH, NY); James, White Mts. (CHI, FH, ILL, K); James, June 1850, Chester Co. (PHIL); James, in 1852, White Mts., Crawford Notch (K, NY); James 59–6, July 1852, Crawford Notch Saco, R. (US); James, Aug. 1853, near Glen (K); James, Aug. 1853, White Mts. (K); James 88, Aug. 1853, Gorham (BM); Kingman, July 3–8, 1912, Mt. Moosilauke (FH); Langdon 18, Aug. 8, 1893, Plymouth (MICH); Lorenz, Aug. 2, 1907, Waterville, alt. 450 m (YALE); G. E. Nichols

(PC); G. E. Nichols, Aug. 12, 1906, White Mts., Crawford Notch (YALE); Rand and Robinson, June 13, 1898, Gap Mt., Troy (FH); W. R. Taylor, Sept. 7, 1916, White Mts., Bretton Woods (MICH, PENN); E. W. Thompson, Aug. 22, 1934, Peterboro, base of Mt. Monadnock (FH); E. W. Thompson, Aug. 26, 1934, Mt. Monadnock (FH); Von Schrenk, June 8, 189–, Shelburne, Cobbat Mt. (MO).

NEW JERSEY: C. F. Austin, Closter (NYSM); C. F. Austin, Pine Barrens (DUKE); H. L. and R. S. Williams, Nov. 30, 1933 (CHI).

NEW YORK: C. F. Austin, Aug. 1861, Bergen Co. (NY); Beals, July 30, 1938, near Ellenville (MICH); Brandegee, Adirondacks (NY); E. G. Britton, Aug. 30, 1894, Adirondacks (NY, WASH); E. G. Britton, Aug. 22 to Sept. 12, 1894, northern New York (MO); E. G. Britton, Sept. 7, 1896, near Lake Placid (NY); E. G. Britton, Sept. 9, 1898, Sunrise Notch (NY); Eaton, Aug. 1855, Fulton Co. (YALE); Greenlach, Oct. 2, 1898, Indian Lake (NY); Grout, Sept. 1931, Cold Spring Harbor (DUKE); Haring, July 8, 1936, Watson Hollow, alt. 510 m (DPU); Haring 4500, Aug. 17, 1946, Ulster Co., Catskill Mts., alt. 750 m (DPU); E. J. Hill 158, 1882, Aug. 11, 1882, Mumford (CHI, ILL); E. J. Hill 123, 1893, July 15, 1893, Genesee Co., Pavilion (ILL); Hulst, Aug. 1899, Lake George (DUKE); Kaiser, Aug. 5, 1910, Catskill Mts., alt. 600 m (ABS, H, PC); Knight, Aug. 1885, Long Isl. (NY); M. F. Miller, June 23, 1900, Shaudaken (ABS, DUKE, NY); Muenscher and Isely, 492 of Winne, Sept. 9, 1941, Ulster Co., near Edgewood, alt. 540 m (DPU, NY); Newman, Sept. 1, 1905, St. Lawrence Co. (FH); Newman 3586, Sept. 1, 1905, northwestern Adirondack Region (PC); Peck, in 1864, Sandlake (NY); Peck, Aug. 1865, Sandlake (NY); Vail, Aug. 1892, Greene Co., near Tannersville (DUKE, NY); Vail, Aug. 22, 1892, Greene Co., Tannersville (DUKE); Warne, Oneida Lake (NY); Winne 293, July 20, 1941, Onondaga Co., Jamesville, alt. 126 m (DPU, NY); Winne 575, Oct. 5, 1941, Essex Co., near Aiden Lair, alt. 510 m (DPU, NY); Winne 576, Oct. 5, 1941, Essex Co., near Minerva, alt. 486 m (DPU, NY); Winne 578, Oct. 7, 1941, Greene Co., alt. 780 m (DPU, NY); Winne 597, Oct. 7, 1941, Greene Co., alt. 600 m (DPU, NY); Winne 601, Oct. 8, 1941, Greene Co., alt. 474 m (DPU, NY); Winne 616, Oct. 8, 1941, Greene Co., alt. 558 m (DPU, NY); Winne 618, Oct. 8, 1941, Greene Co., alt. 510 m (DPU, NY); Winne 621, Oct. 8, 1941, Greene Co., near Hunter, alt. 525 m (DPU, NY); Winne 624, Oct. 8, 1941, Greene Co., near Hunter, alt 492 m (DPU, NY); Winne 630, Oct. 8, 1941, Greene Co., Hunter, alt. 474 m (DPU, NY); Winne 641, Oct. 21, 1941, Greene Co., High Peak Mt., alt. 600 m (DPU, NY); Winne 649, Oct. 21, 1941, Greene Co., High Peak Mt., alt. 810 m (DPU, NY); Winne 670, Oct. 21, 1941, Greene Co., near Plaat Clove, alt. 576 m (DPU, NY); Winne 675, Oct. 21, 1941, Greene Co., near Plaat Clove, alt. 573 m (DPU, NY); Winne 680, Oct. 21, 1941, Greene Co., near Plaat Clove, alt. 570 m (DPU, NY); Winne 688, Oct. 22, 1941, Ulster Co., West Saugerties, alt. 171 m (DPU, NY); Winne 691, Oct. 22, 1941, Ulster Co., West Saugerties, alt. 174 m (DPU, NY); Winne 700, Oct. 22, 1941, Greene Co., Plaat Clove, alt. 573 m (DPU, NY); Winne 707, Oct. 22, 1941, Greene Co., Kaaterskill Junction, alt. 504 m (DPU, NY); Winne and Muenscher, 396 of Winne, Aug. 20, 1941, Essex Co., Mt. Marcy, alt. 672 m (DPU, NY); Winne and Muenscher, 397 of Winne, Aug. 20, 1941, Essex Co., Indian Falls, alt. 1020 m (DPU, NY); Winne, Muenscher, and Shannon, 190 of Winne, June 12, 1941, St. Lawrence Co., South Colton, alt. 276 m (DPU, NY); Winne, Muenscher, and Shannon, 230 of Winne, June 15, 1941, Hamilton Co., alt. 510 m (DPU, NY); Woodward and Beals, July 30, 1938, near Ellenville, alt. 315 m (DPU).

NORTH CAROLINA: L. E. Anderson 428, June 26, 1933, Jackson Co., Soco Falls, near Maggie (PENN); L. E. Anderson 435, June 26, 1933, Jackson Co., near Maggie (DPU, PENN); Blomquist, July 3, 1932, Swain Co. (DPU, DUKE); Blomquist 4020, Oct. 9, 1934, Johnston Co. (DPU, DUKE); F. S. Chapman, Nov. 24, 1920, Alleghany Co., Glade Valley (DPU, DUKE); S. Chapman, Jan. 16, 1923, Forsyth Co., Winston-Salem (DPU, DUKE); Clebsch, June 18, 1944, Graham Co., Unicoi Mts., alt. 1275 m (DPU); Durand 12097, Aug. 2, 1901, Blowing Rock (PC); Fitzgerald 126, May 1881 (FH); F. W. Gray M1636, Oct. 26, 1931, Linnville,

Grandfather Mt., alt. 1200 m (BART); Grout, July 27, 1907 (PC); Schallert 523 (GRI); Schallert 2780, Stokes Co. (BART, CHI); Schallert, Aug. 8, 1921, Stokes Co. (DPU, DUKE); Schallert 46, Aug. 10, 1921, Stokes Co. (WIS); Schallert 341, Aug. 12, 1921, Forsyth Co., Winston-Salem (WASH); Schallert, May 30, 1926 (CAS, S); Schallert, June 1, 1926, Forsyth Co., Winston-Salem (WASH); Schallert, June 10, 1929, Stokes Co. (FH, UC); Schallert, May 30, 1931, Forsyth Co. (DPU); Schallert, Mar. 20, 1934, Forsyth Co. (CHI, NY); A. J. Sharp 107, July 28, 1930, Highlands, Mirror Lake, alt. 1200 m (DPU, TENN); A. J. Sharp 3721, July 4, 1937, Macon Co., Highlands (CHI, DPU, MICH, NY, TENN); Standley and Bollman 10123, Aug. 3, 1913, Montrest (CHI, US); Torrey, Salem (PHIL); K. A. Wagner 1582, June 5, 1949, Jackson Co., Chattooga R. (DPU); K. A. Wagner 1974, June 10, 1949, Jackson Co., Twin Falls, near Cashiers (DPU); K. A. Wagner 1986, June 10, 1949, Jackson Co., Twin Falls, near Cashiers (DPU); Welch 2220, June 17, 1936, Jackson Co., Soco Falls (CHI, DPU); Welch 2221, June 17, 1936, Jackson Co., Soco Falls (CHI, DPU, MICH, NY); Welch 2222 and 2223, June 17, 1936, Jackson Co., Soco Falls (DPU).

OHIO: E. L. Braun, Aug. 13, 1928, Adams Co. (CIN); Rood, July 1, 1935, Trumbull Co. (DPU, OS); A. J. Sharp, May 29, 1937, Adams Co. (DPU, OS); Wareham, June 11, 1937, Adams Co., on dolomite (DPU, OS).

PENNSYLVANIA: C. F. Austin, in 1873 (H); Best, Sept. 13, 1892, Tobyhanna Mills (NY); Boardman, Sept. 16, 1934, Westmoreland Co. (CM); Boardman, Oct. 6, 1934, Laurel Ridge (MT); Boardman, Oct. 6, 1935, Somerset Co. (CM); E. G. Britton, June 8, 1889, Pocono Summit (NY); E. G. Britton, July 3, 1899, Pocono (WASH); Browne, Aug. 26, 1907, Buck Hill Falls (PENN); F. S. Chapman, July 3, 1932, Pike Co. (WYO); Conard 46–122, Sept. 20, 1946, Monroe Co., Pocono Manor (DPU); Conard 46–127, Sept. 20, 1946, Monroe Co., Pocono Manor (DPU); S. K. Eastwood, June 22, 1935, Cameron Co. (CM); S. K. Eastwood, Aug. 4, 1935, Venango Co. (CM); S. K. Eastwood, Sept. 1, 1935, Cameron Co. (CM); Fogg, May 4, 1923, Merwinsburg, Monroe Co. (PENN); Galen (CM); Garber, Nov. 9, 1868, Monroe Co. (NY); Githens 236, July 8, 1946, Sullivan Co. (DPU, PENN); Githens 244, July 13, 1946, near Eaglesmere, Sullivan Co. (DPU, PENN); Glowenke 877, Nov. 5, 1948, Lackawanna Co., Lake Scranton Gorge, near Minooka, alt. 360 m (DPU); Glowenke 907, Nov. 26, 1948, Lackawanna Co., near Moosic, alt. 255 m (DPU); T. Green, Pocono Mt. (CHI); T. Green, July 15, 1857, Monroe Co., Pocono Mt. (CHI); T. Green, June 15, 1859, Pocono Mt. (CHI); James 87, Chester Co. (BM); James, June 1850, Chester Co. (CHI, K); James 85, Sept. 1850, Philadelphia Co. (BM); James, Feb. 6, 1859, Blair Co. (CHI); O. E. Jennings, July 15, 1909, Center Co. (CM); O. E. Jennings, Sept. 22, 1909, Center Co. (CM); O. E. and G. K. Jennings, Aug. 31 to Sept. 1, 1925, Westmoreland Co. (CM); Krout, July 5, 1902, Glenolden, Schuykill Co., alt. 600 m (DUKE); Lewis, July 30, 1927, Monroe Co., Buck Hill Falls (MICH); E. T. Moul 2693, May 11, 1946, Franklin Co. (DPU, PENN); E. T. Moul 5790, June 26, 1946, Pike Co., Milford (DPU, TENN); E. T. Moul 2859, Apr. 1, 1947, Columbia Co., near Red Rock, alt. 345 m (DPU, TENN); E. T. Moul 4386, Sept. 13, 1947, Mifflin Co., near Milroy, alt. 323.7 m (DPU, TENN); E. T. Moul 4418, Sept. 13, 1947, Mifflin Co., near Milroy, alt. 323.7 m (DPU, PENN); E. T. and H. L. Moul 4840, Oct. 26, 1947, Union Co., McKean Spring, alt. 352. 8 m (DPU, PENN); E. T. and H. L. Moul 4881, Oct. 26, 1947, Lycoming Co., near Montgomery, alt. 228 m (DPU, PENN); F. J. Myers, June 8, 1941, Lackawanna Co., near Thornhurst (DPU); T. C. Porter, Lancaster Co. (CHI); T. C. Porter, Oct. 1852, Center Co. (PHIL); T. C. Porter, Oct. 1862, Center Co. (CHI, PHIL); T. C. Porter, Aug. 1, 1867, Pocono, East Branch Lake (CHI, PHIL); Rau (MICH); Rau, Okono Glen (NY); Rau, Pike Co. (NY); Rau, Carbon Co. (OS); Rau 137, Brokoglen (FH); Roberts, Nov. 22, 1924, Center Co., Tassey Mt. (WYO); Schallert, July 3, 1932, Pike Co., Pocono Mts. (DUKE); J. K. Small, Feb. 15, 1890, Penryn (NY, US, WIS); J. K. Small S527, Feb. 15, 1890, near Penryn (NY); Stauffer, Schuylkill Co. (CHI, PHIL); Steere, Mar. 22, 1931, Berwyn (MICH); Wilkens, May 13, 1928, Berks Co. (PHIL); Witz,

Apr. 4, 1937, Rector (CM); Wolle, in 1874, Stony Creek (NY); Wood, July 1860, Monroe Co. (CHI); T. G. Yuncker 10557 and 10559, July 7, 1941, near Bellwood (DPU, NY).

RHODE ISLAND: Burlingame 11152, July 1897, Chepachet (PC); Collins 833, May 9, 1893, East Greenwich (PC); Collins 1840, Apr. 29, 1899 (PC, WIS); Collins 2545, May 30, 1901 (PC).

SOUTH CAROLINA: L. E. Anderson 8349, Aug. 18, 1949, Oconee Co., East Fork, Chattooga R., alt. 840 m (DPU, DUKE).

TENNESSEE: Clebsch, June 22, 1945, Monroe Co., Unicoi Mts., Whigg Cabin Bald, alt. 1500 m (DPU); A. J. Sharp, Sept. 15, 1930, Polk Co., near Isabella, alt. 450 m (DPU, TENN); A. J. Sharp 125, Sept. 15, 1930, Polk Co., near Isabella, alt. 450 m (BART); A. J. Sharp 88, Feb. 8, 1931, Sevier Co., near Elkmont, alt. 840 m (DPU, TENN); A. J. Sharp 126, Mar. 15, 1931, Grainger Co., Church Mt., alt. 450 m (DPU, TENN); A. J. Sharp 33, July 7, 1933, Anderson Co., near Savage's Garden, alt. 300 m (DPU, TENN); A. J. Sharp 133, July 7, 1933, Campbell Co., alt. 330 m (DPU, TENN).

VERMONT: Dutton, June 1909, Goshen, alt. 510 m (BR); Dutton 324, June 24, 1909, Goshen (MO); Dutton 404, June 24, 1910, Rochester, alt. 600 m (CHI, CM, DUKE, FH, MT); Dutton 1307, July 30, 1921, Sherburne (CHI, CM, DUKE, FH, MT); Frost (PC, WIS); Frost, in 1858, near Brattleboro (YALE); Grout, June 5, 1896, Newfane (DUKE, FH); Grout, July 10, 1898, Stratton (DUKE); Grout, Aug. 1900, Stratton (DUKE); Grout, Aug. 1904, Grafton (DUKE); Grout, Aug. 13, 1906, Newfane (FH); Grout, Aug. 26, 1906, Stratton, N. Am. Musci Pl. 286 (CAN, CM, FH, ILL, MICH, MINN, MO, NY, OS, UC, US, WASH, WIS, YALE); Grout, Aug. 3, 1913, near summit of Mt. Hood, Stratton (DUKE); Grout, July 17, 1918, Stratton, alt. 600 m, N. Am. Musci Perf. 22 (apparently a second issue, collected Oct. 17, 1930), (BART, CAS, CHI, CM, DPU, MT, NY, OS, UC, US, UT, WASH, YALE); Grout, July 23, 1921, Newfane, alt. 380 m, N. Am. Musci Perf. 22a (MICH); Kirk, Aug. 21, 1923, Bolton, alt. 600 m (DUKE); N. L. T. Nelson 2661, July 1909, Leicester (S); J. Porter, in 1874, Pownal (FH); Pringle, Oct. 2, 1880, Underhill (CHI, DUKE); A. J. Sharp, July 6, 1932, Stratton (DPU, NY, TENN); E. W. Thompson, July 4, 1935, Newfane (FH); Trask, Habeeb, and Grout 201, July 31, 1943, Newfane (DPU); Trask, Habeeb, and Grout 184, Aug. 7, 1943, Newfane (DPU); Welch 713, Aug. 9, 1932, Newfane (CHI, DPU, NY).

VIRGINIA: H. A. Allard 4602, May 1, 1938, Shenandoah Co., Allegheny Mts. (DPU, WIS); Blomquist 3158, Aug. 3, 1934, Giles Co., Mountain Lake (DUKE, FH); E. G. Britton 238, June 16, 1892, near Marion (PC); E. G. Britton 240, June 17, 1892, near Marion, alt. 690 m (PC); Brown, Hogg, Vail, Timmerman, and E. G. and N. L. Britton, June 2, 1890 (NY, PC); Brown, Hogg, Vail, Timmerman, and E. G. and N. L. Britton, June 3, 1890, Giles Co. (NY); S. Chapman 2064, Apr. 6, 1923, Mission Mt., Nortonsville (DUKE); Fulford, June 1939, Giles Co., Mountain Lake (CIN, DPU); Fulford, June 21, 1939, Giles Co., Mountain Lake (CIN, DPU); D. Parker, July 17, 1938, Giles Co. (DPU, NY); Patterson R-215, May 2, 1943, Roanoke Co., Ft. Lewis Mt., near Salem, alt. 360 m (DPU); Patterson R-509, Sept. 19, 1943, Bedford Co., Peaks of Otter, alt. 840 m (DPU); J. K. Small S377, June 2–4, 1892, Smyth Co., Blue Ridge Mts., alt. 750 m (DPU, NY); J. K. Small, June 17, 1892, Smyth Co., Blue Ridge Mts., alt. 810 m (CIN, DPU, DUKE, NY, PHIL, UC, US, WIS); J. K. Small S14, June 17, 1892, Smyth Co., Blue Ridge Mts., alt. 810 m (CHI, DPU, NY); Vail and Britton 238, May to June 16, 1892 (US); Vail and Britton, June 4, 1892 (MICH, NY); Vail and Britton, June 16, 1892 (NY); Vail and Britton, June 17, 1892 (NY).

WEST VIRGINIA: Ammons, Oct. 7, 1932, Monongalia Co., Morgantown (WVA); A. L. Andrews, Oct. 1903, Tibb's Run (NY); Fox, June 22, 1939, near Nettie (DPU, WVA); Musgrave, June 28, 1932, Monongalia Co., near Uffington (DPU, WVA); J. L. Sheldon 3282, Aug. 6, 1908, Cranesville (NY); J. D. Smith, July 1878, Cheat R. (US).

WISCONSIN: Cheney, Aug. 1893, northern Wisconsin (PC); Cheney 2844, Aug.

2, 1893, Lincoln Co., Merrill (DPU, WIS); Cheney 2848, Aug. 2, 1893, Lincoln Co., Wisconsin R., near Granite Heights (DPU, WIS); Cheney, June 1894, northern Wisconsin (PC); Cheney 3325, June 29, 1894, Marathon Co., near Mosinee (DPU, WIS); Holzinger, Oct. 25, 1907, near Hatfield Dam, La Crosse (PC); Holzinger, Oct. 25, 1907, near Hatfield Dam, La Crosse, Grout, N. Am. Musc. Pl. 286a (BART, CAN, CM, COLO, MINN, MO, NY, OS, UC, US, WASH, WIS, YALE); Holzinger, June 19, 1909, Hatfield (DPU, UT, WYO).

United States without locality: Austin, Musci Appal. 251 (CAN, CAS, DUKE, K, NY, PHIL, US, WIS); Sullivant, Musci Allegh. 189, New England (K, L, MICH, NY, US); Sullivant and Lesquereux, Musci Bor.-Am. Exsic. 229 p.p. (FH, K, MICH, MO, NY, TRT, WIS); Sullivant and Lesquereux, Musci Bor.-Am. Exsic. 342 (CHI, CM, FH, MICH, NY, PHIL, UC, WIS).

Fontinalis dalecarlica is recognized by its slender habit, branches usually close and frequently appearing fasciculate, ends of foliated stems and branches attenuate, median cauline leaves subimbricate to imbricate, erect to slightly spreading, concave, narrowly lanceolate or narrowly ovate-lanceolate, apices usually acute or acuminate, commonly entire, majority of blades 2–4 mm long, 0.5–1.25 mm wide, and 3–3.5 : 1, median cells of leaves linear with ends attenuate, frequently flexuous or subflexuous, alar cells commonly forming distinct auricles, and upper perichaetial leaves narrowed into acute or apiculate apices.

The labels of *F. dalecarlica* var. *curvata* and var. *atra* in the Herbarium of S. O. Lindberg bear the same data. The plants and fruits are very comparable. The specimens which the writer examined in the Lindberg Herbarium are blackish and the smaller branches are frequently curved to one side, conditions which possibly suggested to Lindberg the varietal names, *atra* and *curvata*, respectively. The leaves at the ends of the stems and branches are normal in color, green or yellowish green. The dark color seems to be due to a dark brown deposit on the plants. In experimentation, the deposit was removed, and the usual color of specimens of *F. dalecarlica* was noted. The curved habit of the majority of the younger branches is considered to be one of response to an unusual environmental factor as no comparable plants of *F. dalecarlica* have been seen in the abundance of material examined.

Many of the plants originally regarded as *Fontinalis dichelymoides*, *F. seriata*, *F. microphylla*, and *F. dalecarlica* var. *gracilescens* and var. *baltica* were collected at various depths in lakes. They are considered by the author to be plants of *dalecarlica* influenced by the environmental conditions of their habitats. This conclusion was based upon the observation of transition stages from the plants assumed to be the types, portions of types, or topotypes of the above named to those commonly recognized as *F. dalecarlica*. In some instances numerous phases of transition were seen in the same packet of material.

14. **Fontinalis dalecarlica** Br. and Schimp. var. **Macounii**
Card., in Rev. Bryol. 20: 9. 1893.

Upper perichaetial leaves broadly obtuse; otherwise, plants similar
to those of *F. dalecarlica*. [The varietal name, *Macounii*, honors the
collector of the type, Professor John Macoun.]

Type: Not located; should be in Cardot's collection of Fontinalaceae in the
Herbarium of the Laboratoire de Cryptogamie, Muséum d'Histoire Naturelle,
Paris. Perhaps the type was lost or destroyed during World War I, when Car-
dot's Herbarium was badly damaged. If true, it was not deposited in the Paris
Museum following the death of Cardot. In the instance of the destruction of
Cardot's type, the author has selected to represent the type, the fruiting plants
of Macoun, Sept. 24, 1872, Athabaska Plains, Flora Canadensis 755, in the Her-
barium of the National Museum of Canada, Ottawa.
Type collector: John Macoun, in 1875.
Type locality: Alberta or Saskatchewan: Lake Athabaska.
Distribution: North America: Canada, in Lake Athabaska and Athabaska
Plains.
Additional descriptions: Cardot, in Rev. Bryol. 23: 70. 1896; Welch, in Grout,
Moss Fl. N. Am. 3: 246. 1934.
Specimens examined: NORTH AMERICA: CANADA: ALBERTA or SASKATCHEWAN:
Macoun, Sept. 24, 1872, Athabaska Plains, Fl. Can. 755 (CAN); Macoun, in 1879,
Lake Athabaska (PC).

The author has knowledge of only three collections of *F. dalecarlica*
var. *Macounii*, 1872, 1875, and 1879. The perichaetial leaves on the
1879 specimens are lacerate with age. Those of the two fruits studied
in Macoun, Fl. Can. 755, collected in 1872, are obtuse. Cardot described those of the 1875 collection as round-obtuse. Since the plants
of var. *Macounii* are very comparable to those of the species, *dalecarlica*, the writer is unable to distinguish between the species and the
variety in the absence of complete perichaetial leaves.

15. **Fontinalis novae-angliae** Sull., Musci and Hep. U.S., No.
3b, p. 104, additions and corrections to p. (654) 54. 1856.

Fontinalis Lescurii Sull., Musci and Hep. U.S., No. 5, p. (654) 54, and p. 105,
additions and corrections to p. (654) 54. 1856.
Fontinalis Eatoni Sull., in Sullivant and Lesquereux, Musci Bor.-Am. Exsic.
224c. 1856 [printed Latin description on label]; Cardot, Mon. Font., p. 94.
1892.
Fontinalis Lescurii var. *ramosior* Sull., Icon. Musc., p. 101. 1864.
Fontinalis Lescurii var. Aust., Musci Appal., p. 41. 1870 (nomen nudum).
Fontinalis Delamarei Ren. and Card., in Rev. Bryol. 15: 71. 1888.
Fontinalis squamosa Hedw. subsp. *Delamarei* (Ren. and Card.) Card., in Rev.
Bryol. 18: 83. 1891.
Fontinalis novae-angliae subsp. *Cardoti* Ren. in litt., according to Cardot, in
Rev. Bryol. 18: 83. 1891.
Fontinalis Cardoti (Ren.) Card., in Rev. Bryol. 18: 86. 1891 [in descriptive key];
Cardot, Mon. Font., p. 95. 1892.
Fontinalis Peckii Aust. in herb., according to Cardot, Mon. Font., p. 86. 1892.

Fontinalis Howei Aust. MS. in herb., according to Cardot, Mon. Font., pp. 92 and 93. 1892.
Fontinalis novae-angliae var. *Howei* (Aust.) Card., Mon. Font., p. 93. 1892. [Assumed to be an error in printing as *Hovei* in Rev. Bryol. 20: 9. 1893.]
Fontinalis Frostii Sull. in litt. ad Frost, sec. Eaton, according to Cardot, Mon. Font., p. 117. 1892.
Fontinalis novae-angliae var. *Eatoni* (Sull.) Card., in Rev. Bryol. 20: 9. 1893.
Fontinalis novae-angliae var. *heterophylla* Card., in Nichols, in Rhodora 15: 9. 1913.
Fontinalis novae-angliae var. *Lorenziae* Card., in Nichols, in Rhodora 15: 9. 1913.
Fontinalis novae-angliae var. *Delamarei* (Ren. and Card.) Welch, in Grout, Moss Fl. N. Am. 3: 248. 1934.
Fontinalis novae-angliae var. *Grouti* Welch, in Grout, Moss Fl. N. Am. 3: 249. 1934.

Plants commonly medium in size, occasionally approaching slender or robust, green, yellowish green, brownish green, golden, golden brown, copper color, or brown, younger portions glossy, older parts dull; stems slightly flaccid to almost rigid, up to 40 cm in length, 0.25–0.3 mm in diameter, commonly denuded and dark at base with age, irregularly pinnately branching; branches numerous, erect to erect-spreading, frequently spreading, close to somewhat distant, sometimes appearing fasciculate, up to 11 cm in length, ends of foliated stems and branches usually attenuate, occasionally acute, subacute, subobtuse, or obtuse; median cauline leaves generally somewhat distant, sometimes subimbricate to loosely imbricate, bases 0.5–2 mm apart, blades usually firm, occasionally subflaccid to flaccid, erect to erect-spreading, subconcave to deeply concave, usually ovate-lanceolate, sometimes oblong-lanceolate or narrowly ovate-lanceolate, margins in apical portions generally narrowly involute, occasionally plane; apices commonly subtruncate, truncate, subobtuse, or obtuse, sometimes subacute to acute, at times almost cucullate, leaf tips usually distinctly serrulate, occasionally entire; median cauline blades 2.5–4 mm long, sometimes up to 6 mm in length, 0.75–2 mm wide, sometimes up to 2.5 mm in width, 1.75–4 : 1, occasionally up to 5 : 1; median cells of leaves linear with ends attenuate or narrowly rhomboidal, commonly subflexuous or flexuous, 6.5–12 μ wide, usually 8–15 : 1, rarely up to 20 : 1; alar cells enlarged, subrectangular, subquadrate, or subhexagonal, walls yellowish, brownish, or subhyaline, commonly forming distinct auricles, leaf bases generally short decurrent, up to 0.5 mm, sometimes not decurrent, occasionally subclasping; median branch leaves comparable to median cauline except smaller in size; perichaetial branch 3–5 mm long, perichaetium oblong to cylindric, 0.75–1 mm in diameter; upper perichaetial leaves oval to suborbicular, apices broadly obtuse, truncate and lacerate with age; calyptra long conical, 1.2–1.5 mm long, 0.4–0.5 mm in diameter, occasionally up to 1.85 mm in

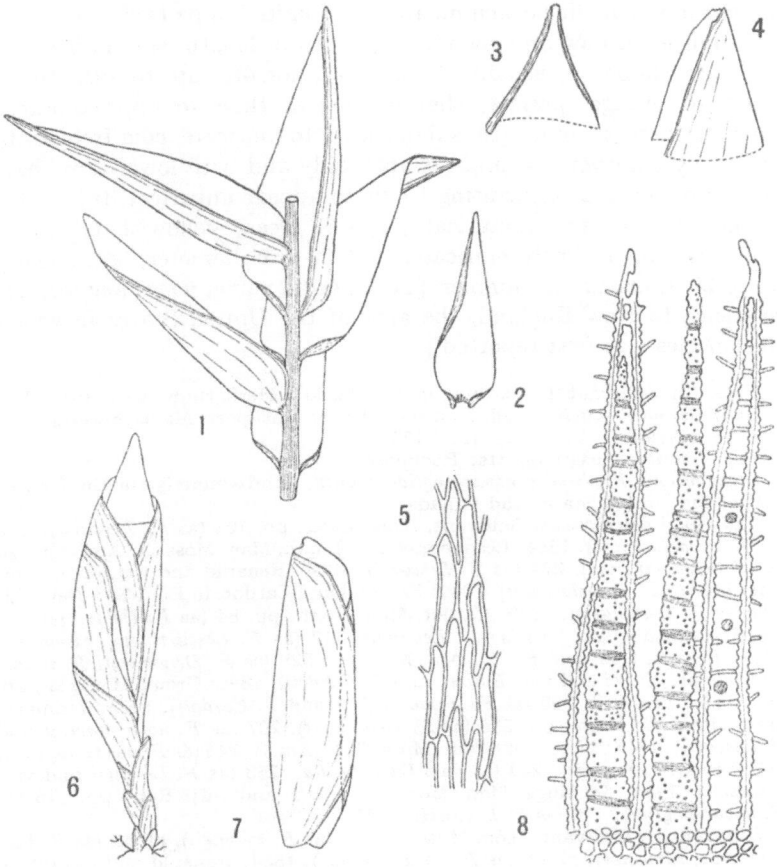

Fig. 14. – *Fontinalis novae-angliae* Sull. – 1.Portion of stem with median cauline leaves. 2.Median cauline leaf. 3 and 4.Leaf apices. 5.Median cells of leaf. 6.Perichaetium containing sporophyte. 7.Perichaetial leaf. 8.Portion of perfect trellis which appears to be imperfect. Nos. 1 and 5 drawn from Sullivant and Lesquereux, Musci Bor.-Am. Exsic. 340; nos. 2–4 from Welch 712, Vermont; nos. 6 and 7 after Sullivant, Icon. Musc., pl. 65 (as *F. novae-angliae*); no. 8 after Sullivant, Icon. Musc., pl. 62 (as *F. Lescurii* var. *ramosior*).

length and 0.75 mm in diameter; operculum immersed to emergent, short conical, obtuse, 0.5–1 mm long, 0.5–0.8 mm in diameter; seta 0.1–0.3 mm long; urn immersed to emergent, subcylindrical to cylindrical, 1.75–2.5 mm long, sometimes up to 3 mm in length, 0.5–1 mm in diameter, 2.6–3.6 : 1, frequently slightly contracted beneath mouth when dry, occasionally not contracted; peristome teeth brownish orange, linear-acuminate, often united in pairs at apex, 0.5–0.75 mm long, now and then up to 1 mm in length, submuricate to muricate, lamellae usually 12–20, occasionally up to 30; trellis brownish orange, perfect, slightly shorter than to approximately equal with teeth in length, submuricate to muricate, cilia frequently united by transverse strands at apex only and with lower cross bars broken or incomplete, causing trellis to appear imperfect, transverse strands smooth to submuricate; spores green, yellowish green, or yellowish brown, finely muricate, 15.3–22 μ in diameter, occasionally up to 28 μ, mature in summer. [The specific name, *novae-angliae*, has reference to New England, the area of the United States in which this species was first reported.]

Type: In Sullivant Herbarium in the Farlow Herbarium, with the label, "*Fontinalis novae-angliae* Sull. In a rivulet near Rockport, Massachusetts."
Type collector: T. P. James, Aug. 1856.
Type locality: Massachusetts: Rockport.
Distribution: Fontinalis novae-angliae occurs, approximately, in the eastern half of the United States and Canada.
Additional descriptions: Sullivant, Icon. Musc., pp. 101 (as *F. Lescurii*), 105 (as *F. novae-angliae*). 1864; Lesquereux and James, Man. Moss. N. Am., pp. 270 (as *F. novae-angliae*), 271 (as *F. Lescurii*). 1884; Renauld and Cardot, in Bot. Gaz. 14: 96 (as *F. Delamarei*). 1889; Renauld and Cardot, in Bull. Soc. Belg. 28: 129 (as *F. Delamarei*). 1889; Cardot, Mon. Font., pp. 84 (as *F. Delamarei*), 91 (as *F. novae-angliae*), 115 (as *F. Lescurii*), 117 (as *F. Lescurii* var. *ramosior*). 1892; Barnes, Gen. and Sp. N. Am. Moss., p. 328 (as *F. Delamarei, F. novae-angliae* var. *Howei* and var. *Eatoni*, and *F. Cardoti*). 1896; Grout, M. H. M., pp. 396 (as *F. Lescurii*), 398 (as *F. novae-angliae* and *F. Cardoti*). 1903; Jennings, Man. Moss. W. Penn., pp. 206 (as *F. Delamarei*), 207 (as *F. novae-angliae* and *F. Lescurii*). 1913; Welch, in Grout, Moss Fl. N. Am. 3: 246 (as *F. novae-angliae*), 247 (as var. *Lorenziae*), 249 (as var. *heterophylla*), 253 (as *F. Lescurii* and var. *ramosior*). 1934; Jennings, Man. Moss. W. Penn. and Adj. Reg., pp. 176 (as *F. novae-angliae*), 178 (as *F. Lescurii*). 1951.
Illustrations: Sullivant, Icon. Musc., pl. 61 (as *F. Lescurii*), pl. 62 (as *F. Lescurii* var. *ramosior*), pl. 65 (as *F. novae-angliae*). 1864; Renauld and Cardot, in Bot. Gaz. 14: pl. 14 (as *F. Delamarei*). 1889; Renauld and Cardot, in Bull. Soc. Bot. Belg. 28: pl. 9 (as *F. Delamarei*). 1889; Grout, M. H. M., pl. 87 (as *F. Lescurii*), fig. 217 and pl. 88 (as *F. novae-angliae*). 1903; Jennings, Man. Moss. W. Penn., pl. 30 (as *F. novae-angliae*). 1913; Welch, in Grout, Moss Fl. N. Am. 3: pl. 76 (as *F. novae-angliae*, var. *heterophylla*, var. *Lorenziae*, and *F. Delamarei*), pl. 78 (as *F. Lescurii* and var. *ramosior*). 1934; Jennings. Man. Moss. W. Penn. and Adj. Reg., pl. 34. 1951. – FIGURE 14.
Specimens examined: NORTH AMERICA: CANADA: CAPE BRETON ISLAND: Macoun, Aug. 6, 1898, Can. Musci 605a (DUKE, FH, US, WASH); G. E. Nichols, July - Aug. 1914 (NY); G. E. Nichols 728b, July – Aug. 1914, Smoky Mt. (YALE); G. E. Nichols, July – Aug. 1915, Valley of Barrasois R. (NY); G. E. Nichols, July – Aug. 1915, Moosehide Lake, mts. west of Ingonish (NY); G. E. Nichols 1481, July – Aug. 1915, valley of Barrasois R. (YALE).
MANITOBA: Hand 527, July 20, 1938, near Rennie (DPU).

MIQUELON ISLAND: Delamare (Label corrected from "*Fontinalis squamosa?*" to "*Delamarei* R. C."), (Assumed to be the TYPE of *F. Delamarei*, PC; S); Delamare (B, DUKE, FH, K, NY, S, US, WIS); Delamare, Renauld and Cardot, Musci Am. Sept. Exsic. 72 (CAN, G-BOIS, NY, PC, YALE); Delamare, Cape Miquelon (B, FH); Delamare, June 1881, Cape Miquelon (NY); Delamare, in 1886 (B); Delamare 119, in 1886, Cape Miquelon (BM); Delamare, in 1887 (K); Delamare, July 1892 (S); Delamare, in 1896, Cape Miquelon (L).

NEW BRUNSWICK: Habeeb, Aug. 28, 1948, McAdam (DPU); W. R. Taylor, Sept. 4, 1922, Grand Manan, near North Head (MICH, YALE).

NEWFOUNDLAND: Long 143, Aug. 25, 1924, Whitbourne (BART, MICH, PHIL); Waghorne, Apr. 12, 1883, Witter's Bay (BM); Waghorne, Apr. 12, 1893, Witter's Bay (FH, MINN, MO, PC, WIS); Waghorne, Apr. 12, 1893, Arslen Bay, New Harbour (G-BOIS); Waghorne, July 10, 1893, New Harbour (G-BOIS, NY).

NOVA SCOTIA: M. S. Brown, Aug. 5, 1907, Yarmouth Co., Brazil Lake (DPU); M. S. Brown, July 16, 1923, Halifax (DPU); M. S. Brown, May 24, 1924 (DPU); M. S. Brown, June 6, 1924, Halifax, Grand Lake (DPU); M. S. Brown, July 5, 1924, Sherbrooke (DPU); M. S. Brown, July 12, 1924, Halifax (DPU); M. S. Brown, Sept. 22, 1931, Jeddore (DPU); M. S. Brown 570b, July 31, 1936, St. Margaret Bay, alt. 90 m (DPU); M. S. Brown 571a, Aug. 1939, Hubbard's, Sawlor Lake (DPU); Fernald, Bartram, and Long 736, July 23, 1921, Yarmouth Co., St. John Wilson's Lake (BART, FH); James Fowler, July 17, 1901, Canso (WIS); Jardin (FH); Jardin, in 1887 (PC); Macoun, Can. Musci 229 p.p. (NY); Macoun, in 1883, Halifax (BM); Macoun, June 18, 1883, Halifax (CAN); Waghorne, July 10, 1883 (PC).

ONTARIO: R. F. Cain, Sept. 4, 1931, Gull Lake Portage, Lake Timagami (DPU, TRT); R. F. Cain, June 26, 1932, Spawning Lake Portage, Lake Timagami (DPU, NY, TRT); R. F. Cain, July 29, 1939, Gull Lake Portage, Lake Timagami (DPU, TRT); R. F. Cain, Aug. 7, 1939, Gull Lake Portage, Lake Timagami (DPU, TRT); R. F. Cain, Aug. 11, 1939, Spawning Lake Portage, Lake Timagami (DPU, TRT); R. F. Cain, Aug. 21, 1939, Madawasca R., Whitney (DPU, TRT); R. F. Cain, Aug. 20, 1940, Costello Lake, Algonquin Park (DPU, TRT); R. F. Cain, Aug. 22, 1940, Little Macaulay Lake, Algonquin Park (DPU, NY, TRT); R. F. Cain, May 21, 1941, Simcoe Co., Sparrow Lake (DPU, TRT); R. F. Cain, May 24, 1941, Simcoe Co., Sparrow Lake (DPU, NY, TRT); R. F. Cain, May 26, 1941, Muskoka District, Kashi R. (DPU, NY, TRT); R. F. Cain, May 28, 1941, Ontario Co., Coopers Falls (DPU, TRT); R. F. Cain, Aug. 18, 1941, Oxbow Lake, near Algonquin Park (DPU, TRT); R. F. Cain, Aug. 8, 1944, Sudbury District, near Espanola (DPU, TRT); R. F. Cain, Sept. 13, 1945, Lake Timagami (DPU, TRT); R. F. Cain, Sept. 13, 1945, Lake Timagami, Gull Lake Portage (DPU, TRT); R. F. Cain, Aug. 23, 1946, Lake Timagami (DPU, TRT); R. F. Cain, Aug. 23, 1946, Lake Timagami, Loon Lake Portage (DPU, TRT); Drexler 120, June 27, 1935, Quetico Park, Lake Kawnipi (DPU); Drexler 158, June 30, 1935, Quetico Park, between McKenzie and Cache Lakes (DPU); Drexler 461, Aug. 23, 1935, Quetico Park, Cairm Lake (DPU); Drexler 491, Aug. 28, 1935, Quetico Park, between Side and Sarah Lakes (DPU); Drexler 1667, Aug. 9, 1938, Quebec Park, Crooked Lake (DPU); Drexler, Aug. 7, 1939, Lake Agness, Quetico Provincial Park, at depth of 15–30 m (DPU); Drexler, Aug. 13, 1939, Quetico Park, Lake Agness (DPU); Hand 469, July 24, 1935, Haliburton District, near Donald (DPU); Hand 594, July 10, 1938, Algonia District, near Cutler (DPU, MICH); Hand 743, Aug. 9, 1941, Muskoka District, Kashi R. (DPU); Hand 739, Aug. 13, 1941, Timiskiming District, near Latchford (DPU, OS); Hand 741, Aug. 13, 1941, Timiskiming District, near Latchford (DPU); Macoun 189, in the Moira (MO, NY); Macoun 100, May 1866, Hastings Co., near Belleville (MO); Macoun 102, July 25, 1868, Peterborough Co., Gull R. (MO); Macoun, in 1869, Gull R. (NY); Macoun 223 p.p. in 1870, North Hastings (BM); Macoun, July 9, 1870, Hastings Co., Scootamata R., Can. Musci 752 p.p. (CAN); Macoun, July 10, 1870, Lake Region, Fl. Can. 2225 (NY); Macoun 341, July 1872, west of Lake Superior (MO, NY); Macoun,

July 4, 1900, Algonquin Park, Cache Lake (TRT); Macoun, July 4, 1900, Algon-
quin Park, Can. Musci 601 (E); Macoun, July 17, 1900, Algonquin Park, Can.
Musci 602a (E, K); Macoun, July 24, 1900 (S); Macoun, July 24, 1900, Algonquin
Park, Can. Musci 604 p.p. (MO, NY); Macoun, July 24, 1900, Algonquin Park,
Can. Musci 605 (FH, K, MO, NY); Macoun 832 July 24, 1900, Algouquin Park,
Catfish Lake (S); Martin 596, in 1937, Algonquin Park, Lake Opeongo (DPU,
TRT); Moxley, May 24, 1936, Bruce Co., Sauble Falls (TRT).

QUEBEC: Allyre 2145, 7–8–1947, Terrebonne Co., Lake Manitou (DPU); Beau-
lac 429, Aug. 10, 1932, Argenteuil Co., Lake Aubin (DPU); Beaulac 536, Aug. 11,
1932, Argenteuil Co., Lake Auby (DPU); Boardman 587, July 18, 1935, Terre-
bonne Co., Lake Mercier (CM); Boardman 588, July 18, 1935, Terrebonne Co.,
Lake Mercier (MT); Deardon, Sept. 10, 1944, Gatineau Park (DPU, TRT); Drolet,
Raymond, and Kucyniak 44–3 and 44–4, Sept. 3, 1944, Quebec Co., Lac des
Roches (DPU, MT); Dupret 118, July 20, 1906, Oka, near Ottawa R. (CAN, DUKE);
Dupret, Aug. 10, 1906, Oka, Ottawa R. (FH, H, PC, PENN); Dupret, Aug. 12,
1906, Oka, Ottawa R. (PENN); Dupret, Sept. 8, 1906, Mt. St. Hilaire (FH, PC);
Dupret 123, Sept. 8, 1906, Mt. St. Hilaire (DUKE); Dupret, Sept. 1906 and 1911,
Mt. St. Hilaire, Bauer, Musci Eur. and Am. Exsic. 1781 (BART, L, NY, WASH);
Dupret, Aug. 1907 (WASH); Dupret 1364 and 1752, Sept. 1907, Rouville Co., Mt.
St. Hilaire (DPU, MT); Fabius 1033, May 10, 1947, Mt. Shefford (DPU); Fabius
1085, May 17, 1947, Mt. Shefford (DPU); Fabius 1086, May 17, 1947, Mt. Shef-
ford, near Granby (DPU); Fabius 1271 and 1273, June 11, 1947, Mt. Shefford
(DPU); Fabius 1280, June 11, 1947, Granby (DPU); Fabius 1522, July 25, 1947,
Waterloo (DPU); Fabius 1531 and 1538, July 26, 1947, Granby (DPU); Fabius
1737, Aug. 21, 1947, West Shefford (DPU); Fabius 1749, Aug. 27, 1947, Granby
(DPU); Fabius 2262 and 2268, Apr. 26, 1948, Mt. Shefford, alt. 300 m. (DPU);
Fabius 2342, May 15, 1948, Mt. Shefford (DPU); Fabius 2467, June 19, 1948,
Mt. Shefford (DPU); Fabius 3125, Nov. 2, 1948, Mt. Shefford (DPU); Fabius 3475,
June 9, 1949, Mt. Yamaska, alt. 300 m (DPU); Fabius 3493 and 3496, June 9,
1949, Mt. Yamaska, alt. 240 m (DPU); Gauthier 2163, July 21, 1934, Berthier
Co., St. Michel des Saints, Lake Richard no. 2 (DPU, MT); Gauthier 2247, Aug. 16,
1934, St. Maurice Co., St. Flore, Lac des Neiges (DPU, MT); Gauthier 11218, July
19, 1938, Laurentides Nat. Park, Lake Hermine (DPU); Kucyniak 42–40, May 9,
1942, Brome Co., Sutton (DPU, MT); Lepage 649, Aug. 29, 1937, Montmagny Co.,
Montmagny (DPU, MT); Lepage 1316 and 1318, Aug. 26, 1939, Rimouski, St.
Paul de la Croix (DPU, MT); Lepage 2778, June 17, 1941, Rimouski Co. (DPU);
Lepage and Dutilly 4366, July 31, 1943, Marten R. (DPU); Lepage and Dutilly
4377, 4–8–1943, Marten R., Camousitchouane Lake (DPU); Lepage and Dutilly
9974, June 26, 1946, Harricanaw R. (DPU); Marie-Anselme 420, Nov. 2, 1935,
La Tuque (DPU, WASH); Marie-Anselme 430, Nov. 9, 1935, La Tuque (DPU);
Marie-Anselme 433, Nov. 9, 1935, near Lake Paradis (DPU, WASH); Marie-
Anselme, May 29, 1936, La Tuque (DPU); Marie-Anselme 759 and 763, July 16,
1936 (DPU); Marie-Anselme 792, July 19, 1936, La Tuque (DPU); Marie-Anselme
742, Aug. 1, 1936 (DPU); Marie-Anselme 782, Aug. 1, 1936, near La Tuque (DPU);
Marie-Anselme 843, Aug. 11, 1936, Waterloo (DPU); Marie-Anselme 839 and
842, Aug. 20, 1936, Waterloo (DPU); Marie-Anselme 1582, 1583, and 1585, July
3, 1937, Waterloo (DPU); Marie-Anselme 1586, July 3, 1937, Lake Waterloo
(DPU); Marie-Anselme 1584, July 7, 1937, Lake Waterloo (DPU); Marie-Anselme
1599, July 14, 1937, Mt. Arford (DPU); Marie-Anselme 1626, July 19, 1937, South
Stukeley (DPU, MT); Marie-Anselme 1627, July 19, 1937, South Stukeley (DPU);
Marie-Anselme 1632, July 22, 1937, Lake Waterloo (DPU); Marie-Anselme
1789, Aug. 29, 1937, Shefford Co., Waterloo (DPU, MT); Marie-Anselme 1638,
Sept. 29, 1937, Lake Leby (DPU); Marie-Anselme 1911, Oct. 23, 1937, Shefford
Co., Waterloo (DPU, MT); Marie-Anselme 3758, July 7, 1942, Wolfe Co., Disraeli
(MT); Marie-Anselme 3757, July 22, 1942, Wolfe Co., Disraeli (DPU, MT); Marie-
Anselme 4026, July 10, 1943, Beauceville (DPU); Marie-Anselme 4034, July 12,
1943, Chaudière R., Beauceville (DPU); Marie-Anselme 4074, July 17, 1943,

Beauceville (DPU); Marie-Jean-Eudes, Aug. 30, 1933, Montcalm Co., Rawdon, Lake McCallum (DPU, MT); Marie-Jean-Eudes 3021a, Aug. 13, 1944, Montcalm Co., Lac à la Truite (DPU); Marie-Victorin 9328, Aug. 1912, Terrebonne Co., Lake Conolly (DPU, MT); Marie-Victorin 9342, in 1914, Portneuf Co., Lake Epinette, Petit-Saguenay (DPU, MT); Marie-Victorin, in 1916, Terrebonne Co., Lake Conolly (DPU, MT); Marie-Victorin 7600, Apr. 1916, Chambly Co., St. Basile (BART, DPU, MT); Marie-Victorin 5–8–1920, Mégantic Co., Lake Bécancour, Flore de la vallée du Saint-Laurent 30 (DPU, FH, MT, NY, PC, US); Marie-Victorin 18274, Aug. 12, 1924, Trois-Lacs, in the Laurentides (DPU, MT); Marie-Victorin and Rolland-Germain 33636, July 13, 1930, Quebec Co., Trois-Lacs (DPU, MT); Raymond and Kucyniak 46–163a, Sept. 21, 1946, Beauce Co., St. Georges, in Chaudière R. (DPU, MT); Victor, May 28, 1937, Lake La Cache, near La Tuque (DPU).

Canada without locality: Macoun, in brooks, Nova Scotia and Ontario, Can. Musci 231 p.p. (NY); Macoun, Sept. 20, 1878, abundant in rivers and small brooks, Can. Musci 227 p.p. (K, NY, US, YALE).

UNITED STATES: ALABAMA: Harvill 5177, July 3, 1949, Walker Co., Duncan Creek, near Curry (DPU); Harvill 6051 & Crawford, Oct. 15, 1949, Cullman Co., near Clifty Creek (DPU); Lesquereux (H, PC); Lesquereux, Falls of Little River, Lookout Mountains, "Fontinalis Lescurii was made on these specimens which are very poor." (TYPE of F. Lescurii, FH); Lesquereux, Falls of Little River, Lookout Mountains, Sullivant and Lesquereux, Musci Bor.-Am. Exsic. 228 (BM, FH, K, MICH, MO, NY, PC, PHIL, TRT, WIS); Mohr, Aug. 1886, Cullman (NY, US); Mohr, Aug. 1889, Cullman (E); R. M. Harper 6, Nov. 24, 1905, DeKalb Co., near Chavies, Town Creek on Sand Mt., coal measures, alt. 345 m (MO, NY, US, YALE); F. Winter, in 1873 (PC).

ARKANSAS: Beyrich (BM); Bush 1454, Mar. 26, 1909, Fulton (MO); Welch 2152, July 5, 1934, Faulkner Co., Cedar Park, near Conway (DPU, NY); Welch 2970, July 5, 1934, Faulkner Co., Cedar Park, near Conway (DPU).

CONNECTICUT: J. A. Allen, Apr. 19, 1878, Woodbridge (NY, PC); J. A. Allen, Nov. 12, 1879, near New Haven (NY); J. A. Allen, Apr. 18, 1880, Hamden (YALE); J. A. Allen, May 15, 1880, Mt. Carmel (YALE); J. A. Allen, Apr. 28, 1881, near Derby (FH, MICH, PC); J. A. Allen, Apr. 30, 1881, near New Haven (NY); J. A. Allen, May 15, 1881, Mt. Carmel (NY); J. A. Allen, May 28, 1881 (NY, PC); O. D. Allen, Apr. 28, 1881, near Birmingham (YALE); Eaton, Killingworth (CAS); Eaton, June 1855, New Haven (YALE); Eaton, Oct. 1855, East Rock, New Haven (YALE); Eaton, Oct. 1856, New Haven (YALE); Eaton, May 11, 1874, New Haven (CAS); Eaton, May 27, 1874, Northfield (CAS); Eaton, May 24, 1875, Hamden (CHI); Eaton, Oct. 25, 1875, Bethany (CHI, UC, US, YALE); Eaton, Nov. 27, 1875 (CAS); A. W. Evans, Oct. 1889, Hamden (YALE); A. W. Evans, Oct. 9, 1889, Hamden (YALE); A. W. Evans, July 1891, Hamden (H); A. W. Evans, May 30, 1907, Beacon Falls (PC); Graves 466, New London Co., Groton (YALE); S. B. Hadley, Canterbury (YALE); S. B. Hadley, Mar. 1907, Canterbury (PC); Lorenz, June 3, 1906, Bloomfield, alt. 180 m (YALE); Lorenz, June 27, 1907, Andover (PC, YALE); Lorenz, May 2, 1913, Bolton Notch, alt. 210 m (YALE); Lorenz, May 11, 1913, Mt. Carmel, Hamden, alt. 180 m (YALE); Merriam, May 23, 1875, Mt. Carmel, near New Haven (CAS); G. E. Nichols, Bear Mt., Salisbury (CIN, YALE); G. E. Nichols, Burlington (YALE); G. E. Nichols, Mar. 29, 1907, Killingworth (NY, PC, YALE); G. E. Nichols, May 3, 1907, East Haven (NY); G. E. Nichols, May 30, 1907, Beacon Falls (CIN, PC, YALE); G. E. Nichols, Apr. 16, 1908, Burlington (YALE); G. E. Nichols, Apr. 17, 1908, Granby (YALE); G. E. Nichols, Apr. 17, 1908, Burlington (YALE); G. E. Nichols, Apr. 17, 1908, Canton (YALE); G. E. Nichols, Dec. 11, 1909, Bethany (CIN, YALE); G. E. Nichols, Apr. 12, 1911, Stonington (YALE); G. E. Nichols, Apr. 14, 1911, North Stonington (PC, YALE); G. E. Nichols, June 11, 1911, North Branford (TYPE of F. novae-angliae var. heterophylla, PC; NY, YALE); Setchell, July 2, 1884, Norwich (UC); J. L. Sheldon C406, Aug. 13, 1901, Central Village (YALE); Weatherby, July 22,

1907, Andover (TYPE of *F. novae-angliae* var. *Lorenziae*, PC; YALE); Young, Oct. 1874, New Haven (CHI, WIS).

DELAWARE: Commons, May 5, 1890, Wilmington (US); Commons, May 15, 1890, Wilmington (PHIL); Commons, May 30, 1890, Wilmington (NY); James 86, June 1851, Brandywine Creek (BM); James 90, June 1851 (BM); Krout, Aug. 15, 1909, Naaman Falls, Brandywine Hundred, alt. 45 m (NY); Maxon, Aug. 15, 1909, Naaman Falls, Brandywine Hundred, alt. 45 m (CHI).

DISTRICT OF COLUMBIA: Holzinger, Rock Creek (NY); Holzinger, in 1892, Blagden's Run, Renauld and Cardot, Musci Am. Sept. Exsic. 185 (BM, CAN, FH, H, NY, PC, YALE); Holzinger, June 25, 1892, Blagden's Run (US); Holzinger, Aug. 6, 1892, Rock Creek (H, US); Holzinger, July 1894, Blagden's Run (S); Lehnert (US); Oldberg, in 1872 (US); Oldberg, Mar. 1873 (S); Oldberg, Jan. 1874 (US); Oldberg, Nov. 18, 1874 (US); Waite, Feb. 1890, Blagden's Run (DPU).

FLORIDA: Gist 109, Aug. 1936, McIntosh (US).

GEORGIA: G. Carroll, Dec. 26, 1940, Dade Co., Lookout Mt. (CIN, DPU); G. Carroll, Apr. 14, 1941, Dade Co., Sand Mt. (DPU); G. Carroll, Apr. 15, 1941, Dade Co., Sulphur Springs (DPU); J. K. Small, July 1893, on granite, Little Stone Mt., alt. 330 m (PC); J. K. Small, July 1893, on granite, Little Stone Mt., alt. 330 m, Mosses of Southern United States 26 (BART, BM, CAN, CM, DPU, FH, MINN, MO, NY, NYSM, OS, UC, US, WASH, WIS); P. Wilson 150, July 27, 1900, Whitfield Co., near Gordon Springs, alt. 292.5 m (BM, K, NY, US).

ILLINOIS: Brendel (H); Brendel, Peoria (BM); Brendel, in 1860, Pulaski Co. (FH); Brendel, in 1869 (B); Grether, Apr. 21, 1948, Grant City Park (DPU); Phinney 1002, Aug. 4, 1944, Pope Co., near Lake Glendale (DPU, TENN).

INDIANA: Deam 58343, Sept. 2, 1937, Crawford Co., near Taswell (CHI, DPU, NY); Welch 5051, July 11, 1937, Perry Co., Blue Wells Hollow (CHI, DPU, MICH, NY); Welch 6368, July 11, 1937, Perry Co., Blue Wells Hollow (DPU); Welch 5050 and 5052, July 14, 1937, Perry Co., Blue Wells Hollow (CHI, DPU, NY).

KENTUCKY: E. L. Braun, May 10, 1937, McCreary Co. (CIN); Daily, Oct. 5, 1940, Wolfe Co., Pine Ridge (CHI); Gleason 2448, Sept. 4, 1939, Madison Co., near Berea (DPU); Shacklette 1957, June 10, 1941, Letcher Co., near Eolia, on sandstone (DPU); Shacklette and Harvill 2175, Sept. 8, 1941, Caldwell Co., near Thaxton, on Pottsville Conglomerate (DPU); Welch 6971 and 6972, Sept. 5, 1941, Wolfe Co., Pine Ridge (DPU, NY); Welch 6973, Sept. 5, 1941, Wolfe Co., Pine Ridge (DPU); Welch 6974, 6975, 6976, 6977, and 6978, Sept. 5, 1941, Wolfe Co., Pine Ridge (DPU, NY); Welch 6979, Sept. 5, 1941, Wolfe Co., Pine Ridge (DPU).

LOUISIANA: D. S. and H. B. Correll 9982, July 31, 1938, Natchitoches Parish, near Perry (DPU, DUKE, FH); Pennebaker, Mar. 17, 1940, Pearl R. (DPU).

MAINE: C. D. Adams, Apr. 27, 1938, Oxford Co., North Hartford (DPU, PSNH); C. D. Adams, July 13, 1938, Oxford Co., Buckfield, Jersey Bog (DPU, PSNH); C. D. Adams, July 14, 1938, Androscoggin Co., Livermore (DPU, DUKE, PSNH); C. D. Adams, July 16, 1938, Androscoggin Co., Livermore (DPU, PSNH); C. D. Adams, Aug. 17, 1940, Oxford Co., Four Ponds (DPU, PSNH); Mrs. M. R. Adams, July 8, 1941, Corinna (DPU); G. Blake, Sept. 1857, Harrison Co. (CHI); J. Blake 109 (FH); J. Blake, in 1876, Harrison (US); J. Blake, Sept. 1876, Harrison (FH); Chamberlain 3589, July 11, 1907, Falmouth (H, PC); Chamberlain and Collins 1684, Aug. 13, 1903, Falmouth (ABS, NY, PC); D. L. Cole, May 25, 1943, Oxford Co., Andover (DPU, PSNH); L. A. Cole 1017 p.p., July 14, 1918, Union (US); Collins 2429 (PC); Fernald 182, Bradley (WIS); Gleason 3204, Mar. 28, 1941, York Co., West Hollis (DPU); R. L. Lowe, July 1930, Upper Dam (DPU, PSNH); R. L. Lowe, July 18, 1939, Aroostook Co., Allagash R. (DPU); R. L. Lowe, Nov. 20, 1939, Isle au Haut, Head Harbor (DPU); R. L. Lowe, May 29, 1940, Falmouth (DPU, PSNH); E. D. Merrill 69 p.p., July 1897, Auburn (NY); E. D. Merrill 70, Aug. 1897 (DUKE, NY); G. K. Merrill 141, July 15, 1914, Rockport (YALE); G. K. Merrill 141, Oct. 16, 1918, Rockport (FH); G. K. Merrill 185, Aug. 17, 1919, Union (DUKE, FH); A. H. Moore, July 8, 1903, Fort Kent, St. John's R. (FH); Northrop, in 1902, Prospect Harbor, The Sands (NY); A. H. Norton, July 4,

1907, Falmouth (PSNH); A. H. Norton, July 26, 1908, Westbrook (PSNH); A. H. Norton, July 16, 1918, Falmouth, Lower Falls (DPU); A. H. Norton, May 26, 1937, York, Mt. Agamenticus (DPU, PSNH); H. Oakes 224, Franklin Co., Rangeley (DPU); H. Oakes, May 10, 1946, Franklin Co., Rangeley (DPU, PSNH); Parlin, Aug. 12, 1932, Oxford Co., Canton (NY); Parlin, July 29, 1936, Androscoggin Co., Livermore Falls (BART, DPU); Parlin, July 14, 1938, Androscoggin Co., Livermore (DPU, PSNH); Wynne and Schnooberger 204, July 12, 1940, Mt. Desert Isl., Acadia Nat. Park (DPU).

MARYLAND: Ammons, Oct. 8, 1932, near Red House (WVA); Boyer, Walford (WASH); Boyer 99, in 1905 (PC); Boyer, May 1909, near Walford (MT); R. G. Brown, Aug. 12, 1946, Garrett Co., New Germany, alt. 741 m (DPU, MD); Drouet 3614, July 22, 1940, Somerset Co., near Princess Anne (DPU); Killip 12928, May 30, 1925, Charles Co., near La Plata (FH, US); Leonard and Stewart 20333, Nov. 7, 1946, Patuxent Research Refuge (DPU); J. B. S. Norton, Aug. 4, 1940, Worcester Co., alt. 7.5 m (DPU, MD); Owens, May 15, 1948, Frederick Co., near Lewiston, alt. 120 m (DPU, MD); Owens, July 5, 1948, Cecil Co., near Elkton, alt. 18 m (DPU, MD); J. D. Smith, July 1876, Garrett Co. (US).

MASSACHUSETTS: E. G. Britton, Sept. 19, 1897, base of Mt. Everett (FH); Coleman, in 1875, Cheshire (US); Farlow, Sept. 1903, Magnolia (PC); Faxon 326, Apr. 7, 1881, Hyde Park (NY); Faxon 331, Sept. 8, 1881, Lynn (NY); Faxon 324, July 3, 1883, Jamaica Plain (NY); Faxon 330, Aug. 4, 1888, West Dedham (NY); Gerritson, May 1, 1906, Waltham, Prospect Hill Park, Grout, N. Am. Musci Pl. 272 (CAN, CM, COLO, DUKE, FH, ILL, MICH, MINN, MO, NY, OS, UC, US, WASH, YALE); Grosvenor, Barre (CHI); Handy, Sept. 1909, Fall R. (PC); Handy, Sept. 23, 1909, Fall R. (H); Handy, June 4, 1910, Fall R. (H); Handy, Sept. 1, 1910, Fall R. (H); Huntington, Feb. 20, 1903, Amesbury (DUKE); Huntington, Mar. 5, 1905, Amesbury (DUKE); Huntington, Jan. 23, 1908, Amesbury (H, PC); James, in 1856, Pigeon Cove (K, MO); James, Aug. 1856, Rockport (TYPE of *F. novaeangliae*, FH); Kingman, May 1, 1908, Melrose (FH); Kingman, Apr. 1, 1912, Andover (FH); Moir, July 5, 1906, Roslindale (NY, PC); Moulton, Nov. 30, 1886, Newburyport (PC, WIS); Rice, July 26, 1941, Nantucket (DPU); Rice, Mar. 29, 1942, Norton (DPU); Rice, May 27, 1942, Foxboro (DPU); Sanford, Apr. 18, 1910, Swansea (FH); Seymour, Oct. 17, 1901, Waltham, Prospect Hill (DPU); W. R. Taylor, July 10, 1917, Cuttyhunk (PENN); Wareham, Sept. 22, 1946, Old North Bridge (DPU).

MICHIGAN: Beeker, June 4, 1933, Galesburg (YALE); Conard 26, Aug. 10, 1937, near Ontonagon and mouth of Floodwood (DPU, GRI); Mich. Dept. Cons. 163, July 24, 1937, Allegon Co., Little Tom Lake (DPU, MICH); Mich. Dept. Cons. 83, July 27, 1937, Baraga Co., Craig Lake (DPU, MICH); Mich. Dept. Cons. 100, Aug. 5, 1937, Marquette Co., Kewayden Lake (DPU, MICH); Mich. Dept. Cons. 106, Aug. 7, 1937, Marquette Co., Indian Lake (DPU, MICH); George Moore, July 24, 1938, Keweenaw Co. (DPU, MICH); G. E. Nichols 215, Aug. 21, 1934, Marquette Co., Upper Mountain Lake, Huron Mts. (DPU, YALE); G. E. Nichols 459, Aug. 24, 1934, Marquette Co., Canyon Lake, Huron Mts. (DPU, MICH, NY, YALE); G. E. Nichols 460, Aug. 24, 1934, Marquette Co., Canyon Lake, Huron Mts. (DPU, YALE); G. E. Nichols, June 1937, Marquette Co., shore of Lake Superior (DPU, MICH, YALE); Povah 129, July 16, 1930, Isle Royale (MICH); Steere, Sept. 1935, Ontonagon Co., Bond Falls (DPU, FH, MICH); Welch 5313, Aug. 31, 1937, Keweenaw Peninsula, Manganese R. Gorge, near Copper Harbor (CHI, DPU, MICH, NY).

MINNESOTA: Holzinger, June 10, 1897, Fall Lake (PC); MacMillan and Lyon, Aug. 27, 1901 (PC); Moyle 3609 and 3622, Lake Co. (MINN); Reif, July 10, 1936, Cook Co., Brule R. (MINN); Reif, July 22, 1936, Temperance R. (MINN); Reif, Aug. 1938, Cook Co. (MINN).

MISSISSIPPI: L. E. Anderson 4555, Sept. 8, 1936, Wayne Co., near Waynesboro (DPU, DUKE); Brenner, Aug. 28, 1940, Stone Co., near Ramsey (MO).

MISSOURI: Bush, Aug. 7, 1889, Pleasant Grove (MO, PC); R. F. Dawson, June 8, 1941, Ironton, Stout's Creek at head of Lake Killarney (DPU, NY); Drew

14738, June 11, 1938, Madison Co., Sam Baker State Park, granitic rocks (UMO); Drexler 655, in 1935, Ozark Mts., near Fredericksburg (DPU); Routien 829, May 5, 1940, St. Genevieve Co., near Farmington, on sandstone (DPU); Russell 54, Sept. 1889, Wayne Co., Williamsville (NY); Russell 55, Aug. 1898, Iron Co., Stone Mt., granitic region (NY); Shoop 536, June 25, 1939, Cedar Co., near Arnica (CHI, DPU, NY); Shoop 572, Aug. 12, 1939, Maries Co., near Paydown (CHI, DPU); Steyermark 13616, July 17, 1934, Polk Co., near Burns (BART, CHI).

NEW HAMPSHIRE: L. A. Carter, July 22, 1901, Gilford (ILL); Clapp, Hancock (DUKE); Croasdale 48, June 24, 1945, Hanover (DPU); Croasdale, Oct. 3, 1948, Lyme (DPU, NJU); Eaton, Cheshire Co. (FH); Eaton 57, Cheshire Co., Hinsdale (Received by Sullivant, Oct. 1856), Sullivant and Lesquereux, Musci Bor.- Am. Exsic. 224c (TYPE of *F. Eatoni*, FH; G-BOIS, K, MICH, MO, NY, PC, PHIL, TRT, WIS); Eaton, Oct. 1856, Cheshire Co., Mt. Wantastiquet (YALE); Faxon 329, June 7, 1881, Mt. Willey (NY); Faxon 473, June 17, 1884, Mt. Willey (NY); Faxon 333, Oct. 14, 1884, Lafayette Co., Franconia, Lonesome Lake (NY); Faxon 334, June 17, 1887, Eagle Lake, Mt. Lafayette (NY); Faxon, July 23, 1891, Lafayette Co., Franconia (PC); Frost, Cheshire Co., Mt. Wantastiquet (K, YALE); Frost, Aug. 1856, Cheshire Co., Mt. Wantastiquet (YALE); Frost, Oct. 1856, Cheshire Co., Mt. Wantastiquet, Indian Road (YALE); Frost, Nov. 1856, Cheshire Co., Mt. Wantastiquet (YALE); Frost, in 1859, Cheshire Co., Mt. Wantastiquet (YALE); Frost and Eaton,[1] in 1858, Cheshire Co., New Hampshire, near Brattleboro, Vermont (TYPE of *F. Lescurii* var. *ramosior* and of *F. Frostii*, FH; NY, PC, YALE),[2] Grout, Oct. 4, 1897, Plymouth (DUKE); Grout, Aug. 1899, Plymouth (FH); Grout, Aug. 1900, Plymouth (DUKE); Huntington, Sept. 8, 1904, Weare (NY); Hutchinson, June 1, 1946, Grafton Co., Brighthollow, alt. 360 m (DPU); Hutchinson, July 15, 1948, Grafton Co., Brighthollow, alt. 273.9 m (DPU); James, White Mts. (K); James, Crawford Notch, White Mts. (PC); Mann, Mt. Hinsdale (NY).

NEW JERSEY: C. F. Austin (H, PC, US); C. F. Austin, Closter (DUKE, NY, NYSM); C. F. Austin, southern New Jersey, Musci Appal. 246 p.p. (CAN, CAS, DUKE, G-BOIS, K, NY in part, PHIL, US, WIS); C. F. Austin, Closter, Sullivant and Lesquereux, Musci Bor.-Am. Exsic. 341 p.p. (NY, US); C. F. Austin, Sept. 1862, Bergen Co. (NY); C. F. Austin, Sept. 1863, Palisades (NY); C. F. Austin, Sept. 1867 (Was *F. Peckii*, PC); C. F. Austin, Sept. 3, 1867, Palisades (WASH); C. F. Austin, Nov. 1868, Tom's R. (NY); C. F. Austin, Sept. 1870, Palisades (NY); C. F. Austin, Sept. 1870, Closter (ILL); C. F. Austin, Apr. 1872, Palisades (NY); C. F. Austin, Oct. 1872, Palisades (NY); C. F. Austin, Sept. 23, 1887, stream to Stag Pond (NY); Best, Jan. 20, 1890, Rosemont (CAN); Best, Nov. 20, 1890, Rosemont (NY); E. G. Britton, June 17, 1886, Lake Hopatcong (NY); N. L. Britton, May 30, 1882, Morris Co., Green Pond (NY); N. L. Britton, June 11, 1886, Morris Co., near Montville (NY); H. L. Fisher, May 23, 1896, near Mount Airy (WIS); H. L. Fisher, Apr. 20, 1897, near Milltown (PHIL); Lawton, Oct. 15, 1950, Sussex Co., Beaver Lake P.O. (DPU); Owens, June 18, 1948, Ocean Co., near Tom's R. (DPU, MD); Rappleye, June 15, 1948, Sussex Co., Kittatinny Mts., High Point State Park, alt. 420 m (DPU, MD).

NEW YORK: Alexander, Apr. 29, 1934, Orange Co. (CHI, NY); Blagg 9–713, June 29, 1929, Long Isl., Cold Spring Harbor, Fish Hatchery (GRI); E. G. Britton, Aug. 30, 1892, Adirondack Mts. (WASH); E. G. Britton, Aug. 30, 1900,

---

[1] Sometimes Frost, alone, is cited as collector; other times, Frost and Eaton are given as the collectors. A letter in the Herbarium of the New York Botanical Garden from David C. Eaton to Mrs. Britton on Dec. 16, 1894, explains that Frost and Eaton collected the plants together, in rivulets of Mt. Wantastiquet, Cheshire Co., N. H., opposite Brattleboro, Vt., and that Sullivant at first thought it new and named it *F. Frostii* in letters to C. C. Frost.

[2] The following note occurs on the type collection in the Sullivant Herbarium, "Tab. 62, Sulliv. Ic. Musc. from these specimens."

Essex Co., Crown Point Gorge, near Chilson Lake (NY); E. G. Britton, Sept. 3, 1900, Essex Co., Gooseneck Falls, near Chilson Lake (NY); E. G. Britton, Sept. 4, 1900, Essex Co., Crown Point Gorge, near Chilson Lake (NY); Burnett, May 31, 1896, Satshaw (G-BOIS); Clinton, Caledonia Creek (CHI); Conard 237, July 2, 1928, Long Isl., Cold Spring Harbor (DPU, GRI); Conard, Aug. 12, 1933, Long Isl., Cold Spring Harbor (DPU, GRI); Conard, in 1934, Long Isl., Smithtown (DPU, GRI, NY); Conard, Aug. 1934, Long Isl., Cold Spring Harbor (DPU, GRI); Conard, Aug. 17, 1934, Long Isl., Smithtown (DPU, GRI); Conard, Sept. 2, 1938, Long Isl., Cold Spring Harbor (CHI, DPU, GRI, MICH, NY); Gerard, Peekskill (US); Grout, Long Isl., Cold Spring Harbor, New York Fish Hatchery (TYPE of *F. novae-angliae* var. *Grouti*, DUKE; DPU); Grout, in 1897, near New York City (DUKE, PC); Grout, Mar. 11, 1906, Long Isl., Jamaica Road (DUKE); Grout, Feb. 20, 1909, Long Isl., New Dorp (DUKE); Grout, Apr. 6, 1913, Staten Isl., Princess Bay (DUKE); Grout, Sept. 1913, Staten Isl., N. Am. Musci Pl. 428 (BART, CAS, CM, COLO, DUKE, H, MICH, MINN, MO, NY, OS, UC, US, WASH, YALE); Grout, May 2, 1914, Staten Isl. (DUKE); Grout, May 4, 1918, Staten Isl. (DUKE); Grout, Jan. 19, 1929, Prince Bay (DPU, DUKE); Grout, July 4, 1929, Long Isl., Cold Spring Harbor (DUKE); Grout, July 5, 1929, Long Isl., Cold Spring Harbor (DUKE); Grout, July 6, 1929, Long Isl., Cold Spring Harbor (DUKE); W. P. Harris, Aug. 20, 1900, Essex Co., Chilson Lake (ABS, DUKE); E. J. Hill 98, 1882, July 24, 1882, Genesee Co., Pavilion (CHI, ILL); E. C. Howe, Poestenkill Mt., near Ft. Edward (Assumed to be the TYPE of *F. Howei*, PC; NY); E. C. Howe, in 1866, Kent Co. (PC); E. C. Howe, Aug. 1866, Rensselaer Co., Grafton Mts. (CHI, DPU, MICH, NY, PC, US); Hulst, June 1898, Long Isl., Flushing (ABS, DUKE, NY); Moldenke 10392, Mar. 27, 1938, Arden (NY); Muenscher, 139 and 142 of Winne, May 30, 1941, Broome Co., Oquaga Lake, alt. 474 m (DPU, NY); Muenscher, 143 of Winne, May 30, 1941, Broome Co., Oquaga Lake (DPU, NY); Muenscher and Isely, 889 of Winne, Sept. 9, 1941, Ulster Co., Ashokan Reservoir, alt. 210 m (DPU, NY); Peck, in 1844, Sandlake (NY); Peck, July 1864, Catskill Mts. (NY); Peck, Aug. 1865, Sandlake (NY); Tees 218, July 31, 1931, Putnam Co. (PHIL); Vail, Aug. 21, 1892, Greene Co., near Tannersville (CIN, DUKE); Vail, Aug. 22, 1892, Greene Co., near Tannersville (DUKE, NY); Winne 256, July 2, 1941, Broome Co., Oquaga Lake, alt. 474 m (DPU, NY); Winne 271, July 5, 1941, Washington Co., Sugarloaf Mt. Range, Pilot Knob Mt., alt. 210 m (DPU, NY); Winne 279, July 5, 1941, Washington Co., Lake George, Pilot Knob, near Kattskill Bay, alt. 330 m (DPU, NY); Winne 418, Aug. 23, 1941, Essex Co., Heart Lake, near North Elba, alt. 642 m (DPU, NY); Winne 477, Sept. 8, 1941, Suffolk Co., Long Isl., Brown R., near Patchogue, alt. 15 m (DPU, NY); Winne 592, Oct. 7, 1941, Greene Co., alt. 612 m (DPU, NY); Winne 595, Oct. 7, 1941, Greene Co. (NY); Winne 596, Oct. 7, 1941, Greene Co. (DPU, NY); Winne 605, Oct. 8, 1941, Greene Co., alt. 564 m (DPU, NY); Winne 612, Oct. 8, 1941, Greene Co., alt. 558 m (DPU); Winne 620, Oct. 8, 1941, Greene Co., near Hunter, alt. 525 m (DPU, NY); Winne 629, Oct. 8, 1941, Greene Co., Hunter, alt. 474 m (DPU, NY); Winne 651, Oct. 21, 1941, near Plaat Clove, on High Peak Mt., alt. 810 m (DPU, NY); Winne 660 and 662, Oct. 21, 1941, Greene Co., alt. 582 m (DPU, NY); Winne 668, Oct. 21, 1941, Greene Co., near Plaat Clove, alt. 576 m (DPU, NY); Winne 679, Oct. 21, 1941, Greene Co., near Plaat Clove, alt. 573 m (DPU, NY); Winne 741, Nov. 13, 1941, Washington Co., Hoosick R., near Eagle Bridge, alt. 102 m (DPU, NY); Winne 757, Nov. 13, 1941, Rensselaer Co., Babcock Lake, alt. 384 m (DPU, NY); Winne 776, Nov. 19, 1941, Schenectady Co., Mariaville Lake alt. 384 m (DPU, NY); Winne and Andrews 129, May 25, 1941, Tioga Co., near Oakley Corners, alt. 427.5 m (DPU, NY); Winne and Muenscher 398, Aug. 21, 1941, Essex Co., Lake Placid, alt. 546 m (DPU, NY); Winne, Muenscher, and Shannon 223, June 14, 1941, Hamilton Co., Lake Pleasant, alt. 516 m (DPU, NY); Winne, Muenscher, and Shannon 228, June 15, 1941, Hamilton Co., Piseco Lake, alt. 498 m (DPU, NY); Winne, Muenscher, and Shannon 241, June 15, 1941, Fulton Co., near Caroga Lake, alt. 411 m (DPU, NY); Winne, Shannon, and

Muenscher 240, June 15, 1941, Fulton Co., near Caroga Lake, alt. 411 m (DPU, NY); Winne, Whitford, and Chase 141, May 30, 1941, Cortland Co., near Cortland, alt. 354 m (DPU, NY); Wynne 2798, June 1943, Orange Co., Palisades Interstate Park, Bear Mt. Section (DPU).

NORTH CAROLINA: L. E. Anderson 906 and 907, July 22, 1933, Macon Co., Dry Falls, Cullasaja R. (DPU, PENN); Atkinson 11958, Sept. 10 and 11, 1901 (PC); Blomquist, July 14, 1922, Durham Co., Durham, Horse Pasture R. (DPU, DUKE); Blomquist, Nov. 16, 1927, Durham Co., Durham (DPU, DUKE); Blomquist, Mar. 24, 1930, Durham Co., Little R., Durham (DPU, DUKE); Blomquist, Apr. 17, 1932, Durham Co., Durham (DPU, DUKE); Blomquist, Apr. 2, 1933, Durham Co., above Laurel Hill (DPU, DUKE); F. S. Chapman, Feb. 20, 1921, Anson Co., near Wadesboro (DPU, DUKE); D. S. Correll 04489, July 8, 1935, Transylvania Co. (DPU, DUKE); D. S. Correll 4439, July 28, 1935, Transylvania Co. (FH); Correll and McDowell 10722, May 24, 1939, Stokes Co. (FH); Durand 12117, Aug. 7, 1901 (PC); Schallert 2987, Apr. 12, 1921, Forsyth Co., Winston-Salem, Nissen Park (BART, WIS); Schallert, July 4, 1923, High Rock (S); Schallert, May 1, 1925, Forsyth Co., Winston-Salem, Bennett's Quarry (DPU, DUKE, MINN); Schallert, June 2, 1925, Forsyth Co., Winston-Salem, Flat Rock (S); Schallert, June 21, 1925, Forsyth Co., Winston-Salem, Flat Rock (DPU, DUKE); Schallert, May 30, 1926, Stokes Co., Cascades (S); Schallert, June 1, 1926, Stokes Co., Cascades (DPU, DUKE); Schallert, Nov. 9, 1926, Columbus Co., near Bug Hill (DPU, DUKE); Schallert, Nov. 9, 1926, Columbus Co., Bug Hill (DPU, DUKE); Schallert, 8–10–1929, Forsyth Co. (CAS); A. J. Sharp 108, July 29, 1930, Cullasaja Gorge, alt. 1050 m (DPU, TENN); Welch 2224, June 16, 1936, between Grandfather Mt. and Blowing Rock (DPU); Welch 2225, June 17, 1936, Dry Falls, Cullasaja R., near Highlands (CHI, DPU, MICH, NY); Welch 2226, June 17, 1936, Dry Falls, Cullasaja R., near Highlands (CHI, DPU, MICH).

OHIO: Bartley and Pontius, July 24, 1935, Jackson Co. (DPU); Bartley and Pontius 232, July 24, 1935, Jackson Co. (DPU, MICH, NY); Bartley and Pontius 531, July 25, 1936, Jackson Co. (DPU); Bartley and Pontius 172, Nov. 8, 1936, Jackson Co. (MICH, NY); Bartley and Pontius, in 1937, Jackson Co. (DPU); E. L. Braun, Oct. 1931, Brown Co., near Williamsburg (CIN); D. M. Brown M39, Aug. 31, 1937, Portage Co., near Garrettsville (OS); Eckfeldt, in 1879 (PENN); Fulford, Mar. 1933, Clermont Co., near Marathon (CIN, DPU); Gordon, May 12, 1929, Hocking Co. (OS); Pontius and Bartley 36, May 15, 1935, Jackson Co. (US); Sullivant, near Columbus, Sullivant and Lesquereux, Musci Bor.-Am. 226b p.p. (PHIL); Sullivant, near Columbus, Sullivant and Lesquereux, Musci Bor.-Am. 337 p.p. (G-BOIS, PHIL); Walters 45, June 7, 1947, Cleveland (DPU); Wareham, Apr. 25, 1936, Licking Co. (DUKE); Wareham, Apr. 25, 1936, Licking Co., Black Hand Gorge (DPU); Wareham, Apr. 13, 1940, Hocking Co. (DPU, OS); Wareham, Apr. 21, 1940, Hocking Co. (DPU, OS); Wareham, July 4, 1942, Hocking Co. (DPU, NY); Welch 9691, Sept. 16, 1944, Hocking Co. (DPU).

OKLAHOMA: De Gruchy 20, May 27, 1936, Medicine Park (NY); J. Engleman, Nov. 1933, near Wilburton (DPU); Norman, July 1, 1947, Lahner Co., Robbers Cave State Park (DPU); A. J. Sharp 12, June 7, 1929, Witchita Nat. Forest (DPU).

PENNSYLVANIA: Ammons, June 22, 1940, Ohiopyle (DPU, WVA); Ammons, June 23, 1940, Ohiopyle (DPU, WVA); Ammons, July 30, 1940, Presque Isle, Erie (CHI, WVA); Bartram 237, Feb. 16, 1919, Pike Co., near Bushkill (BART); Bartram 543, May 7, 1921, near Reseca Falls (BART); Boardman, May 3, 1936, Butler Co. (CM); Boardman, June 23, 1940, Fayette Co. (CM); Boardman, June 23, 1940, Ohiopyle (MT); Burnett, May 9, 1893, Bradford (WASH); Burnett 497, May 9, 1893, McKean Co. (NY); Burnett, May 31, 1896, Bradford (ABS, WASH); Burnett 3110, Aug. 18, 1898, McKean Co. (ABS, FH); Conard 40–1285, June 11, 1940, Ohiopyle (DPU, GRI); Conard 40–1249, June 19, 1940, Ohiopyle (DPU, GRI); Conard 40–1222, 40–1263, and 40–1267, June 23, 1940, Ohiopyle (DPU, GRI); S. K. Eastwood, July 14, 1935, Butler Co. (CM); S. K. Eastwood, Aug. 4, 1935, Venango Co. (CM); S. K. Eastwood, May 30, 1937, Clearfield Co. (CM); Fitzgerald

(H); Githens 242, Sullivan Co., Lake Eaglesmere (DPU, PENN); Gleason 3380, July 22, 1941, Pike Co., Wolf Lake (DPU); Glowenke 598, July 15, 1948, Lackawanna Co., Moosic Mts., near Minooska, alt. 300 m (DPU); Glowenke 908, Nov. 26, 1948, Lackawanna Co., near Moosic, alt. 255 m (DPU); Glowenke 915, Dec. 5, 1948, Lackawanna Co., near Elmhurst, alt. 480 m (DPU); J. A. Graves, Apr. 1897, Susquehanna (MICH); C. Gross, Apr. 30, 1937, between Lewiston and Thurmont (DPU); Henry, Oct. 3, 1925, Butler Co. (CM); Henry, May 3, 1930, Butler Co. (CM); Hepner, Oct. 21, 1933, Somerset Co. (CM); E. J. Hill 92, 1909, Aug. 31, 1909, Crawford Co., Saegertown (CHI, PC); James, Chester Co. (ILL); James, Aug. 1860, Philadelphia, near Germantown (CHI, PHIL); O. E. and G. K. Jennings, Sept. 17, 1909, Westmoreland Co. (CM); Kriebel, Mar. 21, 1927, Hereford (DPU); Krout, Aug. 13, 1907, Delaware Co., Glenolden (DUKE); E. T. Moul 5791 and 5799, Apr. 19, 1946, Adams Co., near Arendtsville (DPU, PENN); E. T. Moul 5798, May 5, 1946, Chester Co., near Daylesford, alt. 120–150 m (DPU, PENN); E. T. Moul 2692a, May 11, 1946, Franklin Co., near St. Thomas (DPU, PENN); E. T. Moul 2692, Sept. 8, 1946, Warren Co., near Russell, alt. 450 m (DPU, PENN); E. T. Moul 2881, Apr. 1, 1947, Luzerne Co., near Red Rock, alt. 660 m (DPU, PENN); E. T. Moul 5192, May 10, 1947, Somerset Co., near Roxbury, alt. 660 m (DPU, PENN); E. T. Moul 3834, Sept. 10, 1947, Wayne Co., Elk Lake, near Waymart, alt. 420 m (DPU, PENN); E. T. Moul 3856, Sept. 10, 1947, Susquehanna Co., Fiddle Lake, near Gelatt, alt. 600 m (DPU, PENN); E. T. Moul 4784, Oct. 19, 1947, Fayette Co., near Marklysburg, alt. 555 m (DPU, PENN); F. J. Myers, June 15, 1941, Pocono Lake (PHIL); F. J. Myers, Sept. 3, 1941, Monroe Co., Naomi Lake (DPU); T. C. Porter, June 15, 1859 (PHIL); Rau, near Bethlehem (MICH); Rau, Aug. 1873, Pike Co. (NY); Rau, in 1874 (NY); Rau, June 1875 (NY); Windle, July 6, 1905, Fayette Co., Ohiopyle (NY); Wolle, Apr. 1874, Bethlehem (US).

RHODE ISLAND: Collins 5900, Scituate, Xoswansicut Reservoir (BART); Congdon, in 1858, North Providence (CAN); Mille 137, Nov. 17, 1946, West Kingston (DPU); Mille 225, May 28, 1947, West Kingston (DPU); Mille 320, Sept. 13, 1947, Kingston (DPU); Mille 401, Apr. 11, 1948, Kingston (DPU); Mille 410, May 1, 1948, Chepachet (DPU); Mille 429, June 20, 1948, Tucker Pond (DPU).

SOUTH CAROLINA: L. E. Anderson 8350 and 8355, Aug. 18, 1949, Oconee Co., East Fork, Chattooga R., alt. 840 m (DPU, DUKE); L. E. Anderson 8488, Aug. 22, 1949, Oconee Co., Lower Falls, Whitewater R., Jocassee, alt. 480 m (DPU, DUKE); McCorkle, Mar. 15, 1950, Lexington Co., Gibson's Pond (DPU); Patterson, Oct. 1931, near Lexington (DPU); Patterson 1179, Aug. 25, 1949, Richlands Co., near Columbia (DPU).

TENNESSEE: A. C. and Floyd Brown, 958 of Clebsch, July 30, 1949, Montgomery Co., near Clarksville (DPU); S. A. Cain 632, Apr. 16, 1935, Sevier Co., Ramsey Fork (DPU, TENN); S. A. Cain 11, Jan. 12, 1936, Morgan Co., near Wartburg (DPU, TENN); S. A. Cain 48, Nov. 22, 1936, Campbell Co., Pine Mt. Narrows (DPU, TENN); Clebsch, July 4, 1947, Coffee Co., near Manchester (DPU); Conard, June 16, 1934, Monteagle (DPU); Jennison and Wilson, Apr. 25, 1934, Grainger Co., near Lea Lakes (DPU, TENN); A. J. Sharp, May 1933, Sevier Co., near Greenbrier (DPU); A. J. Sharp 34103, Jan. 28, 1934, between Rockwood and Harriman, alt. 480 m (DPU, FH, TENN); A. J. Sharp 34348, Apr. 8, 1934, Grainger Co., Lea Lakes, alt. 300 m (CHI, DPU, FH, TENN); A. J. Sharp 34377, Apr. 15, 1934, Johnson Co., Shady Valley, alt. 840 m (DPU, FH, TENN); A. J. Sharp 341116, Oct. 14, 1934, Fentress Co., near Clark Range, alt. 510 m (DPU, FH, MICH, NY, TENN); A. J. Sharp 35223, July 7, 1935, Fentress Co. (FH); A. J. Sharp 38107, June 23, 1938, Blount Co., Montvale, alt. 450 m (DPU); A. J. Sharp 4061, May 5, 1940, Blount Co., Cades Cove, alt. 540 m (DPU, FH, MICH, TENN); A. J. Sharp 4751, Apr. 27, 1947, Polk Co., Hiwassee R. (DPU, TENN); A. J. Sharp 4744, June 29, 1947, Van Buren Co., above Piney Creek Falls, alt. 480 m (DPU, TENN); A. J. Sharp 4763, Aug. 2, 1947, Coffee Co., near Manchester (DPU, TENN); A. J. Sharp and Clebsch, June 29, 1947 Van Buren Co., Piney Creek (DPU, TENN).

VERMONT: Clapp, Aug. 1904, Athens (NY); Dobbin, July 1907, Stratton (DUKE); Dutton 420, July 3, 1910, Bridgewater, in Ottauquechee R. (DUKE, MO); Dutton 2173, Aug. 24, 1924, Goshen (CM, DUKE, FH); Flowers, July 1934, Newfane (UT); Griffin, in 1909, Mullet's Bog (DUKE); Grout, May 29, 1893, Elmore Pond (DUKE, NY, PC); Grout, Aug. 4, 1902, Newfane (DUKE, PC); Grout, Aug. 27, 1903, Newfane, alt. 450 m (DUKE); Grout, Aug. 2, 1906, Newfane, N. Am. Musci Pl. 280 (BART, CAN, CM, DUKE, FH, H, ILL, MICH, MINN, MO, NY, OS, PC, UC, US, WASH, WIS, YALE); Grout, June 2, 1926, Newfane (DPU, DUKE); Mann, in 1863, Brattleboro (FH); Pringle, June 18, 1880, Enosburgh (DPU, FH, PC); E. W. Thompson, Aug. 11, 1934, Newfane (FH); E. W. Thompson, July 4, 1935, Newfane (FH); Trask, Habeeb, and Grout 39, July 16, 1943, Newfane (DPU); Welch 712, Aug. 6, 1932, near Newfane (CHI, DPU, MICH, NY).

VIRGINIA: E. C. Allard 18142, Apr. 27, 1938, Fauquier Co., Bull Run Mt. (US); H. A. Allard 6478, Apr. 23, 1939, Shenandoah Co., Devils Hole Mt., near Edinburgh (DPU, US); E. G. Britton 237, June 15, 1892, near Marion (PC); R. P. Carroll 172, Apr. 26, 1929, Alleghany Co., Eagle Rock (DPU); Fitzgerald (H); Fitzgerald, in 1881, near Bealeton (TYPE of *F. Cardoti*, PC; FH, G-BOIS, K, NY, S, US); Fulford, June 1939, Giles Co., Mountain Lake (CIN, DPU); Gleason 47078, Apr. 21, 1947, Carroll Co., Blue Ridge Mts., alt. 780 m (DPU); Gleason 48022, May 24, 1948, Nansemond Co., near Holland (DPU); Gleason 48087, June 1, 1948, Isle of Wight Co., near Franklin (DPU); Holzinger, May 16, 1891, Arlington (US); Holzinger, Sept. 1891, Arlington (PC); Holzinger, Sept. 11, 1891, Arlington (NY, WIS); Holzinger, Sept. 16, 1891, Arlington (US); Iltis 3794 and 3798, Dec. 26, 1947, Spottsylvania Co., near Fredericksburg (DPU); Leonard 1867, Mar. 26, 1922, Alexandria Reservoir (US); Leonard 2291, July 30, 1922, Dun Loring (US); Leonard 18142, Apr. 27, 1938, Bull Run Mts., near Hopewell (DPU); Leonard 18143, Apr. 27, 1938, Bull Run Mts., near Hopewell (DPU, US); Murrill, Aug. 4, 1896, Mt. Lake (NY); Patterson, Mar. 29, 1937, near Suffolk (DPU); Patterson R-508, Sept. 19, 1943, Bedford Co., Peaks of Otter, alt. 840 m (DPU); Patterson R-581, Oct. 19, 1943, Botetourt Co., Tinker Mt., alt. 300 m (DPU); Patterson R-741, Oct. 14, 1945, Craig Co., Caldwell Mt., alt. 450 m (DPU); Pierce, June 30, 1936, Giles Co., Mountain Lake (CHI); Schnooberger and Wynne 3178, 3181, 5011, 5015, and 5037, June 18, 1944, Page Co., Shenandoah Nat. Park, alt. 900 m (DPU); Schott, May 2, 1857 (CHI); J. K. Small, June 17, 1892, Smyth Co., Blue Ridge, alt. 810 m (DUKE, NY, PC, PHIL, UC, US, WIS); Standley 13151, May 29, 1916, Bluemont (CHI, US); Vail and E. G. Britton, May and June 1892 (MICH); Vail and E. G. Britton 237, June 16, 1892 (NY); Vail and E. G. Britton, June 17, 1892, Smyth Co. (NY, US); Vail and E. G. Britton 241, June 17, 1892 (NY); Vail and E. G. Britton 211, June 27, 1892 (NY).

WEST VIRGINIA: H. A. Allard 6527, May 10, 1939, Hardy Co., Allegheny Mts. (US); H. A. Allard 10949, 10949a, and 10951, Aug. 2, 1943, Tucker Co., Canaan Valley, near Davis, alt. 270–360 m (DPU, US); Ammons, June 29, 1929, Webster Co., Cowen, Gauly R. (DPU, WVA); Ammons, May 27, 1932, Monongalia Co., Morgantown (WVA); Ammons, June 1932, Preston Co., near Aurora (WVA); Ammons, Oct. 23, 1932, Upshur Co., Rock Cave (DPU, WVA); S. E. Bailey, July 25, 1939, Monongalia Co., alt. 633 m (DPU); Bartholomew, July 27, 1939, Monongalia Co., near Morgantown (DPU, WVA); Fox, June 26, 1940, Mercer Co., Princeton (DPU, WVA); Gray M1641, Alderson (BART); Gray M957, May 30, 1928, Greenbank (BART); Gray 1053, July 25, 1928, Pocahontas Co. (BART); Gray M1210, Aug. 1929, Cass, Cheat Mt. (BART); Gray M1552, June 1931, Monroe Co., Flat Top Mt. (BART); Gray 1553, Aug. 28, 1931, Mt. Asbury (BART); Gray M4036, Jan. 22, 1932, Rainelle (BART); Gray M4041, July 1932, Phillippi (BART); Gray M4055, July 1933, Terra Alta (BART); J. Myers, Aug. 4, 1937, Preston Co., near Pisgah (WVA); J. D. Smith, July 1878, Cheat R. (US).

WISCONSIN: Cheney 2690, July 30, 1893, Lincoln Co., Wisconsin R., near Merrill (DPU, WIS); Cheney, June 1894, northern Wisconsin (PC); Cheney 3203, June 25, 1894, Marathon Co., Wisconsin R., Granite Heights (DPU, WIS); Cheney

11133, Apr. 17, 1926, Barron Co. (DPU, WIS); Cheney 12957, July 4, 1929, Barron Co., Pipe Lake, near Barronette (DPU, WIS).
United States without locality: Austin, Musci Appal. 244 (CAN, CAS, DUKE, K, NY, PHIL, US, WIS); Austin, Musci Appal. 247 (CAN in part, CAS, DUKE, G-BOIS, K, NY, PHIL, US, WIS); Sullivant and Lesquereux, Musci Bor.-Am. Exsic. 225, in New England (FH, K, MICH, MO, NY, TRT, WIS); Sullivant and Lesquereux, Musci Bor.-Am. Exsic. 335 p.p., in New England (MICH, NY, PC, PHIL, WIS, YALE); Sullivant and Lesquereux, Musci Bor.-Am. Exsic. 336, in New England (CHI, CM, FH, MICH, NY, PHIL, US, WASH, WIS, YALE); Sullivant and Lesquereux, Musci Bor.-Am. Exsic. 340 p.p., from New England to Alabama (CHI, CM, FH, G-BOIS, MICH, NY, PC, PHIL, WIS, YALE).

Fontinalis novae-angliae may be distinguished by the following combination of characteristics: plants generally medium in size, majority of median cauline leaves firm, erect to erect-spreading, concave, ovate-lanceolate, margins in apical portion of blade usually narrowly involute, apices commonly somewhat truncate, subobtuse, or obtuse, leaf tips frequently serrulate, blades 2.5–4 mm long, sometimes up to 6 mm in length, 0.75–2 mm wide, sometimes up to 2.5 mm in width, 1.75–4 : 1, leaf cells above basal area generally subflexuous to flexuous, auricles commonly distinct, and apices of upper perichaetial leaves broadly obtuse.

In the preparation of the monograph on the Fontinalaceae of North America, the author was not satisfied with the clarity of the characteristics of *F. Lescurii*, and observed the resemblance of plants so determined to those of *F. novae-angliae*. The doubt of the validity of that species has continued through the present study as additional specimens, so labeled, have been examined. Cardot, in his monograph, mentioned the confusion with forms of *F. novae-angliae*. It seems to the writer that Austin was not certain of the distinction as he made *cymbifolia* a variety of *F. Lescurii* with question. The author considers *cymbifolia* to be a variety of *F. novae-angliae*, because leaves true in characteristics to the latter species occur intermixed with the boat-shaped leaves of the variety, *cymbifolia*. Sullivant's *F. Lescurii* var. *ramosior* is regarded by the writer to be *F. novae-angliae*. In exsiccati of Austin and of Sullivant and Lesquereux, distributed as *F. Lescurii*, the author has found plants of *F. novae-angliae* and of *F. novae-angliae* var. *cymbifolia*. In the observation of collections of *F. novae-angliae*, occasionally the writer has found slender plants with flaccid to subrigid stems and flaccid, subconcave to concave, narrowly ovate-lanceolate or oblong-lanceolate leaves, with apices somewhat acuminate, intermixed with specimens which are definitely *novae-angliae*. In a few instances, the above description has applied to branch leaves on plants with median cauline blades true to *novae-angliae* in the majority of diagnostic characteristics. During the examination of plants formerly determined as *F. Lescurii*, the author has concluded that these specimens so determined are in some instances young or not well developed plants of *novae-angliae*, and in other instances the plants are very flaccid specimens of this species, demonstrating the effect of environmental factors. This conclusion is based upon the occurrence of leaves which would be described as those of *Lescurii* on plants with blades which are true to the description of *novae-angliae*.

The plants described by Cardot as *F. novae-angliae* var. *Lorenziae* have been compared with numerous collections of slender plants of *F. novae-angliae*. The habit of branching and the position and characteristics of the leaves are comparable. The author considers the plants formerly determined as var. *Lorenziae* to be slender specimens of *F. novae-angliae*.

For the present treatise the writer has studied a very large number of specimens of *Fontinalis*. During the examination of these plants, observations have been made regarding the influence of environmental factors upon the firmness and flaccidity of the stems and leaves as well as upon the proximity and distance of the branches and blades. Upon this basis the author now considers the specimens which were described previously as *F. novae-angliae* var. *Grouti* as being subflaccid or flaccid plants of the species.

In the Fontinalaceae of North America, the author regarded *F. Delamarei* of Renauld and Cardot as a variety of *F. novae-angliae*, and found difficulty in separating the variety from the species. In the examination of a much larger number of collections in the present study, the search for constant distinguishing characteristics has continued. The presence of leaves formerly regarded as those of *Delamarei* intermixed with those considered as characteristic of *novae-angliae* on the same axis has resulted in the placing of *Delamarei* in the synonymy of *F. novae-angliae*.

16. **Fontinalis novae-angliae** Sull. var. **cymbifolia** (Aust.) Welch, in Grout, Moss Fl. N. Am. 3: 247. 1934.

Fontinalis Lescurii Sull. var. ? *cymbifolia* Aust., Musci Appal., p. 42. 1870; Austin, Musci Appal 248. 1870, [printed Latin description on label].
Fontinalis involuta Ren. and Card., in Cardot, in Rev. Bryol. 18: 86. 1891, [in descriptive key]; Cardot, Mon. Font., p. 96. 1892; Cardot, in Rev. Bryol. 18: 83. 1891 (nomen nudum).
Fontinalis novae-angliae subsp. *involuta* (Ren. and Card.) Kindb., Sp. Eur. and N. Am. Bryin., Part 1: 147. 1896.
Fontinalis Waghornei Card., in Rev. Bryol. 23: 71. 1896.
Fontinalis novae-angliae var. *Waghornei* (Card.) Welch, in Grout, Moss Fl. N. Am. 3: 248. 1934.

Median cauline leaves erect-spreading to spreading, commonly close, bases up to 1 mm apart, some blades comparable in shape and involution to those of *novae-angliae*, others oblong-lanceolate, some suboval-lanceolate, majority deeply concave to subcanaliculate or canaliculate, margins broadly involute in apical portions or almost to base of leaf, blades then appearing sublinear; apices frequently

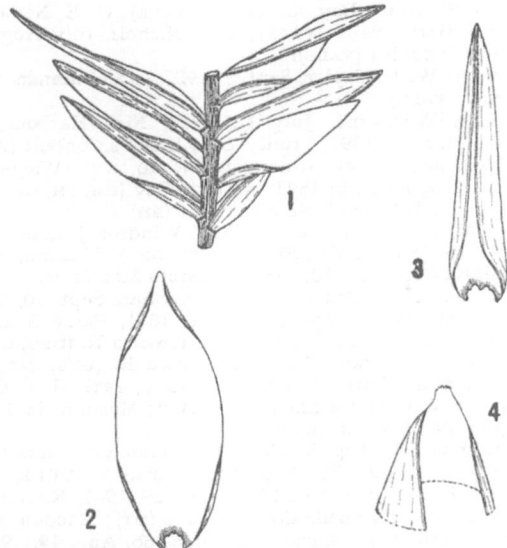

FIG. 15. – *Fontinalis novae-angliae* Sull. var. *cymbifolia* (Aust.) Welch – 1.Portion of stem with median cauline leaves. 2.Median cauline leaf with margins in apical region narrowly involute.   3.Median cauline leaf with margins broadly involute almost to base of blade.   4.Leaf apex. Drawn from Langlois, in 1892, Louisiana, Renauld and Cardot, Musci Am. Sept. Exsic. 186.

subcucullate to cucullate, leaf tips commonly serrulate, occasionally entire; median cauline blades 3–5.5 mm long, 0.5–1.75 mm wide, occasionally up to 2 mm in width, 2.5–4 : 1; otherwise plants, including fruits, similar to those of *F. novae-angliae*. [The varietal name, *cymbifolia*, was used by C. F. Austin to describe the boat-shaped leaves.]

Type: In Herbarium of the New York Botanical Garden. The plants are not so designated, but Austin's notes which accompany them strongly indicate that he used these specimens as a basis for his original description.

Type collector: Coe Finch Austin, Sept. 1865.

Type locality: New Jersey: near "High Point," in a pond on Shawangunk Mountain.

Distribution: Fontinalis novae-angliae var. *cymbifolia* occurs, approximately, in the eastern half of the United States and Canada.

Additional description: Barnes, Gen. and Sp. N. Am. Moss., p. 329 (as *F. involuta*). 1896.

Illustrations: Welch, in Grout, Moss Fl. N. Am. 3: pl. 76 (as vars. *cymbifolia* and *Waghornei*). 1934. – FIGURE 15.

Specimens examined: NORTH AMERICA: CANADA: CAPE BRETON ISLAND: G. E. Nichols, July–Aug. 1914, mts. north of Barrasois R. (NY); G. E. Nichols, July–Aug. 1914, mts. west of St. Ann's Bay (NY); G. E. Nichols, July–Aug. 1914, mts. west of St. Ann's Bay, alt. 300 m (YALE); G. E. Nichols, July–Aug. 1914, mts. south of Barrasois R. (YALE); G. E. Nichols, July–Aug. 1915, Moosehide Lake, west of Ingonish (YALE).

NEW BRUNSWICK: W. R. Taylor, Sept. 9, 1922, Grand Manan, near Eel Lake (DUKE, MICH, PENN, YALE).

NEWFOUNDLAND: Waghorne, July 12, 1890, New Harbour, Broad Cove (CAN); Waghorne, Mar. 30, 1892, Trinity Bay, Heart's Content (Assumed to be TYPE of *F. Waghornei*, PC; FH, G-BOIS, MINN, MO, NY); Waghorne, in 1893, Trinity Bay (NY); Waghorne, in 1893, Witter's Bay (BM, FH, G-BOIS, MINN, MO, NY, PC, S); Waghorne, July 1893, New Harbour (BM).

NOVA SCOTIA: M. S. Brown, July 4, 1923, Windsor Junction (DPU); M. S. Brown, July 5, 1924, Queen's Co., Five Rivers (DPU); Macoun, June 18, 1883, Halifax (CAN); McKay 73, Aug. 10, 1883, Cobequid Mts. (CAN).

ONTARIO: E. G. and N. L. Britton and Timmerman, Sept. 10, 1889, Muskoka Lake, Port Sandfield (NY); Burgess, Aug. 21, 1881, Parry Sound, Mill Lake (TRT); R. F. Cain, Aug. 21, 1939, Whitney, Madawasca R. (DPU, NY, TRT); R. F. Cain, Aug. 21, 1939, Algonquin Park, Madawasca R. (DPU, TRT); R. F. Cain, Sept. 11, 1939, Algonquin Park, Costello Lake (DPU, TRT); R. F. Cain, Sept. 11, 1939, Algonquin Park, Pinetree Lake (DPU, TRT); Macoun, in 1868, Belleville (NY); Macoun, July 24, 1900, Algonquin Park (TRT).

QUEBEC: Boardman 590, Aug. 5, 1936, Terrebonne Co., Lake Barbote, Lake Mercier (CM, MT); Dupret, July 20, 1906, Oka, Ottawa R. (DUKE); Dupret, 5–9–1929, Oka, Ottawa R. (CAS); Fabius 1781, Aug. 29, 1947, Roxton Pond (DPU); Lucien 1292, July 21, 1942, Labelle Co., Bellerive (MT); Macoun, May 16, 1896, Hull, in Beaver Meadow (TRT); Marie-Jean-Eudes 86, Aug. 19, 1930, Montcalm Co., R. Rouge, Rawdon, alt. 135 m (MT); Marie-Victorin 9329, July 1913, Temiscouata Co., Lake St. Hubert (MT); Marie-Victorin 19228, Aug. 23, 1922, Lake St. Jean Co., Kondiaronk (MT).

Canada without locality: Macoun, in brooks, Nova Scotia and Ontario, Can. Musci 231 p.p. (NY).

UNITED STATES: ALABAMA: Harvill 5144, July 2, 1949, Franklin Co. (DPU).

FLORIDA: Fitzgerald, in 1880 (FH, PC).

KENTUCKY: Welch 1086, June 27, 1933, Whitley Co., Cumberland Falls Road, near Corbin (CIN, DPU).

LOUISIANA: Arsène 14665, Aug. 1921, Covington (BART, NY, US); Celestin 1025, Mar. 1893, Covington (US); Drummond, New Orleans, Musci Am. 152 (Assumed to be TYPE of *F. involuta*, PC; BM, FH, G-BOIS, K, NY); Langlois 23c (BM); Langlois 33 (BM); Langlois, in 1882, Covington (US); Langlois 31, in 1882, Covington (US); Langlois, in 1884 (FH); Langlois 31, July 11, 1884; Langlois 745a, 1882–1885, Abita (US); Langlois 262, May 22, 1885, Opelousas (PC, US); Langlois 758, Aug. 1885 (PC); Langlois, in 1891, Abita (K); Langlois 745a, Oct. 5, 1891, Abita (US); Langlois, Oct. 6, 1891, Abita (US); Langlois 745, Oct. 6, 1891, Abita (BART, FH, PC, US); Langlois 745, Oct. 6, 1891, Abita, Grout, N. Am. Musci Pl. 388 (BART, CAN, CIN, CM, COLO, MICH, MINN, MO, NY, OS, UC, US, WASH, WIS, YALE); Langlois 745b p.p., Oct. 6, 1891, Abita (US); Langlois, in 1892, Abita (H); Langlois, in 1892, St. Tammany Co. (US, YALE); Langlois, in 1892, St. Tammany Co., Renauld and Cardot, Musci Am. Sept. Exsic. 186 (CAN, FH, NY, PC, YALE); Langlois 752, Sept. 9, 1892 (PC); Langlois 753, Sept. 14, 1892

PC); Langlois 755, Sept. 17, 1892, St. Tammany Co. (PC, US); Langlois 756, Sept. 17, 1892 (PC); Langlois 757, Sept. 17, 1892 (PC).

MAINE: MacGoron, Aug. 17, 1941, Oxford Co., Hartford (DPU, PSNH); Norton, May 26, 1937, Mt. Agamenticus, York (DPU); Patterson 137, July 6, 1929, Mt. Desert Isl. (DPU).

MASSACHUSETTS: Kingman, Aug. 14, 1911, Wakefield, Lake Inamapowitt (FH); Wolle, Oct. 1874, Fall R. (US).

MICHIGAN: Mich. Dept. Cons. 124, Aug. 11, 1937, Baraga Co., Spruce Lake (DPU, MICH).

MISSISSIPPI: Harvill 7287, Aug. 19, 1950, Jackson Co., near Ocean Springs (DPU); Webster and Wilbur 744, July 12, 1950, Scott Co., near Forest (DPU, MICH).

MISSOURI: Gier 4495, June 15, 1951, St. Genevieve Co., Pickle Spring (DPU, WJC).

NEW HAMPSHIRE: Kennedy, Oct. 13, 1899, Chocorua (FH).

NEW JERSEY: C. F. Austin (NY); C. F. Austin, Musci Appal. 248 (CAN, CAS, DUKE, G-BOIS, H, K, NY, PC, PHIL, US, WIS); C. F. Austin, Oct. 1862 (NY); C. F. Austin, Oct. 1862, Pine Barrens, Manchester (PC); C. F. Austin, in 1863, Ocean Co. (NY); C. F. Austin, in 1865, Closter (NY); C. F. Austin, Sept. 1865, Shawang-unk Mt., near High Point (Assumed to be TYPE of var. *cymbifolia*, NY); C. F. Austin, Nov. 1865, Shawangunk Mts. (NY); C. F. Austin, in 1867 (NY); Mitmann, July 1936, Englewood (NY).

NEW YORK: Conard, Aug. 15, 1934, Long Isl., Smithtown (FH); Conard, Aug. 17, 1934, Long Isl., Smithtown (DPU); W. P. Harris 877, Aug. 20, 1900, Essex Co., Chilson Lake (FH); Hulst, Aug. 1898, Lake George (NY); Hulst, Aug. 1899, Lake George (DUKE); Muenscher and Isely, 511 of Winne, Sept. 9, 1941, Broome Co., Oquaga Lake, alt. 474 m (DPU, NY); Winne 278, July 5, 1941, Warren Co., Lake George, Kattskill or Warner Bay, alt. 96 m (DPU, NY); Winne 314, Aug. 3, 1941, Warren Co., Lake George, bay of Campers' Isl., alt. 96 m (DPU, NY); Winne 401, Aug. 21, 1941, Essex Co., Heart Lake, near North Elba, alt. 642 m (DPU, NY); Winne 564, Sept. 18, 1941, Sullivan Co., Lake Louise Marie, alt. 456 m (DPU, NY); Winne 768, Nov. 19, 1941, Schenectady Co., Mariaville Lake, alt. 384 m (DPU, NY); Winne, Muenscher, and Shannon 215, June 14, 1941, Hamilton Co., Sacandaga Lake, alt. 516 m (DPU, NY); Winne, Shannon, and Muenscher 202, June 13, 1941, Essex Co., Cascade Lake, alt. 609.6 m (DPU, NY).

NORTH CAROLINA: Schallert, Jan. 1, 1934, Columbus Co. (FH, UC); Schallert, Jan. 1, 1934, North China (DPU).

OHIO: Ammons, Aug. 9, 1941, Put-in-Bay (DPU, WVA); Bartley, Sept. 17, 1946, Hocking Co. (DPU).

PENNSYLVANIA: Ammons, July 30, 1940, Presque Isle, Erie (WVA); Eckham, in 1873, Pike Co. (G-BOIS); E. T. Moul 2798, Mar. 31, 1947, Carbon Co., near Nesquehoning, alt. 505.5 m alt. (DPU, PENN); Rau, Aug. 1873, Pike Co. (NY); Rau 134, Pike Co. (FH).

RHODE ISLAND: Collins 885, May 30, 1893, Exeter (PC); Mille 394, Oct. 5, 1947, Tucker Pond (DPU); Mille 341, Oct. 9, 1947 (DPU).

SOUTH CAROLINA: Patterson, Mar. 27, 1931, near Columbia (DPU).

TENNESSEE: S. A. Cain and A. J. Sharp 254, Apr. 19, 1936, Marion Co. (CHI); D. S. and H. B. Correll 8143, Sept. 1, 1937, Roane Co., between Rockwood and Ozone (DUKE, FH); A. J. Sharp 124, Aug. 31, 1930, Morgan Co., Clear Fork R. Rugby, alt. 540 m (BART, DPU, TENN); A. J. Sharp, May 17, 1931, Fentress Co., alt. 480 m (DPU, TENN); A. J. Sharp 34562, May 31, 1934, Morgan Co., Clear Fork R., near Rugby, alt. 510 m (FH, DPU, TENN); A. J. Sharp 34915, July 13, 1934, Marion Co., near Whitwell, alt. 540 m (DPU, TENN); A. J. Sharp 34955, July 15, 1934, Marion Co., Foster Falls, alt. 450 m (DPU, TENN); A. J. Sharp 4761, Aug. 2, 1947, Grundy Co., Altamont, alt. 480 m (DPU, TENN); A. J. Sharp and Clebsch, June 29, 1947, Van Buren Co., Piney Creek (DPU, TENN).

VERMONT: Grout, Aug. 1, 1900, Stratton (DUKE); Grout, Aug. 22, 1910, Stratton, N. Am. Musci Pl. 348 (BART, CAN, CM, COLO, DPU, DUKE, FH, MINN,

NY, PC, UC, US, UT, WASH, WIS, YALE); Grout, July 28, 1927, Stratton (DPU);
Trask, Habeeb, and Grout 250, Aug. 16, 1943, Stratton (DPU); Wynne 1898,
Aug. 1, 1940, Stratton (CHI, DPU).
VIRGINIA: Gleason 3077, Aug. 22, 1940, Nansemond Co., near Holland (DPU).
WEST VIRGINIA: Ammons, Aug. 15, 1929, Hampshire Co., Short Mt. Hanging
Rock (DPU, WVA).

The var. *cymbifolia* may be recognized by the following combination
of characteristics: median cauline leaves erect-spreading to spreading,
many deeply concave to subcanaliculate or canaliculate, margins
frequently broadly involute either in apical portions or approximately to base of blade, and numerous leaves appearing to be sublinear.

In the treatment of this variety by the author in Grout, Moss Fl.
N. Am. 3: 247. 1934, the words *cymbifolia* and *involuta* in the discussion and in the explanation of pl. 76, figs. 26–28, were confused
inadvertently during editing and printing. Also, the type locality of
F. involuta was cited instead of that of var. *cymbifolia*.

In the Fontinalaceae of North America, the author regarded *F.
Waghornei* of Cardot as a variety of *F. novae-angliae*. The characteristics which seemed to separate *Waghornei* from *Delamarei*, *cymbifolia*,
and *novae-angliae* were variable. During the study of additional collections, the search for constant diagnostic characteristics has continued. It now seems that *Waghornei* should be in the synonymy of
var. *cymbifolia*.

17. **Fontinalis novae-angliae** Sull. var. **latifolia** Card., in Nichols,
in Rhodora 15: 9. 1913.

Median cauline leaves usually distant, bases 1–2 mm apart, blades
subflaccid to flaccid, broadly ovate-lanceolate, 4–7.5 mm long, 1.5–
3.5 mm wide, 2–3 : 1; otherwise, plants similar to those of *F. novae-
angliae*. (The varietal name, *latifolia*, has reference to the broad
leaves.)

Type: In Herbarium of Cardot in the herbarium of the Laboratoire de Crypto-
gamie, Muséum National d'Histoire Naturelle, Paris, France.
Type collector: G. E. Nichols, Apr. 16, 1908.
Type locality: Connecticut, Burlington, in a brook.
Distribution: Fontinalis novae-angliae var. *latifolia* occurs, approximately,
in the eastern half of the United States and Canada.
Additional description: Welch, in Grout, Moss Fl. N. Am. 3: 248. 1934.
Illustrations: Welch, in Grout, Moss Fl.N. Am. 3: pl. 76. 1934. – FIGURE 16.
Specimens examined: NORTH AMERICA: CANADA: QUEBEC: Marie-Anselme
222, Oct. 27, 1935, Lake Rivard, near La Tuque (DPU); Marie-Anselme 691,
June 16, 1936, Emmanuel's Lake (DPU, MT); Marie-Anselme 761, June 27,
1936 (DPU).
UNITED STATES: ARKANSAS: Demaree 22722, Mar. 15, 1942, Drew Co.,
Monticello (DPU, FH); Demaree, May 2, 1942, Drew Co., Monticello (DPU, NY, OS).

CONNECTICUT: G. E. Nichols, Apr. 16, 1908, Burlington (TYPE, PC; DUKE, NY, YALE); G. E. Nichols, Apr. 17, 1908 (PC); G. E. Nichols, Apr. 14, 1911, North Stonington (YALE).

INDIANA: Morrison and Dawson, Apr. 15, 1933, Putnam Co., Hoosier Highlands (DPU); Welch 1087, May 7, 1933, Putnam Co., Hoosier Highlands (DPU); Welch 1084, May 24, 1933, Putnam Co., Hoosier Highlands (DPU); Welch 1088, Apr. 23, 1934, Putnam Co., Hoosier Highlands (CHI, DPU, NY).

MAINE: R. L. Lowe, June 22, 1940, Blackstrap Hill (DPU, PSNH); Pier, Sept. 1902, Kinnebunkport (DPU).

MASSACHUSETTS: Ordway, May 2, 1936, Boxford (DPU).

MISSOURI: E. Hall, Vernon Co. (CHI); Steyermark 83, Feb. 15, 1931, Lincoln Co., near Foley (CHI, DPU, MO, NY).

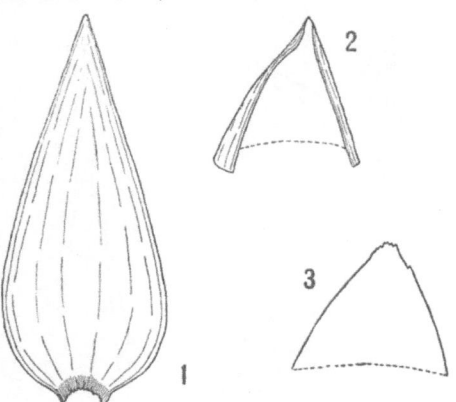

FIG. 16. – *Fontinalis novae-angliae* Sull. var. *latifolia* Card. – 1.Median cauline leaf.   2 and 3.Leaf apices. Drawn from G. E. Nichols, Apr. 16, 1908, Connecitcut.

NEW HAMPSHIRE: Riedeman 337, Apr. 26, 1954, near Durham (DPU).

NEW YORK: Alles, Apr. 8, 1945, White Plains (DPU); Blagg 9–712, July 2, 1929, Long Isl., Cold Spring Harbor (DPU); Haring 134, May 10, 1939, Woodland (CHI, DPU); Lawton 1014, Apr. 22, 1951, Arden (DPU); Palmatier, 86 of Winne, Apr. 27, 1941, Tompkins Co., Connecticut Hill, alt. 510 m (DPU, NY); Winne 868, Aug. 17, 1940, Saratoga Co., Black Lake, near Corinth, alt. 495 m (DPU, NY); Winne 720, Oct. 22, 1941, Kaaterskill Junction, alt. 504 m (DPU).

NORTH CAROLINA: Blomquist, June 12, 1928, Haywood Co., Eagle's Nest, near Lake Junaluska, alt. 930 m (DPU, DUKE); Correll 8658, Feb. 14, 1938, Guilford Co., near Gibsonville (DPU, DUKE).

RHODE ISLAND: Mille 486, July 11, 1948, West Greenwich (DPU).

VIRGINIA: Long 2867, Apr. 18, 1942, Norfolk Co., Dismal Swamp, near Jericho Ditch and margin of Lake Drummond.

WEST VIRGINIA: W. Frye, Nov. 1933, Hampshire Co., Hanging Rock (WVA); W. Sharp, Feb. 6, 1932, Dellslow (WVA).

18. **Fontinalis Bryhnii** Limpr., in Hagen, in Kongl. Norsk. Vid. Selsk. Skrift., No. 3, p. 40. 1908.

Fontinalis antipyretica Hedw. subsp. *Bryhnii* (Limpr.) Podp. Consp. Musc. Eur., p. 507. 1954.

Plants slender to medium in size, yellow green, golden green, olive green, copper brown, dark brown, or black, younger portions glossy, older portions dull; stems subflaccid to flaccid, up to 30 cm in length and 0.3 mm in diameter, generally denuded and dark at base with

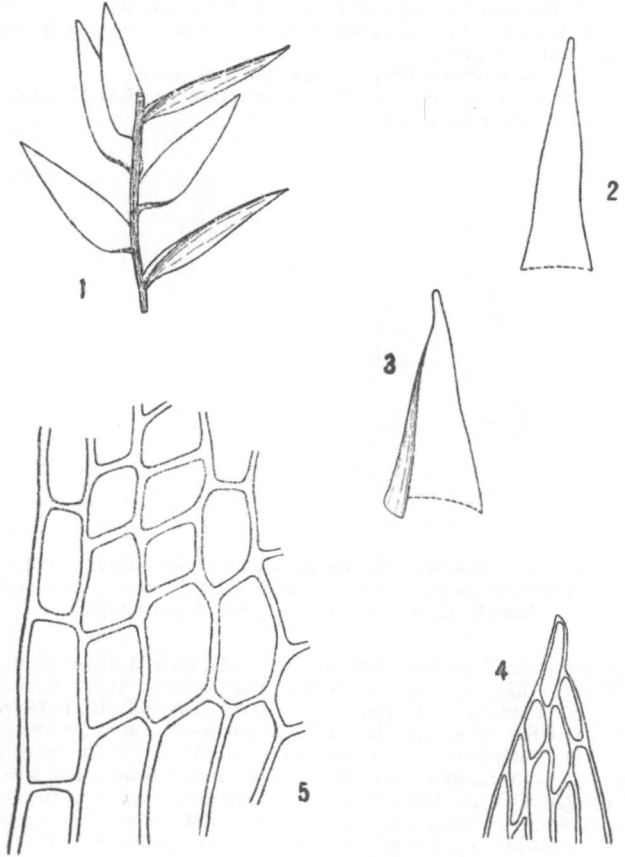

FIG. 17. – *Fontinalis Bryhnii* Limpr. – 1.Portion of stem with median cauline leaves. 2 and 3.Leaf apices. 4.Leaf apex. 5.Portion of alar group. Drawn from N. Bryhn, July 1901, Baegna River, near Sörum, Norway.

age, irregularly pinnately branching; branches numerous, erect-ascending to erect-spreading, short to elongate, up to 11 cm in length, ends of foliated stems and branches attenuate or obtuse; median cauline leaves distant, bases up to 1.5 mm apart, blades firm, erect-spreading, concave throughout or concave at base and plane above,

sometimes loosely folded longitudinally, narrowly ovate-lanceolate,
or ovate-lanceolate, margins gradually tapering from basal fourth or
basal half into the apex; apices commonly acuminate or acute, occasionally narrowly obtuse, now and then twisted one-half turn, leaf
tips usually entire, sometimes subserrulate or serrulate, frequently
terminating in a very prominent hyaline or colored cell; blades 3.5–5
mm long, occasionally up to 5.5 mm in length, 1–1.5 mm wide, occasionally up to 2.2 mm in width, 2.5–5 : 1; median cells of leaves
commonly linear with ends attenuate or obtuse, occasionally narrowly
rhomboidal, 10.2–17 μ wide, 7.5–20 : 1; alar cells enlarged, subquadrate, subrectangular, or subhexagonal, walls subhyaline, hyaline,
golden brown, or brown, cells in longitudinal rows, usually 4 or 5,
occasionally 6, those of the marginal row commonly narrower than
the others, sometimes cells also in horizontal rows, the entire group
rectangular in outline, parallel with leaf margin, and extending upward
from leaf base approximately 0.75 mm, commonly not forming
auricles, or if so, very slight, leaf bases long decurrent, frequently
extending almost to base of adjacent leaf; median branch leaves
similar to median cauline except smaller in size; perichaetium and
sporophyte not seen.[1] (The specific name, *Bryhnii*, honors N. Bryhn,
the collector of the type specimen.)

Type: In Herbarium of Bergens Museum, Bergen, Norway, it is assumed.
Hagen states that *F. Bryhnii* was collected first by Bryhn, Aug. 5, 1885, at
Elverum, Norway, in the Glommen River, growing with *F. seriata*. Limpricht
refers to this collection in Laubm. 2: 669. 1894, in his discussion under *F. baltica*.
Hagen continues with the statement that in July 1901, Bryhn collected *F.
Bryhnii* in Søndre Aurdal, near Sörum, in the Baegna River, and sent these
plants to Limpricht, who then considered them to be a new species which he
wished to publish under the name of *F. Bryhnii*. According to Hagen, Limpricht's death occurred before publication and Bryhn sent specimens from both
localities to him for the preparation of a description. It may be assumed that
Bryhn sent portions of these two collections to Hagen as he did to Limpricht,
and that Hagen used both for his description. The writer has seen duplicates of
the above-mentioned collections in the Herbarium of Naturhistoriska Riksmuseet, Stockholm, and of the July 1901 collection in the Herbarium of the
Botanisches Museum der Universität, Helsinki. All are with the labels of
"Musci Norvegici ex herb. N. Bryhn." It has not been possible, with the data
at hand, to determine the type with certainty.
Type collector: N. Bryhn. Aug. 5, 1885 or July 1901, it is assumed.
Type locality: Norway: either Elverum, in the Glommen River, or Søndre
Aurdal, near Sörum, in the Baegna River, it is assumed.

---

[1] Observations have been made regarding the occurence of the antheridia
and archegonia in the same cluster. Numerous groups of reproductive structures
on plants of *F. Bryhnii* have been examined. The majority contained mature
antheridia and archegonia, some had immature antheridia and mature archegonia, and a few had mature archegonia and no visible antheridia. Ten clusters
were removed from one stem. Six contained both antheridia and archegonia.
Four had archegonia but no evident antheridia.

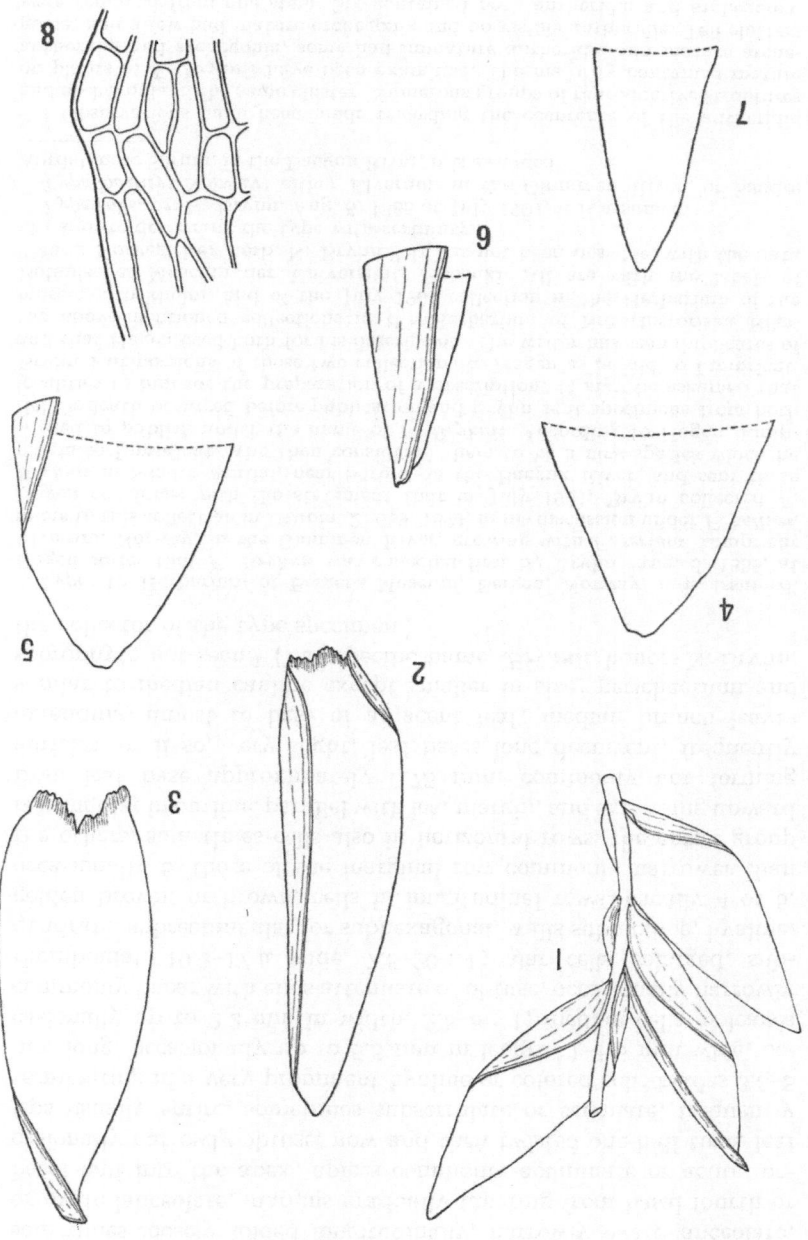

Distribution: Europe: Norway.
Additional descriptions: Brotherus, Laubm. Fenn., p. 396. 1923; Mönkemeyer, Laubm. Eur. Erg.- Bd. 4: 657. 1927; Jensen, Skand. Bladmossfl., p. 379. 1939.
Illustration: FIGURE 17.
Specimens examined: EUROPE: NORWAY: Bryhn, Aug. 1885, Hedemarken Amt, near Elverum, Glommen R., North Lat. 61°, alt. 200 m (s); Bryhn, Aug. 5, 1885, Hedemarken Amt, Elverum, Glommen R. (s); Bryhn, Aug. 1887, Hedemarken Amt, near Elverum, Glommen R. (BG, s); Bryhn, Aug. 1887, Hedemarken Amt, Elverum, Glommen R., North Lat. 61°, alt. 200 m (s); Bryhn, Aug. 1892, Hedemarken Amt, Elverum, Glommen R. (s); Bryhn, Aug. 1899, Buskerud Amt, Hallingdal (s); Bryhn, Aug. 1900, in river, 60° 35', alt. 150 m (DUKE); Bryhn, July 1901, Valdres, near Sörum, Baegna R., North Lat. 60° 45', alt. 150 m (s); Bryhn, July 1901, Valdres, near Sörum, North Lat. 60° 35', alt. 150 m (s); Bryhn, July 1901, Valdres, near Sörum, Baegna R., North Lat. 60° 35', Long. 27° 45', alt. 180 m (H); Bryhn, Aug. 1901, Valdres, near Sörum, Baegna R., North Lat. 60° 35', alt. 150 m (DUKE, MINN, NY, s); Bryhn, Aug. 1904, Valdres, near Sörum, Baegna R., North Lat. 60° 35', alt. 150 m (DPU, H, PC, s); Bryhn, Aug. 1904, Valdres, near Sörum, Baegna R., North Lat. 60° 35', alt. 150 m, Bauer, Musci Eur. Exsic. 491 (BART, PC); Hagen, Sept. 1, 1887, near Sarpsborg (PC); Kaalaas, Aug. 14, 1907, Kristians Amt (s).

Fontinalis Bryhnii is distinguished from other species in the concave group by the following combination of characteristics: majority of median cauline leaves narrowly ovate-lanceolate or ovate-lanceolate, 3.5–5 mm long, 1–1.5 mm wide, 2.5–5 : 1, with apices acuminate or acute, leaf tips entire, ending in a prominent cell, an alar group rectangular in outline, parallel with leaf margin, consisting of 4–6 longitudinal rows of cells, and auricles none or very slight.

19. **Fontinalis bogotensis** Hampe, in Ann. Sci. Nat. (Cinq. Sér.) Bot. 4: 351. 1865; Hampe, in Flora 45: 454. 1862 (nomen nudum).

Plants slender to medium in size, yellowish green, olive green, green, brownish green, or brownish to blackish near base; glossy, especially younger portions, or dull, particularly older and basal leaves; stems flaccid, up to 55 cm in length, 0.25–0.4 mm in diameter, foliated or denuded in basal portions, darker below with age, irregularly pinnately branching; branches usually few, occasionally numerous, erect-ascending to erect-spreading, close to distant, short to elongate, up to 11 cm in length, ends of foliated stems and branches attenuate; median cauline leaves usually distant, bases up to 1 mm apart, blades flaccid to somewhat firm, erect to erect-spreading, commonly subconcave to concave, occasionally plane or nearly so above base, gener-

FIG. 18. *(opposite).* – *Fontinalis bogotensis* Hampe – 1.Portion of stem with median cauline leaves. 2 and 3.Median cauline leaves. 4–7.Leaf apices. 8. Alar cells. Drawn from A. Lindig, April 1860, Colombia, South America.

ally ovate-lanceolate, sometimes subelliptic-lanceolate or oblong-lanceolate, width gradually decreasing from broadest area into short and broad apex; apices obtuse, subobtuse, or truncate, at times incurved, leaf tips usually entire or sinuolate, sometimes slightly serrulate; median cauline blades 3–4.5 mm long, at times up to 6.5 mm in length, 0.75–1.5 mm wide, occasionally up to 2.75 mm in width, 2.5–3.5 : 1, occasionally 2 : 1; median cells of leaves linear with ends attenuate or obtuse, or narrowly rhomboidal, 5–13.5 µ wide, 6–16 : 1; cells in upper portion of leaves often flexuous; alar cells enlarged, subrectangular, subquadrate, or subhexagonal, walls hyaline, sub-hyaline, yellowish, or brownish, frequently forming distinct auricles, leaf bases not decurrent, semiamplexicaul or amplexicaul; median branch leaves similar to cauline except smaller in size; perichaetial leaves, calyptra, and operculum not seen; urn oval or short oblong, 1.25–1.5 mm long, 1 mm in diameter, 1.25–1.5 : 1, not contracted beneath mouth when dry; peristome teeth brownish orange, linear-acuminate, sometimes united in pairs at apex, 0.5 mm long, muricate, lamellae 23–24; trellis brownish orange, perfect, 0.6 mm long, muricate, transverse strands complete and appendiculate; spores yellowish green, smooth, 13.2–19.8 µ in diameter; ripe in summer.[1]

[The specific name, *bogotensis*, refers to Bogota, Colombia, South America, where the type was collected.]

Type: In Herbarium of British Museum of Natural History, London, England. Hampe's Herbarium and Lindig's book of specimens are both deposited in the British Museum of Natural History.

Type collector: A. Lindig, April 1860.

Type locality: Colombia: Bogota, "in rivulis, Novae Granadae, Rio Arzobispo, 2800 meters."

Distribution: South America: Colombia.

Additional description: Cardot, Mon. Font., p. 89. 1892.

Illustration: FIGURE 18.

Specimens examined: SOUTH AMERICA: COLOMBIA: Apollinaire, June 1904, Bogota (H, PC); Apollinaire, June 1905, near Bogota (BART, NY, PC); Apollinaire, Apr. 1906, near Bogota (PC); Ariste-Joseph, in 1909, Bogota, R. Arzobispo (US); Ariste-Joseph, in 1917, near Bogota, R. Arzobispo (NY); Cuervo, Bogota (BM, G-DEL, PC); Holton, Oct. 23, 1852, Bogota (G-BOIS, K, NY, PC); Lindig, Apr. 1860, Bogota, R. Arzobispo, alt. 2800 m (TYPE, BM; K, L, NY, PC); Weir 284, Andes Mts., Bogota, between Tipaquira and Pacho, alt. 2700 m (BM, G-BOIS, H, K, NY).

Fontinalis bogotensis may be distinguished from other species in *Fontinalis* by the following combination of characteristics: leaves flaccid to somewhat firm, commonly subconcave to concave, generally ovate-lanceolate, sometimes subelliptic-lanceolate or oblong-lanceo-

---

[1] The type material is sterile. The two fruits available for study were in the collection made by Cuervo, in Hampe's Herbarium, in the British Museum of Natural History, London, England.

late, apices short and broad, obtuse, subobtuse, or truncate, leaf tips entire or sinuolate, occasionally serrulate, cells in apical portion frequently flexuous, blades usually 3–4.5 mm long, 0.75–1.5 mm wide, 2.5–3.5 : 1, bases semiamplexicaul or amplexicaul, and auricles frequently distinct.

The species with which *F. bogotensis* may be confused are *F. Duriaei* and *F. novae-angliae*. The striking differences between *F. bogotensis* and the former are leaves usually subconcave to concave, especially at bases and apices, commonly ovate-lanceolate, occasionally sub-elliptic-lanceolate or oblong-lanceolate, apices obtuse, subobtuse, or truncate, tips usually entire or sinuolate, and bases semiamplexicaul or amplexicaul in *bogotensis*, and blades commonly plane, broadly ovate-lanceolate or oval-lanceolate, with majority of apices short and broadly acuminate, leaf tips usually acute, serrulate, and bases plane to subconcave in *F. Duriaei*. *F. novae-angliae* may be recognized in contrast by leaves ovate-lanceolate, margins in apical portions usually narrowly involute, and leaf tips generally distinctly serrulate.

20. **Fontinalis missourica** Card., in Rev. Bryol. 23: 69. 1896.

Fontinalis Holzingeri Card., in Holzinger, in Minn. Bot. Stud. 2: 43. 1898; Cardot, in Minn. Bot. Stud. 3: 129. 1903.
Fontinalis Umbachii Card., in Minn. Bot. Stud. 3: 129. 1903.
Fontinalis Nelsoni Card., in Welch, in Grout, Moss Fl. N. Am. 3: 244. 1934 (nomen nudum).

Plants slender to medium in size, yellowish green, green, or brownish green, glossy, especially younger portions, older parts sometimes dull; stems subflaccid to flaccid, up to 20 cm in length, 0.25–0.5 mm in diameter, denuded and darker at base with age, irregularly pinnately branched, occasionally bipinnately divided; branches numerous, erect to erect-spreading, generally close, up to 9.5 cm in length, ends of foliated stems and branches attenuate; median cauline leaves distant, bases up to 2 mm apart, blades usually subflaccid, sometimes flaccid, occasionally firm, erect-spreading, concave throughout or concave below and plane above, commonly lanceolate to narrowly ovate-lanceolate, occasionally broadly ovate-lanceolate; apices long, narrowly to broadly acuminate, leaf tips usually acute, sometimes subobtuse or narrowly obtuse, commonly serrulate, occasionally entire; median cauline blades 4.5–6 mm long, 1–1.75 mm wide, 2.8–5 : 1; median cells of leaves linear with ends attenuate, commonly flexuous, 6.8–13.6 μ wide, 9–17 : 1; alar cells moderately enlarged, subrectangular, subquadrate, or subhexagonal, walls hyaline, yellowish, or yellowish brown, frequently forming slight but distinct auricles; leaf bases not decurrent to briefly so, up to 0.5 mm; median branch leaves comparable to median cauline except smaller in size; perichaetial branches 3.5 mm long; perichaetium cylindrical, 0.75 mm in diameter; upper perichaetial leaves suboval, truncate and lacerate with age; calyptra

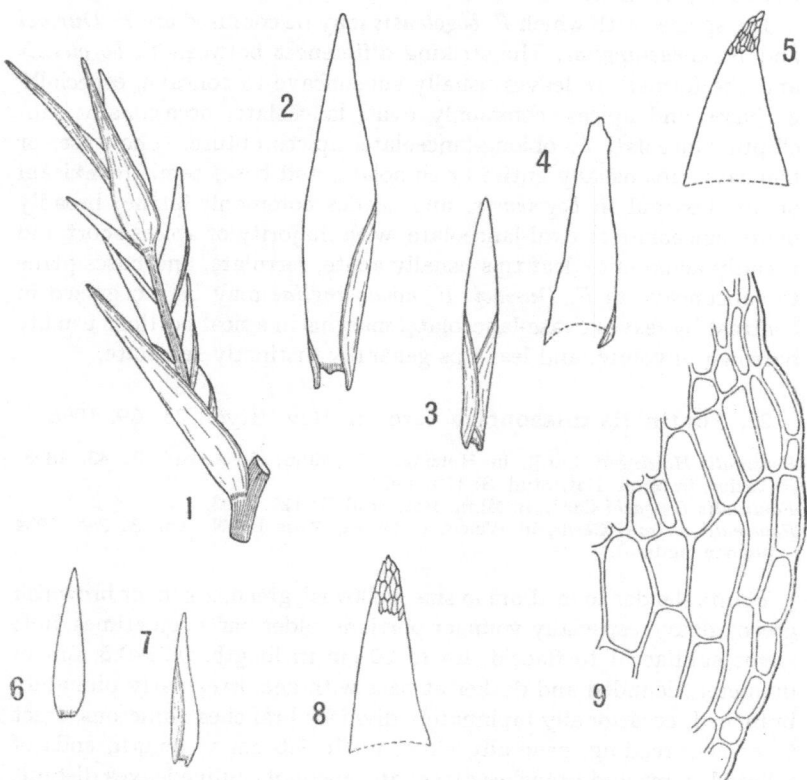

FIG. 19. – *Fontinalis missourica* Card. – 1. Portion of stem with median cauline leaf, bearing in its axil a young branch. 2 and 3.Cauline leaves. 4 and 5. Apices of cauline leaves. 6 and 7.Branch leaves. 8.Apex of branch leaf. 9. Group of alar cells of median cauline leaf. No. 1 drawn from Demetrio, May 29, 1896, Missouri; nos. 2–9 from Demetrio, Sept. 23, 1907, Missouri.

not seen; capsule generally emergent; operculum conical, 0.75 mm long, 0.5 mm in diameter; seta 0.2 mm long; urn immersed to slightly emergent, cylindrical, 2 mm long, 0.6 mm in diameter, 3.3 : 1, slightly contracted beneath mouth when dry; peristome teeth brownish orange, linear-acuminate, often united in pairs at apex, 0.45–0.5 mm in length, muricate, lamellae 11–15; trellis brownish orange, imperfect, cilia slightly longer than teeth, 0.56 mm in length, muricate, united by cross bars into a trellis in upper one-third, transverse strands muricate, not appendiculate; spores yellowish green, green, or yellowish brown, muricate, 13.6–20.4 μ in diameter, ripe in summer. [The specific name, *missourica*, has reference to the state, Missouri, in which the type specimens were collected.]

Type: In Paris Museum, bearing the data: "No. 172. *Fontinalis missourica* Card. sp. nova. Ex Herb. C. H. Demetrio. On rocks, floating in creeks. Near Cole Camp Creek, Benton Co., Missouri."

Type collector: Rev. C. H. Demetrio, July 12, 1894.

Type locality: Missouri: Benton Co., near Cole Camp Creek.

Distribution: North America: in eastern United States and Canada.

Additional descriptions: Hill, in Bryol. 18: 10 (as *F. Umbachii*). 1915; Welch, in Grout, Moss Fl. N. Am. 3: 243 (as *F. missourica*) 1934.

Illustrations: Cardot, in Minn. Bot. Stud. 3: pl. 21 (as *F. Holzingeri*), pl. 22 (as *F. Umbachi*). 1903; Welch, in Grout, Moss Fl. N. Am. 3: pl. 75 (as *F. missourica*) 1934. – FIGURE 19.

Specimens examined: NORTH AMERICA: CANADA: QUEBEC: Cléonique-Joseph 7929, Aug. 21, 1934, St. Anne de Bellevue (DPU, MT).

UNITED STATES: ARKANSAS: Bartram 1513, Jan. 30, 1926, Garland Co., near Hot Springs (BART); Beyrich (BM, K); Beyrich, in 1834 (BM).

ILLINOIS: Hatcher 113: 9, Feb. 11, 1950, Jackson Co., Hickory Ridge (DPU); E. J. Hill, July 21, 1911, Cook Co., Sag Bridge (PC); E. J. Hill 9. 1911, July 21, 1911, Cook Co., Sag Bridge (CHI); E. J. Hill, Aug. 9, 1911, Will Co., Lockport (PC); E. J. Hill 19. 1911, Aug. 9, 1911, Will Co., Lockport (CHI); Umbach, June 1898, Romeo, Des Plaines R. (CHI, DUKE); Umbach, June 18, 1898, Romeo, Des Plaines R. (Assumed to be the TYPE of *F. Umbachi*, PC); Umbach 604, June 18, 1898, Romeo, Des Plaines R. (WIS); Umbach 1104, July 7, 1906, Romeo, Des Plaines R., Grout, N. Am. Musci Pl. 279 (BART, CAN, CM, COLO, DUKE, FH, H, ILL, K, MICH, MINN, MO, NY, OS, PC, S, UC, US, WASH, WIS); Umbach, July 7, 1909, Romeo, Des Plaines R. (PC).

KANSAS: Hall, Aug. 1870, Neosho R. (CHI); Meeker, Dec. 11, 1892, Franklin Co., Ottawa (MO); Meeker, Dec. 26, 1893, Franklin Co., Ottawa (NY); Meeker, May 4, 1894, Franklin Co., Ottawa (FH, WIS); Meeker, in 1900, Franklin Co., Middle Creek (DPU, FH).

MICHIGAN: Nichols and Steere, Aug. 20–27, 1935, Ontonagon Co., shore of Lake Superior (DPU, YALE); Schaffner, in 1895, Washtenaw Co., Ann Arbor (OS); Steere, Aug. 1935, Ontonagon Co., Bond Falls (DPU).

MINNESOTA: Holzinger, in 1897, Granite R. (K); Holzinger and Elftman, June 17, 1897, at the second falls of Granite R., going east from Lake Saganaga (TYPE of *F. Holzingeri*, PC; MINN, MO, NY, S, US, WYO).

MISSOURI: Bush, June 4, 1899, Swan (MO); Bush 187, June 4, 1899, Swan (NY, US); Bush, Sept. 27, 1905, Swan (MO); Conard 47–153, Nov. 9, 1947, Shannon Co., Mountain View, Blue Spring (DPU, GRI); Demetrio (PC); Demetrio, May 24, 1884, Perry Co., near Perryville, Saline Hills (H); Demetrio, in 1894, Benton Co. (FH); Demetrio, July 12, 1894, Benton Co., near Big Cave (NY); Demetrio 172, July 12, 1894, Benton Co., near Cole Camp Creek (TYPE of *F. missourica*, PC); Demetrio, May 29, 1896, Benton Co., near Big Cave (FH, NY, US); Demetrio, May 29, 1896, Benton Co., Cole Camp Creek (PC, S); Demetrio, Sept. 23, 1907, Benton Co., Big Springs, near Cole Camp Creek (FH, WASH); Demetrio, Sept. 1910, Benton Co., near Big Cave, Cole Camp Creek (ILL); Gier 701a, Nov. 28, 1947, Jasper Co. (DPU); Gier 2652 and 2660, Oct. 15, 1949, Camden Co., spring and old mill race Ha Ha Tonka (DPU); Letterman, in 1908, Allenton, Allen's Spring (PC); Letterman, Sept. 21, 1908, Allenton, Allen's Spring

(MO); Metcalf, Morgan Co., near Versailles, Travais Mills (CHI); Metcalf 889, Aug. 23, 1920, Morgan Co., near Versailles, Travais Mills (NY, US); N. L. T. Nelson, Nov. 5, 1905, Irondale (K); N. L. T. Nelson 1020, Nov. 5, 1905, Irondale (DUKE, PC); Steyermark 14419, Aug. 11, 1934, Howell Co., near Bly, Bennet Bayou (BART, CHI); Steyermark 40110 and 40112, June 20, 1941, Taney Co., White R., near Mincy (DPU).

OHIO: Field Crew of Conservation Dept., Dec. 15, 1937, Franklin Co., Big Walnut Creek (DPU, OS).

Fontinalis missourica may be recognized by the following combination of characteristics: median cauline leaves distant, bases up to 2 mm apart, blades subflaccid to flaccid, concave throughout or concave below and plane above, commonly lanceolate or narrowly ovate-lanceolate, apices long, narrowly to broadly acuminate, leaf tips usually acute and serrulate, blades 4.5–6 mm long, 1–1.75 mm wide, 2.8–5 : 1, median cells linear, commonly flexuous, ends attenuate, and alar cells moderately enlarged, frequently forming slight but distinct auricles.

In the Fontinalaceae of North America, the author used the indistinct difference between cauline and branch leaves as one of the distinguishing characteristics of *F. missourica*. In this study the characteristics of the median cauline leaves have been considered as more reliable bases for classification than differences in size and shape of the cauline and branch leaves. These variations seem to be influenced largely by the age of the branch, the blades on the older ones bearing a stronger resemblance to the cauline than those of the younger branches. The writer does not now regard the leaves of *F. missourica* as being dimorphic.

21. **Fontinalis Mac-Millanii** Card., in Rev. Bryol. 23: 71. 1896.

Plants medium in size, yellowish green, green, brownish green, copper color, brown, glossy; stems subrigid to rigid, up to 12 cm in length and 0.5 mm in diameter, commonly denuded and dark at base with age, irregularly pinnately branching; branches numerous, erect to erect-spreading, up to 7 cm in length, varying from close to distant, ends of foliated stems and branches generally attenuate, occasionally obtuse; median cauline leaves loosely imbricate, bases 0.5–1 mm apart, blades generally firm, occasionally subflaccid, erect-spreading, usually concave, sometimes concave at base and plane above, occasionally plane, commonly ovate-lanceolate or oval-lanceolate, sometimes oblong-lanceolate; apices usually acuminate or acute, occasionally subacute or subobtuse, leaf tips frequently twisted one-fourth to one-half turn and occasionally completely, sometimes subcarinate to carinate near one margin, occasionally one side narrowly and briefly involute, sometimes one margin oblique, generally entire, now and then subserrulate; median cauline blades 5–6 mm long, 1–2 mm wide, 2.5–3 : 1, occasionally 5 : 1; median cells of leaves linear with ends attenuate or obtuse, occasionally narrowly rhomboidal, now and then subflexuous, 8.5–17 μ in width, 9–18 : 1; alar cells enlarged, sub-

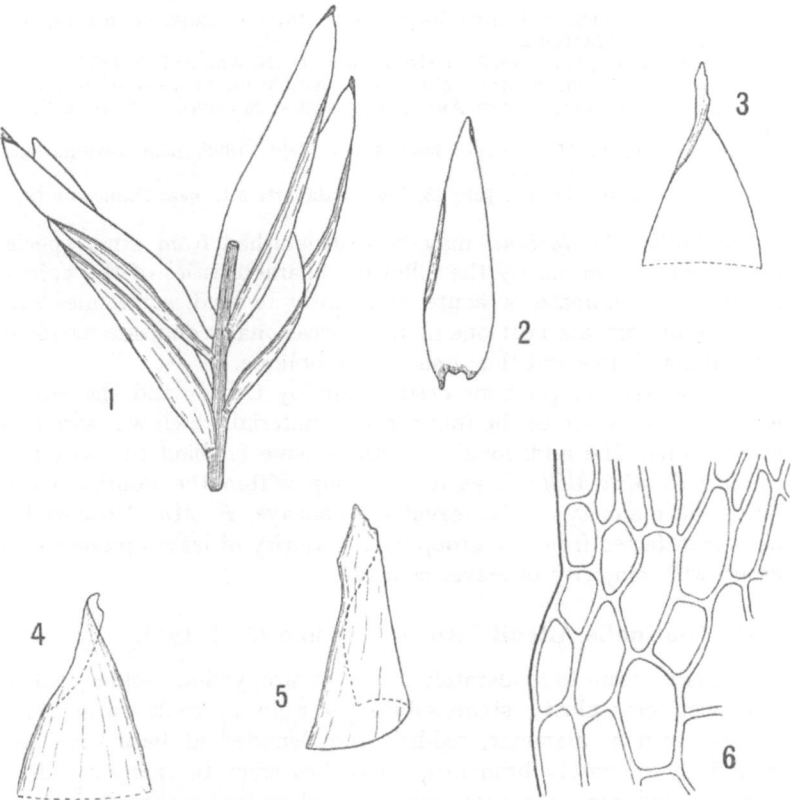

FIG. 20. – *Fontinalis Mac-Millanii* Card. – 1.Portion of stem with median cauline leaves. 2.Median cauline leaf. 3, 4, and 5.Leaf apices. 6.Group of alar cells. Drawn from Conway Mac-Millan, Sept. 1895, Minnesota.

quadrate, subrectangular, or subhexagonal, walls brownish, yellowish, or subhyaline, usually forming distinct auricles, bases not decurrent to very briefly so, up to 0.25 mm; median branch leaves similar to median cauline except smaller in size; perichaetium and sporophyte not seen. [The specific name, *Mac-Millanii*, honors the collector of the type specimen, Professor Conway Mac-Millan.] [French spelling, Mac-Millan; English spelling, MacMillan.]

Type: In Herbarium of Cardot, in Laboratoire de Cryptogamie, Muséum d'Histoire Naturelle, Paris, France.
Type collector: Conway Mac-Millan, Sept. 1895.
Type locality: Minnesota: near International Boundary of the United States and Canada.
Distribution: Northern United States and southern Canada: Minnesota, New Hampshire, and Manitoba.
Additional description: Welch, in Grout, Moss Fl .N. Am. 3: 255. 1934.
Illustrations: Welch, in Grout, Moss Fl. N. Am. 3: pl. 79. 1934. – FIGURE 20.
Specimens examined: NORTH AMERICA: CANADA: MANITOBA: Macoun 26, in 1872 (NY).
UNITED STATES: MINNESOTA: Mac-Millan, Sept. 1895, near International Boundary (TYPE, PC; MINN).
NEW HAMPSHIRE: Faxon, July 23, 1891, Lafayette Mt., near Franconia (NY).

Fontinalis Mac-Milanii may be distinguished from other species in the concave group by the following characteristics of the apices: commonly acuminate or acute, frequently twisted, sometimes subcarinate or carinate near one margin, occasionally one side narrowly involute, and now and then one margin oblique.

The variance of previous descriptions by Cardot and the author with this one is due to the fragmentary material which was available at the time. The additional collections have enabled the writer to classify *F. Mac-Millanii* as to its group within the Fontinalaceae. Since the majority of the leaves are concave, *F. Mac-Millanii* has been transferred from the group with majority of leaves plane to the group with majority of leaves concave.

22. **Fontinalis Allenii** Card., in Rhodora 15: 8. 1913.

Plants medium to moderately robust in size, yellow, golden yellow, or copper color, glossy; stems subrigid to rigid, up to 26 cm in length and 0.5 mm in diameter, reddish and denuded at base with age, irregularly pinnately branching; branches erect to erect-spreading, short to elongate, attenuate; median cauline leaves subimbricate to imbricate, bases up to 1 mm apart, blades firm, erect to erect-spreading, subconcave to concave, ovate-lanceolate, width decreasing gradually from basal fourth or half into the broad apex; apices short, frequently subcarinate or carinate, sometimes concave, subconcave or plane, broadly obtuse, sometimes subtruncate, occasionally subobtuse to subacute, leaf tips subserrulate to serrulate, occasionally entire; blades 5–5.5 mm long, occasionally up to 7 mm in length, 1.5–2 mm wide, occasionally up to 2.5 mm in width, 2.4–3 : 1; median cells of leaves linear, ends attenuate, 6.8–17 μ wide, 10–24 : 1; alar cells enlarged, rectangular, quadrate, or subhexagonal, subhyaline, hyaline, yellow, golden brown, or brown, commonly forming distinct auricles, leaf bases decurrent, approximately 0.5 mm; median branch

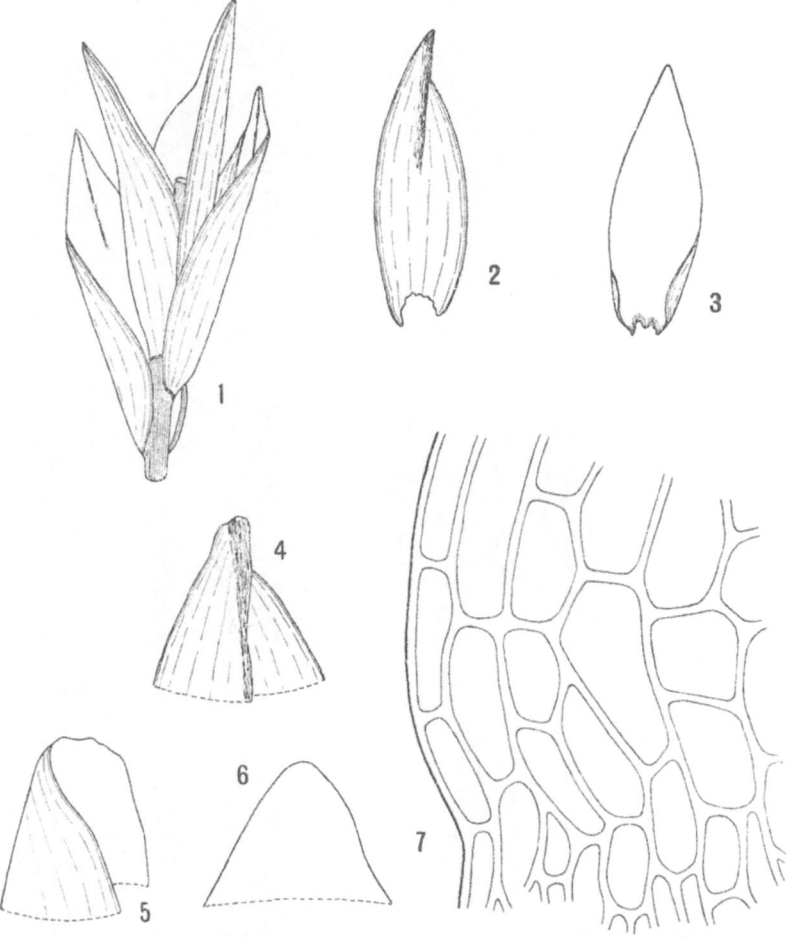

FIG. 21. – *Fontinalis Allenii* Card. – 1.Portion of stem with median cauline leaves. 2.Median cauline leaf, concave below and carinate above. 3.Median cauline leaf, concave below and plane above. 4, 5, and 6.Leaf apices, carinate, concave, and plane, respectively. 7.Group of alar cells. Drawn from J. A. Allen, Apr. 22, 1880, Connecticut.

leaves comparable to median cauline except smaller in size; perichaetium and sporophyte not seen. [The specific name, *Allenii*, honors the collector of the type specimen, J. A. Allen.]

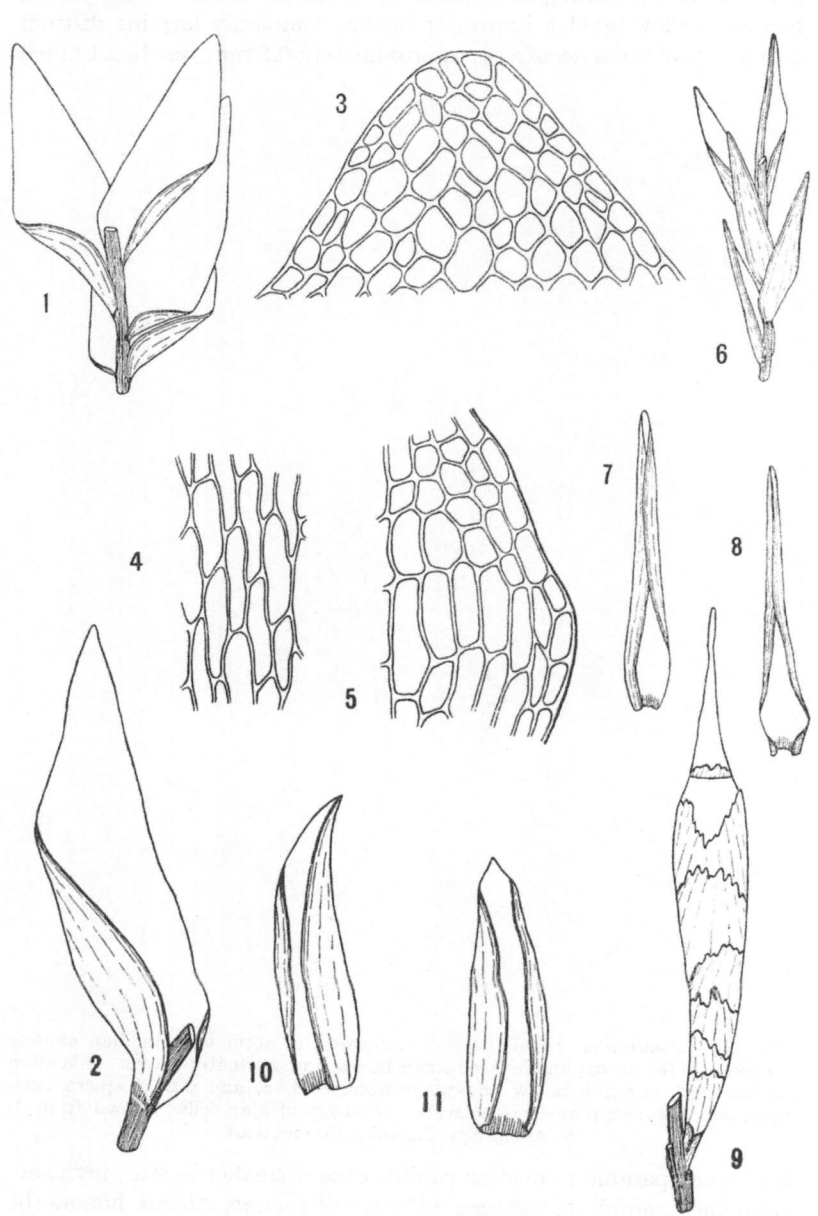

Type: In the Cardot Herbarium, in the herbarium of the Laboratoire de Cryptogamie, Muséum d'Histoire Naturelle, Paris, France.
Type collector: J. A. Allen, Apr. 22, 1880.
Type locality: Connecticut: Hamden, Mount Carmel.
Distributions: Northeastern United States: Connecticut, Massachusetts, and Pennsylvania.
Additional description: Welch, in Grout, Moss Fl. N. Am. 3: 249. 1934.
Illustrations: Welch, in Grout, Moss Fl. N. Am. 3: pls. 76 and 78. 1934. – FIGURE 21.
Specimens examined: NORTH AMERICA: UNITED STATES: CONNECTICUT: J. A. Allen, Apr. 22, 1880, Mount Carmel, Hamden (TYPE, PC; NY, YALE); J. A. Allen, Oct. 22, 1880, Mount Carmel, Hamden, Grout, N. Am. Musci Pl. 395 (Topotype; BART, CAN, CM, COLO, DUKE, H, MINN, MO, NY, UC, US, WASH, WIS); G. E. Nichols, Apr. 12, 1911, Stonington.
MASSACHUSETTS: H. C. Dunham, May 6, 1905, Weston (NY); Seymour, May 8, 1913, Waverly (DPU).
PENNSYLVANIA: Rau, July 1872, Pike Co. (NY).

Fontinalis Allenii is distinguished from other species in the concave group by the broadly obtuse, subcarinate to carinate leaf apices, and from species in the carinate-conduplicate group by leaves subconcave to concave, and subcarinate to carinate in leaf apices only.

23. **Fontinalis biformis** Sull., Musci and Hep. U. S., p. (654) 54. 1856; Sullivant, in Sullivant and Lesquereux, Musci Bor. – Am. Exsic. 226. 1856, [printed Latin description on label].

Fontinalis disticha Hook. and Wils. var. Sullivant, Musci Allegh., p. 46. 1846.
Pilotrichum distichum C. Müll., p.p., Syn. Musc. 2: 150. 1850.
Pilotrichum sphagnifolium C. Müll., Syn. Musc. 2: 150. 1850.
Fontinalis biformis Sull. f. *vernalis* Sull., Icon. Musc., p. 99. 1864.
Fontinalis biformis Sull. f. *aestivalis* Sull., Icon. Musc. p. 99. 1864.

Plants medium in size in spring and approaching slender in summer and autumn, with two distinct kinds of leaves during one growing period, appearance changing from early spring to summer, vernal stage yellowish green, green, or brownish green, slightly glossy to dull, aestival phase brownish green or blackish, dull; stems flaccid to subflaccid when young, subrigid to rigid when older, up to 67.5 cm

---

FIG. 22. (*opposite*). – *Fontinalis biformis* Sull. – 1.Portion of stem with median cauline leaves (vernal). 2.Cauline leaf (vernal). 3.Sphagniform apical cells of leaf (vernal). 4.Median cells of leaf (vernal). 5.Alar cells of leaf (vernal). 6.Portion of stem with median cauline leaves (aestival). 7 and 8.Cauline leaves (aestival). 9.Calyptra, perichaetial branch, andspo rophyte within the perichaetium. 10 and 11.Perichaetial leaves. Nos. 1, 3, 4, and 5 drawn from Sullivant and Lesquereux, Musci Bor.-Am. Exsic. 337; no. 6 from Sullivant and Lesquereux, Musci Bor.-Am. Exsic. 338; nos. 2, 7, 8, and 9 after Sullivant, Icon. Musc., pl. 60; nos. 10 and 11 after Sullivant, Icon. Musc., pl. 59.

in length, 0.35–0.5 mm in diameter, denuded and dark with age, irregularly pinnately branching; branches numerous, erect-spreading, distant to close, often appearing fasciculate, up to 10 cm in length, ends of foliated stems and branches attenuate; bases of vernal median cauline leaves up to 1.5 mm apart, blades flaccid, erect-spreading, concave, subconcave to concave at base and plane above, or plane, usually broadly ovate-lanceolate, sometimes lanceolate, width decreasing rapidly from basal fourth or from middle of blade into the short or long, broadly acuminate apex, occasionally blade abruptly narrowed above basal two-thirds; leaf tips usually narrowly to broadly obtuse, sometimes subobtuse, subacute or acute, commonly subserrulate to serrulate, sometimes entire; median cauline leaves 4–6.5 mm long, rarely up to 7.5 mm in length, 1.25–2.5 mm wide, occasionally up to 3.5 mm in width, 2.5–4 : 1; median cells of leaves linear, sometimes subflexuous, ends attenuate or obtuse, 10.2–17 μ wide, 2–6 : 1, occasionally up to 10 : 1; apical cells rhomboidal, rhombic, subquadrate, subrectangular, or subhexagonal, somewhat sphagniform; alar cells enlarged, oblong, subrectangular, subquadrate, or subhexagonal, walls hyaline, subhyaline, or yellowish brown, auricles distinct, leaf bases occasionally subclasping, not decurrent or briefly so, up to 0.5 mm; median branch leaves comparable to cauline except smaller in size; aestival median cauline leaves very different in appearance from the former, occurring with the vernal and eventually composing the dominant foliage of the plants for a period, occasional young or vernal leaves sometimes present, bases up to 1.5 mm apart, blades firm, erect-spreading, usually concave, canaliculate or subtubular, sometimes concave below and plane above, occasionally plane, margins involute to convolute, blades narrowly ovate-lanceolate; apices long acuminate, leaf tips acute, generally serrulate, sometimes entire; median cauline leaves 2–3.5 mm long, 0.4–0.8 mm wide, 3–5 : 1, occasionally 7.5 : 1; median cells of leaves linear, sometimes subflexuous, ends attenuate to obtuse, 6.5–13.6 μ wide, 5–15 : 1; apical cells rhombic, rhomboidal, subquadrate, subrectangular, or subhexagonal, somewhat sphagniform; alar cells enlarged, subrectangular, subquadrate, or subhexagonal, walls hyaline, subhyaline, or yellowish brown, auricles distinct; leaf bases not decurrent or briefly so, up to 0.5 mm; median branch leaves comparable to cauline except smaller in size; perichaetial branch 4.5 mm long; perichaetium oblong to subcylindrical, 1.5 mm in diameter; upper perichaetial leaves suborbicular or suboval, apices obtuse, truncate and lacerate with age; calyptra long conical; capsule immersed to emergent; operculum long conical, 0.75–1.5 mm long, 0.75

mm in diameter; seta 0.25 mm long; urn slightly oblong in outline or subcylindric, 2–2.5 mm long, 0.75–1 mm in diameter, 2.5–2.6 : 1, contracted beneath mouth when dry; peristome teeth brownish orange, linear-lanceolate, acuminate, 0.67–0.75 mm in length, occasionally united in pairs at apex, submuricate to muricate, lamellae 18–20; trellis brownish orange, imperfect, slightly longer than teeth, cilia muricate, transverse strands connecting cilia in the upper two-thirds into a lattice; spores yellowish green, green, or yellowish brown, usually finely muricate, occasionally smooth, 10–20.4 μ in diameter, ripe in summer. [The specific name, *biformis*, has reference to the two kinds of leaves.]

Type: In Sullivant Herbarium in the Farlow Herbarium.
Type collector: William S. Sullivant; aestival phase in 1851, and vernal phase. in 1852.
Type locality: Ohio: 4 miles west of Columbus, in woodland brooks.
Distribution: United States: Wisconsin, Illinois, Indiana, and Ohio.
Additional descriptions: Sullivant, Icon. Musc., p. 99. 1864; Lesquereux and James, Man. Moss. N. Am., p. 270. 1884; Cardot, Mon. Font., p. 72. 1892; Grout, M. H. M., p. 400. 1903; Jennings, Man. Moss. W. Penn., p. 205. 1913; Welch, in Grout, Moss Fl. N. Am. 3: 241. 1934; Jennings, Man. Moss. W. Penn. and Adj. Reg., p. 175. 1951.
Illustrations: Sullivant, Icon. Musc., pls. 59 and 60. 1864; Brotherus, in Engler and Prantl, Nat. Pflanz.: Bryophyta 2: fig. 545. 1905; Brotherus, in Engler and Prantl, Nat. Pflanz., Musci 11 (2): fig. 475. 1925; Welch, in Grout, Moss Fl. N. Am. 3: pl. 75. 1934. – FIGURE 22.
Specimens examined: NORTH AMERICA: UNITED STATES: ILLINOIS: Vasey (CHI); Wolf, Fulton Co., Canton (CHI, ILL, UT).
INDIANA: L. Allen, Apr. 20, 1907, Putnam Co., near Limedale (DPU); Banker, Mar. 23, 1907, Putnam Co., near Greencastle (NY); Naylor, in 1906, Putnam Co., near Greencastle (DPU); Naylor, Mar. 1906, Putnam Co., near Greencastle (PC); Naylor, Mar. 23, 1907, Putnam Co., near Greencastle (DPU); Welch 7391, Apr. 24, 1942, Putnam Co., Geode Gorge, near Greencastle (DPU); Welch 7392 and 7393, Apr. 25, 1942, Putnam Co., Geode Gorge, near Greencastle (DPU); Welch 7393, Apr. 25, 1942, Putnam Co., Geode Gorge, near Greencastle, Grout, N. Am. Musc. Pl. Suppl. 71 (CM, DPU, MINN, NY, UC); Welch 7394, 7395, 7396, 7397, 7398, 7399, 7400, 7401, 7402, and 7403, Apr. 25, 1942, Putnam Co., Geode Gorge, near Greencastle (DPU); Welch 7515, July 3, 1942, Owen Co., Eel R., Cataract Falls, southwest of Greencastle (DPU, NY); Welch 9133, 9134, 9135, 9136, 9137, 9138, 9139, 9140, and 9141, May 9, 1948, Putnam Co., Geode Gorge, near Greencastle (DPU).
KENTUCKY: H. Bishop, May 1, 1941, Cedar Creek (DPU, UL).
OHIO: Bartley, June 10, 1946, Adams Co., Beaver Pond (DPU); Bartley, June 12, 1946, Adams Co., near Beaver Pond (DPU); Bartley and Pontius, Mar. 19, 1938, Highland Co., near Fort Hill (DPU, OS); E. L. Braun, Apr. 15, 1932, Adams Co., Bundle Run (CIN); Hopkins, Aug. 20, 1909, Portage Co. (CM); Hopkins, Aug. 25, 1909, Portage Co. (US); Hopkins, in 1910, Portage Co. (NY); Hopkins, Aug. 13, 1912, Portage Co., Woodworth's Glen (CM, MO, NY); Lesquereux (H, PC); Schrader, in 1862 (H, L); Schrader, in 1862 and 1863 (H); Sullivant, Columbus (K, NY); Sullivant, Big Miami, near Middletown (US); Sullivant, Columbus, Austin, Musci Appal. 245 (CAN, CAS, CM, DUKE, G-BOIS, K, NY, PHIL, US, WIS); Sullivant, near Columbus, Sullivant and Lesquereux, Musci Bor.-Am. Exsic. 226 (Latin description of *F. biformis* printed on label), (K, NY, TRT); Sullivant,

near Columbus, Sullivant and Lesquereux, Musci Bor.- Am. Exsic. 226b p.p. (DUKE, FH, K, MICH, MO, NY, TRT, WIS); Sullivant, near Columbus, Sullivant and Lesquereux, Musci Bor.-Am. Exsic. 226c (DUKE, FH, K, MICH, MO, NY, PHIL, TRT, WIS); Sullivant, near Columbus, Sullivant and Lesquereux, Musci Bor.-Am. Exsic. 337 p.p. (CHI, CIN, CM, FH, G-BOIS, K, MICH, NY, US, WIS, YALE); Sullivant, near Columbus, Musci Bor.-Am. Exsic. 338 p.p. (CHI, CM, FH, G-BOIS, K, MICH, NY, PHIL, US, WASH, WIS, YALE); Sullivant 47 (G-BOIS, K, PC); Sullivant, in 1842 (PHIL); Sullivant 33, in 1842 (NY); Sullivant 34, in 1842 (NY); Sullivant 62, in 1842 (K, PC); Sullivant 63, in 1842 (G-BOIS, K); Sullivant 25, in 1843, Columbus (G-BOIS); Sullivant 90, Apr. 1849 (NY); Sullivant, in 1851, near Columbus (TYPE of F. biformis ("status aestivalis"), FH); Sullivant, in 1851, near Columbus, Musci Allegh. 192 ("status aestivalis"), (Assumed to be a portion of the type collection, FH; K, MICH, NY, US); Sullivant, in 1852, near Columbus (TYPE of F. biformis ("status vernalis"), FH); Sullivant, in 1852, near Columbus, Musci Allegh. 191 ("status vernalis"), (Assumed to be a portion of the type collection, FH; K, L, MICH, NY, US).

WISCONSIN: Barnes, in 1859, Wamatosa (PC); Cheney 10737, Sept. 16, 1925, Waukesha Co., North Lake (DPU, WIS); Cheney 10737a, Sept. 16, 1925, Waukesha Co., North Lake (WIS); Lapham (K, WIS); Lapham, Milwaukee (WIS); Lapham 57 (FH); Lapham, May 7, 1859, Milwaukee (K, PC, WIS).

United States without locality: Sullivant and Lesquereux, Musci Bor.-Am. Exsic. 340 p.p., from New England to Alabama (K, PC).

Fontinalis biformis is distinguished from other species of the genus by the distinct contrast in leaves, and thus appearance, during different periods of the year, the large, flaccid, concave to plane blades being produced in the earlier months and the concave to tubular ones later. The vernal leaves are distinct because of the following combination of characteristics: blades flaccid, concave to plane, usually broadly ovate-lanceolate, apices generally narrowly to broadly obtuse, leaves 4–6.5 mm long, rarely up to 7.5 mm in length, 1.25–2.5 mm wide, occasionally up to 3.5 mm in width, 2.5–4 : 1, median cells 2–6 : 1, occasionally up to 10 : 1, and apical cells rhombic, rhomboidal, subquadrate, subrectangular, or subhexagonal, somewhat sphagniform. The aestival leaves are characterized by the following combination of characteristics: firm, erect-spreading, majority concave, canaliculate, or subtubular, narrowly ovate-lanceolate, margins involute to convolute, apices long acuminate, leaf tips acute, usually serrulate, blades 2–3.5 mm long, 0.4–0.8 mm wide, majority 3–5 : 1, median cells 5–15 : 1, and apical cells rhombic, rhomboidal, subquadrate, subrectangular, or subhexagonal, somewhat sphagniform. Some aestival leaves are frequently present at base of plants in which the vernal leaves are dominant, and the latter, either complete or as remnants, commonly occur in early summer with the dominant aestival leaves.

24. **Fontinalis Langloisii** Card., in Rev. Bryol. 18: 86. 1891, [in descriptive key]; Cardot, Mon. Font., p. 126. 1892; Cardot, in Rev. Bryol. 18: 84. 1891 (nomen nudum).

Fontinalis involuta Ren. and Card. f. *angustifolia* Card. (?), in Welch, in Grout, Moss Fl. N. Am. 3: 257. 1934.

Plants slender, yellowish, yellowish green, or brownish green, glossy in upper portions, dull near base; stems usually subrigid to rigid, usually up to 12 cm in length, occasionally up to 20 cm long, 0.25–0.4 mm in diameter, denuded and darker below with age, irregularly pinnately branching, frequently bipinnately and often tripinnately divided, occasionally to the fourth division, producing a fruticose appearance; branches numerous, usually erect-spreading

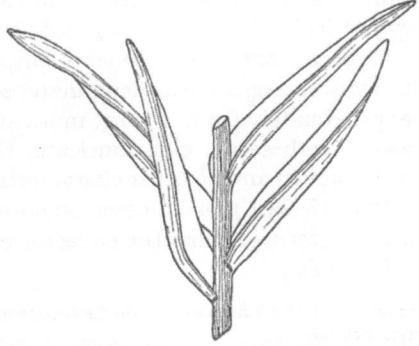

FIG. 23. – *Fontinalis Langloisii* Card. – Portion of stem with median cauline leaves. Drawn from A. B. Langlois, Oct. 1, 1890, Flora Ludoviciana 648, Louisiana.

but sometimes spreading, close to distant, up to 5 cm in length, frequently appearing distichous, ends of foliated stems and branches generally attenuate, sometimes obtuse because of loosely imbricated leaves; bases of median cauline leaves up to 1 mm apart, blades usually firm when moist, sometimes subflaccid to flaccid, erect-spreading, generally plane throughout but sometimes subconcave, or slightly concave to concave at base and plane above or occasionally subconcave at apex, now and then subtubular to tubular, narrowly lanceolate, width gradually decreasing from basal fourth or third into broad apex; apices obtuse or abruptly narrowed into subacute, acute, or obtuse tips, occasionally subtruncate, leaf tips generally serrulate, rarely entire; median cauline blades 3.5–5.5 mm long, 0.9–1.4 mm wide, 3.5–4.5 : 1; median cells of leaves linear or narrowly rhomboidal, ends attenuate or obtuse, 6.8–15.3 μ wide, 9–16 : 1; alar cells moderately enlarged to enlarged, subrectangular, subquadrate, or subhexagonal, walls hyaline, subhyaline, or yellowish brown, auricles none to slight but distinct, leaf bases not decurrent to very briefly so, up to 0.25 mm; median branch leaves frequently plane or nearly so as in cauline but sometimes deeply concave from base into subtubular or tubular apex, the latter usually applicable to basal and apical leaves of the branches, 3–5 mm long, 0.5–0.65 mm in width, 5.36–10 : 1; perichaetial branches 3.25–4.5 mm long; perichaetium cylindrical, 0.5–0.75 mm in diameter; upper perichaetial leaves suboval, apices obtuse, truncate and lacerate with age; calyptra long conical, 1.5 mm long, 0.25 mm in diameter; capsule immersed to slightly emergent; operculum conical, obtuse, approximately 0.5 mm long and 0.5 mm in diameter; seta 0.15–0.3 mm long; urn narrowly cylindrical to cylindrical, 1.5–2.4 mm long, 0.5–0.65 mm in diameter, 2.5–4.5 : 1, slightly contracted beneath mouth when dry; peristome teeth brownish orange, linear-acuminate, sometimes united in pairs at apex, approximately 0.5 mm long, muricate, lamellae more than 14 (only broken teeth seen); cilia muricate, the one available inner peristome badly broken and no other characteristics discernible); spores yellowish brown, 17–22.1 μ in diameter, smooth, ripe in summer. [The specific name, *Langloisii*, honors the collector of the type specimens, Rev. A. B. Langlois.]

Type: In the herbarium of the Laboratoire de Cryptogamie, Muséum d'Histoire Naturelle, Paris, France.

Type collector: A. B. Langlois, Oct. 1, 1890.

Type locality: Louisiana: Ravine-aux-Cannes, near Mandeville, "sur un bois pourri submergé dans une mare."

Distribution: United States: Louisiana and Mississippi.

Additional descriptions: Cardot, Mon. Font., p. 126. 1892; Barnes, Gen. and Sp. N. Am. Moss., p. 331. 1896; Welch, in Grout, Moss Fl. N. Am. 3: 257. 1934.

Illustrations: Welch, in Grout, Moss Fl. N. Am. 3: pl. 79. 1934. – FIGURE 23.

Specimens examined: NORTH AMERICA: UNITED STATES: LOUISIANA: Celestin, 816 of Langlois, Dec. 1892, St. Tammany Co. (NY, PC, US); Langlois, Ravine-aux-Cannes, near Mandeville (FH); Langlois 648, Oct. 1, 1890, Ravine-aux-Cannes, near Mandeville (Assumed to be the TYPE of *F. Langloisii*, PC; NY, US, YALE); Langlois 745b p.p., Oct. 6, 1891, Abita (Originally *F. involuta* f. *angustifolia*, PC, US); Langlois, in 1892, Ravine-aux-Cyprès, near Bonfouca (H, US); Langlois, in 1892, Ravine-aux-Cannes (K, PC); Langlois, in 1892, St. Tammany Co. (K); Langlois, in 1892, Renauld and Cardot, Musci Am. Sept. Exsic. 229 (CAN, FH, H, MICH, NY, PC, YALE); Langlois 754, in 1892, Renauld and Cardot, Musci Am. Sept. Exsic. 231 (CAN, FH, H, MICH, NY, PC, YALE); Langlois, Sept. 9, 1892, Ravine-aux-Cannes (NY, US); Langlois, Sept. 9, 1892, Grout, N. Am. Musci Pl. 387 (BART, CAN, CM, H, MICH, MO, NY, OS, UC, US, WASH, WIS, YALE); Langlois 751, Sept. 9, 1892 (PC); Langlois, Sept. 14, 1892, Ravine-aux-Cyprès (NY, US); Langlois, Sept. 27, 1892, Tchiffonté (US); Langlois, Dec. 1892, Bayou Bonfouca (ABS); Langlois 518, Dec. 1892 (US); Langlois 814, Dec. 1892, near Bayou Bonfouca (NY, US); Langlois 814, Dec. 1892, near Bonfouca, Renauld and Cardot, Musci Am. Sept. Exsic. 229 (H, MICH, PC); Langlois 815, Dec. 1892, near Bonfouca (PC); Langlois 816², Dec. 1892, near Bonfouca (NY, US); Langlois 818, Dec. 1892 (PC).

MISSISSIPPI: Penfound 109, Oct. 4, 1936, Pearlington (DPU).

Fontinalis Langloisii may be distinguished from other species in the genus by the following combination of characteristics: plants slender, commonly rather short, up to 12 cm in length, appearing fruticose because of the frequent bipinnately and sometimes tripinnately divided axes, stems usually rigid or subrigid, median cauline leaves narrowly lanceolate, width gradually decreasing from basal fourth or third into broad, obtuse, or abruptly narrowed, subacute to acute, usually serrulate apex, blades 3.5–5.5 mm long, 0.9–1.4 mm wide, 3.5–4.5 : 1, and median branch leaves narrower than median cauline and varying from plane throughout to deeply concave with subtubular or tubular apices.

The assignment of *F. Langloisii* to a group is rather difficult because of the range in leaf form from plane to tubular. Cardot placed this species in the *Solenophyllae* on the basis of the numerous subtubular to tubular and occasionally convolute branch leaves, it is assumed. In 1934, the author did likewise. Since in this study the emphasis is placed upon the median cauline leaves, *Langloisii* has been placed in the group with majority of leaves concave because the blades are sometimes subconcave to concave or subtubular to tubular even though usually plane, and generally firm when moist although sometimes subflaccid to flaccid. The plane, flaccid leaves suggest the group with majority of leaves plane, but due to the numerous concave to tubular blades on the branches, this placement does not appear to be correct, because in this group the branch leaves are comparable, with exception of smaller size, to the median cauline blades.

The author regrets the inclusion in the 1934 study of the name and description of *F. involuta* f. *angustifolia* from a packet in Cardot's herbarium as continued search for the publication of this form has not been successful. Langlois 745 and 745a are *F. novae-angliae* var. *cymbifolia* and 745b is in part *F. novae-angliae* var. *cymbifolia* and in part *F. Langloisii*. Upon these data, it seems possible that the term, *angustifolia*, was applied to the plants of *F. Langloisii*, assuming that the leaves were narrow blades of *F. novae-angliae* var. *cymbifolia*. Thus, the form is included in the synonymy of *F. Langloisii* although 745, 745a, and a portion of 745b are *F. novae-angliae* var. *cymbifolia*. The author considers *F. involuta* Ren. and Card. to be a synonym of the var. *cymbifolia*.

25. **Fontinalis disticha** Hook. and Wils., in Drummond, Musci Am. (Southern States) No. 151. 1841, [printed Latin description on label].

Pilotrichum distichum C. Müll., p.p., Syn. Musc. 2: 150. 1850.
Fontinalis Lescurii Sull. var. *gracilescens* Sull., Icon. Musc., p. 101. 1864.

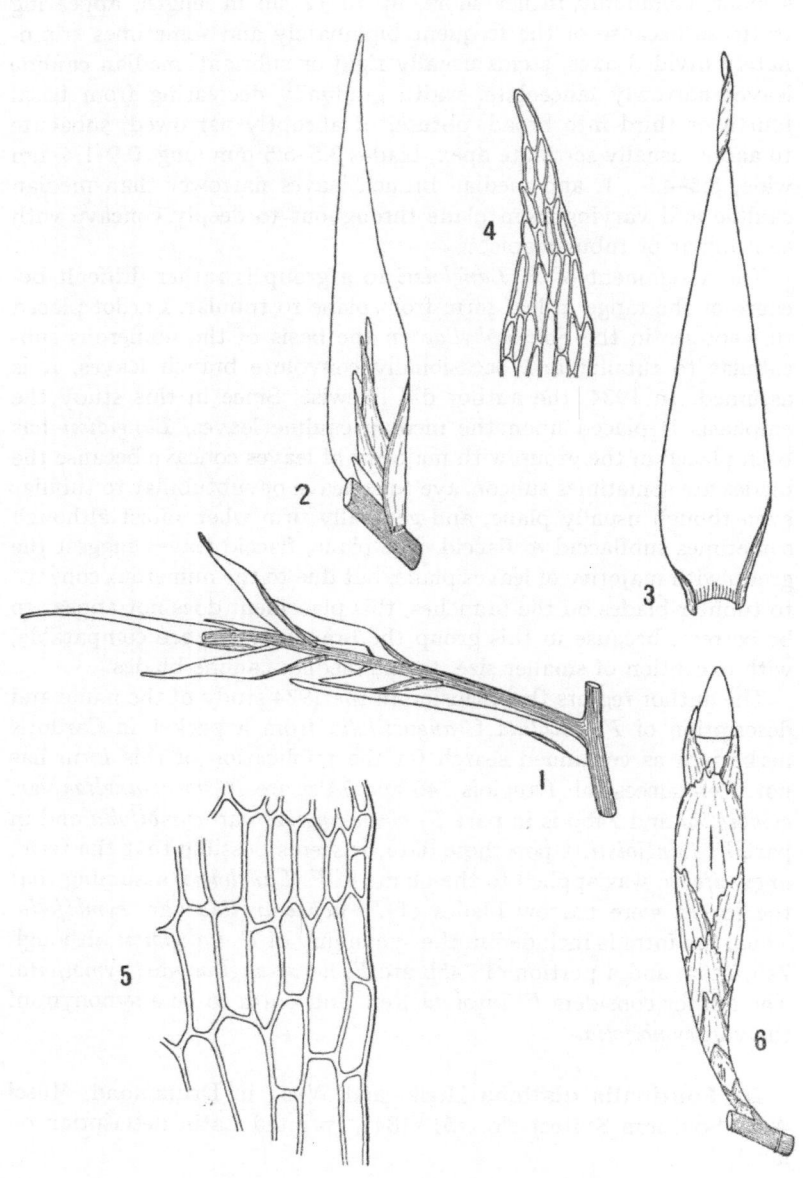

Fontinalis Sullivanti Lindb. in Öfvers. Finska Vet.-Soc. Förh. 12: 77. 1869.
Fontinalis Renauldi Card., in Rev. Bryol. 18: 85. 1891, [in descriptive key].
Fontinalis microdonta Ren. in litt., according to Cardot, Mon. Font., p. 120.
1892.
Fontinalis Sullivanti Lindb. f. *viridis* Card., in Welch, in Grout, Moss Fl. N.
Am. 3: 243. 1934.
Fontinalis Sullivanti Lindb. var. *microdonta* (Ren.) Welch, in Grout, M. Fl. N.
Am. 3: 243. 1934.

Plants slender in size, yellowish green, green, or brownish green, younger parts glossy, older dull; stems subrigid to rigid, up to 25 cm in length, 0.25–0.4 mm in diameter, denuded and darker at base with age, usually irregularly pinnately branched, sometimes bipinnately divided; branches generally rigid or nearly so, few to numerous, commonly spreading, frequently erect-spreading, occasionally recurved, usually close, occasionally distant, appearing to be distichous,[1] short, up to 5 cm long, ends of foliated stems and branches attenuate; median cauline leaves distant, bases up to 2 mm apart, blades generally firm, sometimes subflaccid, occasionally flaccid, commonly erect to erect-spreading, spreading when bearing spreading branches in axils, concave throughout, or subconcave at base and plane above, occasionally canaliculate to subtubular, narrowly lanceolate, width gradually decreasing from basal fourth or half into long and narrowly acuminate apices; acumen sometimes slightly twisted, leaf tips commonly acute or subacute, at times subobtuse or narrowly obtuse, usually serrulate or subserrulate, occasionally entire; median cauline leaves usually 4–6.5 mm long, sometimes up to 9 mm in length, 0.75–1.5 mm wide, 3.7–8 : 1; median cells of leaves linear, ends attenuate, 6.5–11.9 μ wide, 8–16 : 1; alar cells enlarged, subrectangular, subquadrate, or subhexagonal, walls hyaline, subhyaline, yellowish, or yellowish brown, auricles usually distinct, leaf bases commonly not decurrent, occasionally briefly so, sometimes up

---

[1] Branches, although tristichous, appear to be distichous in herbarium specimens because of the twisting of the stem to the right and then to the left, or vice versa, and occasionally in a long, complete twist. However, when the plants remain in water a short time, the tristichous arrangement becomes evident.

FIG. 24. (*opposite*). – *Fontinalis disticha* Hook. and Wills. – 1.Portion of stem with median cauline leaf, bearing in its axil a portion of a well developed branch. 2.Part of stem with median cauline leaf, bearing in its axil a young branch. 3.Median cauline leaf. 4.Apex of median cauline leaf. 5.Group of alar cells of median cauline leaf. 6.Perichaetial branch and sporophyte within the perichaetium. No. 1 drawn from Sullivant, in Sullivant and Lesquereux, Musci Bor.-Am. Exsic. 339, Alabama; nos. 2, 3, and 6 after Sullivant, Icon. Musc., pl. 63; nos. 4 and 5 from Sullivant, in Sullivant and Lesquereux, Musci Bor.-Am. Exsic. 227, Alabama.

to 0.5 mm; median branch leaves usually similar to median cauline except smaller in size, youngest ones frequently canaliculate to subtubular; perichaetial branches 3.5–4.5 mm long; perichaetium subcylindrical to cylindrical, 0.5–0.75 mm in diameter; upper perichaetial leaves suboval, apices obtuse, truncate and lacerate with age; calyptra long conical, acuminate; capsule immersed or emergent; operculum long conical, usually emergent, occasionally immersed, 0.8–1 mm long, 0.5 mm in diameter; seta 0.25 mm long; urn immersed or emergent, narrowly cylindrical, 1.7–2.5 mm long, 0.5–0.75 mm in diameter, 3.3–4 : 1, usually contracted beneath mouth when dry, occasionally not; peristome teeth generally brownish orange, sometimes yellowish brown, linear-acuminate, occasionally united in pairs at apex, 0.6–0.65 mm long, submuricate to muricate, lamellae 12–26; inner peristome generally brownish orange, sometimes yellowish brown, imperfect, cilia about length of teeth or slightly longer, up to 0.75 mm in length, submuricate to muricate, united in approximately the upper half or two-thirds by transverse strands into imperfect trellis, cilia occasionally subappendiculate below, lower cross bars broken or incomplete, smooth to submuricate; spores yellowish green, green, or yellowish brown, usually muricate, sometimes smooth to submuricate, 11.9–19.8 μ in diameter, ripe in summer. [The specific name, *disticha*, has reference to the two-ranked or distichous appearance of the arrangement of the branches in herbarium specimens.]

Type: In British Museum of Natural History, London, England, indicated as "Drummond, New Orleans, No. 23," in Herb. Musci W. Wilson, and distributed as Thomas Drummond, Musci Americani (Southern States) No. 151. 1841.
Type collector: Thomas Drummond.
Type locality: Louisiana, New Orleans.
Distribution: North America: in, approximately, the eastern half of the United States.
Additional descriptions: Sullivant, Musci and Hep. U. S., p. (654) 54. 1856; Sullivant, Icon. Musc., p. 103. 1864; Lesquereux and James, Man. Moss. N. Am., pp. 271 (as *F. Lescurii* var. *gracilescens*), 272 (as *F. disticha*). 1884; Cardot, Mon. Font., pp. 75 (as *F. disticha*), 76 (as *F. Sullivanti*). 1892; Barnes, Gen. and Sp. N. Am. Moss., p. 330 (as *F. microdonta*). 1896; Grout, M. H. M., p. 400 (as *F. Sullivanti*). 1903; Jennings, Man. Moss. W. Penn., p. 205 (as *F. Sullivanti*). 1913; Welch, in Grout, Moss Fl. N. Am. 3: 242 (as *F. disticha* and as *F. Sullivanti*). 1934; Jennings, Man. Moss. W. Penn. and Adj. Reg., p. 175 (as *F. Sullivanti*). 1951.
Illustrations: Sullivant, Icon. Musc., pl. 63. 1864; Welch, in Grout, Moss Fl. N. Am. 3: pl. 74 (as *F. Sullivanti* and var. *microdonta*), pl. 75 (as *F. disticha*). 1934. – FIGURE 24.
Specimens examined: NORTH AMERICA: UNITED STATES: ALABAMA: B. Andrews, Nov. 1931, Utah Road swamp, near Tuscaloosa (DUKE); Sullivant, Mobile (US); Sullivant, near Mobile, Musci Allegh. 190 (FH, K, MICH, NY, US); Sullivant, near Mobile, Sullivant and Lesquereux, Musci Bor.-Am. Exsic. 227 (DUKE, FH, G-BOIS, K, MICH, MO, NY, PC, PHIL, TRT, WIS); Sullivant, near Mobile, Sulli-

vant and Lesquereux, Musci Bor.-Am. Exsic. 339 p.p.[1] (BM, CM, MICH, NY, PC, PHIL, WIS, YALE); Sullivant, in 1845, Mobile (NY).

ARKANSAS: Bush, Aug. 31, 1897, Varner (MO); Bush, in 1898, Varner (H); Bush, Apr. 25, 1898, Varner (NY, PC); Demaree 26541, Oct. 2, 1947, Craighead Co., Jonesboro (DPU).

DELAWARE: James 91, Wilmington (BM, ILL); James, in 1851, Wilmington (K).

FLORIDA: Ford and Redfearn, Feb. 8, 1950, Alachua Co., Hatchet Creek, Bry. Fla. 179 (DPU); H. Kurz, 2513 of Schornherst, Jan. 14, 1950, Gadsden Co. (DPU, FLSU); J. K. Small and E. T. Wherry 11660, Apr. 5, 1925, Leon Co., swamp near Ocklockonee R., near Tallahassee (NY).

GEORGIA: Schornherst 872, Feb. 25, 1939, Grady Co., Ocklockonee R., near Cairo (DPU, MICH, NY).

INDIANA: Welch 9191, May 1931, Knox Co., Little Cypress Swamp (DPU).

LOUISIANA: Drummond, New Orleans (K); Drummond 23, New Orleans (TYPE of F. disticha, BM, K), Musci Am. 151 p.p. (FH, K, NY, PC, WASH); Pennebaker, June 7, 1939, East Feliciana Parish, near Jackson (DPU).

MASSACHUSETTS: C. E. Faxon, Nov. 1886, West Roxbury (DPU, WELC).

MICHIGAN: Ehlers, Aug. 7, 1937, Cheboygan Co. (DPU, YALE).

NEW JERSEY: C. F. Austin, Closter, Musci Appal. 249 (CAN, CAS, CIN, DUKE, FH, G-BOIS, H, K, NY, OS, PC, PHIL, US, WIS, YALE); C. F. Austin, "New Jersey and Pennsylvania," Musci Appal., Suppl. 1, 524 (CAN, CHI, FH, NY, PHIL, US, YALE); C. F. Austin, Closter, Sullivant and Lesquereux, Musci Bor.-Am. Exsic. 341 (TYPE of F. Sullivanti Lindb., FH, bearing the note, "On these specimens Lindberg made his Sullivanti"; CHI, CM, G-BOIS, MICH, NY, PC, PHIL, WIS, YALE); C. F. Austin, Sept. 1862, Bergen Co. (NY); C. F. Austin, in 1878, Closter (PC, WIS); C. F. Austin, Sept. 1878, Closter (YALE).

NEW YORK: C. F. Austin, Palisades (NYSM); T. Green, in 1865, Long Isl., Montauk Point (CHI, PHIL); Peck, Sandlake (NYSM).

NORTH CAROLINA: L. E. Anderson 5722, May 8, 1937, Gates Co., near Corapeake (DPU, DUKE); Wetherby, Aug. 1896, Roan Mt. (DPU, DUKE).

OHIO: Emmitt 1060a, Aug. 21, 1949, Wood Co., Portage Twp. (BGSU, DPU).

TENNESSEE: A. J. Sharp 5013, June 17, 1950, Grundy Co., near Pelham (DPU, TENN).

United States without locality: Lesquereux, in 1883 (TYPE of F. microdonta, PC; FH).

Fontinalis disticha may be recognized by the following combination of characteristics: plants slender in size, stems and branches subrigid to rigid, branches generally spreading, frequently erect-spreading, usually close, appearing to be distichous in herbarium specimens, short, up to 5 cm long, median cauline leaves distant, bases up to 2 mm apart, blades commonly firm, sometimes subflaccid to flaccid, usually erect-spreading, spreading if bearing branches in axils, concave throughout or subconcave at base and plane above, narrowly lanceolate with long and narrowly acuminate apices, usually 4–6.5 mm long but sometimes up to 9 mm in length, 0.75–1.5 mm wide, 3.7–8 : 1.

In the Sullivant Herbarium in Farlow Herbarium, with *F. Lescurii* var. *gracilescens*, Sullivant and Lesquereux, Musci Bor.–Am. Exsic.

---

[1] The two localities cited on labels of No. 339 are Alabama, near Mobile, and southern Kentucky. According to Cardot, Mon. Font., p. 15, the plants of *F. disticha* were collected in Alabama and those of *F. filiformis* in Kentucky.

341, is Sullivant's notation, "On these specimens Lindberg made his *F. Sullivanti.*" In his Icones Muscorum, Sullivant states that *F. Lescurii* var. *gracilescens* resembles *F. disticha* but has a much shorter capsule. The following exsiccati are cited in the Icones Muscorum with *F. disticha* Hook. and Wils.: Drummond, Musci Am. 151, Sullivant, Musci Allegh. 190, and Sullivant and Lesquereux, Musci Bor.–Am. Exsic. 227. The author is in accord with Sullivant regarding the above citations. However, it seems to the writer that the plants of Sullivant and Lesquereux, Musci Bor.–Am. Exsic. 341 are similar to those in the other exsiccati cited above. Thus, *F. Sullivantii* and *F. Lescurii* var. *gracilescens* are reduced to the synonymy of *F. disticha* Hook. and Wills.

In the Fontinalaceae of North America, the author used for the basic distinction between *F. disticha* and *F. Sullivantii* the difference between length ratio of cauline and branch leaves. In the present study, the contrast in leaf length is not considered to be of taxonomic significance because the size and frequently the shape of the branch leaves have been found to differ on the same plant, especially with age, the median blades on the older branches bearing a strong resemblance to the corresponding cauline leaves. It now seems that the characteristics of the median cauline leaves are the more reliable bases for classification.

The Marie-Victorin specimens from St. Basile, Chambly Co., Quebec, reported by Kucyniak (1944) as *F. disticha*, have been determined more recently as *F. novae-angliae*. The former determination was changed after additional material was available for study.

26. **Fontinalis filiformis** Sull. and Lesq., in Austin, Musci Appal., p. 42. 1870; Austin, Musci Appal. 250. 1870, [printed Latin description on label].

Fontinalis disticha Hook. and Wils. var. *tenuior* Sull., Icon. Musc., p. 103. 1864.
Fontinalis filiformis Sull. and Lesq. var. *tenuifolia* Card., Mon. Font., p. 126. 1892.

Plants very slender in size, filiform in appearance, yellowish, yellowish brown, yellowish green, green, brownish, reddish brown, or golden brown, younger portions glossy, older parts dull; stems usually rigid, occasionally subrigid, up to 20 cm long, 0.25–0.35 mm in diameter, denuded and darker at base with age, irregularly pinnately branching; branches few to numerous, commonly erect-spreading, sometimes spreading, close to distant, appearing to be distichous, up to 7 cm in length, ends of foliated stems and branches attenuate; median cauline leaves distant, bases up to 2 mm apart, blades firm, commonly erect-

spreading, sometimes erect, occasionally spreading, concave or deeply
concave to convolute-tubulose, very narrowly lanceolate; apices long
acuminate, subulate or nearly so, tips entire, subserrulate, or serrulate; median cauline blades 3–6 mm long, 0.35–0.5 mm wide,
occasionally up to 0.75 mm in width, 7–10 : 1; median cells of leaves
linear or narrowly rhomboidal, ends attenuate or obtuse, 6.5–12 μ
wide, 6–15 : 1; alar cells slightly enlarged, subrectangular, sub-

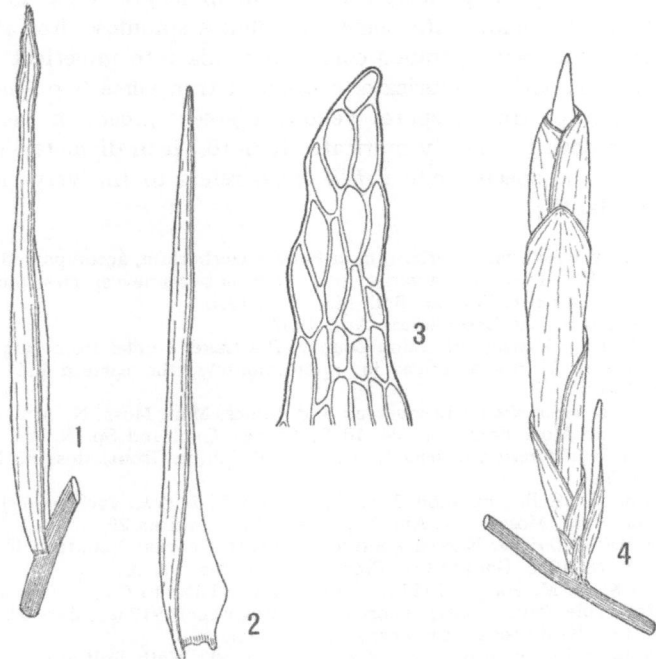

FIG. 25. – *Fontinalis filiformis* Sull. and Lesq. – 1.Portion of
stem with median cauline leaf.   2.Median cauline leaf.   3.Leaf
apex.   4.Portion of stem with perichaetial branch and sporophyte within the perichaetium. Nos. 1, 2, and 4 drawn after
Sullivant, Icon. Musc., pl. 64; no. 3 from Lesquereux, in southern
Kentucky.

quadrate, or subhexagonal, walls hyaline, subhyaline, yellowish, or
yellowish brown, commonly forming distinct auricles, leaf bases not
decurrent; median branch leaves similar to median cauline except
smaller in size; perichaetial branches 4–4.5 mm long; perichaetium
narrowly cylindrical, 0.5 mm in diameter; upper perichaetial leaves
suboval to almost oblong, apices usually obtuse, occasionally subapiculate, lacerate with age; calyptra long conical, acuminate, 1–1.5

mm long, 0.25–0.3 mm in diameter; capsule immersed or emergent; operculum long conical, immersed or emergent, 0.5–1 mm long, 0.3–0.5 mm in diameter; seta 0.2–0.25 mm long; urn immersed, narrowly cylindric, 1.5–2.5 mm long, 0.35–0.5 mm in diameter, 4–5 : 1, slightly contracted beneath mouth when dry; peristome teeth brownish orange, linear-acuminate or subulate, occasionally united in pairs at apex, 0.5–0.6 mm in length, muricate, lamellae 12–20; inner peristome brownish orange, imperfect, 0.4–0.6 mm in length, cilia muricate, occasionally appendiculate, sometimes almost spinulose, free at base, near apex transverse strands connecting cilia into imperfect trellis, cross bars smooth to muricate, occasional transverse bars join cilia near center of trellis; spores yellowish green, green, or yellowish brown, smooth to slightly muricate, 13.6–18.7 μ in diameter, ripe in summer. [The specific name, *filiformis*, refers to the very slender, filiform plants.]

Type: In the Sullivant Herbarium in Farlow Herbarium, accompanied by the following note: "This is not a variety of *F. disticha* but a new species which may be named *Fontinalis filiformis* Sull. MS., Feb. 1870."

Type collector: Leo Lesquereux, Apr. 1857.

Type locality: Kentucky: Union County, Big Lake, 4 miles from Caryride.

Distribution: North America: in, approximately, the eastern half of the United States.

Additional descriptions: Lesquereux and James, Man. Moss. N. Am., p. 271. 1884; Cardot, Mon. Font., p. 124. 1892; Barnes, Gen. and Sp. N. Am. Moss., p. 331 (as *F. filiformis* var. *tenuifolia*). 1896; Welch, in Grout, Moss Fl. N. Am. 3: 256. 1934.

Illustrations: Sullivant, Icon. Musc., pl. 64 (as *F. disticha* var. *tenuior*). 1864; Welch, in Grout, Moss Fl. N. Am. 3: pl. 79. 1934. – FIGURE 25.

Specimens examined: NORTH AMERICA: UNITED STATES: FLORIDA: Redfearn 2023, Apr. 18, 1956, Gadsen Co., Pronto Springs (DPU, FLSU).

GEORGIA: R. M. Harper 2151a, Apr. 26, 1904, Tattnall Co., in Ohoopee R., near Reidsville (NY, YALE); Thorne and Muenscher 9173a, Mar. 25, 1949, Calhoun Co., Keel Creek, near Leary (DPU, IOWA).

ILLINOIS: Schneck, Aug. 1881, Wabash Co. (NY); Wolf, Fulton Co., Canton (CHI, FH, ILL, NY, PC, US, UT).

KENTUCKY: Lesquereux (BM, BR, H, NY, PC); Lesquereux, Austin, Musci Appal. 250 (CAS, CIN, DUKE, K, NY, PHIL, US, WIS, YALE); Lesquereux, Sullivant and Lesquereux, Musci Bor.-Am. Exsic. 339 p.p.[1] (CHI, CIN, CM, FH, G-BOIS, K, MICH, NY, PC, PHIL, WIS, YALE); Lesquereux, in 1857 (K); Lesquereux, in 1857, Union Co., Big Lake, 4 miles from Caryride (TYPE, FH); Shacklette 1760, Sept. 3, 1940, Union Co., Rockford Slough (DPU); Shacklette 2043, Aug. 26, 1941, Union Co., Chalybeate Springs (DPU, MICH).

LOUISIANA: Langlois, in 1892, near St. Martinsville (US); Langlois, in 1892, near St. Martinsville, Renauld and Cardot, Musci Am. Sept. Exsic. 230 (CAN, FH, NY, PC, YALE); Langlois 746, Jan. 23, 1892, at base of tree in a lagoon, Abbé's Forest, near St. Martinsville (Assumed to be TYPE of *F. filiformis* var. *tenuifolia*, PC; NY, US, WIS); Langlois 746, Jan. 28, 1892, near St. Martinsville (PC, US); Langlois, Aug. 6, 1892, Abbé's Forest, near St. Martinsville (US); Langlois 750, Aug. 6, 1892, near St. Martinsville (PC); Langlois 750, Aug. 6, 1892,

---

[1] See page 165, note 1.

near St. Martinsville, Renauld and Cardot, Musci Am. Sept. Exsic. 230 (BM, PC);
Langlois, Mar. 14, 1893, Abbé's Forest, near St. Martinsville (NY, US); Langlois,
Mar. 15, 1898, Abbé's Forest, near St. Martinsville (US); Pennebaker, July 27,
1940, Livingston Parish, Amite R. (DPU).

MAINE: Harvey, Orono (US).

MARYLAND: Leonard and Killip 845, July 17, 1921, Charles Co., Tompkins-
ville (FH, US).

MICHIGAN: Schnooberger 1960, Oct. 29, 1938, Gratiot Co. line and East
Superior (DPU).

MISSOURI: Demetrio, July 1886, Perry Co., "Bois-brûlé Bottom" (DUKE, FH,
H, ILL); Demetrio, Aug. 1886, Perry Co., Mississippi Valley (PC).

NEW JERSEY: Delisle (K).

SOUTH CAROLINA: Kennedy, Mar. 25, 1899, Summerville, Dorchester Road,
at Six Bridges (FH).

TEXAS: Thurow, in 1892, Harris Co., Hockley (PC); Thurow 27, Mar. 25, 1893,
Waller Co. (WIS).

VIRGINIA: D. S. Correll 11562, Oct. 14, 1941, Greensville Co., Fontaine Creek
(DPU, DUKE); Oldberg, Aug. 10, 1877, Loudoun Co., Snikkir's Gap (US).

As far as now known, *Fontinalis filiformis* is the most slender
species in the genus. In addition to its filiform appearance, it is
further characterized by stems subrigid to rigid, median cauline
leaves distant, bases up to 2 mm apart, blades firm when moist,
deeply concave to convolute-tubulose, very narrowly lanceolate,
apices long acuminate, subulate or nearly so, majority of blades 3–6
mm long, 0.35–0.5 mm wide, 7–10 : 1, and median branch leaves
comparable to median cauline except smaller in size.

GROUP 3: LEAVES PLANE

Leaves generally plane, but intermixed with these blades may be those which are subconcave to concave at base.

27. Fontinalis hypnoides C. J. Hartm., Handb. Skand. Fl., p. 434. 1843.[1]

Fontinalis squamosa Hedw. var. *tenella* Br. and Schimp., Bry. Eur., Fasc. 31, 5: 7. 1846. [Type not seen.]
Pilotrichum Strömbäckii C. Müll., Syn. Musc. 2: 150. 1850.
Fontinalis Ravani Hy, in Mém. Soc. Agr., Sci. et Arts Angers 24: 136. 1882.
Fontinalis hypnoides var. *Ravani* (Hy) Card., in Rev. Bryol. 18: 83. 1891.
Fontinalis hypnoides subsp. *tenella* Card., in Rev. Bryol. 18: 83. 1891 (nomen nudum).
Fontinalis tenella (Card.) Card., in Rev. Bryol. 18: 85. 1891, [in descriptive key]; Cardot, Mon. Font., p. 105. 1892.
Fontinalis Lescurii Sull. var. E. G. Britton, in Cardot, Mon. Font., p. 105. 1892.
Fontinalis hypnoides var. *pungens* Von Klinggr., Leber- u. Laubm. West- u. Ostpr., p. 228. 1893.
Fontinalis hypnoides var. *angustifolia* Warnst., in Roth, Eur. Laubm. 2: 286. 1904; in Jaap, in Verh. Bot. Ver. Brand. 43: 66. 1901 (nomen nudum).
Fontinalis hypnoides var. *Adlerzii* Card., in Adlerz, Bladmossfl. Sver., p. 26. 1907.
Fontinalis hypnoides f. *pungens* (Von Klinggr.) Mönkem., in Pascher, Süsswasserfl. 14: 106. 1914.

Plants slender in size, delicate, pale green, yellow green, green, or brown, younger portions glossy, older sometimes dull; stems usually filiform, flaccid, up to 30 cm in length and 0.25 mm in diameter, frequently denuded and darker below with age, irregularly pinnately branching; branches few to numerous, erect, erect-spreading, or spreading, close to distant, short to elongate, up to 8 cm in length, ends of foliated stems and branches attenuate; median cauline leaves usually distant, bases up to 2 mm apart, blades flaccid, erect-spreading

---

[1] With regard to whether C(arl) J(ohan) Hartman or R(obert Wilhelm) Hartman described *Fontinalis hypnoides*, the author was referred by Dr. Herman Persson to Dr. Olle Mårtensson, Uppsala, Sweden. Dr. Mårtensson replied that C. J. Hartman (1790–1849) described several bryophytes. He was the author of the first five editions of Handbok i Skandinaviens. Since *F. hypnoides* was described in the fourth edition (1843), the writer has concluded that the correct citation is *F. hypnoides* C. J. Hartm. R. W. Hartman lived from 1827–1891. (Also, see Krok, Bib. Bot., 1925.)

to spreading, commonly plane, occasionally subconcave at base, generally narrowly ovate-lanceolate or narrowly lanceolate, occasionally oblong-lanceolate, width gradually decreasing from basal fourth or half into the acumen; majority of apices long and narrow but occasionally on same axis short and broad, leaf tips commonly acute but now and then subacute, usually entire, sometimes serrulate; median cauline blades 3–5.5 mm long, occasionally 6 mm in length, 0.75–1.5 mm wide, occasionally up to 2.5 mm in width, generally 4–5.5 : 1, sometimes 2.5–3 : 1; median cells of leaves linear with ends attenuate, or narrowly rhomboidal, 8.5–15.5 μ wide, 5–16 : 1; alar cells usually somewhat enlarged, subquadrate, subrectangular, or subhexagonal, walls hyaline, subhyaline, yellowish, or brownish, alar area occasionally indistinct, generally subrectangular to rectangular in outline, parallel with leaf margin, frequently extending upward along the margin farther than elsewhere in the blade, auricles usually none, occasionally very slight, leaf bases commonly decurrent, up to 0.5 mm; median branch leaves similar to median cauline except smaller in size; perichaetial branch 3.5–4.75 mm in length; perichaetium oval to oblong, 0.75–1.5 mm in diameter; upper perichaetial leaves oval, broadly oval, or suborbicular, apices obtuse or subobtuse, sometimes with a short obtuse tip, truncate and lacerate with age; calyptra long conical, 1–1.5 mm long, 0.5 mm in diameter; operculum short conical, 0.5–1 mm long, 0.5 mm in diameter; seta 0.25–0.5 mm in length; urn immersed or emergent on same plant, suboval to oval, 1.75–2.5 mm long, 1–1.25 mm in diameter, 1.4–2 : 1, generally slightly contracted beneath mouth when dry; peristome teeth brownish orange, linear-acuminate, often united in pairs at apex, 0.75–0.8 mm long, muricate, sometimes densely so, lamellae 15–32; trellis brownish orange, perfect, approximately the length of the teeth, submuricate to densely muricate, subspinulose to spinulose, transverse bars complete, appendiculate, subspinulose to spinulose; spores green, yellowish green, or yellowish brown, submuricate to muricate, 12–20.4 μ in diameter, ripe in summer. [The specific name, *hypnoides*, indicates that Hartman saw some semblance between these plants and Hypnum.]

Type: In the Herbarium of the Botanical Museum of the Institute of Systematic Botany, University of Uppsala, Uppsala, Sweden.

Type collectors: C. and E. A. Strömbäck and C. Hartman, Jr., Aug. and Sept. 1842.

Type locality: Sweden: Gästrikland, Hille Parish, in the swampy meadow, Flyet, by the factory at Oslättfors.

Distribution: Canada, northern United States, Europe, and northern Asia.

Additional descriptions: Bruch and Schimper, Bry. Eur., Fasc. 31, 5: 8. 1846; Schimper, Nya Moss., p. 167. 1848; Schimper, Syn. Musc., pp. 457 (as *F. squa-*

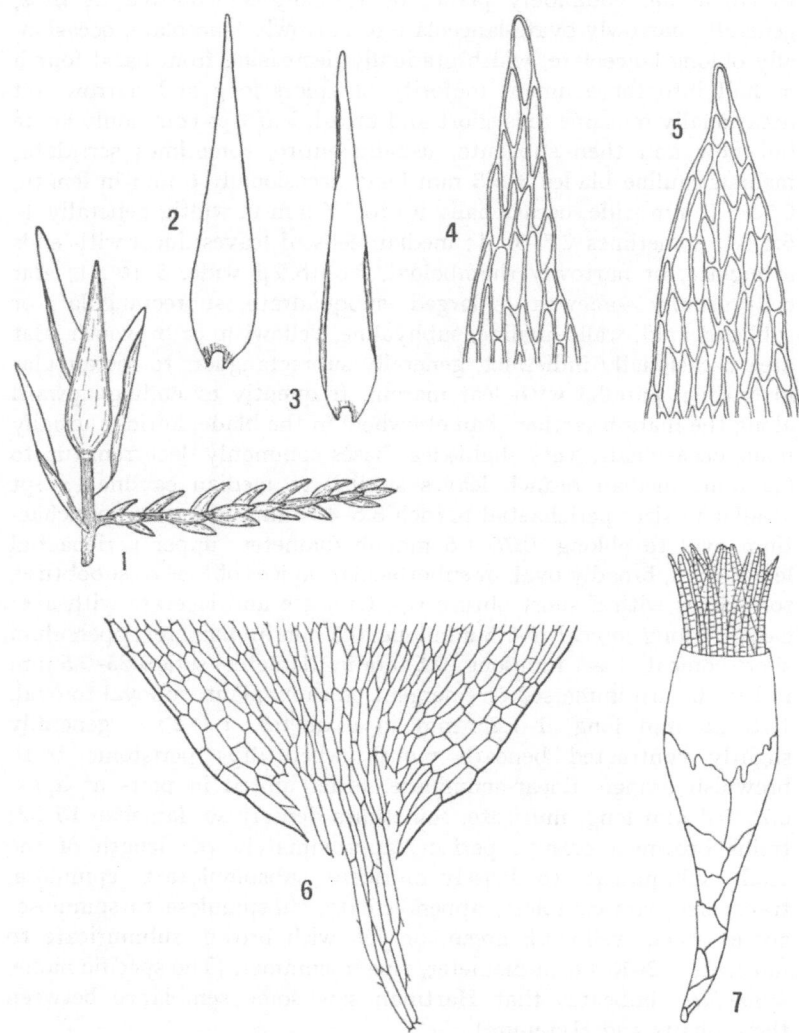

FIG. 26 – *Fontinalis hypnoides* C. J. Hartm. – 1.Portion of stem with median
cauline leaves and a branch.   2 and 3.Cauline leaves.   4 and 5. Leaf apices.
6.Leaf base.   7.Perichaetial branch, urn, peristome teeth and perfect trellis.
Nos. 1–3 and 7 drawn after Bruch and Schimper, Bry. Eur., pl. 432; nos. 4–6
from Leiberg 137, in 1888, Idaho.

mosa var. *tenella*), 458 (as *F. hypnoides*). 1860; Milde, Bry. Siles., p. 277. 1869; Schimper, Syn. Musc., p. 556. 1876; Hy, in Mém. Soc. Agr., Sci. et Arts Angers 24: 135. 1882; Lesquereux and James, Man. Moss. N. Am., p. 272. 1884; Cardot, Mon. Font., p. 98. 1892; Husnot, Musc. Gall., p. 287. 1892; Farneti, in Atti Ist. Bot. Univ., Pavia (Nuov. Ser.) 3: 68 (as var. *Ravani*). 1894; Limpricht, Laubm. 2: 663 (as *F. hypnoides*), 664 (as var. *pungens*). 1894; Barnes, Gen. and Sp. Moss. N. Am., p. 329 (as *F. tenella*). 1896; Roth, Eur. Laubm. 2: 285 (as *F. hypnoides*), 286 (as var. *pungens*). 1904; Braithwaite, Br. Moss Fl., p. 214 (as *F. seriata*). 1905; Warnstorf, Laubm., pp. 632 (as *F. hypnoides*), 633 (as var. *pungens*), 634 (as var. *angustifolia*). 1905; Mönkemeyer, in Pascher, Süsswasserfl. 14: 106. 1914; Hj. Möller, in Ark. Bot. 17 (14): 70 (as *F. hypnoides*), 74 (as var. *Adlerzii*). 1922; Brotherus, Laubm. Fenn., pp. 398 (as *F. hypnoides*), 399 (as var. *Adlerzii*). 1923; Jensen, Danm. Moss., p. 190. 1923; Dixon, Handb. Br. Moss., p. 394 (as *F. seriata*). 1924: Mönkemeyer, Laubm. Eur. Erg.-Bd. 4: 664 (as *F. hypnoides*, and as f. *pungens*). 1927; Welch, in Grout, Moss Fl. N. Am. 3: 250 (as *F. hypnoides*), 252 (as *F. tenella*). 1934; Jensen, Skand. Bladmossfl., pp. 382 (as *F. hypnoides*), 383 (as var. *Adlerzii*). 1939; Jennings, Man. Moss. W. Penn. and Adj. Reg., p. 176. 1951.

Illustrations: Bruch and Schimper, Bry. Eur., Fasc. 31, 5: pl. 432. 1846; Schimper, Nya Moss., pl. 15. 1848; Hy, in Mém. Soc. Agr., Sci. et Arts Angers 24: 137, fig. a (as *F. Ravani*), fig. c (as *F. hypnoides*). 1892; Husnot, Musc. Gall., pl. 81 (as *F. hypnoides* and *F. Ravani*). 1892; Farneti, in Atti Ist. Bot. Univ., Pavia (Nuov. Ser.) 3: pl. 24 (as var. *Ravani*). 1894; Roth, Eur. Laubm. 2: pl. 30. 1904; Braithwaite, Br. Moss Fl., pl. 123 (as *F. seriata*). 1905; Mönkemeyer, in Pascher, Süsswasserfl. 14: fig. 32. 1914; Györffy, in Magy. Bot. Lap. 15: pl. 6. 1916; Hj. Möller, in Ark. Bot. 17 (14): figs. 32–34 (as *F. hypnoides*), fig. 35 and pl. 9[17] (as var. *Adlerzii*). 1922; Brotherus, Laubm. Fenn., fig. 68. 1923; Jensen, Danm. Moss., pl. 8. 1923; Dixon, Handb. Br. Moss., pl. 49 (as *F. seriata*). 1924; Mönkemeyer, Laubm. Eur. Erg.-Bd. 4: figs. 143 and 145. 1927; Welch, in Grout, Moss Fl. N. Am. 3: pl. 77 (as *F. hypnoides* and *F. tenella*). 1934. – FIGURE 26.

Specimens examined: ASIA: JAPAN: Faurie 8691, Sept. 10, 1892, Yezo, Kushiro R. (PC); Jakenchi 1867, May 20, 1921, Yezo, base of Mt. Tarumai (H); Miyabe 333, May 1893, Sapporo (FH, H, K, PC); Sioda, Oct. 16, 1921, Mino (G-DEL).

UNION OF SOVIET SOCIALIST REPUBLICS: Arnell, May 27, 1876, Siberia, Ob R. (H, PC); Arnell, June 29, 1876, Siberia, Yenisei R. (PC); Gordiagin 16, Aug. 2, 1901, Siberia, Akmolinsk (H); Gorodkov, Aug. 14, 1913, Siberia, Tobolsk (H); Gorodkov, Aug. 12, 1914, Siberia, Tobolsk (H); Kuzeneva, July 23, 1914, Siberia, Amur R. (H); Martianov, July 1882, Siberia, Minusinsk (H); Shalosubov, May 26, 1906, Siberia, Tobolsk (H).

EUROPE: AUSTRIA: Baumgartner, Vienna, Danube R., alt. 160 m, Krypt. Exsic. 2095 (K, NY, US); Baumgartner, Feb. 11, 1912, near Vienna, alt. 160 m, Bauer, Musci Eur. Exsic. 1156 (BART, L, WASH); Baumgartner, Nov. 12, 1916, near Vienna, alt. 150 m, Bauer, Musci Eur. Exsic. 1395 (BART, NY, WASH); Roemer, in 1865, near Brünn (H).

DENMARK: Hesselbo, July 1906 (DPU, S); Hesselbo, July 1908, Sjaelland (S).

ENGLAND: Binstead, Feb. 26, 1896, Herefordshire, near Winforton, R. Wye (BM, DPU, PC); Binstead, Apr. 9, 1896, Herefordshire, near Winforton, R. Wye (K).

FINLAND: Bomansson, Alandia (PC); Bomansson, Aug. 1872, Alandia (S); Bomansson, in 1855, Alandia (NY); Bomansson, July 1885 (YALE); V. F. Brotherus, July 7, 1869, Tavastia borealis, Viitasaari (H); V. F. Brotherus, Aug. 7, 1869, Tavastia borealis, Viitasaari (H, NY); Buch, May 12, 1904, Nylandia, Helsinki (H, UC); Buch, Aug. 11, 1911, Savonia australis, near Savitaipale, in Lake Karhijärvi, V. F. Brotherus, Bryoth. Fenn. 360b (CHI, L); Hult, May 2, 1880, Nylandia, Helsinki (S); Kihlman, Sept. 6, 1878, Nylandia, Helsinki, V. F. Brotherus, Musci Fenn. Exsic. 199 (K, PC); Kihlman, May 2, 1881, Nylandia, Helsinki, Stansvik (DPU); Kotilainen, Feb. 6, 1925, Ostrobothnia, near Tervola (H); Kotilainen, July 6, 925, Ostrobothnia (DPU, WASH); Kotilainen, July 9,

1926, Ostrobothnia borealis (s); Kotilainen, July 10, 1926, Ostrobothnia bore-
alis (CAS); Kotilainen, July 23, 1926, Ostrobothnia (DUKE); Kotilainen, July 23,
1926, Ostrobothnia, near Tervola (FH, s); Kotilainen, Aug. 21, 1934, Ostro-
bothnia borealis (s, US); Lång, Oct. 1898 (s); Lång, Oct. 1898, Tavastia australis
(CAS, DUKE, UCE); Lång, in 1899, Tavastia australis (PC); Lång, Apr. 27, 1899,
Tavastia australis (DPU, H, s); Lång, Apr. 27, 1899, Tavastia australis, V. F.
Brotherus, Bryoth. Fenn. 171 (CHI, DPU, L, YALE); H. Lindberg, July 30, 1891,
Åbo lán (s); H. Lindberg, Oct. 22, 1898, Tavastia australis (G-DEL, s); Malm-
berg, June 27, 1866 (H); Norrlin, Sept. 1863 (US); Norrlin, Sept. 2, 1863, Tavastia
(K, NY); Norrlin, Sept. 1864, Tavastia (K); Palmén, Aug. 1865, Tavastia (K);
Sola, May 12, 1904, Nylandia, Helsinki, Degerö (s); Sundvik, May 18, 1904,
Nylandia, near Helsinki, Degerö, V. F. Brotherus, Bryoth. Fenn. 360a (DUKE,
L, YALE).

FRANCE: Bioret, May 1919, Maine-et-Loire, Société Française, 1919, Exsic.
3235 (DPU, G-DEL); Boulay, in 1874, Fréjus (PC); Bureau and Camus, Oct. 4,
1891, Loire Inférieure (PC); Camus, Sept. 12, 1891, Loire Inférieure (PC); Hy,
Rochefort-sur-Loire (MINN); Hy, Loire R., near Angers, Husnot, Musci Gall.
776 (CHI, FH, K, PC, WASH, WIS); Hy, Maine-et-Loire, Isl. of St. Jean-de-la-Croix,
near Angers (PC); Hy, "F. Ravani Hy sp. nov. 1882! France: in fossis aqua
stagnante repletis secus fluvium Ligerim, in agro Andegavensi" (Assumed to
be portion of TYPE of F. Ravani, PC); Hy, Mar. 21, 1882, Rochefort-sur-Loire
(MINN), (Also letters from Hy, written Feb. 27, 1882, and Mar. 3, 1882, concerning plants which he named F. Ravani); Hy, Apr. 3, 1882, Isl. of St. Jean-de-
la-Croix, Loire R., near Angers (CHI, PC); Hy, in 1884, Loire R., near Angers
(H); Hy, June 1885, Loire R. (H); Hy, Sept. 1888, Loire R. (PC); Hy, Oct. 1888,
Maine-et-Loire (K); Migault, Bureau, and Camus, Sept. 1891 (PC); Ravain, Apr.
4, 1882, Isl. of St. Jean-de-la-Croix, near Angers (PC).

GERMANY: A. Braun, Mark Brandenburg, near Berlin (H); A. Braun, in 1855,
Mark Brandenburg, near Weissensee, near Berlin (K); A. Braun, June 1855,
Mark Brandenburg, Weissensee, near Berlin (s); A. Braun, July 1855, Mark
Brandenburg, Weissensee, near Berlin (s); A. Braun, July 27, 1855, Mark Brandenburg, Weissensee, near Berlin (B, K, s); A. Braun, July 1899, Mark Branden-
burg, Berlin (H); Caspary, Aug. 22, 1881, West Prussia, Wandsburger Lake,
near Flatow (H); Caspary, Sept. 7, 1882, West Prussia, Batlewo Lake, near
Kulm (H); Caspary, July 21, 1883, West Prussia (H); Freiberg, Brandenburg,
alt. 30 m, Bauer, Musci Eur. Exsic. 1158 (BART, L, WASH); Freiberg, Sept. 4,
1909, West Prussia, Torfsee, near Wahlendorf, alt. 154 m. (DUKE, s); Freiberg,
July 31, 1911, East Prussia, Brandenburg, Tilsit (s); Freiberg, Aug. 18, 1911,
Brandenburg, Tilsit, alt. 12 m, Bauer, Musci Eur. Exsic. 1157 (BART, L, WASH);
Freiberg, July 31, 1912, East Prussia (G-DEL); Grebe, Aug. 1884, East Prussia
(US); Jaap, Sept. 3, 1900, Brandenburg, in an old marl quarry (s); Jaap, Sept.
3, 1901, Brandenburg, in an old marl quarry, Heide, "Triglitz i.d. Prignitz"
(Perhaps a portion of the TYPE of F. hypnoides var. angustifolia, s), (Label was
"F. hypnoides Hartm. var. angustifolia Warnst. n. var. !."); Kalmus, Aug. 8,
1907, East Prussia, Elbing (PC); Kern, May 6, 1888, Breslau (PC); Kolb, near
Wildbad (ILL); Mougeot, in 1814 (G-DEL); Ruthe, Waltersdorf, near Bärwalde
in Neumark (H, K, L, NY); Ruthe, Oderwiesen, near Bärwalde, Rabenhorst,
Bryoth. Eur. 1313 (FH, NY, UC); Ruthe, June 1870, Neumark (BR); Ruthe, Aug.
1872, Neumark, Bärwalde, Rabenhorst, Bryoth. Eur. 1228 (FH, MICH, NY);
Ruthe, Sept. 1872, near Bärwalde, Märkische Laubmoose 161 (CAS); Ruthe, in
1873, Brandenburg, Neumark, near Bärwalde (H, K, L, s); Ruthe, July 1874,
Neumark, Bärwalde (R); Ruthe, in 1875, Neumark, Bärwalde (G-BOIS); Ruthe,
in 1876, Neumark, Bärwalde (H, K); Ruthe, Aug. 1879, Neumark, near Schönfliess (CHI, FH, NY, US); Ruthe, Aug. 20, 1879, Neumark, near Schönfliess
(UCLA); Ruthe, Aug. 25, 1879, Neumark, near Schönfliess (CHI, G-DEL, PC);
Ruthe, in 1887, Neumark, near Bärwalde (G-DEL); Sanio, in 1886, East Prussia,
near Lyck (PC); Sanio, Aug. 28, 1886, East Prussia, near Lyck (Perhaps a portion
of the TYPE of F. hypnoides var. pungens, PC).

ITALY: Anzi, Mar. 1881, Lake Como, Erb. Critt. Ital. (Ser. II) 1103 (CHI, G-DEL, NY, PC, UC); Artaria, Apr. 7, 1895, Pavia (H); Artaria, Jan. 16, 1898, Pavia (PC); Artaria, Jan. 19, 1898, Pavia, alt. 80 m (CAN); Artaria, Jan. 19, 1898, Pavia, alt. 80 m, Fleischer and Warnstorf, Bryoth. Eur. Merid., Cent. III, 259 (PC); Artaria, Dec. 11, 1898, Pavia, alt. 80 m (PC); Artaria, Aug. 16, 1899, Pavia (s).

LATVIA (LIVLAND): Mikutowicz 15297, June 5, 1902, Kurland, Talsen, Bryoth. Balt. 296 (H, US); Mikutowicz 15299, June 5, 1902, Kurland, Talsen (E, H, US).

NORWAY: Kaalaas, July 15, 1895, Bratsberg Amt, alt. 160 m (H, s); Kaalaas, July 16, 1895, Bratsberg Amt (DPU, s); Ryan, in 1888, Smaalenenes Amt (s); Sörensen, Apr. 30, 1911, Akershus Amt, alt. 122 m (G-DEL, s).

SWEDEN: Åberg 139 and 467, July 17, 1912, Jämtland (s); Åberg 1601, Sept. 2, 1920, Jämtland (s); Adlerz, Aug. 8, 1904, Närke, Kvistbro (Assumed to be portions of the TYPE of F. hypnoides var. Adlerzii, DPU, PC, s); Adlerz, Aug. 14, 1906, Närke, Kvistbro (s); Adlerz, June 16, 1908, Närke (s); Adlerz, June and Aug. 1904–1908, Närke, Kvistbro, Bauer, Musci Eur. Exsic. 564 (BART, PC, s); Ångström, Uppland, Uppsala (CHI, PC, s); Bergman, Aug. 1855, Bohuslän (K); Berudes, in 1868, Stockholm (CHI); Ekstrand, Aug. 25, 1878, Södermanland (s); Hartman, Gästrikland (L, MINN, NY); Hartman, Uppland (NY, s); Hartman, in 1850, Uppland, Uppsala (H); Hartman, July 1891, Gästrikland (PC); C. Hartman, in 1852, Gästrikland, Oslättfors (s); C. and R. Hartman, Sept. 3, 1842, Gästrikland, Oslättfors (K); C. Hartman and E. A. Strömbäck, June 1845, Uppland, Uppsala (s); R. Hartman, Gästrikland, Flyet, Oslättfors (DPU, s); R. Hartman, Sept. 3, 1842, Gästrikland, Oslättfors (NY); R. Hartman, Sept. 5, 1842, Gästrikland, Oslättfors (TRT); R. Hartman, in 1847, Gästrikland, Oslättfors (DPU, H, s); R. Hartman, Aug. 2, 1847, Gästrikland, Oslättfors (s); R. Hartman, Aug. 1852, Gästrikland, Oslättfors (CHI); R. Hartman, July 1854, Gästrikland, Oslättfors (DPU, s); R. Hartman, in 1858, Gästrikland, Oslättfors, Flyet, Bry. Scand. Exsic. 85 (K, s); R. Hartman, July 1859, Gästrikland, Oslättfors (K); R. Hartman, in 1871, Gästrikland, Oslättfors (s); Hellsing, Oct. 6, 1893, Uppland, near Uppsala (DPU, s); Hellsing, May 1896, Uppland, Danmark (DPU, s); Hellsing, Aug. 14, 1899, Norrbotten, Öfver-Torneå (s); Hovgard, May 20, 1926, Skåne (s); Jäderholm, Oct. 1894, Uppland (s); Kindberg, July 1869 (CAN); Kjellmark, Aug. 27, 1897, Närke, Axberg (s); Z. M. T. Lange, Aug. 4, 1854, Närke (NY); S. O. Lindberg, Stockholm (PC); S. O. Lindberg, Stockholm, Rabenhorst, Bryoth. Eur. 629 (NY, PC, s, UC); S. O. Lindberg, in 1852, Stockholm (K); S. O. Lindberg, Oct. 1853, Stockholm (DPU, s); S. O. Lindberg, Nov. 1853, Stockholm (K, YALE); S. O. Lindberg, July 1854, Dalarna (YALE); S. O. Lindberg, June 1862, Stockholm (H); S. O. Lindberg, in 1893, Stockholm (PC); Hj. Möller, June 1910, Dalarna, Ore (DPU, s); Hj. Möller, June 25, 1910, Dalarna, Ore, Oresjön (DPU, s); Hj. Möller, July 1920, Södermanland, Nacka (DPU, DUKE, FH, WASH); Hj. Möller, July 29, 1920, Södermanland, Nacka (DPU, DUKE, MICH, s, US); Hj. Möller, Apr. 12, 1921, Södermanland, Nacka (DPU, s); Nyman, June 1896, Uppland, Söderfors (DPU, H, s); Schimper, Gästrikland, Oslättfors (NY, US); Schimper, in 1844, Gästrikland (G-DEL, PC); Schimper, in 1856, Gästrikland (G-BOIS); Sillén 47, Aug. 1875, Gästrikland, Oslättfors (DPU, H, s); Sillén 48, Aug. 1875, Gästrikland, Oslättfors (H); C. and E. A. Strömbäck and C. Hartman, Jr., Aug. and Sept. 1842, Gästrikland, Hille, Oslättfors, Flyet (TYPE of F. hypnoides, UPS; DPU); E. A. Strömbäck, Sept. 9, 1844, Gästrikland, Oslättfors, Flyet (s); E. A. Strömbäck, Sept. 17, 1852, Uppland, Uppsala (s); Sundén, in 1855, Dalarna, Torsång (H); Swartz, in 1807, Tarna (K); Swartz 801, in 1807, Dalecarlia (s); Tärnlund, Aug. 20, 1923, Stockholm, Nacka (DPU); H. Thedenius, in 1863, Gästrikland, Oslättfors, Flyet (s); H. Thedenius, in 1885, Södermanland, Stockholm, Nacka (s); H. Thedenius, Aug. 1888, Södermanland, Stockholm, Nacka (DPU, s); Vetterhall, Aug. 25, 1878, Södermanland, Stockholm, Nacka (s); Westling, Aug. 1863, Gästrikland, Oslättfors (CHI, s); Westling 196,

Aug. 1863, Gästrikland, Oslättfors (NY); Wikström, Lule Lappmark (DPU, s); Zetterstedt, July 30, 1859, Uppsala (PC, s); Zetterstedt, Aug. 13, 1859, Uppsala (G-DEL).

UNION OF SOVIET SOCIALIST REPUBLICS: Anufriev, July 28, 1914, Russia (H); Kihlman, Feb. 3, 1897, Russia (H); Koclmakov 1355, June 30, 1895, Russia (H); Krylov, Sept. 18, 1883, Russia (PC); Ziedendrath, June 23, 1895, Russia (H).

NORTH AMERICA: CANADA: ALBERTA: Macoun, Sept. 16, 1872, Athabaska Plains, Fl. Can. 2226 (NY); Macoun, Aug. 9, 1904, Battle Creek (FH); Macoun, Aug. 9, 1906, Battle R. (E, FH, K).

BRITISH COLUMBIA: Brinkman, Oct. 19, 1908, Shuswap Lake, Can. Moss. 674 (CM, FH, US); Brinkman, Nov. 11, 1908, Shuswap Lake, Can. Moss. 719 (CM, E, FH, K); Brinkman, June 11, 1910, Nicola R. District, near Spence's Bridge, alt. 1170 m, Can. Moss. 219 (ABS, DPU, DUKE, FH, MACF, MICH); Brinkman, June 18, 1910, Nicola District, near Spence's Bridge, alt. 1140 m, Can. Moss. 231 (ABS, CM, E, FH); Brinkman, July 10, 1910, Cougar Lake, West Savonas, alt. 1320 m, Can. Moss. 247 (CM, E, FH, US); Brinkman, July 12, 1910, Nicola R. District, near Cougar Lake, alt. 1320 m, Can. Moss. 247 (ABS, FH, MACF); MacFadden, New Denver, Slocan Lake (DPU, TRT); MacFadden, June 27, 1922, Slocan Lake (DPU, MACF); MacFadden, Sept. 2, 1925, New Denver, Slocan Lake (DPU, MACF); Macoun, June 8, 1875, Telegraph Trail (CAN); Macoun, July 3, 1889, Sicamous, Can. Musci 602 (DUKE, FH, US, WASH); Macoun, July 7, 1889 and July 3, 1890, Sicamous, Can. Moss. 432 (K); Macoun, July 8, 1889, Sicamous, Can. Musci 432 (CIN, MINN, MO, WIS, YALE); Macoun, July 11, 1889, Sicamous, on base of tree trunks 0.9 m from ground, Can. Musci 432 (FH, NY); Macoun, July 17, 1889, Sicamous (CAN); Macoun, July 17, 1889, Sicamous, Can. Musci 432 (FH, H, MINN, NY, S, TRT, UC, US); Macoun, July 17, 1889, Sicamous, Can. Moss. 207 (BM, FH, K, MO, MT, NY, US); Macoun, May 31, 1890, Revelstoke, Can. Musci 206 (CAN); Macoun 604, May 31, 1890 (US, WASH).

CAPE BRETON ISLAND: G. E. Nichols, July–Aug. 1914, South Ingonish (NY, YALE).

MANITOBA: Dudley, Sept. 1, 1938, Whiteshell R. (MINN).

NORTHWEST TERRITORIES: Macoun, Can. Musci 232 (CIN, FH, K, MINN, MO, NY, PC, US, WIS, YALE).

NOVA SCOTIA: Macoun, Can. Musci 231 p.p. (NY).

ONTARIO: H. H. Brown 887, Aug. 26, 1939, Peel Co., near Summerville (DPU, TRT); Conard 8–122, June 22, 1938, Owen Sound, Bruce Peninsula (DPU, GRI); Hand 468, July 24, 1935, Haliburton District, near Donald (DPU); Hand 611, June 1938, Bruce Co., near Hepworth (DPU); Macoun, Hastings Co. (NY); Macoun, Can. Musci 231 p.p. (NY); Macoun, Sept. 1870, near Belleville (TRT); Macoun, Oct. 1870, Lake Region, Fl. Can. 2227 (NY); Macoun, May 24, 1871, Belleville, Can. Musci 604a (FH, PC, US, WASH); Macoun, Sept. 26, 1900, Golden Lake, Can. Moss. 206 p.p. (FH, US); Macoun, Sept. 27, 1900, Golden Lake (K); Macoun, Oct. 27, 1900, Britannia, Can. Moss. 206 p.p. (BM); Macoun, July 3, 1907, Kettle Creek, near St. Thomas (CAN); Macoun, July 8, 1907, St. Thomas (K, S); Moxley, July 10, 1927, Bruce Co., Oliphant Beach (CM, FH); Schnooberger 1373a, June 23, 1938, Bruce Co., near Oliphant Beach (DPU).

QUEBEC: Cléonique-Joseph 10177, July 2, 1938, St. Jean-d'Iberville Co., Sabrevois (DPU, MT); Lepage 2005, July 21, 1940, Matane Co., Lake Malfait, St. Leandre (DPU).

SASKATCHEWAN: Macoun, July 18, 1880, Thunder Creek, Can. Musci 604 p.p. (DUKE).

VANCOUVER ISLAND: Macoun, May 1875, Victoria (NY); Macoun, May 8, 1875, Fl. Can. 257 p.p. (CAN).

UNITED STATES: COLORADO: C. F. Baker 1, July 28, 1896, Larimer Co., alt. 2550 m (PC); C. F. Baker 631, July 28, 1901, Gunnison Watershed Region, Doyle's, alt. 2430 m (BART, DUKE, FH, K, MINN, NY, PC, UC, US, WYO); Conrad, July 16, 1941, Lake Co. (DPU, GRI, NY, OS); Rodeck, Aug. 8, 1938, Boulder Co., alt. 3150 m (DPU).

CONNECTICUT: Clark, May 30, 1931, Taconic (DPU, NY); Lorenz, Nov. 1902, near New Haven, Cedar Hill Cemetery (DUKE); Wright, June 30, 1882, Hartford (FH).

IDAHO: Leiberg, Kootenai Co., Lake Pend d'Oreille (FH, PC, US); Leiberg 137, Lake Pend d'Oreille (Assumed to be the TYPE of *F. tenella*, PC; WIS); Leiberg, in 1888, Kootenai Co., Lake Pend d'Oreille (S); Leiberg 137, in 1888, Kootenai Co., Lake Pend d'Oreille (CAN, NY, S, WIS, YALE); Leiberg 137, April–Dec. 1889, Kootenai Co., Lake Pend d'Oreille (CAN, CHI, FH, NY, WIS, YALE); Leiberg 137, July 1890, Kootenai Co., Lake Pend d'Oreille (MINN); Leiberg 256, July 31, 1890, Ellisport Bay, Lake Pend d'Oreille (NY); Leiberg 256, Aug. 1891, Kootenai Co., Lake Pend d'Oreille (WIS); Leiberg 137m1, Oct. 1893, Kootenai Co., Lake Pend d'Oreille, alt. 620 m (NY); Röll 1242 p.p., Kootenai Co., Lake Pend d'Oreille (PC).

ILLINOIS: E. Hall, in 1860, Marion Co., ponds along Illinois R. (CHI); E. Hall, in 1861, Havana, Illinois R. (CHI, PC, WIS); Wolf, Canton (CHI); Wolf, Fulton Co. (ILL).

MAINE: R. Lowe, July 7, 1934, Haleb (DPU, PSNH).

MASSACHUSETTS: Githens 1050, July 31, 1951, Berkshire Co., at outlet of Lake Averis (PHIL).

MICHIGAN: Purpus, Oct. 1891, Clarks Lake (PC).

MINNESOTA: Holzinger, Aug. 1894, near Lamoille Cave (MINN); Holzinger and Elftman, June 20, 1897, between North and Little Gunflint Lakes (DUKE, FH, MINN, MO, NY, US, WYO); Macmillan, Brand, and Lyon, Aug. 20, 1901 (PC); Moyle 3624, July 2, 1940, St. Louis Co. (MINN); Tilden, Minneapolis (PC); Welch 9697, 10360, and 10419, Aug. 3, 1950, Pipestone Co., Pipestone National Monument (DPU).

MISSOURI: Bush, May 11, 1894, Montier (MO, PC); Bush, Aug. 9, 1899, Pleasant Grove (MO, PC).

MONTANA: Röll 1432–1434, Deer Lodge (CHI, FH, NY, PC); R. S. Williams 284, Aug. 12, 1897, Two-Medicine Lake (K, MINN, MO, NY).

NEW YORK: Clinton, Caledonia Creek (MICH).

OREGON: Lyall, in 1861, Oregon Boundary Commission, from Fort Coville to Rocky Mts. (K).

SOUTH DAKOTA: A. C. McIntosh, Rapid City, Cleghorn Springs (DPU).

UTAH: Flowers 2324, Aug. 10, 1940, Summit Co., near Mirror Lake, 3090 m (DPU, UT).

WISCONSIN: Coll.?, Apr. 11, 1859, Wamutosa (WIS).

WYOMING: Kelley 541, July 20, 1926, Albany Co., Nash Canyon (MICH, NY); A. Nelson 8002, Aug. 7, 1900, Albany Co., Centennial (CHI, CM, FH, ILL, K, MINN, MO, NY, US, WASH, WYO). A. Nelson 9671, Aug. 12, 1912, Albany Co., Medicine Bow Mts. (FH, MINN, MO, NY, US, WYO); Röll 1554, Yellowstone Nat. Park (FH, PC); Röll 1582, Yellowstone Nat. Park, Grand Canyon, alt. 2100 m (FH, PC); Röll 1583, Yellowstone Nat. Park, Grand Canyon, alt. 2100 m (FH, PC); Röll 1502 and 1503, Sept. 2, 1888, Yellowstone Nat. Park, Grand Canyon (MICH); Weed, Sept. 11, 1889, Yellowstone Lake, margin of hot spring, temp. of water 32° C. (DUKE, NY).

Plants of *Fontinalis hypnoides* show great variation in vegetative and fruiting structures. On some plants, there occur leaves which resemble those of *F. Duriaei* and others which are true to the description of *F. hypnoides*. On other plants in the same collection or in different collections, all leaves are typical of *F. hypnoides*. Some branch leaves on plants of *F. Duriaei* resemble median cauline leaves of *F. hypnoides*. It is very important that median cauline blades of well developed or mature plants be used for accurate determination.

Occasionally it is difficult to name the species with certainty. However, plants which are distinctly *F. hypnoides* and those which are *F. Duriaei* without question have leaves which are definitely different and give cause for retaining the two species. The author has used the following combination of characteristics for the determination of *F. hypnoides:* plants slender, stems and blades flaccid, majority of median cauline leaves distant, bases up to 2 mm apart, blades plane or nearly so, narrowly ovate-lanceolate or narrowly lanceolate, width decreasing gradually from basal fourth or half into long and narrowly acuminate apices, leaf tips entire, majority of blades 3–5.5 mm long, 0.75–1.5 mm wide, 4–5.5 : 1, alar group of cells subrectangular in outline, parallel with leaf margin, and frequently but not always extending upward along margin farther than elsewhere in blade, auricles none or occasionally slight, leaf bases frequently decurrent, and trellis perfect, subspinulose to spinulose.

In the author's treatise of the Fontinalaceae of North America in 1934, the trellis of *F. tenella* was considered as being imperfect. In the examination of additional collections during the present study, including those in Herb. Renauld and in Herb. Cardot, a number of fruits of plants determined as *F. tenella* have been seen. Perfect trellises were found in several capsules. Also, in these collections there was evidence of the median cauline leaves being very similar to those of *F. hypnoides*. Leaves from the type of *F. tenella* were compared with blades from the type of *F. hypnoides*. No constant differences in size, shape, median cells, or alar cells were detected. Auricles were commonly absent in both. The two alar groups of one leaf base and the cells of these groups show variation in shape and size. A slight auricle sometimes occurs on one side, while on the opposite side of the same leaf the alar cells do not form an auricle. Since the author can find no characteristics by which *F. tenella* can be separated from *F. hypnoides*, the former has been placed in the synonymy of the latter.

28. **Fontinalis Duriaei** Schimp., Syn. Musc., p. 555. 1876.

Fontinalis fasciculata Lindb., in Öfvers. Finska Vet.-Soc. Förh. 12: 76. 1869 (nomen dubium). [Erroneously spelled as *F. fascicularis* in Schimper, Syn. Musc. Eur., p. 555. 1876, and in Bescherelle, Cat. Mouss. Algérie, p. 30. 1882.]
Fontinalis nitida Lindb. and Arn., in Kongl. Svenska Vet.- Akad. Handl. 23 (10): 161. 1890.
Fontinalis hypnoides C. J. Hartm. subsp. *nitida* (Lindb. and Arn.) Card., in Rev. Bryol. 18: 83. 1891.
Fontinalis Bovei Card., Mon. Font., p. 110. 1892.
Fontinalis Duriaei f. *latifolia* Card., Mon. Font., p. 114. 1892.
Fontinalis hypnoides var. *Duriaei* (Schimp.) Husn., Musc. Gall., p. 287. 1892.

Fontinalis hypnoides var. *ramosa* Farn., Atti Ist. Bot. Univ., Pavia (Nuov. Ser.)
3: 69. 1893. [On basis of description; type not seen.]
Fontinalis Camusi Card., in Rev. Bryol. 22: 53. 1895.
Fontinalis antipyretica Hedw. subsp. *Duriaei* (Schimp.) Kindb., Sp. Eur. and
N. Am. Bryin., Part. 1: 149. 1896; Kindberg, in Can. Rec. Sci. 6: 75. 1894.
Fontinalis hypnoides var. *japonica* Card., in Rev. Bryol. 24: 34. 1897.
Fontinalis amblyphylla Card. var. *pungens* Card., in Rev. Bryol. 24: 36. 1897.
Fontinalis obscura Card., in Holzinger, in Minn. Bot. Stud. 3: 120. 1903.
Fontinalis subcarinata Card., in Bot. Gaz. 37: 376. 1904.
Fontinalis Duriaei var. *pungens* Roth and Zodda, in Hedwigia 49: 220. 1910.
Fontinalis mesopotamica Schiffn., in Annal. K. K. Naturhist. Hofm. 27: 498. 1913.
Fontinalis antipyretica var. *thermalis* Boros, in Bauer, Musci Eur. et Am. Exsic.
1930; Schedae und Bemerkungen zur 39 Serie, nos. 1901–1950, p. 8. 1927.
Fontinalis seriata Lindb. var. *pseudofastigiata* P. de la Varde, in Werner, in
Rev. Bryol. et Lichénol. (N. S.) 5: 227. 1932 (nomen nudum).[1]
Fontinalis Duriaei var. *integra* Trab., in Maire and Werner, in Bull. Soc. Hist.
Nat. Afr. Nord. 25: 56. 1934. [On basis of description; type not seen.]
Fontinalis nitida var. *angustiretis* Card., in Welch, in Grout, Moss Fl. N. Am.
3: 251 and 252. 1934 (nomen nudum).
Fontinalis Fiorii Tong., in Nuov. Giorn. Bot. Ital. (N. S.) 45: 401. 1938.
Fontinalis antipyretica subsp. *vulgaris* (Mönkem.) Giacom. var. *thermalis* Boros,
in Podpěra, Consp. Musc. Eur., p. 506. 1954.
Fontinalis hypnoides Györffy, according to Podpěra, Consp. Musc. Eur., p. 506.
1954; I. Györffy, in Magyar Bot. Lapok 15: 235. 1916, as *F. hypnoides* R.
Hartm.
Fontinalis Györffii Boros, in Podpěra, Consp. Musc. Eur., p. 506. 1954 (nomen
nudum).
Fontinalis hypnoides C. J. Hartm. f. *ramosa* (Farn.) Podp., Consp. Musc. Eur.,
p. 510. 1954.
Fontinalis Trabutii Card., in Jelenc, Musc. Afrique Nord, p. 119. 1955.
Fontinalis intermedia Card., in Jelenc, Musc. Afrique Nord, p. 119. 1955.
Fontinalis Durieui var. *intermedia* Card., from herbarium label. In Jelenc, Musc.
Afrique Nord, p. 119. 1955.

Plants slender to medium in size, sometimes rather delicate, pale
green, yellowish green, olive green, green, brownish green, or
brownish, often blackish near base, occasionally entire plant black,
glossy, especially younger portions, or dull, particularly older and
basal leaves; stems flaccid, up to 30 cm in length and 0.25 mm in
diameter, occasionally up to 0.5 mm in diameter, foliated or denuded
in basal portions, darker below with age, irregularly pinnately
branching; branches few to numerous, erect-spreading to spreading,
close to distant, short to elongate, up to 12 cm in length, ends of
foliated stems and branches attenuate; median cauline leaves usually
distant, bases commonly up to 2 mm but occasionally up to 2.5 mm
apart, blades flaccid to somewhat firm, erect-spreading to spreading,
commonly plane, occasionally subconcave at base, sometimes with

[1] *Fontinalis seriata* Lindb. var. *pseudo-fasciculata* P. de la Varde, in Gatte-
fosse et Werner, Catalogus Bryophytum Maroccanorum, 1932, (nomen nudum)
seems to have been used interchangeably with *F. seriata* Lindb. var. *pseudofastigiata* P. de la Varde (nomen nudum). The author's correspondence with
Mr. Robert Potier de la Varde confirms this opinion.

one to two slight longitudinal folds, generally broadly ovate-lanceolate or oval-lanceolate, sometimes oblong-lanceolate, width decreasing
either gradually or somewhat abruptly from the approximate middle
of blade into the apex; majority of apices short and broadly acuminate, leaf tips usually acute, occasionally subobtuse, commonly
serrulate, often entire; median cauline blades 3–5 mm long, occasionally up to 6 mm in length, rarely up to 7 mm, 1–2.5 mm wide, majority

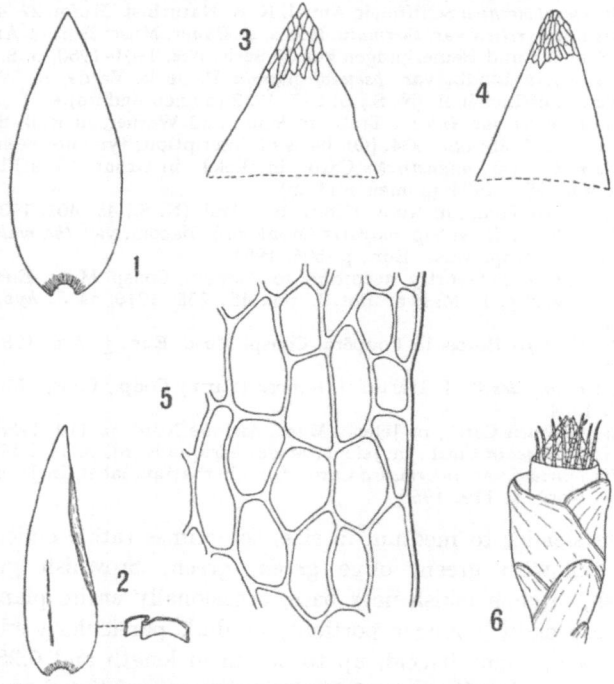

FIG. 27. – *Fontinalis Duriaei* Schimp. – 1.Median cauline leaf.
2.Median cauline leaf with longitudinal fold. 3 and 4.Leaf apices.
5.Alar cells. 6.Perichaetium, emergent urn, and peristomes. Nos.
1–5 drawn from Holzinger, July 28, 1902, Minnesota; no. 6 after
Husnot, Musc. Gall., pl. 81, fig. 14.

2–3.5 : 1, sometimes up to 5.5 : 1; median cells of leaves usually
linear with ends attenuate, sometimes narrowly rhomboidal, 8.5–17 μ
wide, 6–18 : 1; alar cells enlarged, subrectangular, subquadrate, or
subhexagonal, walls hyaline, subhyaline, yellowish, or brownish,
vertical rows of cells generally 5–7, group of alar cells subrectangular
in outline and parallel with margin of blade if auricles absent, suboval
if auricles slight, or suborbicular if auricles distinct, auricles of some
degree commonly present, leaf bases frequently rounded, not de-

current to very briefly so, up to 0.5 mm; median branch leaves similar to median cauline except smaller in size; perichaetial branch 3.5–5.25 mm long, perichaetium oval to oblong, 0.75–1.5 mm in diameter; upper perichaetial leaves suboval to suborbicular, apices usually broadly obtuse, sometimes with a short obtuse tip, truncate and lacerate with age; calyptra long conical, 1.2–1.5 mm long, 0.5–0.6 mm in diameter; operculum obtuse conical, 0.8–1.25 mm long, 0.7–1.25 mm in diameter; seta 0.25 mm in length; urn usually immersed, occasionally slightly emergent, immersed and emergent on same plant, oval, suboval, or oblong, 1.5–2.5 mm long, 1–1.5 mm in diameter, 1.7–2 : 1, generally not contracted beneath mouth when dry, but occasionally so; peristome teeth brownish orange, linear-acuminate, often united in pairs at apex, 0.75–1 mm long, muricate, lamellae 15–35; trellis brownish orange, perfect, approximate length of the teeth, muricate, transverse strands complete, lower ones appendiculate; spores green, yellowish green, or yellowish brown, finely muricate to smooth, 13.6–20 μ in diameter; ripe in summer. [The specific name, *Duriaei*, honors Durieu de Maisonneuve, the collector of the type specimens.]

Type: No specimen so designated has been seen. It is assumed that the plants collected by Durieu, Sept. 13, 1840, in Algeria, Africa, near La Calle, shore of Lake El Hout, and sent to Schimper, served the latter as the basis of his original description. These plants are in the Kew Herbarium and are assumed by the author to be the type of *Fontinalis Duriaei* since Schimper's Herbarium is in the Herbarium of the Royal Botanic Gardens of Kew.

Type collector: Durieu de Maisonneuve, in 1840.

Type locality: Algeria: near La Calle, shore of Lake El Hout.

Distribution: Africa, Asia, Europe, North America north of Mexico, and South America in Brazil.

Additional descriptions: Hy, in Mém. Soc. Agr., Sci. et Arts Angers 24: 135. 1882; Cardot, Mon. Font., pp. 103 (as *F. nitida*), 108 (as *F. fasciculata*), 111 (as *F. Duriaei*). 1892; Husnot, Musc. Gall., p. 287. 1892; Limpricht, Laubm. 2: 661. 1894; Barnes, Gen. and Sp. N. Am. Moss., pp. 329 (as *F. nitida*), 330 (as *F. Duriaei*). 1896; Grout (as *Editor*), Moss Fl. Upp. Minn. R. by J. Holzinger, in Bryol. 7: 11 (as *F. obscura*). 1904; Roth, Eur. Laubm. 2: 284 (as *F. Duriaei*), 288 (as *F. Camusi*). 1904; Cardot and Thériot, in Bryol. 9: 7 (as *F. subcarinata*). 1906; Mönkemeyer, Laubm. Eur. Erg.-Bd. 4: 661 (as *F. Camusii*), 664 (as *F. Duriaei*), 665 (as *F. nitida*). 1927; Welch, in Grout, Moss Fl. N. Am. 3: 251 (as *F. nitida*), 252 (as *F. Duriaei*), 255 (as *F. subcarinata*). 1934; Jennings, Man. Moss. W. Penn. and Adj. Reg., p. 177 (as *F. nitida* and as *F. Duriaei*). 1951.

Illustrations: Hy, in Mém. Soc. Agr., Sci. et Arts Angers 24: 137, figs. b, 1, 3, 4. 1882; Husnot, Musc. Gall., pl. 81. 1892; Farneti, in Atti Ist. Bot. Univ., Pavia (Nuov. Ser.) 3: pl. 24 (as *F. hypnoides* var. *ramosa*). 1893; Cardot, in Holzinger, in Minn. Bot. Stud. 3: pl. 22 (as *F. obscura*). 1903; Cardot, in Bot. Gaz. 37: pl. 23 (as *F. subcarinata*). 1904; Roth, Eur. Laubm. 2: pl. 30 (as *F. Duriaei*), pl. 32 (as *F. Camusi*). 1904; Roth and Zodda, in Hedwigia 49: pls. 7 and 8. 1910; Schiffner, in Ann. K. K. Naturhist. Hofm. 27: 498, figs. 88–91 (as *F. mesopotamica*). 1913; Mönkemeyer, Laubm. Eur. Erg.-Bd. 4: fig. 145. 1927; Welch, in Grout, Moss Fl. N. Am. 3: pl. 77 (as *F. nitida* and *F. Duriaei*), pl. 79 (as *F. sub-*

carinata). 1934; Tongiorgi, in Nuov. Giorn. Bot. Ital. (N. S.) 45: 402, fig. 3 (as
F. Fiorii). 1938. – FIGURE 27.

Specimens examined: AFRICA: ALGERIA: Bové, Mar. 1839, Algiers (Assumed
to be the TYPE of *F. Bovei*, PC; BR, G-BOIS, K, NY, S, US); Durieu (K); Durieu, La
Calle (PC); Durieu, Feb. 1840, "bassin de la fontaine du Café maure de Kaddous"
(Assumed to be TYPE of *F. fasciculata* Lindb., H; BM, K, PC); Durieu, Sept. 13,
1840, near La Calle, shore of Lake El Hout (Assumed to be the TYPE of *F.
Duriaei*, K; BM); Durieu, Nov. 13, 1840, La Calle, near Lake El Hout (PC);
Trabut (PC); Trabut, Oued Reghaïn (PC); Trabut, Oued Saoula (PC); Trabut,
Mar. 1902 (PC); Trabut, in 1907 (H).
ERITREA: Fiori, Apr. 2, 1909, Hamasen, "Valle Mergat-Feres presso Uochì,"
alt. 2500 m. (Assumed to be TYPE of *F. Fiorii*, FI; DPU).
MOROCCO: Gattefossé, May 1932, Tinghir on the Todra, alt. 1300 m [The
name, *F. seriata* var. *pseudofastigiata*, was based upon this collection, Herb.
Potier de la Varde; DPU]; Pitard 280, Feb. 1911, Tangier (PC).
ASIA: JAPAN: Arimoto, July 7, 1903, Yezo, Sapporo (FH); Faurie (Assumed
to be TYPE of *F. amblyphylla* var. *pungens*, PC); Faurie 9073 and 9073a, Apr. 3,
1893, Yezo, Sapporo (PC); Faurie 9073b, Apr. 3, 1893, Yezo, public garden of
Sapporo (Assumed to be COTYPE of *F. hypnoides* var. *japonica*, PC); Faurie
12305, Apr. 27, 1894, Yezo, Sapporo (Assumed to be TYPE of *F. hypnoides* var.
japonica, PC); Faurie 12645, May 25, 1894, Sambongi (Assumed to be COTYPE
of *F. hypnoides* var. *japonica*, PC); Faurie 3203, in 1905, Sambongi (H, NY);
Okamura 959, Aug. 1910, Mino, Ōgaki (H); Okamura 960, Feb. 12, 1911, Aki,
Kabe (H); Shiota, Oct. 16, 1921 (FH).
MESOPOTAMIA: Von Handel-Mazzetti 348, June 20, 1910, near Arran, in
Chabur R., volcanic substratum, alt. 400 m, Mesopot.-Exped. nat.-wissensch.
Orinetv. in Wien 1690 (Assumed to be TYPE portion of *F. mesopotamica*, H).
UNION OF SOVIET SOCIALIST REPUBLICS: Arnell, May 28, 1876, Siberia,
Neolevka, Ob and Irtish Rivers (Assumed to be TYPE of *F. nitida*, H; G-BOIS,
PC, WASH); Arnell, July 27, 1876, Siberia, Yenisei (H, PC); Arnell, Aug. 26, 1876,
Siberia, Yenisei, Tolstoinos (Assumed to be COTYPE of *F. nitida*, H; L, PC);
Arnell, Sept. 1876, Yenisei, Potkamina Tunguska (PC); Arnell, Sept. 28, 1876,
Siberia, Yenisei (H); Arnell, Sept. 30, 1876, Siberia, Yenisei, Vorogova (H, K,
PC, S); Gorodkov, Aug. 15, 1915, Siberia, Tobolsk (H); Gorodkov, Aug. 16, 1915,
Siberia, Tobolsk (H); Gorodkov, Aug. 18, 1915, Siberia, Tobolsk (H); Kirillov,
July 29, 1914, Siberia, Institutum Cryptogamicum Horti-Botanici Petropoli-
tani 1535 (H); Korotky, Lebadeff, and Okusebko, June 15, 1912, Transbaikalia
(H); Korotky, Lebadev, and Okusebko, June 16, 1912, Transbaikalia (H);
Kuznetsov, July 29, 1914, Siberia, Institutum Cryptogamicum Horti-Botanici
Petropolitani 1578 (H); Sahlberg, Sept. 30, 1876, Yenisei, Vorogova (H); Slerar-
drov, Aug. 26, 1911, Siberia, Irkutsk (H); Tohle, July 30, 1913, western Siberia
(PC); Tohle, Aug. 30, 1913, Siberia, Tobolsk (H); Waldburg-Zeil, June 10, 1876,
western Siberia, Sarajgor (K, PC); Waldburg-Zeil, Sept. 1876, Western Siberia,
Sarajgor (B, K).
EUROPE: CORSICA: Philibert, in 1877 (H).
FRANCE: Bizot 1102, Oct. 13, 1928, Côte-d'Or (MICH); Boulay, in 1873, Nîmes
(PC); Bureau, in 1894, Loire-Inférieure, Maine R., near Aigrefeuille (Assumed
to be portion of COTYPE of *F. Camusi*, H, K, MINN); Bureau, Apr. 1, 1894, Loire-
Inférieure, Rousselin, Sèvre R., near Boussay (Assumed to be COTYPE of *F.
Camusi*, PC); Bureau, Apr. 15, 1894, Loire-Inférieure, Maine R., Aigrefeuille
(Assumed to be COTYPE of *F. Camusi*, PC); Bureau, Apr. 15, 1894, Loire-In-
férieure, Maine R., Aigrefeuille, Dismier, Bryoth. Gall. 293 (BART, YALE, WASH);
Bureau, in 1894 and 1895, Loire-Inférieure, Maine R., at Aigrefeuille (Assumed
to be COTYPE of *F. Camusi*, PC); Bureau, June 16, 1895, Loire-Inférieure, Tréli-
tière, Maine R., near Aigrefeuille (Assumed to be COTYPE of *F. Camusi*, PC);
Bureau and Camus, Loire-Inférieure, Sèvre-Nantaise and Maine Rivers, Boussay and Aigrefeuille (Assumed to be COTYPE of *F. Camusi*, PC; CHI); Camus,

Aug. 21, 1890, Loire-Inférieure, Chaudron, Sèvre-Nantaise R., near Boussay (Assumed to be the TYPE of *F. Camusi*, PC); Camus, Apr. 1, 1894, Loire-Inférieure, Loire-Nantaise R., Rousselin, near Brussière (Vendée) and Boussay (Loire-Inférieure), (Assumed to be COTYPE of *F. Camusi*, PC); Camus and Bureau, Loire-Inférieure, Sèvre-Nantaise and Maine Rivers, at Boussay and Aigrefeuille, Husnot, Musci Gall. 933 (CHI, DPU, DUKE, FH, K, PC, WIS); Cardot, Aug. 1917, Loire R. (PC); Crozals, Mar. 1902, Hérault, near Vias, alt. 10 m (PC); Dismier, Aug. 10, 1922, Ardèche (WASH); Renauld, Feb. 1887, at Cohors (PC).

GERMANY: Milde, Breslau (H).

GIGLIO: Bottini (NY); Bottini, Apr. 15, 1887 (H, PC); Bottini, Apr. 1888 (BM, DPU, G-BOIS, PC).

GREECE: Bory de Saint-Vincent, in 1829 (PC).

HUNGARY: Boros, Apr. 3, 1926, in thermal lake, "Malomto," in Tapolca, alt. 120 m (S); Boros, Aug. 1926, in thermal lake, "Malomto," in Tapolca, alt. 120 m, water temperature 16° C., Fl. Hungar. Exsic. 930 (BM, BR, CHI, FH, ILL, MICH, UC); Boros, Aug. 6, 1926, in thermal lake, "Malomto," in Tapolca, alt. 120 m, temperature of water 16° C. (CAS, FH); Boros, Aug. 6, 1926, in thermal lake, "Malomto," in Tapolca, alt. 120 m, water temperature 16° C., Bauer, Musci Eur. et Amer. Exsic. 1930 (Assumed to be portion of TYPE of *F. antipyretica* var. *thermalis*, BART, NY, S).

ITALY: Artaria, Apr. 7, 1895, Pavia (WASH); Gibetti, in 1876, Guiglia (Assumed to be TYPE of *F. Duriaei* f. *latifolia*, PC).

MINORCA ISLAND: Hegelmaier, Apr. 3, 1873 (K).

PORTUGAL: P. Allorge, May 1931, at Bussaco (PC); Ervideira, Nov. 1924, Lisbon, Jamor R., at Cruz-Quebrada, Bauer, Musci Eur. et Amer. Exsic. 1780 (BART, L, NY, WASH); Luisier, Dec. 1908, Bellas, near Lisbon (G-DEL, S); A. Moller, Lusitania (H); Welwitsch 256, June 1847, Algarve, near Silves (K).

SARDINIA: Canepsa, Nov. 10, 1867 (PC); Fleischer (PC); Fleischer, Mar. 9, 1894, Nuoro, Fleisch. and Warnst., Bryoth. Eur. Merid. 74 (PC, WASH); Fleischer, Apr. 9, 1894, Nuoro (PC); Marcucci, Guspini (PC); Marcucci, in 1866, Guspini (G-BOIS, NY, PC).

SICILY: Zodda 444, July 1907, in the Alcantara, at Francavilla, near Messina, alt. 400 m (Assumed to be portion of TYPE of *F. Duriaei* var. *pungens*, S); Zodda, July 24, 1907, in the Alcantara, at Francavilla, near Messina, alt. 400 m (Assumed to be portion of TYPE of *F. Duriaei* var. *pungens*, K, PC).

SPAIN: Casares-Gil, Madrid (FH); Casares-Gil, July 1902, Barcelona (H); Hegelmaier, in 1878, Hispania, near Ronda, alt. 600 m. (B); Hegelmaier, July 22, 1878, Hispania, near Ronda (B); Reuter, July 1841, Pardo, near Madrid (G-BOIS, PC).

UNION OF SOVIET SOCIALIST REPUBLICS: Voronichin, May 5, 1916, Lenkoran, Girdani (H); Voronichin, May 5, 1916, Lenkoran, Azerbaijan (H); Voronichin, May 8, 1916, Lenkoran, Burupsali (H).

NORTH AMERICA: ALASKA: Stair 5029, July 18, 1945, Yakutat, Ophir Creek (DPU); Stair 5029f, July 18, 1945, Yakutat, Ophir Creek, 1.54 m above high water mark, on branch of Salix, "It can probably be explained by the high rainfall at Yakutat which amounts to approximately 350 cms per year and that mostly in the form of a drizzle." (DPU).

CANADA: ALBERTA: Brinkman 4309, July 2, 1929, Lesser Slave Lake District, near Atauwau R. (MICH); Macoun, Sept. 20, 1872, Athabaska Plains, Fl. Can. 750 (CAN); Macoun 2613, Aug. 23, 1879, Great Plains, Red Deer R. (CAN); Macoun, June 14, 1897, Jumping Pound Creek, Rocky Mts. (CAN); Macoun 987, June 14, 1897, Jumping Pound Creek, Rocky Mts. (K); Macoun, June 14 and July 7, 1897, Bragg's Creek and Jumping Pound Creek, foothills of Rocky Mts. (CAN); Macoun, Aug. 4, 1897, in Old Man R. at discharge of Crow's Nest Lake (CAN); Macoun, Aug. 6, 1897, in discharge of Crow's Nest Lake, Rocky Mts. (CAN).

ANTICOSTI ISLAND: Macoun, Aug. 16, 1883, Can. Moss. 205 p.p. (US); Marie-Victorin 19086 (14), Aug. 20, 1917, Grand Ruisseau (DPU, MT).

BRITISH COLUMBIA: Brinkman 145, Nov. 27, 1909 Deadman R., Savonas, alt. 540 m (US); MacFadden, Apr. 24, 1926, mouth of Bonanza Creek, Slocan Lake (CHI, DPU, MACF, NY); Macoun, Oct. 1872, Peace R., Fl. Can. 2615 (CAN); Macoun 311, July 17, 1889, Sicamous (PC).

MANITOBA: Gillett 2293, July 28, 1948, Knife R., Knife Lake, near Churchill (DPU, MICH); Hand 1021, Aug. 6, 1946, Riding Mt. Nat. Park (DPU); Macoun, June 15, 1881, Manitoba House, Can. Musci 752 p.p. (CAN).

NEW BRUNSWICK: Habeeb 449 and 449a, June 18, 1944, Grand Falls (DPU); Habeeb, Aug. 27, 1944, Grand Falls, Musci Novi Bruns. 51 (DPU).

NOVA SCOTIA and ONTARIO: Macoun, Can. Musci 231 p.p. (FH, K, NY, PC, UC, WIS, YALE).

ONTARIO: H. H. Brown 821, Aug. 7, 1939, Victoria Co., Oak Hill, near Head Lake (TRT); R. F. Cain, May 1, 1941, Wentworth Co., near Cambellville (DPU, TRT); R. F. Cain, June 22, 1941, Bruce Co., Sauble Falls (DPU, TRT); R. F. Cain, Aug. 11, 1942, Halton Co., near Milton (DPU, MT, NY, TRT); R. F. Cain, Dec. 24, 1943, Brant Co., near Hatchley (DPU, TRT); Grassl 7676, Sept. 14, 1936, Lavigne (DPU); Hand 586, Oct. 9, 1938, Grey Co., Oxenden (DPU); Hand 646, July 22, 1939, Grey Co., near Rock Mills (DPU); Hand 750, Aug. 14, 1941, Timiskiming District, near New Liskeard (DPU); Kucyniak 42–127, Aug. 5, 1942, Carleton Co., Lake Constance (DPU, MT); Lepage and Dutilly 6199, June 22, 1944, Moosonee (DPU); Lepage and Dutilly 8956, June 27, 1945, Moosonee, James Bay (DPU); Macoun, in 1864, Hastings Co., Moira R., Belleville (NY); Macoun, July 1864, near Belleville (K); Macoun 118, Aug. 1864, Moira R., near Belleville (K); Macoun, in 1866, near Belleville (NY); Macoun 99, in 1868, Hastings Co. (MO); Macoun 223 p.p., in 1870, North Hastings (BM); Macoun, in 1871, near Belleville (CHI); Macoun, May 1871, near Belleville, Can. Musci 759b (CAN); Macoun 256, in 1875 (K); Macoun, in 1876, Colpoy's Bay, Lake Huron (TRT); Macoun, in 1876, Fl. Can. 257 p.p. (K); Macoun, July 8, 1884, Nipigon R. (CAN); Macoun, Sept. 21, 1900, Golden Lake, Can. Musci 604 p.p. (FH); Macoun, Sept. 26, 1900, Golden Lake (CAN); Macoun 874, Sept. 26, 1900, Golden Lake (H); Macoun, Sept. 26, 1900, Golden Lake, Can. Moss. 206 p.p. (CAN, MO, MT, NY, US); Macoun, Oct. 21, 1900, Britannia, Can. Moss. 206 p.p. (K); Macoun, Oct. 27, 1900, Can. Moss. 206 p.p. (B, BM); Moxley, Aug. 18, 1928, Grey Co., Oxenden (DPU, TRT); Moxley, Aug. 17, 1929, Bruce Co., near Colpoy's Bay (DUKE); Moxley, Aug. 4, 1935, Bruce Co., near Wiarton (UT); Roy 54, Owen Sound (E).

QUEBEC: Beaulac 416, July 10, 1932, Abitibi Co., Makamik (MT); Dupret, Oka, Ottawa R. (PC); Dupret, at Como (FH); Dupret, July 12, 1905, Como, Ottawa R. (FH, PC); Dupret, Oct. 1905, Como, Ottawa R. (DPU); Dupret 46, Oct. 30, 1905, Como, Ottawa R. (FH, H); Dupret, July 19, 1906, Como, Ottawa R. (CAN, DUKE, PC, WASH); Dupret, July 19, 1906, Como, Ottawa R., Grout, N. Am. Musci Pl. 297 p.p. (BART, CAN, CM, COLO, DUKE, FH, G-DEL, H, ILL, MINN, MO, NY, OS, PC, S, UC, US, WASH, WIS); Dupret, July 20, 1906, Como and Oka, Ottawa R. (CAN, DUKE); Dupret, Aug. 10, 1906, Oka, Ottawa R. (FH, MO, PENN, YALE); Dupret 1365 p.p., Aug. 10, 1906, Deux-Montagnes Co., Oka, Ottawa R. (BART, MT); Dupret 1758, Aug. 10, 1906, Deux-Montagnes Co., Oka, Ottawa R. (MT); Dupret, June 16, 1913, Cartierville, Montreal Isl. (DPU, MT, NY, US); Dupret 3705, June 16, 1913, Cartierville, Montreal Isl. (MT); Dupret, Jan. 30, 1915, Oka, Point Boileau, Ottawa R. (CAS); Fabius 1055 and 1056, May 10, 1947, Mt. Shefford (DPU); Fabius H–62, May 30, 1948, St. Dominique (DPU); Kucyniak 39–1, Oct. 1939, David Bridge, Montreal Isl. (DPU, MT); Lepage 1870 and 1882, July 5, 1940, Lac à l'Anguille, St. Anaclet, Rimouski (DPU); Lepage 1879, July 5, 1940, Lac à l'Anguille, St. Anaclet, Rimouski (DPU, MT); Lepage 1922, July 15, 1940, Témiscouata Co., Lake Légaré (DPU); Lepage 3233, Aug. 16, 1941, Rimouski Co., Lake St. Mathieu (DPU); Lepage 3383, June 29, 1942, Rimouski Co., Shaw Lake, St. Narcisse (DPU); Lepage 3959, Aug. 26, 1942, Rimouski Co.,

Lake Croche, St. Guy (DPU); Lepage and Dutilly 6434, July 8, 1944, Vieux Comptoir, lat. 52° 33' (DPU); Lepage and Dutilly 6721, Aug. 27, 1944, Vieux Comptoir, lat. 52° 33' (DPU); Lepage and Dutilly 12147 and 12151, July 7, 1946, Harricanaw R. (DPU); Marie-Anselme 66, Sept. 8, 1932, Saguenay R., Chicoutime (DPU); Marie-Anselme 843, Aug. 14, 1936, Lake Waterloo (DPU); Marie-Anselme 912, Sept. 14, 1936, Lake Waterloo (DPU); Marie-Anselme 1643, July 3, 1937, Mt. Shefford (DPU); Marie-Anselme 1624 and 1625, July 19, 1937, South Stukebey (DPU); Marie-Victorin 1, in 1908, Chambly Co., Longueuil, St. Antoine (DPU, MT); Marie-Victorin 1335 and 1740, June 1909, Chambly Co., Longueuil (DPU, MT); Marie-Victorin 9330, in 1912, Deux-Montagnes Co., St. Eustache, Ottawa R. (DPU, MT); Marie-Victorin 3715, Aug. 1912, Deux-Montagnes Co., St. Eustache, Ottawa R. (DPU, MT); Marie-Victorin 19164, Sept. 1916, Chambly Co., Longueuil (DPU, MT); Marie-Victorin 65, July 20, 1921, Lake St. Jean Co., St. Prime Bay (DPU, MT); Marie-Victorin 66, July 29, 1921, Lake St. Jean Co., Roberval (DPU, MT); Marie-Victorin and Rolland-Germain, Aug. 11, 1940, Laval Co., St. Rose (DPU, MT); Marie-Victorin, Rolland-Germain, and Boivin 10999, Aug. 11, 1940, Laval Co., St. Rose, Milles-Iles R. (DPU, MT); Marie-Victorin, Rolland-Germain, and Brunel 45306, July 6, 1933, Labelle Co., Lake Gatineau (DPU, MT); Rhéole 1917, July 1917, St. Jérôme region (DPU, MT).

QUEEN CHARLOTTE ISLANDS: Spreadborough, July 1, 1910, Skidegate (B, CAN, CM, MO).

SASKATCHEWAN: Bourgeau (NY); Bourgeau, 1857–1858 (K); Bourgeau, in 1859 (PC); Macoun, July 19, 1880, Moose Jaw Creek (CAN); Macoun, July 19, 1880, Moose Jaw Creek, Can. Musci 759b (CAN); Macoun, July 4, 1895, Cottonwood Coulee, Milk R. (CAN); Macoun, July 4, 1895, Cypress Hills, Battle Creek (E, H, K, S); Macoun, July 4, 1895, Cypress Hills, Battle Creek, Can. Musci 603a (CAN, E, FH, K, MO, NY, S); Macoun 264, July 4, 1895, Cypress Hills, Battle Creek (Assumed to be TYPE of *F. subcarinata*, PC).

VANCOUVER ISLAND: Macoun, May 8, 1875, Fl. Can. 257 p.p. (CAN).

Canada without locality: Macoun, Moss. of Can. 190 (MO).

UNITED STATES: ALABAMA: Worthington, Feb. 10, 1877, Lookout Mts. Little River Falls (OS).

ARIZONA: Pringle, May 6, 1881, Santa Rita Mts. (DUKE, NY, PC).

CALIFORNIA: Mrs. R. M. Austin, in 1880, Plumas Co., Feather R. (UC); Baker and Nutting, May 28, 1894, Burney Valley (CAS, NY, UC); Barnes (PC); Bolander 79 p.p. (FH, NY, US, WIS).

COLORADO: C. F. Baker 42, July 13–19, 1895, Larimer Co., Chambers Lake, alt. 2700 m (BART, BM, CIN, COLO, FH, MINN, NY, PC, US); C. F. Baker 2, Aug. 1, 1896, Larimer Co., Chambers Lake, alt. 2700 m (PC); C. F. Baker, Sept. 1, 1896, Larimer Co., Chambers Lake (DUKE); Brandegee (NY); Brandegee, 1874–1878, within 100 miles of Canyon City (MICH); Craft 832, Sept. 1951, Hinsdale Co., Lake City (DPU); Emmitt 1613, Oct. 28, 1950, Arapahoe Nat. Forest, near Echo Lake, alt. 2880 m (DPU); McCaskey 201, July 31, 1939, Boulder Co., Shadow Lake, alt. approximately 3000 m (DPU); Rodeck, Aug. 8, 1938, Boulder Co., Shadow Lake, alt. 3150 m (DPU); Spencer, Aug. 1, 1938, La Plata Co., alt. 2070 m (DPU).

CONNECTICUT: J. A. Allen, Apr. 20, 1880, Mill R., New Haven (CIN, FH, NY, PC, YALE); G. E. Nichols, June 11, 1911, North Branford (PC, YALE); G. E. Nichols, June 30, 1911, Lakeville (CIN, PC, YALE); G. E. Nichols, July 3, 1911, Salisbury (NY, PC, YALE).

IDAHO: T. C. Frye, Sept. 4, 1929, Custer Co., Mackay (DPU, WASH); Larsen and Mercer, Mar. 10, 1945, Cassia Co., Burley, Burley Irrigation District, "Submersed in canals, on hard bottoms, catches on most anything and holds silt, forming mounds in the canals. It is hard to kill, and costs considerable to keep it out." (UT); Leiberg 114–230 p.p., Kootenai Co., near Lake Pend d'Oreille (DUKE, PC).

INDIANA: Deam 49117, July 18, 1930, Pulaski Co., Tippecanoe R., near Winamac (BUT, DPU); Deam 55177, July 18, 1934, Elkhart Co., St. Joseph R., Bristol (CHI, DPU, MICH, NY); Deam 57125, Aug. 2, 1926, Fulton Co., Tippecanoe R., near DeLong (DPU); Deam 59827, Aug. 21, 1940, Elkhart Co., bayou of St. Joseph R., near Bristol, "Very common over an acre or two in the bed of a bayou of the St. Joseph River just below the bridge at Bristol. Attached to rocks of all sizes and shapes." (CHI, DPU, NY); Hull, in 1934, Lake Co., Hobart Road, between Miller and East Gary (DPU).

IOWA: Conard 42–12, June 24, 1942, Emmet Co., Estherville, Ft. Defiance State Park (CHI, DPU, NY, US, UT, WASH); Linder, Johnson Co. (DPU); Wolden, Apr. 19, 1925, Emmet Co., Estherville, Ft. Defiance State Park (DPU).

MAINE: Collector ?, July 7, 1934, Holeb (DUKE).

MASSACHUSETTS: Kennedy, June 9, 1905, Bishops Brook (FH).

MICHIGAN: Conard 27, Aug. 16, 1937, near mouth of Iron R., west of Ontonagon (DPU, GRI); Conard 22, Aug. 18, 1937, Cranberry R., near Ontonagon (DPU, GRI); Darlington 346, Aug. 1935, Leelanau Co., Colen Lake (MICH); Ehlers, June to Aug. 1920, Chippewa Co., Scotty Bay Creek (MICH, NY, YALE); Ehlers, June to Aug. 1920, Emmet and Cheboygan Counties, between Levering and Carp Lake (MICH, NY); Ehlers 2072, June to Aug. 1920, Emmet Co., Carp Lake (YALE); Ehlers 2075, June to Aug. 1920, Chippewa Co., Scotty Bay Creek (YALE); Ehlers 136, July 18, 1920, Emmet Co., near Carp Lake (MICH, WASH); Ehlers 2074, July 30, 1920, Chippewa Co., Scotty Bay Creek (MICH); Ehlers, Aug. 26, 1920, Mackinac Co., Prentis Bay Creek (MICH); Ehlers, July 30, 1931, Cheboygan, Nigger Creek (DPU, WVA); Ehlers and Blinks, June to Aug. 1920, near Burt Lake (MICH, NY, YALE); Gleason 2337, Aug. 5, 1939, Cheboygan Co. (DPU); Hermann, July 20, 1936, Keweenaw Co, near Copper Harbor (MICH); G. E. Nichols, June to Aug. 1920, Emmet and Cheboygan Counties, Maple R., near Pellston (MICH, NY); G. E. Nichols, June to Aug. 1920, near Pellston (YALE); G. E. Nichols, July to Aug. 1921, Emmet and Cheboygan Counties, Carp Creek, near Burt and Douglas Lakes (MICH, NY, YALE); G. E. Nichols, Aug. 1921, Carp Creek, near Douglas Lake (DPU, YALE); G. E. Nichols, Aug. 1926, Carp Creek, near Douglas Lake (DPU, YALE); G. E. Nichols, June to Aug. 1930, near Mackinaw City (YALE); G. E. Nichols, Aug. 1935, Tahquamenon (DPU, YALE); G. E. Nichols, June 1937, Marquette Co., Huron Mt., Mountain Lake (DPU, YALE); G. E. Nichols and Steere, Aug. 1935, Ontonagon Co., Porcupine Mts. (DPU, MICH); G. E. Nichols and Steere, Aug. 1935, Ontonagon Co., Porcupine Mts., pool on Lake Superior beach (DPU, MICH); G. E. Nichols and Steere, Aug. 20–27, 1935, Ontonagon Co., Porcupine Mts., pool along shore of Lake Superior (DPU, MICH, YALE); Phinney 321, July 22, 1941, Cheboygan Co., Maple R. (CHI); Schaffner, Apr. 27, 1895, Geddes (OS); Schnooberger 1960a, Oct. 29, 1938, Gratiot Co. Line and East Superior Road (DPU); Schnooberger 3732, Oct. 27, 1940, Gratiot Co. (DPU, OS); Shacklette 2026, July 21, 1941, Cheboygan Co., near Lake Huron (DPU); Steere, Aug. 1935, Ontonagon Co., Porcupine Mts., pool along Lake Superior Shore (MICH); Steere, Sept. 1, 1935, Ontonagon Co., Bond Falls (DPU, MICH); Tarzwell 253, Aug. 6, 1936, Calhoun Co., Rice Creek (DPU, MICH); Tinney, July 6, 1933, Cheboygan Co., Carp Creek (DPU); Wynne 1310, Aug. 22, 1939, Ogemaw Co., Prior Creek (DPU).

MINNESOTA: Cahn 720 and 740, in 1935, near Burntside Lake, near Ely (DPU); Drexler, Aug. 5, 1948, Cottonwood Co., Pipestone Nat. Monument (DPU); Holzinger 77, Sept. 1895, Minneapolis (FH, NY, WIS); Holzinger, Aug. 1896, near Minneapolis (CIN, H, PC); Holzinger, Aug. 1896, near Minneapolis, Grout, N. Am. Musci Pl. 190 (CAN, CM, DUKE, FH, ILL, MINN, MO, NY, OS, PC, UC, US, WASH, WIS, YALE); Holzinger, June 8–10, 1897, Ball Lake, near Ely (MINN); Holzinger, June 10, 1897, Fall Lake (PC); Holzinger, in 1901, Minnesota R., near Granite Falls (H); Holzinger, July 10–15, 1901, Yellow Medicine Co., Minnesota R., near Granite Falls (DUKE, FH, H, K, MINN, MO, NY, PC, US, WYO); Holzinger, July 10–15, 1901, Minnesota R., near Granite Falls, Renauld and Cardot, Musci Am. Sept. Exsic. 382 (BM, CAN, FH, MICH, NY, PC, S, YALE); Holzinger, July 12, 1901,

Minnesota R., at Granite Falls (Assumed to be TYPE of *F. obscura*, PC); Holzinger, in 1902, near Grand Marais, Renauld and Cardot, Musci Am. Sept. Exsic. 383 (CAN, FH, NY, PC, YALE); Holzinger, July 28, 1902, Cook Co., Rosebush Falls and Rosebush Creek, near Grand Marais (DUKE, NY, PC, WASH); Holzinger, Aug. 5, 1902, Cook Co., near Grand Marais (PC); Holzinger and Elftman, June 17, 1897, second falls of Granite R., east of Saganaga (WASH); Holzinger and Elftman, June 21, 1897, northern Minnesota (PC); Linnaean Club, Univ. of Minnesota, No. 67, May 21, 1939, Rice Co. (MINN); MacMillan, Brand, and Lyon, Aug. 23, 1901, Gunflint Lake (PC); Moyle 2627, Sept. 28, 1935, St. Louis Co. (MINN); Moyle 3335, July 26, 1939, Benton Co. (MINN); Moyle 3330, Sept. 7, 1939, Morrison Co. (MINN); Moyle 3336, Sept. 11, 1939, Morrison Co. (MINN); Moyle 3618, July 8, 1940, St. Louis Co. (MINN); Moyle 3586, July 10, 1940, St. Louis Co. (MINN); W. A. Wheeler 1145, Sept. 14, 1901, Rock Co., Minnesota Springs (MINN).

MISSOURI: Bush, June 17, 1897, Eagle Rock (DUKE, MO, NY, PC, US); Bush, May 24, 1898, Eagle Rock (MO); Conard, Apr. 5, 1947, Shannon Co., Round Spring State Park (DPU, GRI); Conard 47–5, 47–6, 47–7, and 47–9, Apr. 5, 1947, Shannon Co., Round Spring State Park (DPU, GRI); Conard 47–52 and 47–54, Apr. 6, 1947, Shannon Co., Blue Spring (DPU, GRI); Conard 47–59, 47–62, and 47–65, Apr. 6, 1947, Shannon Co., Powder Mill Spring (DPU, GRI); Conard 47–28, 47–29, 47–30, 47–31, and 47–32, Apr. 7, 1947, Carter Co., Big Spring State Park (DPU, GRI); Conard 47–68, Apr. 10, 1947, Shannon Co., Cove Spring (DPU, GRI); Davis, Mar. 28, 1915, Pike Co. (MO, OS, UC); Drew 7438, June 9, 1938, Shannon Co., Round Spring State Park, Round Spring (UMO); E. J. Palmer 2260, June 19, 1909, Joplin (CHI, MO, NY); Steyermark 26914, June 4, 1939, Shannon Co., Blue Spring, Jack's Fork, Current R., near Montier (CHI).

MONTANA: T. C. Frye, Sept. 7, 1928, Ronan (DPU, WASH); T. C. Frye, July 21, 1934, St. Ignatius, near St. Mary's Lake (DPU, WASH); Maguire and Piranian 5330, July 4, 1934, Glacier Nat. Park, near Divide Lake (UC); R. S. Williams 82, Apr. 12, 1887 (CHI, YALE); T. G. and E. C. Yuncker 10905, July 25, 1937, Glacier Nat. Park, Camas Creek, alt. 1050 m (DPU).

NEW YORK: Brewer, May 7, 1858, Ovid (NY); Clinton (MICH); Cook, Sept. 1888, Canandaigua (ILL, NY); E. J. Hill 157. 1882, Aug. 11, 1882, Mumford, Spring Creek (CHI, ILL); Sartwell, Yates Co. (CHI); Winne 867, Aug. 16, 1935, Saratoga Co., Ballston Lake, alt. 96 m (DPU, NY); Winne 864, Aug. 11, 1940, Warren Co., bay of Lake George, Assembly Point, alt. 960 m (DPU, NY); Winne 866, Sept. 15, 1940, Albany Co., Colonie, Rudd's Pond, alt. 90 m (DPU, NY); Winne 309, Aug. 2, 1941, Cortland Co., Tioughnioga R., Cuyler, near DeRuyter, alt. 354 m (DPU); Winne 336, Aug. 10, 1941, Lewis Co., Deer R., Copenhagen, alt. approximately 352 m (DPU, NY); Winne 724, Nov. 7, 1941, Albany Co., Watervliet Reservoir, Norman's Kill, alt. 78 m (DPU, NY); Winne 840, Dec. 5, 1941, Saratoga Co., Ballston Lake, alt. 75 m (DPU, NY); Winne and Muenscher 249, June 29, 1941, Oswego Co., Oneida Lake, Three Mile Bay, near Constantia, alt. 108 m (DPU, NY).

NORTH DAKOTA: Brenckle (PC); Brenckle, June 1910, Kulm, Grout, N. Am. Musci Pl. 404 (BART, CAS, CM, COLO, DUKE, FH, MINN, MO, NY, OS, UC, US, WASH, YALE).

OHIO: Moseley, in 1911, Castalia (CM); Sterki, Apr. 1910, near Geneva (CM); Sterki, Aug. 24, 1912, Summit Co., Long Lake (CM).

OREGON: J. A. Allen, in 1912, Willamette R., Rockspur, near Portland (DUKE, YALE); Coville and Applegate 1142, Aug. 22, 1898, Linn Co., Clear Lake, bottom of lake in about 6 m of water, alt. 660 m (NY).

PENNSYLVANIA: Bartram 233, Feb. 12, 1921, Monroe Co., Bushkill (BART); Jennings, Aug. 4, 1909, Crawford Co. (CM); T. C. Porter (K); T. C. Porter, Pocono (NY); T. C. Porter, Huntingdon Co. (PHIL); T. C. Porter, in 1863, Huntingdon (CHI); Rau, Monocacy R., springs along bank (NY); Rau, Bethlehem (OS); Rau 135, Bethlehem (FH); Wolle and Rau, Monocacy R., Bethlehem (NY).

SOUTH DAKOTA: A. C. McIntosh, Dec. 20, 1926, Rapid City (DPU); Over 15457, Aug. 18, 1923, Roberts Co., Jim Creek (NY, US).

TEXAS: Whitehouse 23109, Mar. 27, 1950, Polk Co., near Corrigan (DPU, SMU).

UTAH: Flowers 2525, June 22, 1927, Cache Co., Logan, alt. 1410 m (DPU, UT); Flowers 2524, Apr. 10, 1938, Wasatch Co., Midway Fish Hatchery, alt. 1680 m (DPU, UT); Flowers 2527, Sept. 24, 1940, Murray, alt. 1290 m (DPU, UT); Flowers 2528, Oct. 2, 1943, Salt Lake City, alt. 1290 m (DPU, UT); M. E. Jones, July 12, 1886, Coalville (US).

VERMONT: Bureau of Fisheries, Aug. 15, 1909, Groton (CHI); Dutton 439, July 31, 1910, Brandon (DUKE); Dutton, July 11, 1911, Ferrisburg (ABS); Dutton 741, July 11, 1911, Ferrisburg (CM); Dutton 900, July 26, 1915, Brandon (CHI, DUKE, FH, MO, MT); Faxon 328, June 18, 1881, Ferrisburg (NY); Faxon 327, July 27, 1884, Ferrisburg (NY); Grout, in 1909, Mullets Bog (DUKE); Kennedy, Sept. 16, 1898, Willoughby (FH); Kennedy, Oct. 27, 1898, Willoughby (FH); Kirk 198, July 8, 1911, Ferrisburg (ABS); Pringle, June 19, 1880, Willoughby Lake (DUKE, NY).

WASHINGTON: A. S. Foster 1821 p.p., Aug. 11, 1911, Clallam Co., Crescent Lake (US); Sereno Watson, 10–9–1880. Okanagan R. (FH, NY).

WISCONSIN: Cheney 638, June 23, 1893, Vilas Co., Wisconsin R. (DPU); Cheney 639 and 640, June 23, 1893, Vilas Co., Wisconsin R. (DPU, WIS); Cheney 196a, June 26, 1893, Vilas Co., between Conover and Eagle R. (WIS); Cheney 796a, June 26, 1893, Vilas Co., Wisconsin R., near Conover (DPU); Cheney 841, June 26, 1893, Vilas Co., Wisconsin R., near Eagle R. (DPU, WIS); Cheney 875, June 27, 1893, Vilas Co., Wisconsin R., near Eagle R. (DPU, WIS); Cheney 880 and 881, June 27, 1893, Vilas Co., Wisconsin R., near Eagle R. (DPU); Cheney 976, June 29, 1893, Oneida Co., Wisconsin R., near Rainbow Rapids (DPU, WIS); Cheney, July 1893, northern Wisconsin (PC); Cheney 1730, July 16, 1893, Oneida Co., Wisconsin R., near Rhinelander (DPU); Cheney 1853 and 1867, July 19, 1893, Oneida Co., Wisconsin R., Hat Rapids, near Rhinelander (DPU, WIS); Cheney 2015, July 20, 1893, Lincoln Co., Wisconsin R., Whirlpool Rapid (DPU, WIS); Cheney 2505, July 29, 1893, Lincoln Co., Wisconsin R., Grandfather Bull Falls, near Merrill (DPU, WIS); Cheney 2843, Aug. 2, 1893, Lincoln Co., Wisconsin R., Merrill (DPU, WIS); Cheney 2849, Aug. 2, 1893, Lincoln Co., Wisconsin R., near Granite Heights (DPU, WIS); Cheney 3375, June 30, 1894, Portage Co., Wisconsin R., between Knowlton and Stevens Point (DPU, WIS); Cheney 3477, July 3, 1894, Portage Co., Wisconsin R., Stevens Point (DPU, WIS); Cheney, June 1896, northern Wisconsin (PC); Cheney 4066, June 22, 1896, Bayfield Co., White R., Drummond (DPU, WIS); Cheney 7993, Aug. 1, 1897, Douglas Co., near St. Louis R. (DPU); Cheney 9721, Jan. 7, 1925, Barron Co., Barron (DPU, WIS); Cheney 10273, July 10, 1925, Waukesha Co., North Lake (DPU); Cheney 10273a, July 10, 1925, Waukesha Co., North Lake (WIS); Cheney 10893, Oct. 6, 1925, Jefferson Co., Palmyra (DPU, WIS); Cheney 10920, Oct. 7, 1925, Waukesha Co., near Eagle (DPU, WIS); Cheney 10927, Oct. 8, 1925, Waukesha Co., near Eagle (DPU, WIS); Cheney 13086, July 3, 1930, Burnett Co., Roosevelt Township (DPU, WIS); Cheney 13143, in 1931, Douglas Co. (DPU, WIS); Cheney 13130, May 22, 1931, Barron Co., Barron Township (DPU, WIS); Cheney 13144, July 16, 1931, Douglas Co. (DPU, WIS); Cheney 13170, June 25, 1933, Barron Co., Doyle Township (DPU, WIS); Cheney 13207a, June 15, 1935, Polk Co. (DPU, WIS); Cheney 13207b, June 15, 1935, Polk Co. (DPU); Gillman, July 6, 1866, White Fish Bay (NY); Schallert, June 22, 1926, Bark R., Rome (WASH); L. R. Wilson 2041, Aug. 29, 1931, Douglas Co., Lucius Lake (WIS).

WYOMING: Conard, in 1924, Camp Roosevelt (NY); Dennings 1092, July 17, 1941, Centennial (DPU); Aven Nelson 69, July 8, 1896 (PC); Aven Nelson 2248, July 8, 1896, Crook Co., Moorcroft (DUKE, E, FH, MO, NY, WYO); Aven Nelson, July 17, 1897, Sweet Water Co. (PC); C. L. Porter 656, Oct. 5, 1930, Little Laramie R., near Laramie, alt. 2100 m (WYO); C. L. Porter 1072, June 20, 1932, Albany Co., Laramie Mts., alt. 2400 m (WYO); C. L. Porter 1370, July 16, 1933,

Albany Co., near Centennial, alt. 2250 m (DPU, FH, NY, WYO); T. G. and E. C. Yuncker 12370, July 15, 1946, Medicine Bow Nat. Forest, near Centennial, alt. about 2400 m (DPU); T. G. and E. C. Yuncker 12371, July 16, 1946, Medicine Bow Nat. Forest, near Centennial, alt. 2400 m (DPU).

SOUTH AMERICA: BRAZIL: E. M. Reineck, Feb. 19, 1898, Rio Grande do Sul, Santa Cruz (PC).

Plants of *F. Duriaei* show great variation in vegetative structures. On the same plants, leaves occur which resemble those of *F. hypnoides* along with blades which are true to the description of *F. Duriaei*, or leaves with bases resembling the former and apices similar to those of the latter species. Some branch blades on plants of *F. Duriaei* resemble median cauline leaves of *F. hypnoides*. The outline of the alar group on one side of the base may be suborbicular and on the opposite, rectangular. An auricle may be distinct on one side of the blade and absent on the other. It is very important that median cauline blades of well developed or mature plants be used for accurate determination. Occasionally it is difficult to name the species with certainty. The author feels justified in retaining *Duriaei* in specific rank because a large number of plants bear leaves which are distinctly true to this species. The writer has used the following combination of characteristics for the determination of *F. Duriaei:* plants slender to medium in size, stems and blades commonly flaccid, leaves sometimes somewhat firm, majority of median cauline blades distant, bases up to 2 mm apart, leaves plane or nearly so, broadly ovate-lanceolate or oval-lanceolate, sometimes oblong-lanceolate, width decreasing either gradually or somewhat abruptly from approximate middle into broad and short apices, leaf tips generally acute, serrulate or entire, blades usually 3–5 mm long, 1–2.5 mm wide, 2–3.5 : 1, alar group subrectangular, suboval, or suborbicular in outline, rarely extending up margin farther than elsewhere in the blade, auricles frequently present, slight to distinct, leaf bases often rounded, usually not decurrent but sometimes briefly so, and trellis perfect and muricate.

Cardot in his monograph on Fontinalaceae refers to *F. nitida* as a regional race of *F. hypnoides*, but retained it as a species. The writer in the Fontinalaceae of North America treated *F. nitida* as a species. Having examined numerous additional collections of specimens determined as *F. nitida*, *F. hypnoides*, and *F. Duriaei*, including the types, the author has observed in the present study that the plants which have been named *F. nitida* have characteristics in common with both *F. hypnoides* and *F. Duriaei*. The leaf width and the shape of base are comparable to *hypnoides*, but the apices, alar groups, and auricles are usually similar to those of *Duriaei*. The median branch leaves and the smaller cauline blades of plants of *Duriaei* often duplicate many characteristics of the median cauline leaves of plants formerly considered as *F. nitida*. The smaller leaf measurements of plants called *F. Duriaei* coincide with those of the larger leaves of specimens of *F. nitida*. Thus, it now seems to the author that plants which have been determined previously as *F. nitida* may be either young or small specimens of *F. Duriaei*, or plants which have been so influenced by the environmental factors of their habitats that the leaves do not attain the size which is commonly characteristic of *F. Duriaei*.

Cardot published *F. Camusi* as a member of the *Heterophyllae* (cauline and branch leaves unlike). Since the leaves are plane, the author has placed the species in the group with majority of leaves plane. According to a note in Cardot's Herbarium, he later regarded these plants as belonging to the latter group (Cardot's *Malacophyllae*). Cardot described the trellis as imperfect. If that is true, the writer may be in error in placing this species in the synonymy of *F. Duriaei*. A few fruits, with broken trellises, were available for study. The few remaining cilia were examined. One had long fragments of cross bars throughout its length. In other peristomes, basal fragments of cilia attached to the urn were joined by cross bars. Thus, the author assumes that originally the trellis may have been perfect and that it has appeared to be imperfect because of the broken condition.

Plants of *F. Duriaei* are sometimes confused with *F. antipyretica* in determinations because flaccid leaves of *antipyretica* often appear plane, especially those of the branches, and frequently blades of *Duriaei*, because of their flaccidity, become folded lengthwise, equally or unequally, seeming to be conduplicate. Also, halves of leaves of *antipyretica*, formed by the splitting of the keel, sometimes resemble complete blades of *Duriaei*. Under these circumstances, the plants of *F. antipyretica* may be distinguished by stems of greater diameter, commonly 0.3–0.5 mm and majority of leaf tips subobtuse or obtuse and alar groups with more numerous cells, vertical rows up to 13, in contrast with *F. Duriaei*, stems commonly approximately 0.25 mm in diameter, and majority of leaf tips acute and alar groups with fewer cells, 5–7 vertical rows.

The plants of *F. fasciculata* in the herbarium in Helsinki, collected by Durieu in Algeria and sent to Lindberg by J. Lange, have been studied, along with portions of the same collection, it is assumed, in herbaria in London, Kew, and Paris. The specimens at Kew are from the Schimper Herbarium, and the others are from the Herbarium of Bescherelle. As indicated by Lindberg in his description, the plants are in very poor condition. The stems and older branches are either denuded or the remaining leaf fragments are badly lacerated. Lindberg, in his selection of name and his emphasis in description, indicates that the fasciculate condition of the branches was the basis for his new species. Fasciculation has been noted occasionally by the author in the examination of *Fontinalis* specimens and has been regarded as a response to an environmental factor, which may have destroyed the meristematic tissue at the ends of the stems and branches, with the resultant development of new branches. It is assumed that the plants upon which Lindberg based this species are abnormal specimens. It does not seem probable that the fasciculation in *Fontinalis* is hereditary, and thus the writer does not consider the fascicles of branches as diagnostic characteristics sufficient for the retention of the species, *F. fasciculata*. If, to the contrary, *F. fasciculata* should be assumed to be a distinct species, the author would hesitate to describe the specimens in the absence of complete median cauline blades. The assignments to synonymy can not be made with certainty in the absence of complete median cauline leaves. Branch leaves usually show greater variation than median cauline blades. The canaliculate leaves described by Lindberg are assumed to be the deeply concave blades of some of the branches in the fascicles. On the longer and more normal branches the leaves are plane and resemble in several characteristics those of *F. Duriaei*. Also, the cauline fragments are suggestive of *F. Duriaei*. If the determination of these plants could be made definitely as *F. fasciculata*, the publication date of *F. fasciculata*, 1869, would have priority over that of *F. Duriaei*, 1876. Since the type specimens of *F. Duriaei* show the major characteristics of that species, and since the type material of *F. fasciculata* appears to be composed of abnormal plants and is so poor that the specific characteristics are indefinite, the author considers *F. fasciculata* as a nomen dubium and as a synonym of *F. Duriaei*.

29. **Fontinalis flaccida** Ren. and Card., in Bot. Gaz. 13: 201. 1888.

Fontinalis denticulata Kindb., in Röll, in Hedwigia 36: 61. 1897.
Fontinalis flaccida Ren. and Card. f. *minor* Card., in Welch, in Grout, Moss Fl. N. Am. 3: 254. 1934 (nomen nudum).

Plants slender in size, yellowish, yellowish green, green, brownish green, or brownish, glossy, especially in younger portions, older portions glossy or dull; stems usually flaccid, occasionally slightly rigid to rigid, up to 40 cm in length, 0.2–0.25 mm in diameter, blackish and denuded at base with age, irregularly pinnately branching; branches few to numerous, usually erect-spreading, occasionally

spreading, close to distant, up to 6.5 cm in length, ends of foliated
stems and branches attenuate; median cauline leaves distant, bases

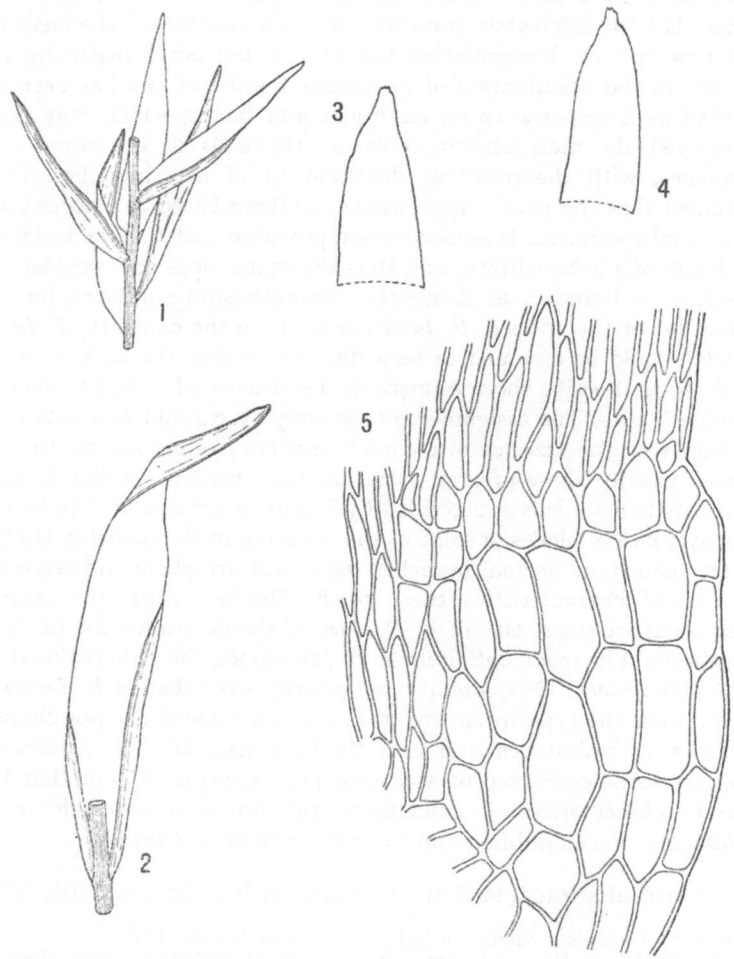

Fig. 28. – *Fontinalis flaccida* Ren. and Card. – 1.Portion of stem with
median cauline leaves. 2.Median cauline leaf. 3 and 4.Leaf apices.
5.Alar cells. Nos. 1–4 drawn from Langlois 25, Mar. 24, 1886, Louisiana;
no. 5 after Renauld and Cardot, in Bot. Gaz. 13: pl. 19.

up to 1.5 mm apart, blades flaccid, erect-spreading to spreading,
generally plane throughout, sometimes subconcave to concave at
base, occasionally very slightly concave at apex, elongate, narrowly

ovate-lanceolate, oblong-lanceolate, or lanceolate; apices long, either narrowly or broadly acuminate, usually abruptly narrowed into truncate, obtuse, or subobtuse ends, leaf tips occasionally acute, generally distinctly serrulate with a few prominent cells, rarely entire, 4–7 mm long, 0.5–1.5 mm wide, 3–6 : 1; median cells of leaves linear, ends attenuate, 6.5–11.9 µ wide, usually 10–15 : 1, occasionally up to 20 : 1; alar cells generally much enlarged, subrectangular, subquadrate, or subhexagonal, walls hyaline, yellowish, or brownish, frequently in vertical rows, up to 8, and sometimes in horizontal rows, up to 5, forming very conspicuous auricles, leaf bases not decurrent to briefly so, up to 0.5 mm; median branch leaves similar to median cauline except smaller in size; perichaetial branch 4–5 mm long; perichaetium subcylindrical to cylindrical, 0.75–1 mm in diameter; upper perichaetial leaves broadly ovate, apices obtuse, lacerate and truncate with age; calyptra not seen; operculum long conical; seta 0.2–0.25 mm long; urn immersed, subcylindrical, 2–2.25 mm long, 0.5–0.75 mm in diameter, 2.66–3 : 1, either contracted or not contracted beneath mouth when dry; peristome teeth brownish orange, linear-acuminate, often united in pairs at apex, 0.65–0.75 mm in length, muricate, lamellae 18–20; trellis brownish orange, imperfect, approximate length of teeth, muricate; spores yellowish green, green, or yellowish brown, finely muricate, 17–22.1 µ in diameter, ripe in summer. [The specific name, *flaccida*, has reference to the flaccid stems and leaves.]

Type: Not previously designated. In Cardot's Herbarium in Herbarium, Laboratoire de Cryptogamie, Muséum d'Histoire Naturelle, Paris, France, the specimens under the following label are assumed to be the type material: "*Fontinalis flaccida* n. sp. No. 140. Langlois, 24.3.86. La. orientale: – sur branches et racines boignant dans l'eau du Bayou Bonfouca." On the back of the packet Cardot wrote the description.

Type collector: A. B. Langlois, Mar. 24, 1886.

Type locality: Louisiana: Bayou Bonfouca.

Distribution: North America: eastern United States and Canada.

Additional descriptions: Renauld and Cardot, in Bull. Soc. Roy. Bot. Belg. 27: 134. 1888; Cardot, Mon. Font., p. 118. 1892; Renauld and Cardot, in Bull. Herb. Bois. 4: 18. 1896; Barnes, Gen. and Sp. N. Am. Moss., p. 330. 1896; Welch, in Grout, Moss Fl. N. Am. 3: 254 (as *F. flaccida* and as *F. denticulata*). 1934; Jennings, Man. Moss. W. Penn. and Adj. Reg., p. 177. 1951.

Illustrations: Renauld and Cardot, in Bot. Gaz. 13: pl. 19. 1888; Renauld and Cardot, in Bull. Soc. Bot. Belg. 27: pl. 9. 1888; Welch, in Grout, Moss Fl. N. Am. 3: pl. 77 (as *F. denticulata*), pl. 78 (as *F. flaccida*). 1934. – FIGURE 28.

Specimens examined: NORTH AMERICA: CANADA: CAPE BRETON ISLAND: G. E. Nichols, July to Aug. 1914, Mountain Lake, near Dingwall (NY, YALE).

NOVA SCOTIA: Jardin, in 1861 (PC); Macoun, July 17, 1883, Yarmouth (CAN); Macoun 553, Aug. 17, 1899, Sable Isl. (K).

ONTARIO: Macoun, July 1869, Lake Superior, near Michipicoten, Moss. Can. 192 (MO, NY); Macoun, June 1886, Belleville (CAN).

QUEBEC: Marie-Anselme 727, July 2, 1936, Lake Bourgeois (DPU, YALE); Marie-Anselme 11, Apr. 7, 1937, Waterloo (DPU).

UNITED STATES: ALABAMA: C. F. Baker, May 26, 1897, Athens (DUKE, MINN, NY, PC); Harvill 6841, Apr. 23, 1950, Mobile Co. (DPU).

ARKANSAS: Bush 5414, Mar. 26, 1909, Fulton (CHI, FH, NY, US); Bush 5414a, Mar. 26, 1909, Fulton (FH, NY, US); Engelmann, Mar. 7, 1837, Little Rock (K); Scully 1349, June 2, 1939 (DPU); Scully 1321, July 23, 1939, Hot Springs (DPU).

CONNECTICUT: J. A. Allen, Apr. 19, 1880, East Haven (Was determined by Cardot as F. *flaccida* f. *minor*, YALE; CIN); Eaton, Oct. 1855, New Haven (PC); Lorenz, Apr. 6, 1913, South Mountain, Bristol, alt. 240 m (NY, S, YALE); G. E. Nichols, Apr. 13, 1911, North Stonington (NY, PC); G. E. Nichols, Apr. 14, 1911, North Stonington (YALE); G. E. Nichols, Aug. 15, 1911, Monroe (NY, PC, YALE).

DELAWARE: Commons, June 11, 1890, Townsend (NY, PHIL); James, in 1851, Wilmington (K).

FLORIDA: Hood, near Orange City (DUKE); Hood, Apr. 29, 1911, near Lake Helen (ABS, DUKE, NY); Hood, Apr. 28, 1912, near Lake Helen, Grout, N. Am. Musci Pl. 407 (BART, CAS, CM, COLO, FH, H, MINN, MO, NY, OS, UC, US, WASH, YALE); Murrill, Feb. 20, 1938, Alachua Co., Gainesville (FLAS); Murrill, Feb. 20, 1938, Alachua Co., Fairbanks (FLAS).

GEORGIA: R. M. Harper 2185b, May 4, 1904, Glynn Co., near Brunswick (NY, YALE); J. K. Small 9663 and S43/9663, July 28–30, 1895, Altamaha R. Swamp, near Jessup (DPU, NY).

KENTUCKY: Lesquereux (PC).

LOUISIANA: Dr. Hale, Red River (NY, PC); Langlois, Bayou Bonfouca (NY); Langlois, in 1886, Bayou Bonfouca (US); Langlois 25, Mar. 24, 1886, Bayou Bonfouca, Slidell (FH); Langlois 140, Mar. 24, 1886, Bayou Bonfouca (Assumed to be TYPE of F. *flaccida*, PC; S, US); Langlois, Mar. 18, 1891, Bayou Bonfouca (YALE); Langlois 925, Nov. 26, 1891 (PC); Langlois, Dec. 1892, Bayou Bonfouca (ABS); Langlois, May 1, 1893, Ravine aux Cannes (US); Penfound 105, Mar. 1936, Indian Village (DPU); Penfound 37–3, Jan. 2, 1937, Indian Village (DPU); Penfound 25, Mar. 14, 1937, Indian Village (DPU); Penfound and Pennebaker, Mar. 24, 1940, Sangpoursang, near Natchitoches (DPU, NO); J. K. Small, Apr. 13, 1931, New Orleans (NY); J. B. Wallace, Feb. 1, 1933, New Orleans (FLAS, NY).

MAINE: Norton, July 3, 1907, Oxford (DPU, PSNH); Norton, July 11, 1907, Falmouth (DPU, PSNH); Norton, June 25, 1933, Saco (DPU, PSNH).

MARYLAND: Boyer, Walford (E, MT, S); Boyer 50, in 1906, Walford (PC); Boyer, Jan. 1906, Walford (FH, NY); Boyer, Dec. 1907, Walford (NY); Leonard 19738, Mar. 12, 1944, near Bladensburg (DPU, US).

MASSACHUSETTS: Doty 6632, July 1946, Duke Co., Pasque Isl. (DPU); Doty 6625, July 10, 1946, Barnstable Co., near Woods Hole (DPU); Drouet, Patrick, and Hodge 3599, July 16, 1940, Woods Hole (CHI, DPU, DUKE, MT); E. M. Dunham, June 17, 1905, Worcester Co., Holden (FH, NY); Farlow, Aug. 1903, near Magnolia (FH); Farlow, Sept. 1903, Magnolia (PC); Faxon 332, Apr. 11, 1881, Dedham (NY); Huntington, Dec. 9, 1899, Essex Co., Amesbury (FH, NY); Huntington, Nov. 1900, Essex Co., Amesbury (DUKE, MO); Huntington, Nov. 18, 1900, Essex Co., Amesbury, Grout, N. Am. Musci Pl. 73 (BART, CAN, CAS, CM, DUKE, FH, ILL, MICH, MINN, MO, NY, OS, PC, UC, US, WASH, WIS, YALE).

MICHIGAN: Mich. Dept. Cons. 141, Aug. 18, 1937, Houghton Co., Bob Lake (DPU, MICH).

MISSOURI: Engelmann, near St. Louis (Assumed to be TYPE of F. *denticulata*, S; B, NY); Engelmann, Mar. 1837 (K, MO).

NEW HAMPSHIRE: Huntington, Hampton Falls (FH); James, White Mts. (K); James, July 1853, Crawford Notch, White Mts. (K, MO).

NEW JERSEY: C. F. Austin (?), Musci Appal. 246 p.p. (FH, NY); C. F. Austin, Oct. 1862, Pines (WIS); C. F. Austin, Nov. 1868, Tom's R. (NY, PC, S, US, YALE); Kaiser, June 20, 1914, Mays Landing (DPU, DUKE, MT, US).

NEW YORK: Latham 54, May 31, 1914, Long Isl., Greenport (DUKE); Latham 7850, May 10, 1937, Long Isl., Orient (NY).

NORTH CAROLINA: Schallert, Aug. 15, 1921, Stokes Co. (DPU); Van Vlesk, Salem Creek (PHIL).
PENNSYLVANIA: Krout, May 17, 1910, Chester Co., Falls Creek (PENN); J. K. Small, Sept. 1894, Lebanon Co., near Cold Spring (NY).
RHODE ISLAND: Collins 5897, Providence Co., Foster (BART).
TENNESSEE: Jennison, Mar. 7, 1936, Blount Co., Cades Cove, alt. 540 m (DPU, TENN); A. J. Sharp 341, Dec. 1934, Blount Co., Cades Cove, alt. 540 m (MICH, NY); Shields, Dec. 1934, Blount Co., Cades Cove (DPU, TENN).
VIRGINIA: Patterson 1700, Mar. 5, 1954, Sussex Co., near Grizzard (DPU, HOL); Patterson 1882, Mar. 18, 1954, Essex Co., in Mount Landing Creek (DPU, HOL).
United States without locality: Engelmann (E, S); Sullivant and Lesquereux, Musci Bor.-Am. Exsic. 338 p.p. (G-BOIS, PC); Sullivant and Lesquereux, Musci Bor.-Am. Exsic. 340 p.p. (FH, NY, PHIL).

Fontinalis flaccida may be distinguished from other species in the genus by the following combination of characteristics: stems and leaves usually flaccid, majority of blades plane throughout, sometimes subconcave to concave at base, occasionally very slightly concave at apex, elongate, narrowly ovate-lanceolate, oblong-lanceolate, or lanceolate, apices long, either narrowly or broadly acuminate, usually abruptly narrowed into truncate, obtuse, or subobtuse ends, occasionally acute, leaf tips distinctly serrulate with a few prominent cells, blades usually 4–7 mm long, 0.5–1.5 mm wide, 3–6 : 1, median cells linear with ends attenuate, and alar cells much enlarged, forming very conspicuous auricles.

Brachelyma

Brachelyma Schimp., Syn. Musc., p. 557. 1876.

Neckera, Sect. 9 *Dichelyma*, Subsect. 2 *Cryphaeadelphus* C. Müll., Syn. Musc. 2: 145. 1850.
Cryphaeadelphus (C. Müll.) Card., in Rev. Bryol. 31: 6. 1904.

Plants submerged, floating, slender to medium in size, youngest portions conspicuously three-angled; stems up to 20 cm long, 0.2–04. mm in diameter; leaves tristichous, carinate-conduplicate, costate, unfolded blades subulate, oblong-lanceolate, sublanceolate, lanceolate, narrowly ovate-lanceolate, or elliptic-lanceolate; apices acute to obtuse, occasionally subcucullate to cucullate, leaf tips subserrulate to serrulate, sometimes almost entire; blades 2–4 mm long, 0.5–1.5 mm wide, 2–7 : 1; costa percurrent or disappearing a short distance below leaf tip; median cells of leaves subrhombic, subrhomboidal, subhexagonal, or linear with ends attenuate; marginal cells in *B. subulatum* forming a border which becomes indistinct or disappears near leaf apex; alar cells not enlarged to very slightly so, subquadrate or subrectangular; auricles none or very slight; dioecious; perichaetial branch 4.5–7.5 mm long; perichaetium subcylindrical to cylindrical, 0.5–1.25 mm in diameter, brownish green or brownish yellow; perichaetial leaves ovate-lanceolate, elliptic-lanceolate, or linear-lanceolate, apices long acuminate or acute, usually entire, occasionally serrulate; calyptra dimidiate, long conical, acuminate, covering operculum only, very fugacious, 2–2.8 mm in length; capsule briefly pedicellate, completely immersed, brownish yellow or brown when mature; operculum long conical, rostrate, beak oblique; seta short, 0.75–1.5 mm long; urn immersed, oval, 1.5–2.25 mm long, 0.75–1 mm in diameter; peristome teeth brownish yellow, linear, 0.4–0.5 mm long, finely muricate, with 8–10 lamellae; cilia of inner peristome brownish yellow, up to 0.6 mm in length, finely muricate, nodulose, or appendiculate, free or united at apex by transverse strands; spores 13.5–23.8 μ in diameter. [The word, *Brachelyma*, is composed of the Greek words meaning short veil, with reference to the calyptra in this genus of plants covering the operculum only.]

Key to the Species of Brachelyma

1. Stems subflaccid to flaccid, 0.2–0.25 mm in diameter; leaves 0.5–1.2 mm
   wide, 3–7 : 1; apices acute to obtuse; median cells subrhombic, subrhomboidal, or subhexagonal, commonly 2–4 : 1; marginal cells linear, 10–15 : 1,
   forming a distinct border . . . . 1. *B. subulatum* (p. 197)
2. Stems rigid, 0.25–0.4 mm in diameter; leaves 1–1.5 mm wide, 2–3 : 1; apices
   narrowly to broadly obtuse; median cells linear-rhombic, linear-rhomboidal,
   or linear, 6–9 : 1; marginal cells linear, 10–13 : 1, not forming a distinct
   border . . . . 2. *B. robustum* (p. 200)

1. **Brachelyma subulatum** (Palis. Beauv.) Schimp., Syn. Musc., p. 557. 1876.

Fontinalis subulata Palis. Beauv., Prodr. AEtheog., p. 58. 1805.
Dichelyma subulatum (Palis. Beauv.) Myr., in Act. Reg. Acad. Sci. Holm. 1832:
    281. 1833.
Cryphaea inundata Nees, in v. Wied, Musci Frond., p. 27. 1841.
Neckera subulata (Palis. Beauv.) C. Müll., Syn. Musc. 2: 145. 1850.
Cryphaeadelphus subulatus (Palis. Beauv.) Card., in Rev. Bryol. 31: 6. 1904.

Plants slender in size, yellowish green, green, brownish green, or
brown, usually dull, youngest portions sometimes glossy, conspicuously three-angled; stems subflaccid to flaccid, up to 20 cm long,
0.2–0.25 mm in diameter, darker and denuded at base with age,
regularly or irregularly pinnately branched, sometimes bipinnately
divided; branches numerous, usually spreading, sometimes erect-
spreading, up to 7 cm in length, ends of foliated stems and branches
obtuse; median cauline leaves imbricate, bases up to 0.5 mm apart,
blades firm, erect-spreading, carinate-conduplicate, subulate, oblong-
lanceolate, sublanceolate, or lanceolate, width gradually decreasing
from base into obtuse or acute apices; keel straight to moderately
curved, frequently abruptly curved near apex; leaf tips usually serrulate, sometimes subserrulate, occasionally entire, margins of apices
frequently serrulate; blades 2–4 mm long, 0.5–1 mm wide, occasionally up to 1.2 mm in width, 3–7 : 1; costa brownish yellow or golden,
22.5–52.5 μ wide in conduplicate leaves, percurrent or disappearing
a short distance below leaf tip; median cells of leaves subrhombic,
subrhomboidal, or subhexagonal, 5–8.5 μ wide, generally 2–4 : 1,
sometimes 6–7 : 1; marginal cells linear with ends attenuate, 6.8–8.5 μ
wide, 10–15 : 1, forming a border of 4–5 rows of cells, becoming indistinct or disappearing near apex; alar cells subquadrate or subrectangular, auricles none or very slight, leaf bases briefly decurrent, up to
0.25 mm; median branch leaves similar to cauline except smaller in
size; perichaetial branch 4.5–7.5 mm long; perichaetium subcylindrical, 1–1.25 mm in diameter, brownish green or brownish yellow;
perichaetial leaves ecostate, the upper ovate-lanceolate, elliptic-

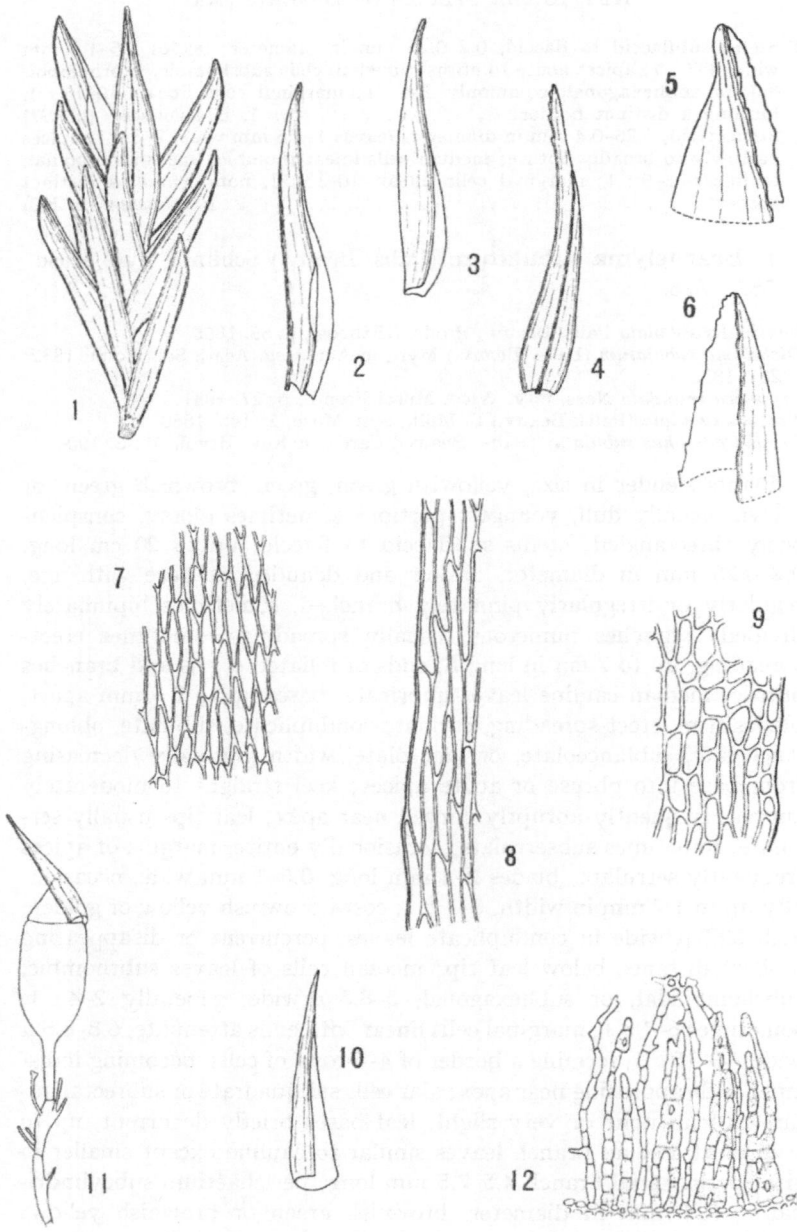

lanceolate, or linear-lanceolate, apices long acuminate or acute, entire; calyptra dimidiate, long conical, acuminate, up to 2 mm in length, 0.4 mm in diameter, covering only the operculum, very fugacious; capsule completely immersed, brownish yellow or brown; operculum long conical, 1–1.2 mm long, 0.7–0.75 mm in diameter, rostrate, beak oblique; seta short, 0.75–1.5 mm long; urn immersed, oval in outline, 1.5–2 mm long, 0.75–1 mm in diameter, 1.75–2.66 : 1, contracted or not contracted beneath mouth when dry; peristome teeth brownish yellow, linear, frequently split along the divisural line, occasionally almost the complete length of the tooth, 0.4–0.5 mm long, finely muricate, lamellae 8–10; cilia brownish yellow, linear, usually longer than the teeth, up to 0.6 mm in length, finely muricate, nodulose, or appendiculate, free or united by transverse strands at apex only; spores yellowish green, green, or yellowish brown, almost smooth or slightly muricate, 13.5–18.7 μ in diameter, ripe in summer. [The specific name, *subulatum*, has reference to the awl-shaped leaves.]

Type: According to Lasègue (1845), in Musée Botanique de M. Benjamin Delessert, p. 70, Delessert purchased, in 1820, the collections made by Palisot de Beauvois in some parts of the United States of America. The author did not see the Palisot de Beauvois collection of *Brachelyma subulatum* in the Delessert Herbarium. In the Herbarium of the New York Botanical Garden, there is a collection accompanied by the following label: "*Fontinalis subulata* P. Beauv. Savannah, Georgia. Ex Hb. Delessert. (type !)."

Type collector: Palisot de Beauvois. Date ?

Type locality: Georgia: Savannah.

Distribution: North America: approximately the southeastern portion of the United States.

Additional descriptions: Bridel, Musc. Recent. Suppl. 3: 110 (as *Fontinalis subulata*). 1817; Bridel, Bryol. Univ. 2: 661 (as *F. subulata*). 1827; Swartz, Adnot. Bot., p. 178 (as *F. subulata*). 1829; Bruch and Schimper, Bry. Eur., Fasc. 16, 5: 9 (as *Dichelyma subulatum*). 1842; Sullivant, Musci and Hep. U. S., pp. (655) 55 (as *D. subulatum*), (656) 56 (as *Cryphaea inundata*). 1856; Lesquereux and James, Man. Moss. N. Am., pp. 274 (as *D. subulatum*), 413 (as *C. inundata*). 1884; Cardot, Mon. Font., p. 131. 1892; Welch, in Grout, Moss Fl. N. Am. 3: 258. 1934.

Illustrations: Myrin, in Act. Reg. Acad. Sci. Holm. 1832: pl. 7b (as *D. subulatum*). 1833; Bruch and Schimper, Bry. Eur., Fasc. 16, 5: pl. 434 (as *D. subulatum*). 1842; Brotherus, in Engler and Prantl, Nat. Pflanz.: Bryophyta 2:

FIG. 29 (*opposite*). – *Brachelyma subulatum* (Palis. Beauv.) Schimp. – 1. Portion of stem with median cauline leaves. 2, 3, and 4. Median cauline leaves. 5 and 6. Leaf apices. 7. Median cells of leaf. 8. Marginal cells of leaf. 9. Alar and basal cells of leaf. 10. Perichaetial leaf. 11. Calyptra, capsule, seta, and portion of perichaetial branch. 12. Portion of peristome. Nos. 1 and 5–10 drawn from Small 5079, Georgia; 2, 3, 4, 11, and 12 after Bruch and Schimper, Bry. Eur., pl. 434.

fig. 547 (as *Cryphaeadelphus subulatus*). 1905; Brotherus, in Engler and Prantl, Nat. Pflanz., Musci 11 (2): fig. 477. 1925; Welch, in Grout, Moss Fl. N. Am. 3: pl. 79. 1934. – FIGURE 29.

Specimens examined: NORTH AMERICA: UNITED STATES: ALABAMA: R. M. Harper 104, June 22, 1906, Geneva Co., Pea R., near Geneva (MO, NY, US, YALE); R. M. Harper, May 27, 1933, Tuscaloosa Co. or Hale Co., Keaton Lake, Warrior R. bottoms (NY); R. M. Harper 3065, May 27, 1933, Tuscaloosa Co. or Hale Co., Keaton Lake, Warrior R. bottoms (DPU, NY, US); Mohr, Mobile (US); Sullivant, near Mobile, Sullivant and Lesquereux, Musci Bor.-Am. Exsic. 339 p.p. (K).

FLORIDA: A. W. Chapman, Gladsden Co., Rocky Comfort Swamp (NY).

GEORGIA: Mohr, River Junction, Appalachicola (US); Mohr, River Junction, Chattahoochee R. (US); Palisot de Beauvois, Savannah (TYPE of *Fontinalis subulata*, G-DEL; NY, PC); Palisot de Beauvois (B, K, PC); J. K. Small 5068, July 28–30, 1895, Altamaha R. Swamp, near Jessup (DUKE, NY); J. K. Small 5079, July 28–30, 1895, Altamaha R. Swamp, near Jessup (CHI, NY); J. K. Small S184/5079, July 28–30, 1895, Altamaha R. Swamp, near Jessup (DPU).

ILLINOIS: Von Wied, Wabash R. (BM, H, PC); Von Wied, Wabash, Fox, and Black Rivers (K); Von Wied, in 1832, Black R. (NY); Von Wied, in 1832, Wabash, Fox, and Black Rivers (B); Von Wied 130, in 1832, Wabash, Fox, and Black Rivers (PC); Von Wied, Dec. 1832, Wabash R. (BR, PC).

LOUISIANA: Drummond (K); Drummond, New Orleans, Musci Am. 151 p.p. (K); Drummond, Musci Am. 153 (BR, FH, NY, PC, WASH); Drummond 62 (BM, K); Langlois, July 9, 1893, Calcassieux R. (US); Langlois 1003, July 9, 1893 (PC); Pennebaker, Mar. 25, 1939, Bogue Chitto Bush, near Covington (CHI, DPU, MICH, NY); Pennebaker, July 27, 1940, Amite bottomlands, near Denham Springs (DPU).

NORTH CAROLINA: Conard, Oct. 20, 1949, Greene Co., Snow Hill (DPU); Conard, Sept. 22, 1950, Greene Co., Snow Hill (DPU); Gleason, Aug. 26, 1941, Gates Co., Chowan R., Wynoak (DPU); F. W. Gray M607, Oct. 10–20, 1927, Charlotte (ABS, BART); Schallert, 11–9–1926, Waccamaw R., Bug Hill (MINN, S); Schallert 514, 11–8–1927 (DPU).

SOUTH CAROLINA: Swails 28, Williamsburg Co., near Black R. (DPU, USC).

TEXAS: Whitehouse 27834, Sept. 28, 1953, Harrison Co., Caddo Lake State Park (DPU, SMU).

VIRGINIA: R. P. Carroll 86, May 15, 1932, Dinwiddie Co., Curtis Pond (DPU); Correll 11567, Oct. 14, 1941, Greensville Co., Fontaine Creek (DUKE).

Brachelyma subulatum differs from *B. robustum* in the following combination of characteristics: plants slender, stems flaccid, leaves narrow, 0.5–1 mm wide, occasionally up to 1.2 mm in width, median cells of leaves subrhombic, subrhomboidal, or subhexagonal, usually 2–4 : 1, occasionally 6–7 : 1, and marginal cells in distinct contrast with median, linear with ends attenuate, 10–15 : 1, forming a border of 4–5 rows of cells, becoming indistinct or disappearing near apex.

2. **Brachelyma robustum** (Card.) E. G. Britt., in Bryol. 7: 48. 1904.

Cryphaeadelphus robustus Card., in Rev. Bryol. 31: 7 and 8. 1904.

Plants slender to medium in size, yellowish green, brownish green, or brown, usually dull, youngest portions sometimes glossy, conspicuously three-angled; stems rigid, up to 20 cm long, 0.25–0.4 mm in diameter, darker and denuded near base with age, commonly pinnately branched, occasionally bipinnately divided; branches numerous, usually spreading, occasionally erect-spreading, close to distant, up to 6 cm in length, ends of foliated stems and branches obtuse; median cauline leaves subimbricate to imbricate, bases up to 0.5 mm apart, blades firm, erect-spreading, carinate-conduplicate, oblong-lanceolate, elliptic-lanceolate, narrowly ovate-lanceolate, or sublanceolate, width gradually decreasing from base into narrowly or broadly obtuse apices, ends of leaves sometimes subcucullate to cucullate; keel straight to moderately curved, frequently abruptly

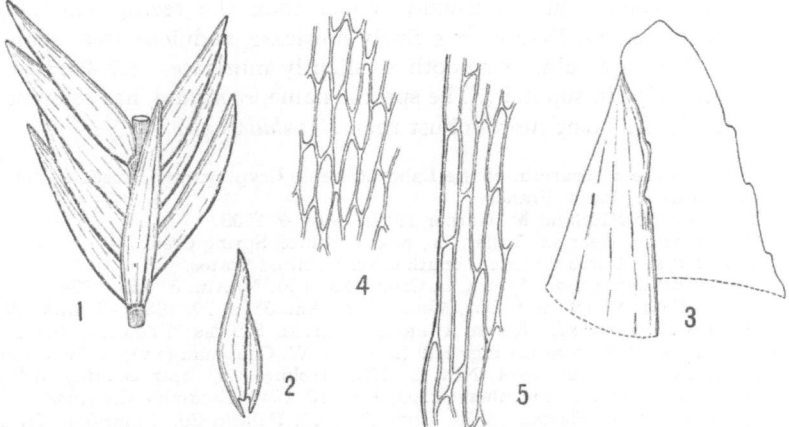

Fig. 30. – *Brachelyma robustum* (Card.) E. G. Britt. – 1.Portion of stem with median cauline leaves. 2.Median cauline leaf. 3.Leaf apex. 4.Median cells of leaf. 5.Marginal cells of leaf. Drawn from Harper 1919a, Georigia.

curved near apex; leaf tips usually serrulate, sometimes subserrulate, occasionally entire, margins of apices frequently serrulate; blades 3–4 mm long, 1–1.5 mm wide, 2–3 : 1; costa brownish green or golden brown, 22.5–52.5 μ wide in conduplicate leaves, percurrent or disappearing a short distance below leaf tip; median cells of leaves linear-rhombic, linear-rhomboidal, or linear with ends attenuate, 6.5–8.5 μ wide, 6–9 : 1; marginal cells linear, ends attenuate, 10–13 : 1; alar cells subquadrate or subrectangular, auricles none or very slight, leaf bases briefly decurrent, up to 0.25 mm; median branch leaves similar to cauline except smaller in size; perichaetial branch 5–6.5 mm long; perichaetium cylindrical, 0.5–1 mm in diameter, brownish green, brownish yellow, perichaetial leaves ecostate, the upper ovate-lanceolate, elliptic-lanceolate, or linear-lanceolate, apices long acuminate or acute, usually entire, occasionally serrulate, lacerate with age; calyptra long conical, 2.5–2.8 mm long; capsule completely immersed, brownish yellow or brown; operculum conical, beak tip absent,[1] remains of operculum 0.75 mm long, 0.75 mm in diameter, rostrate, beak oblique; seta 0.75–0.85 mm long; urn immersed, oval, 1.75–2.25 mm long, 0.75–1 mm in diameter, 2.25–2.33 : 1, contracted or not contracted beneath mouth when dry; peristome teeth brownish yellow, linear, frequently split along the divisural line, occasionally almost the complete length of the tooth, remaining fragments up to 0.26 mm long, finely muricate, lamellae 8; cilia brownish yellow, linear, usually longer than the teeth, remaining fragments up to 0.33 mm long, finely muricate, nodulose, free; spores brownish yellow, almost smooth or slightly muricate, 18.7–23.8 μ in diameter, ripe in summer. [The specific name, *robustum*, has reference to this species being more robust than *B. subulatum*.]

Type: In the Herbarium of the Laboratoire de Cryptogamie, Muséum d'Histoire Naturelle, Paris, France.
Type collector: Roland M. Harper 1919a. Aug. 4, 1903.[2]
Type locality: Georgia: Miller Co., near Colquitt, Spring Creek.
Distribution: North America: southeastern United States.
Additional description: Welch, in Grout, Moss Fl. N. Am. 3: 258. 1934.
Illustrations: Welch, in Grout, Moss. Fl. N. Am. 3: pl. 79. 1934. – FIGURE 30.
Specimens examined: NORTH AMERICA: UNITED STATES: FLORIDA: Brinker 498, Aug. 20, 1941, near Cottage Hill (DPU); A. W. Chapman (NY); A. W. Chapman 39 (K); McFarlin 1694, Nov. 8, 1937, Holmes Co., near Bonifay, along Holmes Creek (DPU); Schornherst 2503, June 10, 1949, Escambia Co. (DPU).
GEORGIA: R. M. Harper 1377a, June 27, 1902, Pulaski Co., Limestone Creek Swamp (Assumed to be COTYPE, PC; BM, NY, US); R. M. Harper, in 1903, Miller Co., near Colquitt, Spring Creek Swamp (K); R. M. Harper 1919a, Aug. 4, 1903, Miller Co., near Colquitt, Spring Creek Swamp (Assumed to be TYPE, PC; E,

---

[1] Several perichaetial branches and two old capsules have been available for study. Even though parts were incomplete, the measurements have been included because the author has found no other description of the sporophytes of *B. robustum*.
[2] Apparently Cardot based his original description upon two collections: (1) R. M. Harper 1377a, June 27, 1902, on base of trunk of *Nyssa uniflora* in swamp of Limestone Creek, Pulaski Co., Georgia, and (2) R. M. Harper 1919a, Aug. 4, 1903, on bases of trees in muddy swamp of Spring Creek near Colquitt, Miller Co., Georgia. In his description Cardot refers to one perichaetium and old capsule. Since only one of the above collections can be considered as the type, the author assumes it is the material bearing the sporophyte, Harper 1919a.
    In the Fontinalaceae of North America, the author cited the 1902 collection as the type. No plants with sporophytes were available for study, and since Cardot does not state in his original description in which collection he observed the perichaetium and capsule, the earlier date was used as the basis of selection of type. During the examination of the plants in the Paris Museum for the present study, the perichaetial leaves were observed on the 1903 collection.

Brachelyma robustum differs from *B. subulatum* in the following combination of characteristics: plants slender to medium in size, stems rigid, leaves narrow, 1–1.5 mm wide, median cells of leaves linear-rhombic, linear-rhomboidal, or linear with ends attenuate, 6–9 : 1, and marginal cells usually not in distinct contrast with median in shape and length, linear, ends attenuate, 10–13 : 1, not forming a border.

Dichelyma

Dichelyma Myr., in Act. Reg. Acad. Sci. Holm. 1832: 274. 1833.

Neckera, Sect. 9 *Dichelyma*, Subsect. 1 *Eudichelyma* C. Müll., Syn. Musc. 2: 143. 1850.

Plants submerged or emerged; stems up to 20 cm in length, 0.2–0.35 mm in diameter, ends of foliated stems and branches commonly curved, frequently uncinate; leaves tristichous, carinate-condupli-cate, costate, unfolded blades oblong-lanceolate, narrowly lanceolate to lanceolate-subulate, often subsecund or secund, frequently falcate to uncinate, some straight along keel, others with keel moderately to strongly curved; apices subobtuse, obtuse, subulate, or acuminate, leaf tips entire, subserrulate, serrulate, or sinuolate; median cauline blades 2.75–7 mm long, 0.4–1.4 mm wide, 3–11 : 1; costa subper-current, percurrent, or briefly to long excurrent; median cells of leaves linear-rhomboidal or linear with ends attenuate; alar cells indistinct, not enlarged to very slightly so; auricles none; dioecious; perichaetial branch 4–10 mm long; perichaetium narrowly cylindri-cal, 0.25–0.65 mm in diameter; inner perichaetial leaves linear, narrowly lanceolate, or narrowly ovate-lanceolate, convolute to tubular, subspirally to spirally enveloping the seta, apices acute to long acuminate, entire, subserrulate, or serrulate; calyptra long conical, dimidiate, enveloping the capsule, when young clasping seta by base, usually free when mature, 1.75–6.5 mm in length; capsule pedicellate, erect or oblique, immersed, emerging laterally from perichaetium or from end of perichaetium, or surpassing it, brownish yellow or brownish orange when mature; operculum conical, acumi-nate, or rostrate, beak straight, oblique, or curved; seta 3–15 mm long, occasionally up to 21.5 mm in length, almost to completely hidden in perichaetial leaves or surpassing them; urn suboval, oval, oval-oblong, subcylindrical, or cylindrical, 0.65–3 mm long, 0.35–1 mm in diameter; peristome teeth 0.26–1 mm long, lamellae 8–23; cilia of inner peristome 0.45–1 mm in length, cilia free or united by cross bars into imperfect or perfect trellis, transverse strands not appendiculate; spores 10–20 μ in diameter. [The word, *Dichelyma*, is composed of Greek words meaning veil in two, with reference to the calyptra being cleft on one side in this genus of plants. In contrast with *Brachelyma*, the calyptra covers the entire capsule in *Dichelyma*.]

KEY TO THE SPECIES OF DICHELYMA

(Specimens sterile)

1. Costa subpercurrent to briefly excurrent.
   2. Costa freqently briefly excurrent; leaf apices subulate or acuminate . . .
   . . . . 1. *D. falcatum* (p. 205)
   2. Costa not excurrent; leaf apices often subobtuse or obtuse.
      3. Leaves 0.5–0.8 mm wide, bases 0.25–0.5 mm apart . . . .
      . . . . 3. *D. pallescens* (p. 217)
      3. Leaves 0.8–1.4 mm wide, bases up to 0.25 mm apart . . . .
      . . . . 2. *D. japonicum* (p. 215)
1. Costa long excurrent.
   2. Median cauline leaves commonly falcate, 4–5 mm long . . . .
   . . . . 4. *D. uncinatum* (p. 221)
   2. Median cauline leaves commonly erect-ascending, usually straight to moderately curved along keel, 5–7 mm long . . . .
   . . . . 5. *D. capillaceum* (p. 227)

(Specimens fertile)

1. Capsule emerging laterally from perichaetium; seta almost to completely hidden in perichaetial leaves, 3–6.5 mm long; cilia free throughout or united at apices into imperfect trellis.
   2. Costa subpercurrent to percurrent . . . . . 3. *D. pallescens* (p. 217)
   2. Costa long excurrent . . . . 5. *D. capillaceum* (p. 227)
1. Capsule emerging from end of perichaetium; seta 4–21.5 mm in length; trellis perfect.
   2. Costa long excurrent, leaves 0.4–0.7 mm wide   4. *D. uncinatum* (p. 221)
   2. Costa percurrent to briefly excurrent, leaves 0.7–1.4 mm wide.
      3. Costa briefly excurrent, leaf apices subulate or acuminate . . . . . .
      . . . . 1. *D. falcatum* (p. 205)
      3. Costa percurrent, leaf apices subobtuse or obtuse . . . .
      . . . . 2. *D. japonicum* (p. 215)

1. **Dichelyma falcatum** (Hedw.) Myr., in Act. Reg. Acad. Sci. Holm. 1832: 274. 1833.

Fontinalis falcata Hedw., Sp. Musc., p. 299. 1801.
Neckera falcata C. Müll., Syn. Musc. 2: 143. 1850.
Dichelyma falcatum f. *chrysea* Hj. Möll., in Ark. Bot. 17 (14): 9. 1922.
Dichelyma falcatum f. *atra* Hj. Möll., in Ark. Bot. 17 (14): 9. 1922.

Plants medium to moderately robust in size, yellowish green, green, grayish green, copper-colored, golden, golden brown, brownish yellow, brownish green, or blackish green, younger portions usually glossy, older parts commonly dull; stems rigid, up to 15 cm in length, 0.2–0.3 mm in diameter, darker and denuded at base with age, usually irregularly branched, occasionally pinnately divided; branches few to numerous, erect-ascending, erect-spreading, or spreading, close

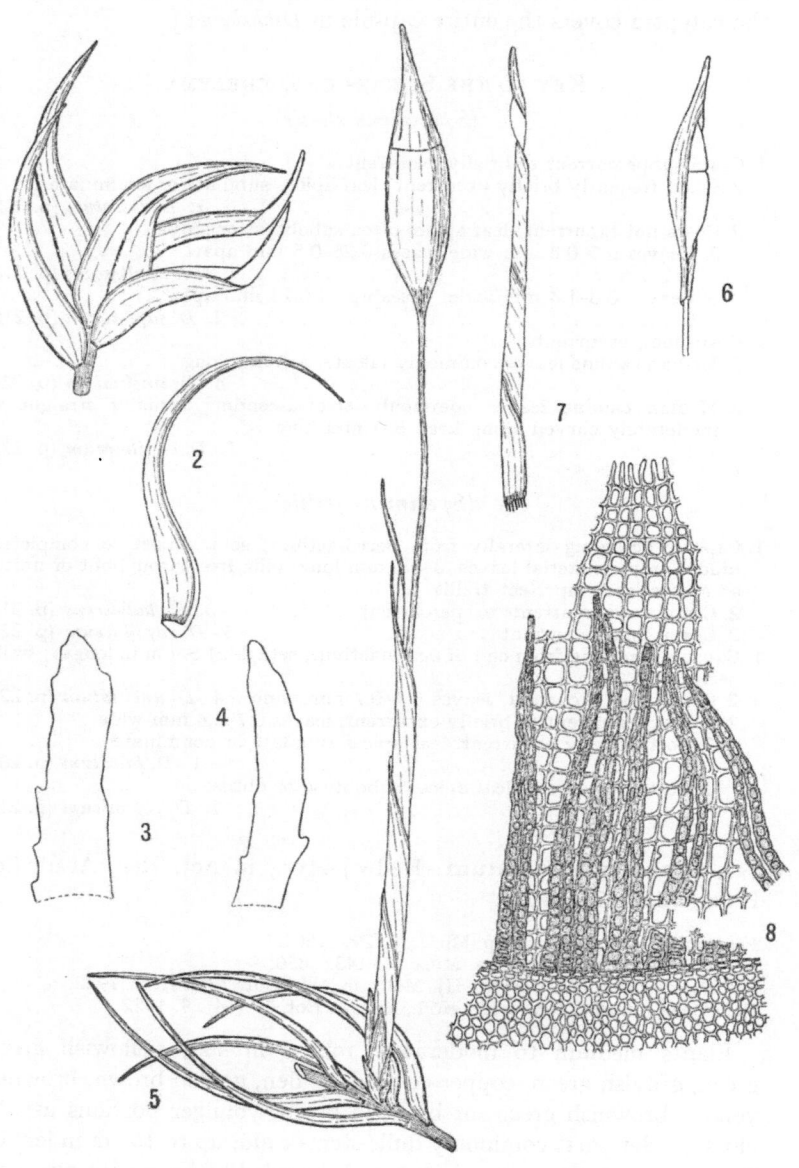

to distant, up to 4 cm in length, ends of foliated stems and branches obtuse, frequently uncinate with falcate-secund leaves; median cauline leaves usually imbricate at base, occasionally subimbricate, bases up to 0.25 mm apart, blades firm, erect-spreading, carinate-conduplicate, usually falcate-secund, some straight along keel, others with keel moderately curved, narrowly lanceolate, width gradually decreasing from base to apex, margins entire in lower two-thirds or three-fourths of blade and serrulate above; apices subulate or acuminate, serrulate; median cauline blades 3–5 mm long, 0.7–1.35 mm wide, 3–5 : 1; costa brownish green, brownish yellow, brownish red, or brown, 30–45 μ wide in conduplicate leaves, generally briefly excurrent, frequently percurrent or nearly so; median cells of leaves linear, ends attenuate, 5–6.5 μ wide, 10–20 : 1; alar cells indistinct, slightly enlarged, usually quadrate or rectangular, occasionally subhexagonal, walls yellowish or brownish yellow, auricles none, leaf bases not decurrent to briefly so, up to 0.25 mm; median branch leaves similar to median cauline except smaller in size, frequently falcate-secund; perichaetial branch 6–9 mm long; perichaetium narrowly cylindrical, 0.4–0.5 mm in diameter; perichaetial leaves ecostate, the inner narrowly lanceolate, long acuminate, apices entire or subserrulate, blades convolute, twisted; calyptra clasping seta by base when young, long conical, dimidiate, 3.5–6.5 mm long, 0.2–0.75 mm in diameter; capsule erect, emerging from end of perichaetium, usually surpassing the perichaetial leaves; operculum conical, acuminate, 0.5–1 mm long, 0.5–0.75 mm in diameter, beak oblique; seta erect, brownish orange, 5–15 mm long, commonly surpassing the perichaetial leaves; urn erect, oval, oval-oblong, subcylindric, or cylindric, brownish yellow or brownish orange when mature, sometimes darkening with age, rounded at base, 1–2 mm long, 0.5–0.9 mm in diameter, 1.6–3 : 1; frequently contracted beneath mouth when dry, sometimes not contracted; peristome teeth brownish yellow or brownish orange, linear-lanceolate, 0.4–0.73 mm long, muricate, lamellae 11–23; trellis perfect, brownish yellow or brownish

FIG. 31 (opposite). – Dichelyma falcatum (Hedw.) Myr. – 1.Portion of stem with median cauline leaves. 2.Median cauline leaf. 3 and 4.Leaf apices. 5.Portion of stem with median cauline leaves, perichaetium, and sporophyte. 6.Capsule and dimidiate calyptra clasping, at its base, the seta. 7.Perichaetial leaf. 8.Upper portion of urn, bearing the outer peristome of teeth and the inner of cilia, united into a perfect trellis by transverse bars. Nos. 1, 3, and 4 drawn from J. A. Allen, July 21, 1880, New Hampshire, Grout, N. Am. Musci Pl. 396; nos. 2 and 5–8 after Bruch and Schimper, Bry. Eur., Fasc. 16, 5: pl. 433.

orange, cilia longer than the teeth, 0.6–0.86 mm in length, muricate; spores yellowish green, green, or yellowish brown, smooth to muricate, 10–15 μ in diameter, ripe in summer. [The specific name, *falcatum*, is indicative of the falcate leaves.]

Type: In the Herbarium of Hedwig, in the Boissier Herbarium, Geneva, Switzerland. Label: "*Fontinalis falcata* Hedw. St. Cr. Vd. III. p. 57, t. 24." The author saw no collection designated as the type. The specimens of *D. falcatum* in the Hedwig Herbarium are fragmentary and lack data as to collector, date, and locality. The plants with the above label are assumed to be the type because of the reference to the original description and plate, Hedwig, Musc. Frond. 3: 57, pl. 24. 1792. No sporophytes were seen on the fragment.

Type collector: Swartz, according to Hedwig, Musc. Frond. 3: 58. 1792.

Type locality: Sweden: "Locus: in Sueciae aquis repertum misit Cl. Swartz," according to Hedwig, Musc. Frond. 3: 58. 1792.

Distribution: North America, Europe, and Asia: Alaska, Canada, United States, Europe, and Siberia.

Additional descriptions: Hedwig, Musc. Frond. 3: 57 (as *Fontinalis falcata*). 1792; Bridel, Musc. Recent. 2 (3): 161 (as *F. falcata*). 1803; Schwaegrichen, Hedwig, Sp. Musc., p. 308 (as *F. falcata*). 1816; Bridel, Musc. Recent. Suppl. 3: 109 (as *F. falcata*). 1817; Hartman, Handb. Skand. Fl., p. 434 (as *F. falcata*). 1820; Bridel, Bry. Univ. 2: 659 (as *F. falcata*). 1827; Bruch and Schimper, Bry. Eur., Fasc. 16, 5: 6. 1842; Hartman, Handb. Skand. Fl., p. 433. 1843; Hartman, Handb. Skand. Fl., p. 341. 1849; Hartman, Handb. Skand. Fl., p. 371. 1854; Sullivant, Musc. and Hep. U. S., p. (655) 55. 1856; Schimper, Syn. Musc., p. 459. 1860; Milde, Bry. Siles., p. 278. 1869; Schimper, Syn. Musc., p. 557. 1876; Lesquereux and James, Man. Moss. N. Am., p. 273. 1884; Cardot, Mon. Font., p. 135. 1892; Husnot, Musc. Gall., p. 288. 1892; Limpricht, Laubm. 2: 674. 1894; Grout, M. H. M., p. 402. 1903; Roth, Eur. Laubm. 2: 294. 1904; Warnstorf, Laubm., p. 635. 1905; Mönkemeyer, in Pascher, Süsswasserfl. 14: 108. 1914; Hj. Möller, in Ark. Bot. 17 (14): 6. 1922; Brotherus, Laubm. Fenn., p. 399. 1923; Mönkemeyer, in Laubm. Eur. Erg.–Bd. 4: 666. 1927; Welch, in Grout, Moss Fl. N. Am. 3: 259. 1934; Jensen, Skand. Bladmossfl., p. 383. 1939.

Illustrations: Hedwig, Musc. Frond. 3: pl. 24 (as *Fontinalis falcata*). 1792; Bridel, Musc. Recent., pl. 4 (as *F. falcata*). 1798; Bridel, Bry. Univ. 2: pl. 10 (as *F. falcata*). 1827; Myrin, in Act. Reg. Acad. Sci. Holm. 1832: pl. 6. 1833; Bruch and Schimper, Bry. Eur., Fasc. 16, 5: pl. 433. 1842; Husnot, Musc. Gall., pl. 81. 1892; Limpricht, Laubm. 2: figs. 327 and 328. 1894; Roth, Eur. Laubm. 2: pl. 30. 1904; Brotherus, in Engler and Prantl, Nat. Pflanz.: Bryophyta 2: fig. 548. 1905; Warnstorf, Laubm., p. 638, fig. 6. 1905; Mönkemeyer, in Pascher, Süsswasserfl. 14: fig. 33. 1914; Brotherus, Laubm. Fenn., fig. 69. 1923; Brotherus, in Engler and Prantl, Nat. Pflanz., Musci 11 (2): fig. 478. 1925; Mönkemeyer, Laubm. Eur. Erg.-Bd. 4: fig. 146. 1927; Welch, in Grout, Moss Fl. N. Am. 3: pl. 79. 1934; Noguchi, in Journ. Jap. Bot. 22: fig. 34. 1948. – FIGURE 31.

Specimens examined: ASIA: UNION OF SOVIET SOCIALIST REPUBLICS: Arnell, June 20, 1876, Siberia, Yenisei (H); Arnell, June 30, 1876, Siberia, Yenisei (PC, S).

EUROPE: CZECHOSLOVAKIA: Schiffner, June 13, 1886, Bohemia, Riesengebirge (US).

FINLAND: Axelson, June 21, 1901, Lapponia kemensis, near Salla (H); Backman, June 13, 1903, Ostrobothnia (H); Boldt, Sept. 4, 1892, Regio aboënsis (H); Bomansson, July 12, 1864 (S); Bomansson, Sept. 5, 1887, Alandia (H); Bomansson, Aug. 1890, Alandia (H); V. F. Brotherus, Oct. 10, 1868, Nylandia, Helsinki (UC); V. F. Brotherus, Aug. 1869, Tavastia (NY); V. F. Brotherus, July 12, 1883, Kuusamo (H); V. F. Brotherus, July 1887, Lapponia murmanica (G-BOIS, PC, S); V. F. Brotherus, July 4, 1887, Lapponia murmanica (H); V. F. Brotherus,

July 1916, Nylandia (MICH, S); Buch, Aug. 20, 1904, Savonia australis (H); Buch, Aug. 22, 1911, Savonia australis, V. F. Brotherus, Bryoth. Fenn. 257 (CHI, H, S, YALE); Collin, Aug. 26, 1883, Satakunta (S); Elfving, Aug. 1873, Regio aboënsis (H); Hjelt, Aug. 5, 1874, Satakunta, V. F. Brotherus, Musci Fenn. Exsic. 132 (B, K); Hjelt and Hult, Aug. 13, 1877, Lapponia kemensis (S); Hjelt and Hult, Aug. 17, 1877, Ostrobothnia borealis (S); Högman, June 25, 1906, Regio aboënsis (H); Hult, July 5, 1880, Lapponia inarensis (S); Kihlman, July 25, 1882, Lapponia imandrensis (H); Kihlman, July 25, 1887, Lapponia imandrensis (H, S); Lackström, Sept. 1871 (UC); H. Lindberg, Oct. 1884, Nylandia, Helsinki (S); H. Lindberg, July 2, 1893, Savonia borealis (S); H. Lindberg, Oct. 23, 1893, Isthmus karelicus, Muola (H); H. Lindberg, Oct. 27, 1893, Isthmus karelicus (S); H. Lindberg, June 13, 1894, Karelia (S); H. Lindberg, July 8, 1895, Karelia (S); H. Lindberg, Nov. 5, 1900, Nylandia (S); H. Lindberg, July 30, 1901 (G-DEL); S. O. Lindberg, Tavastia (MINN); S. O. Lindberg, July 1856 (K); S. O. Lindberg, June 30, 1868, Nylandia, Helsinki (H); S. O. Lindberg, Oct. 11, 1868, Nylandia, Helsinki (H, K, S); S. O. Lindberg, Oct. 1877, Nylandia, Helsinki (NY); Linkola, Aug. 2, 1906, Savonia borealis (UC); Linkola, July 6, 1928, Lapponia enontekiensis (H); Linnaniemi, Nov. 17, 1907, Karelia ladogensis (H); Niklander, Aug. 11, 1855, Regio aboënsis (H); Pesola, July 24, 1915, Karelia (H); Silén, in 1861, Tavastia borealis (UC); Söyrinki, Jan. 9, 1934, Tavastia australis (H).

GERMANY: Baenitz, Aug. 15, 1894, Silesia, Riesengebirge, alt. 1200 m (CHI, E, PC, S, US); Bauer, Aug. 27, 1839, Riesengebirge (S); Freiberg, Aug. 1, 1920, Riesengebirge, Sudeten Mts. (S); Fritze, July 1865, Riesengebirge (NY); Hampe, in 1846, Silesia, Sudeten Mts. (BM, K); Hise, July 25, 1862, Riesengebirge, Rabenhorst, Bryoth. Eur. 779 (FH, G-BOIS, H, NY, PC, S); Hovgard, June 8, 1926, Riesengebirge (S); Hübener, Sudeten Mts. (G-BOIS); Kern, July 1877, Silesia, Riesengebirge (ILL); Kern, July 1885, Silesia, Riesengebirge, alt. 1300 m (R); Kern, July 26, 1886, Silesia, Riesengebirge, alt. 1200 m (CAN, FH, G-BOIS); Krieger, July 28, 1908, Riesengebirge (MT); Kühn, Riesengebirge, Rabenhorst, Bryoth. Eur. 1132 (G-BOIS, UC); Kurz, in 1853, Riesengebirge (G-BOIS); Limpricht, Aug. 1, 1860, Isergebirge (S); Limpricht, July 28, 1865, Riesengebirge (S); Limpricht, July 23, 1866, Riesengebirge, Bryoth. Siles. 34 (NY, PC, S); Limpricht, Aug. 1, 1868, Isergebirge (S); Milde, Riesengebirge (K, NY); Milde, in 1860, Riesengebirge (CHI, DPU, S); Milde, Aug. 1866, Silesia, Riesengebirge (K); Milde, in 1868, Riesengebirge (DUKE); Reinsch, in 1872, Sudeten Mts. (S); Schiffner, June 13, 1886, Riesengebirge (CHI, DPU, S); Schimper, Riesengebirge (PC); Schulze, Aug. 27, 1839, Riesengebirge (BM); Schulze, Aug. 1868, Riesengebirge (NY); Schulze, Sept. 9, 1878, Silesia (NY); Sendtner, Sept. 15, 1838, Riesengebirge (K); Stübner, in 1844, Sudeten Mts. (BM); Voigt, Riesengebirge (S); Voigt, in July, Sudeten, Dresden (NY); Von Klinggraeff, June 1861 (S); D. Winter, Sept. 1901, Riesengebirge (DUKE); Ferd Winter, Riesengebirge (DUKE); Zimmermann, July 31, 1865, Riesengebirge (S); Zimmermann, July 29, 1868, Riesengebirge (S).

NORWAY: Adlerz, July 24, 1897, Hedemarkens Amt (S); Berggren, June 1865, Kristians Amt (S); Berggren, July 8, 1865 (S); Blanchet, in 1832 (NY); Blytt, near Christiania (NY); Blytt, in 1826, Akershus Amt (S); Bryhn, June 1876, Akershus Amt (S); Bryhn, July 1878, Akershus Amt (S); Bryhn, Aug. 13, 1899, Buskeruds Amt (S); Bryhn, July 1901, Valdres (MINN); Bryhn, Aug. 1904, Kristians Amt (S); Bryhn, Aug. 1904, Valdres, near Sörum, alt. 150 m, Bauer, Musci Eur. Exsic. 566 (B, H, K, PC, S); Conradi, July 18, 1899, Hedemarkens Amt (S); De Notaris, in 1839 (G-BOIS); Ferguson, June 1880, Vola Vand (E); Hagen, July 30, 1891, Nedenes Amt (S); Hagen, July 17, 1894 (S); Hagen, Aug. 19, 1900, Søndre Trondhjems Amt, Opdal (CAN, H, S); Holmgren, July 27, 1868 (S); Kaalaas, Aug. 13, 1882, Akershus Amt (S); Kaalaas, June 11, 1886, Akershus Amt (S); Kaalaas, Aug. 10, 1890, Buskeruds Amt (S); Kaalaas, May 15, 1892, Akershus Amt (CHI, S); Kaalaas, July 16, 1895, Bratsberg Amt (S); Kaalaas, July 27, 1903, Kristians Amt (s); Kaalaas, Sept. 13, 1914, Akershus Amt (s); Kiaer, Oct. 31, 1866, Christiania (H, MINN); Kiaer, Aug. 9, 1867, Kristians Amt, Dovre (NY, s); Kiaer, July 9, 1869, Kristians Amt (s); Kiaer, Aug. 2, 1869, Kristians Amt (s); Kiaer, Aug. 4, 1869, Lesjaverk (NY); Kiaer 40, Sept. 29, 1878, near Christiania (G-BOIS, K, MINN, NY); Kiaer, Aug. 6, 1884, Jarlsberg and Larviks Amt (s); Kiaer, Aug. 6, 1891, Akershus Amt (s); Kindberg, in 1853, Hedemarkens Amt (s); Kindberg, July 1879, Kristians Amt, Dovre (s); Kindberg, July 26, 1879, Søndre Trondhjems (s); Kindberg, July 24, 1897, Hedemarkens Amt (s); Lénstrom, Aug. 1875, Søndre Trondhjems (s); Nicholson, July 1900, Saera Maristuen (PC); Ryan, May 29, 1886, Smaalenenes Amt (s); Ryan, Oct. 1888, Smaalenenes Amt (s); Ryan, July 17, 1894, Finnmarkens Amt (s); Schimper, in 1844 (PC); Sørensen, June 5, 1911, Akershus Amt (G-DEL, s); Wahlstedt, Aug. 1867, Hedemarkens Amt (s); Zetterstedt, Aug. 13, 1870, Hedemarkens Amt (s); Zetterstedt, July 31, 1868 (s); Zetterstedt, in 1870, Søndre Trondhjems Amt, Dovre (UC); Zetterstedt 174a, July 26, 1870, Søndre Trondhjems Amt (s); Zetterstedt, in 1872 (MINN); Zetterstedt, July 26, 1878, Søndre Trondhjems Amt, Dovre (PC); Zetterstedt and Wickbom, July 26, 1870, Søndre Trondhjems Amt (CHI, FH, MINN, PC, s).

RUMANIA: Simkovics, Aug. 15, 1874, Transylvania (NY); Simkovics, Aug. 15, 1874, Transylvania, alt. 2100 m, Schultz Herb. Normale (Nov. Ser.) Cent. 17, 1698 (FH, G-DEL, PC, s).

SWEDEN: Åberg 135, Aug. 17, 1912, Jämtland (s); Åberg 136, Aug. 19, 1912, Jämtland (s); Åberg 841, Aug. 18, 1916, Jämtland (s); Åberg 1013, June 11, 1917, Jämtland (s); Åberg 1014, July 8, 1917, Jämtland (s); Åberg 1015, Sept. 18, 1917, Jämtland (s); Åberg 1602, Sept. 13, 1920, Jämtland (s); Åberg 1904, Aug. 19, 1921, Värmland (s); Åberg 1905, Sept. 27, 1921, Värmland (s); Åberg 3046, Aug. 18, 1925, Värmland (s); Åberg 3329, July 21, 1926, Jämtland (s); Angströn (CHI, K); Arnell, June 1873, Ångermanland (H); Arnell, Aug. 19, 1873, Ångermanland (s); Arnell, Aug. 8, 1874, Ångermanland (s); Arnell, July 10, 1881, Ångermanland (s); Arnell, Aug. 1883, Medelpad, Husnot, Musci Gall. 736 (CHI, DPU, FH, K, PC, s, WASH); Arvén, Aug. 19, 1892 (CHI); Arvén, June 1900, Södermanland (DPU, s); Arvén, May 19, 1901, Dalarna (FH); Arvén, July 16, 1910, Dalarna (DPU, s); Arvén, Aug. 9, 1912, Dalarna (s); Arvén, July 14, 1914, Ångermanland (s); Arvén, July 16, 1916, Ångermanland (DPU, s); Arvén, Aug. 2, 1920, Hälsingland (s); Arwidsson, Oct. 10, 1931, Åsele Lappmark (s); Aurell, Oct. 1, 1889, Närke (s); Bergström, Sept. 9, 1918, Dalsland (s); Brundin, Aug. 17, 1889, Uppland (s); Brundin, Aug. 1892, Uppland (s); Collinder, June 1873, Hälsingland (s); Collinder, Aug. 14, 1873, Jämtland (s); Dusén, in 1870, Östergötland (CAN); Dusén, Aug. 17, 1870, Östergötland (US); Dybeck, Aug. 17, 1833, Västmanland (s); Ehrhart, Pl. Cr. Exsic. 205, Uppland, Uppsala (G-DEL, K); Elmqvist, in 1873, Närke, Örebro (DPU, s); Elmqvist, June 27, 1873, Närke (s); Elmqvist, July 1873, Närke (s); Elmqvist, July 1879, Östergötland (s); Florin, June 25, 1924, Hälsingland (s); Florin, July 17, 1931, Härjedalen (s); Florin, July 29, 1932, Jämtland (s); Grape, July 1875, Västerbotten (s); Halle, July 28, 1928, Härjedalen (s); Halle, July 29, 1928 (s); Halle, Aug. 16, 1936, Härjedalen (s); Hamnström, Aug. 1875, Östergötland (s); Hartman (BR, PC); Hartman, in 1851, Gästrikland (G-BOIS); Hartman, July 1859, Gästrikland (H); Hartman, in 1873 (G-DEL); C. Hartman, Gästrikland, Oslättfors (TRT); C. Hartman, Sept. 23, 1870, Närke (s); C. Hartman, in 1873 (K); C. Hartman, Aug. 11, 1873, Närke (s); C. J. Hartman, Sept. 16, 1830, Södermanland (s); R. Hartman, in 1858, Gästrikland, Bry. Scand. Exsic. 86 (BM, K, s); R. Hartman, July 1859, Gästrikland (H); R. Hartman, Oct., 1870, Gästrikland (NY, US); Hellbom, in 1867, Härjdalen, Un. Itin. Cr., 1867, 33 (G-BOIS, NY, PC, s); Hellsing, June 22, 1893, Dalarna (s); Hellsing, Aug. 15, 1899, Norrbotten (s); Hellsing, Oct. 2, 1914, Bohuslän, Musci Scandinavici 267 (s); Hellsing, June 11, 1918, Gästrikland (s); Hellsing, June 15, 1918, Hälsingland (s); Hellsing 107, Aug. 4, 1918 (s); Hellsing, Sept. 10, 1918, Lule Lappmark (s); Hellsing, Sept. 10, 1918, Lule

Lappmark, Musci Scandinavici 194 (s); Hellsing, June 13, 1920, Värmland, Musci Scandinavici 335 (s); Holmgren, July 5, 1867, Lule Lappmark (s); Holmgren, Aug. 1, 1867, Lule Lappmark (s); Indebetou, June 6, 1877, Värmland (s); Indebetou, July 1879, Dalarna (s); Indebetou, Sept. 1880, Dalarna (s); Jäderholm, in 1855, Småland (s); Jäderholm, July 23, 1893, Dalarna (s); Jäderholm, July 1909, Småland (s); Jäderholm, July 14, 1914, Torne Lappmark (s); Jäderholm, Aug. 10, 1916, Torne Lappmark (s); Jensen and Arnell, Aug. 1, 1902, Lule Lappmark (s); H. E. Johannson, July 13, 1918, Pite Lappmark (DPU, s); H. E. Johansson, Sept. 12, 1918, Lule Lappmark (s); H. E. Johansson, Aug. 16, 1926, Värmland (s); H. E. Johansson, Aug. 25, 1926, Värmland (s); H. E. Johansson, Aug. 28, 1926, Värmland (s); H. E. Johansson, Oct. 4, 1926, Värmland (s); H. E. Johansson, Oct. 6, 1926, Värmland (s); H. E. Johansson, July 26, 1927, Värmland (s); H. E. Johansson, Aug. 24, 1927, Värmland (s); H. E. Johansson, July 27, 1928, Värmland (s); H. E. Johansson, Sept. 25, 1928, Värmland (s); Kjellmark, Oct. 24, 1920, Småland (s); Laestadius, July 5, 1874, Västerbotten (s); Laestadius, Aug. 1, 1874, Västerbotten (s); Laestadius, Aug. 16, 1874, Västerbotten (s); Laestadius, Aug. 21, 1874 (s); J. Lange, Aug. 1849, Gästrikland (s); S. O. Lindberg, in 1854, Dalarna (B, PC); S. O. Lindberg, June 1854, Dalarna (s); S. O. Lindberg, June and July 1854, Dalarna (H); S. O. Lindberg, July 1854, Dalarna (DPU, H, K, s, US, YALE); S. O. Lindberg, July 1854, Dalarna, Rabenhorst, Bryoth. Eur. 628 (FH, G-BOIS, NY, PC, s); S. O. Lindberg, June 1859, Västergötland (s); S. O. Lindberg, Oct. 1868 (H, s); Lindman, July 13, 1871, Blekinge (s); Malmström, July 1924, Västerbotten (s); S. Medelius, Dec. 1909, Småland (MICH, s); Hj. Möller, Oct. 7, 1908, Dalarna (s); Hj. Möller, July 13, 1909, Dalarna (s); Hj. Möller, July 24, 1909, Dalarna (s); Hj. Möller, Aug. 27, 1910, Dalarna (s); Hj. Möller, Oct. 15, 1911, Dalarna (s); Hj. Möller, June 16, 1912, Västerbotten (DPU, s); Hj. Möller, June 20, 1912, Norrbotten (s); Hj. Möller, July 15, 1912, Torne Lappmark (DPU, s); Hj. Möller, July 24, 1912, Torne Lappmark (s); Hj. Möller, May 26, 1913, Gästrikland (s); Hj. Möller, Aug. 5, 1914, Jämtland (H, s); Hj. Möller, Aug. 5, 1914, Jämtland, Bauer, Musci Eur. et Am. Exsic. 2280a (BART); Hj. Möller, July 5, 1916, Åsele Lappmark (s); Hj. Möller, July 6, 1916, Lycksele Lappmark (s); Hj. Möller, July 18, 1916, Lycksele Lappmark, Bauer, Musci Eur. et Am. 2280b (BART); Hj. Möller, July 13, 1918, Pite Lappmark (s); Hj. Möller, July 28, 1918, Pite Lappmark (s); Hj. Möller, July 19, 1919, Lule Lappmark (DPU, s); Hj. Möller, July 1, 1921, Lule Lappmark (s); Hj. Möller, July 12, 1921, Lule Lappmark (CHI, DPU, s); Hj. Möller, Aug. 26, 1921, Dalarna (DPU, s); Hj. Möller, July 1, 1923, Lule Lappmark (s); Hj. Möller, Aug. 1923, Lule Lappmark (DPU, s); Hj. Möller, Aug. 2, 1926, Åsele Lappmark (DPU, s); Mosén, Aug. 12, 1868, Södermanland (DPU, s); Myrin, Aug. 10, 1833, Uppland (s); Nyman, July 28, 1890, Östergötland (H); Nyman, July 23, 1891, Lule Lappmark (s); Nyman, Oct. 4, 1891, Uppland (s); Nyman, Sept. 5, 1893, Lule Lappmark (s); Nyman, June 29, 1896, Uppland (DPU, s); Ohlsson, Värmland (NY); Oldberg, Aug. 1868, Gästrikland (US); Olsson, June 1855, Värmland (DPU, s); Olsson, in 1856, Värmland (s); Östman, May 14, 1904, Härjedalen (s); Östman, Oct. 1, 1909, Härjedalen (s); E. Persson, July 21, 1888, Närke, Örebro (US); Scheutz, Småland (NY); Scheutz, in 1869, Småland (DPU, s); Scheutz, in 1879, Bohuslän (US, s); Schimper, Gästrikland (BM, G-BOIS, MINN, US); Schimper 185 (UC); Schimper, in 1844, Gästrikland (G-BOIS); Schimper, in 1845 (PC); Schimper, in 1856 (G-BOIS); Seth, in 1879, Medelpad (s); Sillén, Sept. 1834, Gästrikland (s); Stenholm, Aug. 4, 1918, Jämtland (BART, MT); Stenholm, Aug. 6, 1918, Jämtland (BART); Stenholm, Aug. 20, 1918, Jämtland (MICH); Stenholm, Aug. 16, 1926, Pite Lappmark (BART); Stenholm, Aug. 17, 1926, Pite Lappmark (WASH); Swartz, (according to Hedwig, Musc. Frond. 3: 58. 1792), (Assumed to be TYPE of *Fontinalis falcata*, G-BOIS); Swartz, in 1795 (s); Swartz, in 1807, Dalarna (s); Tärnlund, June 1890, Västmanland (DPU, s); H. Thedenius, July 1888, Gästrikland (FH, s, UC); H. Thedenius, Aug. 1888, Gästrikland (CHI, DPU, s, UCLA, US); K. F. Thedenius, Härjedalen (s); K.

F. Thedenius, Aug. 23, 1834, Gästrikland (s); K. F. Thedenius, Sept. 14, 1834, Gästrikland (s); K. F. Thedenius, in 1837, Gästrikland (κ); K. F. Thedenius, July 1837, Gästrikland (FH, s, UC); Tolf, July 1870, Småland (NY); Tolf, June 1888, Småland (s); Vestergren, July 10, 1901, Lule Lappmark (s); Vetterhall, Aug. 1874, Småland (s); Von Post, in 1847, Västmanland (CHI, DPU, s); Vrang, July 27, 1896, Närke, near Örebro, Renauld and Cardot, Musci Eur. Exsic. 186 (CHI, DUKE, FH, MINN, PC, YALE); Wahlenberg (s); Wahlstedt, in 1867 (PC); Westerberg, Dec. 22, 1897, Östergötland (s); Westerberg, in 1907, Östergötland (s); Westling, in 1863, Gästrikland (s); Westling, Aug. 1863, Gästrikland, Oslättfors (CHI); Zetterstedt, Aug. 18, 1853, Västergötland (s); Zetterstedt, Oct. 1, 1855, Uppland (s); Zetterstedt, July 25, 1865, Småland (s); Zetterstedt, July 27, 1865, Småland (FH); Zetterstedt, June 10, 1875, Västergötland (s).

SWITZERLAND: J. Weber, July 18–21, 1878, Albulaquelle, alt. 2300 m (G-DEL).

UNION OF SOVIET SOCIALIST REPUBLICS: Sahlberg, July 28, 1869, Russia, Lapponia, Arctic Ocean Bay (H).

NORTH AMERICA: ALASKA: R. S. Williams, May 15, 1898, Yukon Territory, Lake Lindeman (B); R. S. Williams 692, May 15, 1898, Yukon Territory, Lake Lindeman (κ, MO, NY, US); R. S. Williams 692, May 22, 1898, Yukon Territory, Lake Lindeman (CHI, NY).

CANADA: ALBERTA: MacFadden, July 30, 1926, Jonquin Valley, Jasper Nat. Park (MO, NY, UC, WASH, WYO); MacFadden, Aug. 1, 1926, Jonquin Valley (CHI, DUKE).

ANTICOSTI ISLAND: Marie-Victorin 19081, Aug. 22, 1917, Small Rat Lake (MT).

BRITISH COLUMBIA: Macoun, in 1875, Fl. Can. 261 (κ).

LABRADOR: J. A. Allen, in 1882 (MICH); J. A. Allen, July 27, 1882, Esquimaux R. (NY, YALE); J. A. Allen 25, July 27, 1882, Esquimaux R. (NY); Waghorne, June 29, 1892, Battle Harbor (BM, NY); Waghorne 108, June 29, 1892, Battle Harbor (G-BOIS, PC); Waghorne 25, Aug. 3, 1894, near Forteau (CAN, CHI, FH, G-BOIS, MINN, MO, PC, S, US).

MANITOBA: Ritchie 1834, 6–7–1956, Seal R., near Great Island (CAN, DPU).

NEWFOUNDLAND: Waghorne, Mar. 28, 1891, New Peslican (NY).

ONTARIO: R. F. Cain, Aug. 30, 1939, Algonquin Park, Norway Lake (DPU, TRT); R. F. Cain, Aug. 10, 1940, Algonquin Park, Little Macaulay Lake (DPU, DUKE, NY, TRT); R. F. Cain, Aug. 16, 1940, Algonquin Park, Little Macaulay Lake (DPU, TRT); R. F. Cain, Aug. 15, 1945, Lake Timagami, Bear Isl. (TRT); Drexler 188, Quetico Park, Russell Lake (DPU); Hand 744, Aug. 13, 1941, Timiskiming Dist., near Latchford (DPU, OS); Macoun, in 1869, north of Michipicoten (PC); Macoun, in 1869, Lake Superior (PC); Macoun 193, July 1869, Lake Superior (MO, NY); Macoun, July 27, 1869, lake region (NY); Macoun, July 27, 1869, near Michipicoten, Can. Musci 761 p.p. (CAN); Macoun, July 27, 1869, Lake Superior, Fl. Can. 257 p.p. (CAN); Macoun 71, July 27, 1869, near Michipicoten (MO, NY); Macoun 890, July 27, 1869, near Lake Superior (PC, s); Macoun 224, Aug. 16, 1874 (BM).

QUEBEC: Boivin 3000 and 3001, Aug. 27, 1939, Charlevoix Co., Lac à Ange, alt. 800 m (MT); Lepage and Dutilly 4445, Jan. 8, 1943, Marten R. (DPU); Marie-Anselme, Oct. 20, 1935, La Tuque (DPU); Marie-Anselme 385, Oct. 26, 1935, La Tuque (NY); Marie-Anselme 706, June 26, 1936, La Tuque (CHI, DPU); Marie-Anselme 793, July 19, 1936, La Tuque (CHI, DPU); Marie-Anselme 1708, Aug. 12, 1937, St. Rivière (NY); Marie-Victorin 3571 p.p., June 1910, Terrebonne Co., Lake Conolly (DPU, MT).

UNITED STATES: COLORADO: McCaskey 202, July 31, 1939, Boulder Co., near Gentian Lake, alt. 3000 m (DPU); Sayre 801, July 1939, Boulder Co., Rock Lake, alt. 3060 m (DPU); Searle, Aug. 15, 1938, Roosevelt Nat. Forest, near Rainbow Lakes, alt. 3030 m (DPU); Vasey, in 1868 (NY).

MAINE: C. D. Adams, July 14, 1938, Livermore (DUKE); C. D. Adams, Aug. 2, 1939, Livermore, Grout, N. Am. Musci Perf. 375 (BART, CAS, CHI, CM, DPU, MICH, MINN, MT, NY, UC, WASH, YALE); H. B. Bailey, Aug. 13, 1902, Mt. Pleasant,

Denmark (NY); E. M. Dunham, Sept. 12, yr. ?, Moosehead (FH); Kennedy, Feb.
23, 1901, Grand Lake (FH); Lorenz, Aug. 6, 1913, Franklin Co., Round Mt. Lake
(NY); H. E. Oakes, May 10, 1946, Franklin Co., Rangeley (DPU, PSNH); Parlin,
Nov. 6, 1938, South Hartford (DPU, PSNH).

MICHIGAN: Holt, July 1905, Isle Royale, Rock Harbor (NY); G. E. Nichols,
Aug. 1934, Marquette Co., Cliff R. (MICH, NY, YALE); Nichols and Steere, Aug.
1935, Ontonagon Co., Porcupine Mts., near Lake Superior (CHI, DPU, MICH).

MINNESOTA: Holzinger, July 16 to Aug. 7, 1902, Cook Co., Grand Marais
(DUKE, NY, PC, WASH); MacMillan and Lyon, Aug. 27, 1901 (PC).

NEW HAMPSHIRE: J. A. Allen, July 21, 1880, Jackson, Grout, N. Am. Musci
Pl. 396 (BART, CAN, CIN, CM, COLO, DUKE, MINN, MO, NY, OS, UC, US, WASH, YALE);
J. A. Allen, July 21, 1885 (YALE); Eaton, Aug. 1856, Wantastiquet (YALE);
Farlow, Sept. 1897, Shelburne (FH); Farlow, Sept. 1, 1897, White Mts., Shel-
burne (PC); Farlow 579, Oct. 1897, Shelburne (G-BOIS, ILL, MICH, NY, UC, US,
WIS, WYO); Farlow, Oct. 1899, Shelburne (DUKE, FH); Farlow, July 4, 1905,
Shelburne (CHI, FH); James, White Mts., Saco R. (ILL); James, July 1852, White
Mts., Crawford Notch, Saco R. (PHIL); James 43, July 1852, White Mts., Craw-
ford Notch (BM); James, July 1855, White Mts., Crawford Notch (MO); James,
July 1857, White Mts., Crawford Notch, Saco R. (CHI, NY); James 60–1, July
1857, White Mts., Crawford Notch, Saco R. (US); James, Aug. 1866, White Mts.,
Crawford Notch (K, S).

NEW YORK: Eckfeldt, in 1868 (PENN); Greenalch, Sept. 10, 1898, Indian Lake
(NY); Peck, July 1865, Catskill Mts. (NY); Winne 583, Oct. 7, 1941, Greene Co.,
East Kill, alt. 780 m (DPU, NY).

UTAH: Flowers 2139, Sept. 6, 1927, Summit Co., Uintah Mts., Bald Mt., alt.
3000 m (UT); Flowers 1705, July 10, 1939, Summit Co., Henry's Fork, alt.
2100 m (UT).

WISCONSIN: Cheney 2554, July 29, 1893, Lincoln Co., Wisconsin R., Grandfather Falls, near Tomahawk (DPU, WIS); Cheney 2862, Aug. 3, 1893, Marathon
Co., Wisconsin R., near Granite Heights (DPU, WIS); Cheney 2885, Aug. 3, 1893,
Marathon Co., Wisconsin R., Granite Heights (DPU, WIS); Cheney 2967, June
22, 1894, Marathon Co., Wisconsin R., Granite Heights (DPU, WIS); Cheney
5440, July 25, 1896, Bayfield Co., Houghton Quarries (DPU, WIS); Cheney 5944
and 5945, Aug. 6, 1896, Apostle Islands, Presque Isle, near Bayfield (DPU, WIS);
Cheney 7565, July 20, 1897, Douglas Co., near Lake Superior (DPU, WIS);
Cheney, Nov. 17, 1902, Rusk Co. (WASH); Cheney 9635, Nov. 17, 1902, Rusk
Co., near Old Murray (DPU); Cheney 9932, Apr. 9, 1925, Rusk Co. (WIS).

WYOMING: T. G. and E. C. Yuncker 12369, July 25, 1946, Medicine Bow Nat.
Forest, near Centennial, alt. 2700 m (DPU).

United States without locality: Sullivant and Lesquereux, Musci Bor.-Am.
Exsic. 226 (G-BOIS); James, Sullivant and Lesquereux, Musci Bor.-Am. Exsic.
229b, New England (FH, K, MICH, MO, NY, PC, TRT, WIS); James, Sullivant and
Lesquereux, Musci Bor.-Am. 343, New England (CHI, CM, FH, MICH, NY, PC,
PHIL, UC, WIS, YALE).

Dichelyma falcatum may be distinguished by the following combination of characteristics: plants medium to moderately robust in
size, median cauline leaves carinate-conduplicate, erect-spreading,
usually falcate-secund, oblong-lanceolate or narrowly lanceolate,
nearly imbricate to imbricate at base, bases up to 0.25 mm apart,
blades 3–5 mm long, 0.7–1.35 mm wide, 3–5 : 1, apices subulate or
acuminate, costa generally briefly excurrent but frequently percurrent or nearly so, seta usually passing at considerable length
beyond the perichaetial leaves, 5–15 mm in length, and trellis perfect.

Fig. 32. – *Dichelyma japonicum* Card. – 1. Portion of stem with median cauline leaves. 2. Median cauline leaf. 3. Perichaetium and sporophyte. Drawn from Faurie 3049, Japan.

2. **Dichelyma japonicum** Card., in Bull. Soc. Bot. Genève (Sér. 2) 1: 132. 1909.

Dichelyma Hatakeyamae Okam., in Bot. Mag., Tokyo 25: 137. 1911.
Dichelyma japonicum Card. var. *Hatakeyamae* (Okam.) Nog., in Journ. Jap. Bot. 22: 84. 1948.

Plants medium in size, green, yellowish green, grayish green, or brownish green, younger portions glossy, older parts dull; stems subrigid to rigid, up to 15 cm long, 0.2–0.35 mm in diameter, darker and denuded at base with age, irregularly divided; branches few to numerous, erect-ascending to erect-spreading, rather close to somewhat distant, up to 1.5 cm in length, usually straight, ends of foliated stems and branches obtuse, occasionally uncinate with falcate-secund leaves; median cauline leaves subimbricate to imbricate at base or slightly distant, bases 0.25–0.5 mm apart, blades firm, erect-spreading, carinate-conduplicate, not secund, subsecund, or secund, some straight along keel, others with keel moderately curved, many subfalcate or falcate, narrowly lanceolate, width gradually decreasing from base to apex, margins entire below apical portion; apices subobtuse or obtuse, leaf tips serrulate; median cauline blades 2.75–4.5 mm long, 0.8–1.4 mm wide, 3–3.75 : 1; costa brownish green, 37.5–45 μ in width in conduplicate leaves, percurrent; median cells of leaves linear, ends attenuate, 5–10.2 μ wide, 10–21 : 1; alar cells indistinct, slightly enlarged, usually quadrate or rectangular, occasionally subhexagonal, walls yellowish or brownish yellow, auricles none, leaf bases not decurrent to briefly so, up to 0.25 mm; median branch leaves similar to median cauline except smaller in size; perichaetial branch 5.5–9.5 mm long; perichaetium narrowly cylindrical, 0.4–0.5 mm in diameter; perichaetial leaves ecostate, the inner narrowly lanceolate, apices acuminate or acute, tips entire or serrulate, blades convolute, twisted; calyptra clasping seta by base when young, long conical, dimidiate, 3–5 mm in length, 0.5–0.7 mm in diameter; capsule erect, usually surpassing the perichaetial leaves; operculum conical, 0.5–1.25 mm long, 0.5–0.75 mm in diameter, beak straight or slightly oblique; seta erect, brownish orange, 8–11.5 mm long, commonly surpassing the perichaetial leaves; urn erect, subcylindrical to cylindrical, brownish yellow when mature, rounded or slightly attenuate at base, 2–3 mm long, 0.7–1 mm in diameter, 1.16–3.3 : 1, usually slightly contracted beneath mouth when dry, occasionally not; peristome teeth brownish yellow or brownish orange, linear-lanceolate, 0.5–0.7 mm long, muricate, lamellae 8–15; trellis perfect, brownish yellow or brownish orange, usually longer than the teeth, 0.5–0.75 mm in length, muricate; spores yellowish green, green, or yellowish brown, muricate, 12–20 μ in diameter, ripe in summer or autumn. [The specific name, *japonicum*, is significant of the country, Japan, or Japon in French, in which the species was first collected.]

Type: Assumed to be in Cardot's Herbarium in the Herbarium of the Laboratoire de Cryptogamie, Muséum d'Histoire Naturelle, Paris, France. The writer was not successful in locating it there. Perhaps it was destroyed in World War I. Isotypes or portions of the type, with labels, "Herb. J. Cardot," occur in herbaria in Helsingfors, Geneva, and in the New York Botanical Garden.
Type collector: Faurie 3049, in 1904.
Type locality: Japan: Nayoro, at the base and on the roots of trees, at the margin of the Teshiogava.
Distribution: Japan.
Additional descriptions: Noguchi, in Journ. Jap. Bot. 22: 84. 1948.
Illustrations: Okamura, in Bot. Mag., Tokyo 25: fig. 5 (as *D. Hatakeyamae*). 1911; Noguchi, in Journ. Jap. Bot. 22: 85, fig. 33 (as *D. japonicum*), 86, fig. 34 (as *D. japonicum* var. *Hatakeyamae*). 1948; Brotherus Herbarium, in Herbarium of Botanisches Museum der Universität, Helsinki, Finland, a printed postal card with drawings, to scale, of plants in fruit, leaves, perichaetium, sporophyte, calyptra, peristome (teeth and trellis), and a spore, accompanied by numerous Japanese symbols (as *D. Hatakeyamae*). – FIGURE 32.
Specimens examined: JAPAN: Faurie 3049, in 1904, Nayoro (TYPE, PC ?; G-BOIS, H, NY); Hatakeyama, Oct. 10, 1909, Echigo, Mt. Odo (Assumed to be portions of TYPE or COTYPE of *D. Hatakeymae*, H, PC, S); Sakurai 399, Aug. 2, 1911, Twasiro, Okugawa, alt. 1680 m (H); Tameki 1532, July 6, 1912, Twasiro, Okugawa (H, L); Tameki, Sept. 8, 1912, Twasiro, Okugawa (WASH).

Dichelyma japonicum may be distinguished by the following combination of characteristics: plants medium in size, median cauline leaves carinate-conduplicate, not secund, subsecund, or secund, almost straight, moderately curved, subfalcate, or falcate, narrowly lanceolate, apices subobtuse or obtuse, serrulate, blades 2.75–4.5 mm long, 0.8–1.4 mm wide, 3–3.75 : 1, costa percurrent, seta 8–11.5 mm long, usually surpassing the perichaetial leaves, and trellis perfect.

Dichelyma japonicum resembles *D. pallescens* and *D. falcatum*, but differs from them in the following characteristics: in *D. japonicum*, leaves 0.8–1.4 mm in width, bases up to 0.25 mm apart, blades not secund, subsecund, or secund, some with keels almost straight or moderately curved, others subfalcate or falcate, apices subobtuse to obtuse, costa percurrent, seta usually surpassing perichaetial leaves, and trellis perfect, in contrast with *D. pallescens*, leaves 0.5–0.8 mm wide, bases 0.25–0.5 mm apart, blades secund, subfalcate to falcate, apices obtuse, subobtuse, acute, or acuminate, costa subpercurrent to percurrent, seta almost to entirely hidden in the perichaetial leaves, and trellis none or imperfect, cilia free or united into trellis at apex with 2–4 rows of transverse strands; and in contrast with *D. falcatum*, leaves usually falcate-secund, apices subulate or acuminate, and costa commonly briefly excurrent but frequently percurrent or nearly so.

3. **Dichelyma pallescens** Br. and Schimp., Bry. Eur., Suppl. 1, Fasc. 31, 5: 2. 1846.

Dichelyma capillaceum Myr., in Act. Reg. Acad. Sci. Holm. 1832, p. 278. 1833, not Bruch and Schimper, 1846.
Neckera leucoclada C. Müll., Syn. Musc. 2: 144. 1850.
Dichelyma obtusulum Kindb., in Macoun, Cat. Can. Pl., Part 6: 159. 1892.
Dichelyma novae brunsviciae Kindb. in litt., according to Cardot, Mon. Font., p. 143. 1892 (nomen nudum); *D. novae-brunsviciae*, Can. Musci No. 534 (nomen nudum), fide Macoun, Cat. Can. Pl., Part 6: 159. 1892 (nomen nudum).

Plants slender in size, green, yellowish green, brownish yellow, or brown, younger parts slightly glossy to glossy, older parts usually dull; stems subrigid to rigid, up to 10.5 cm in length, 0.2–0.25 mm in diameter, darker and denuded at base with age, irregularly or subpinnately divided; branches few to numerous, erect-spreading to spreading, close to distant, up to 4 cm in length, ends of foliated stems and branches obtuse, often slightly to distinctly curved, or uncinate with falcate-secund leaves; median cauline leaves subimbricate to imbricate at base or slightly distant, bases 0.25–0.5 mm apart, blades firm, erect-spreading, carinate-conduplicate, secund, commonly subfalcate or falcate, narrowly lanceolate, width gradually decreasing from base to apex, margins entire below and serrulate in apical portions; apices acuminate, acute, subobtuse, or obtuse, leaf tips serrulate, rarely entire; median cauline blades 3–4 mm long, 0.5–0.8 mm wide, 4–7 : 1; costa brownish yellow or greenish yellow, 22.5–60 µ in width in conduplicate leaves, subpercurrent to percurrent; median cells of leaves linear-rhomboidal, or linear, ends attenuate, 5.1–8.5 µ wide, 8–15 : 1, alar cells indistinct, slightly enlarged, usually quadrate or rectangular, occasionally subhexagonal, walls yellowish or brownish yellow, auricles none, leaf bases not decurrent to very briefly so, up to 0.25 mm; median branch leaves similar to median cauline except smaller in size, frequently falcate-secund; perichaetial branch 4–10 mm long; perichaetium narrowly cylindrical, 0.25–0.4 mm in diameter; perichaetial leaves ecostate, the inner linear, apices long acuminate, tips entire, blades convolute, twisted; calyptra clasping seta by base when young, long conical, dimidiate, 3–4 mm in length, 0.25–0.5 mm in diameter; capsule erect or oblique, emerging laterally from the perichaetium or slightly surpassing it; operculum conical, acuminate, 0.5–1 mm long, 0.3–0.6 mm in diameter; seta erect, brownish yellow, brownish orange, orange, or reddish, 4–6.5 mm long, almost to entirely hidden in the perichaetial leaves; urn emergent, suboval or subcylindrial, brownish, brownish orange, or orange, rounded or slightly attenuate at base,

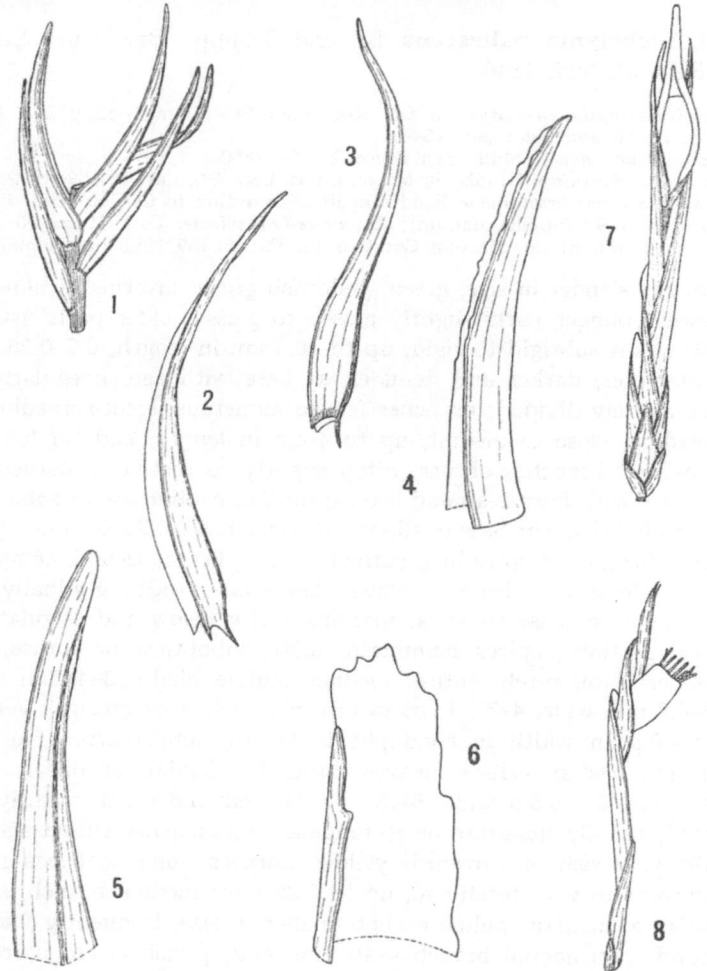

FIG. 33. – *Dichelyma pallescens* Br. and Schimp. – 1.Portion of stem with median cauline leaves.   2 and 3. Median cauline leaves.   4 and 5.Leaf apices.   6.Leaf tip.   7.Perichaetium, calyptra, and sporophyte.   8.Perichaetium and laterally emergent capsule. Nos. 1, 4, 5, and 6 drawn from Barron, in Austin, Musci Appal., Suppl. 1, 525, New York; nos. 2, 3, 7, and 8 after Bruch and Schimper, Bry. Eur., pl. 435.

0.65–1.8 mm long, 0.35–0.9 mm in diameter, 1.9–2.5 : 1, usually slightly contracted beneath mouth when dry, occasionally not;

peristome teeth yellowish, brownish yellow, or brownish orange, linear, 0.33–0.4 mm long, muricate, lamellae 8–12; trellis none or imperfect, cilia yellowish, brownish yellow, or brownish orange, usually longer than the teeth, up to 0.45 mm in length, muricate, nodulose, entirely free or united at the apex with 2–4 transverse strands; spores brownish yellow, brown, 10–15.3 μ in diameter, smooth to muricate, ripe in summer. [The specific name, *pallescens*, was applied to the plants by Bruch and Schimper to signify the pale color of the specimens which they described.]

Type: In Herbarium of Naturhistoriska Riksmuseet, Stockholm, Sweden. The plants are not marked as type but are so regarded.

Type collector: Thomas Drummond.

Type locality: North America. [Assumed to be Canada; perhaps Ontario, Holland Landing.]

The writer has examined Myrin's type material of *Dichelyma capillaceum* Myr. and has determined it as *D. pallescens* Br. and Schimp. The data with the plants are: "*Dichelyma capillaceum* M. In America Boreali, legit Drummond, Hooker misit 1832"; and "*Dichelyma capillaceum* Myrin. *Fontinalis capillacea.* N. Am. The best I have. Drummond. Hooker misit 1832." It appears that Myrin thought he was transferring the species *capillacea* from the genus *Fontinalis* to his genus *Dichelyma*, because he cites as a synonym, *Fontinalis capillacea* Dicks., Crypt. Fasc. 2, p. 1. 1790. Since Dickson included in the synonymy of *F. capillacea* Dicks, *Fontinalis capillacea, calycibus stili instar cuspidatis* Dill., Hist. Musc., p. 260. 1741, and since Dillenius cites as locality, Pennsylvania, and as collector, Bartram, it is logical that Myrin would cite the collector and the type locality of *D. capillaceum* Myr. as given in the original description of *F. capillacea* Dill. Thus it seems necessary to rely upon the labels of his type material for the type locality and collector of *Dichelyma pallescens* Br. and Schimp., which they based upon *Dichelyma capillaceum* Myr. Myrin, in Act. Reg. Acad. Sci. Holm. 1832, pl. 7a, 1833, is considered by the writer as illustrating *D. pallescens* because the costa is shown as subpercurrent to percurrent as it is in *D. pallescens* and not long excurrent as it is in *D. capillaceum* (With.) Myr. emend. Br. and Schimp.

Distribution: North America: northeastern United States and southeastern Canada.

Additional descriptions: Bruch and Schimper, Bry. Eur., Fasc. 16, 5: 7 (as *D. capillaceum* Myr., excl. var. *subulifolium*, p. 8). 1842; Sullivant, Musci and Hep. U. S., p. (655) 55. 1856; Lesquereux and James, Man. Moss. N. Am., p. 274. 1884; Cardot, Mon. Font., p. 142. 1892; Grout, M. H. M., p. 402. 1903; Jennings, Man. Moss. W. Penn., p. 209. 1913; Welch, in Grout, Moss Fl. N. Am. 3: 261. 1934; Jennings, Man. Moss. W. Penn. and Adj. Reg., p. 179. 1951.

Illustrations: Myrin, in Act. Reg. Acad. Sci. Holm. 1832, pl. 7 (as *D. capillaceum* Myr.). 1833; Bruch and Schimper, Bry. Eur., Fasc. 16, 5: pl. 435, excl. var. *subulifolium*, 1, 2, 2a (as *D. capillaceum* Myr.). 1842; Grout, M. H. M., fig. 220. 1903; Jennings, Man. Moss. W. Penn., pl. 31. 1913; Welch, in Grout, Moss Fl. N. Am. 3: pl. 79. 1934; Jennings, Man. Moss. W. Penn. and Adj. Reg., pl. 35. 1951. – FIGURE 33.

Specimens examined: NORTH AMERICA: CANADA: NEW BRUNSWICK: Fowler, in 1878, Fredericton (PC); Fowler, July 13, 1878, Fredericton (NY); Fowler, July 13, 1878, Bass R., Macoun, Can. Musci 534 (Assumed to be portion of TYPE of *D. obtusulum*, BM, FH, K, MINN, MO, NY, TRT, UC, US, WIS); Fowler, July 13, 1878, Bass R., Macoun, Can. Musci 609 (DUKE, FH, PC, US, WASH); Fowler, July 13, 1878, Fredericton, Macoun, Can. Musci 994 (CAN); Fowler, in 1890,

Bass R. (as *D. novae brunsviciae* Kindb. in litt., s; H, PC); Habeeb and Davidson 13, Oct. 7, 1944, Fredericton (DPU); Moser, in 1892, Queens Co. (CAN); Moser, in 1894, Canaan Forks (s).

NEWFOUNDLAND: Fowler, Sept. 18, 1879, Grand Lake (CAN).

NOVA SCOTIA: James (B).

ONTARIO: R. F. Cain, Aug. 30, 1941, Frontenac Co., Silver Lake (DPU, TRT); R. F. Cain, Sept. 1, 1941, Frontenac Co., Silver Lake (DPU, DUKE, MT, NY, TRT); R. F. Cain, Aug. 14, 1945, Lake Timagami, Bear Isl. (DPU); R. F. Cain, Aug. 13, 1946, Lake Timagami, Long Point (DPU, TRT); Drexler 1706, Aug. 9, 1936, Crooked Lake, Quebec Park (DPU); Drummond, Holland Landing, Musci Am. 234 p.p. (B, BM, G-BOIS, K, MICH, NY, s); Hand 612, July 9, 1938, Algoma Dist., near Iron Bridge (DPU); Hand 612, July 17, 1938, Algoma Dist., near Iron Bridge (DPU); Ibbatson, in 1847, Holland Landing (NY); Macoun, Ottawa R., Ioness Isl., near St. Andrews (H); Macoun 227, Apr. 10, 1884, Ottawa (BM); Macoun 247, Sept. 17, 1887, Ottawa (B, H); Macoun, in 1889, Ottawa, Can. Musci 235 (BM, FH, H, K, MINN, MO, NY, PC, s, TRT, UC, US, WIS, YALE); Macoun, Nov. 9, 1896, Ottawa, Can. Moss. 210 (BM, CAN, E, FH, K, MO, MT, NY, US).

QUEBEC: Beaulac, Apr. 19, 1931, Montreal (MT); Beaulac 657, July 3, 1934 (DUKE); Cléonique-Joseph 9223, Sept. 19, 1936, Deux-Montagnes Co., Grande Baie (DPU, MT); Cléonique-Joseph 10176 p.p., July 2, 1938, St. Jean-d'Iberville Co. (DPU, MT); Dupret, Sault-au-Récollet (MT); Dupret, July 12, 1906, Sault-au-Récollet (FH); Dupret, July 16, 1906, Tessichee (PC); Dupret, July 19, 1906, Ottawa R., Como, Grout, N. Am. Musci Pl. 297 p.p. (G-DEL); Dupret, Aug. 22, 1906, Sault-au-Récollet (DUKE, FH); Dupret, Aug. 22, 1906, near Montreal (YALE); Dupret, Sept. 19, 1907, Montreal (PENN, s, WASH); Dupret, Aug. 22, 1916, Sault-au-Récollet (CAN); Dupret 4107, July 15, 1924, near Montreal (DUKE); Dupret, Oct. 10, 1924, Montreal (H, WASH); Dupret, Nov. 28, 1924, near Montreal (DUKE); Marie-Anselme, July 11, 1934, Iberville (DUKE); Marie-Anselme, Feb. 21, 1935, Iberville (DUKE); Marie-Anselme 112, Mar. 7, 1935 (DPU); Marie-Anselme 500, Feb. 1936, La Tuque (DPU); Marie-Anselme 1446, June 6, 1937, Shefford Co., Waterloo (DPU); Marie-Anselme 1816, Sept. 5, 1937, Shefford Co., Waterloo (MT); Marie-Anselme 1909, Oct. 23, 1937, Shefford Co., Lake Waterloo (CHI, DPU, TRT); Marie-Anselme 1912, Oct. 23, 1937, Shefford Co., Lake Waterloo (CHI, DPU, MT); Marie-Victorin 3571 p.p., June 1910, Terrebonne Co., Lake Conolly (MT); Marie-Victorin and Jacques Rousseau 50843, Sept. 28, 1941, Pontiac Co., Calumet Isl. (DPU, MT); Scott 73, Nov. 1890, Hull (TRT); Urban, June 23, 1858, Rouge R. (CAN).

Canada without locality: Drummond, British North America (H, K); Drummond, Canada (K).

UNITED STATES: MAINE: J. Blake, Oct. 1878, Harrison (FH, NY, US); Crockett, May 28, 1903, Megunticook Lake, Camden (DUKE); Fernald 139, July 8, 1891, Orono (FH, MINN); Kennedy, July 4, 1900, Penobscot R. (FH); H. D. Merrill 117, Oct. 1898, Pea Cove (DUKE); Norton, Dec. 10, 1923, Brownfield (DPU, PSNH); Parlin, July 14, 1938, Androscoggin Co., Livermore (DPU, PSNH).

MASSACHUSETTS: Cummings, South Natick (FH); Cummings, Oct. 1883, South Natick (US); Cummings, Oct. 31, 1883, South Natick (B, CAN, CHI, CM, FH, NY, OS, PC, S, WIS).

MICHIGAN: G. E. Nichols, Aug. 1934, Marquette Co., near Canyon Lakes (MICH, NY, YALE); G. E. Nichols, Aug. 1937, Keweenaw Co., Keweenaw Peninsula (YALE); G. E. Nichols and Steere, Aug. 1935, Ontonagon Co. (DPU, MICH, YALE); G. E. Nichols and Steere, Aug. 20–27, 1935, Ontonagon Co., shore of Lake Superior (DPU, MICH, YALE); Schnooberger 3718, Oct. 12, 1940, Roscommon Co., Muskegon and Wolf Rivers (DPU, WASH); Steere, Aug. 1934, Alger Co., near Au Train (BART); Steere 702, Aug. 1934, Alger Co., near Au Train (FH, MICH); Steere 3199, July 3, 1941, Cheboygan Co., Blake Lake (DPU, FH, MICH); Welch 5312, Aug. 31, 1937, Keweenaw Peninsula, near Center (CHI, DPU, MICH, NY).

MINNESOTA: Holzinger, in 1897, Fall Lake (FH); Holzinger, June 8–10, 1897, Fall Lake, near Ely (DUKE, FH, H, MINN, MO, NY, PC, S, US, WYO).

NEW HAMPSHIRE: A. W. Evans, July 17, 1890, Jackson (YALE); James (K); James, Aug. 1869, White Mts. (K).

NEW YORK: Barron, Oneida Lake (FH, US); Barron, Oneida Lake, Austin, Musc. Appal., Suppl. 1, 525 (CAN, CHI, DUKE, FH, NY, OS, US, YALE); Burnett, Riverside Park (CIN); Burnett 1768, May 31, 1896, Cattaraugus Co., Latishaw (ABS); Burnett, Aug. 19, 1896, Cattaraugus Co., Riverside Park (NY); Burnett 2065, Aug. 19, 1896, Cattaraugus Co., Riverside Park, Grout, N. Am. Musci Pl. 121a (CAN, CHI, CM, DUKE, FH, ILL, MICH, MINN, MO, NY, OS, PC, PENN, UC, US, WASH, YALE); Burnett, Oct. 18, 1897, Cattaraugus Co., Riverside Park (CM, NY); Burnett 2768, Oct. 18, 1897, Cattaraugus Co., Riverside Park, Grout, N. Am. Musci Pl. 121 (ABS, BART, CAN, CIN, CM, COLO, DUKE, FH, ILL, MINN, MO, NY, OS, PC, PENN, PHIL, UC, US. WASH, YALE); Pratt, Sept. 17, 1924, Wayne Co. (WASH); Pringle, Oct. 21, 1880, Saranac Lake (CHI, DUKE, PC); Rau, Oct. 21, 1880, Saranac Lake (NY); Warne and Barron, Nov. 1, 1877, Oneida Lake (NY); Winne 823, Nov. 26, 1941, Chautauqua Co., Bear Lake, alt. 390 m (DPU, NY); Winne 843, Dec. 5, 1941, Saratoga Co., Ballston Lake, alt. 75 m (DPU, NY); Winne, Shannon, and Muenscher 222 p.p., June 14, 1941, Hamilton Co., Sacandaga Lake, alt. 522 m (DPU, NY).

PENNSYLVANIA: Burnett, Bradford (WASH); James, Oct. 1869, Monroe Co., Tobyhanna (CHI, PHIL); Porter (NY).

VERMONT: Kirk, May 1910, Rutland (DUKE); Kirk 215, May 5, 1911, Rutland (ABS); N. L. T. Nelson 2719, May 1910, Rutland (S).

WISCONSIN: Cheney 1862, July 19, 1893, Oneida Co., Wisconsin R., Hat Rapids, near Rhinelander (DPU, WIS); Cheney 2504, July 29, 1893, Lincoln Co., Grandfather Falls, near Tomahawk (WIS); Cheney 2509, July 29, 1893, Lincoln Co., Wisconsin R., Grandfather Falls, near Tomahawk (DPU); Cheney 2818, Aug. 1, 1893, Lincoln Co., Wisconsin R., Merrill (DPU); Livingstone, Oct. 6, 1928, Outagamie Co., Appleton (DPU, WIS).

Dichelyma pallescens may be recognized by the following combination of characteristics: plants slender in size, median cauline leaves carinate-conduplicate, secund, commonly subfalcate or falcate, oblong lanceolate or narrowly lanceolate, apices obtuse, subobtuse, acuminate, or acute, margins usually serrulate in apical portion, blades 3–4 mm long, 0.5–0.8 mm wide, 4–7 : 1, costa subpercurrent to percurrent, seta 4–6.5 mm long, almost to entirely hidden in the perichaetial leaves, capsule emerging laterally from the perichaetium or slightly surpassing it, and trellis none or imperfect, cilia free or united into trellis at apex with 2–4 rows of transverse strands.

4. **Dichelyma uncinatum** Mitt., in Journ. Linn. Soc. 8: 44. 1865.

Dichelyma cylindricarpum Aust., in Bot. Gaz. 2: 111. 1877.
Dichelyma uncinatum Mitt. var. *cylindricarpum* (Aust.) Card., Mon. Font., p. 139. 1892.

Plants slender in size, yellowish green, green, brownish green, or brown, younger parts glossy, older parts dull or glossy; stems subrigid to rigid, up to 12 cm long, 0.2–0.25 mm in diameter, brownish or blackish and denuded near base with age, irregularly or regularly

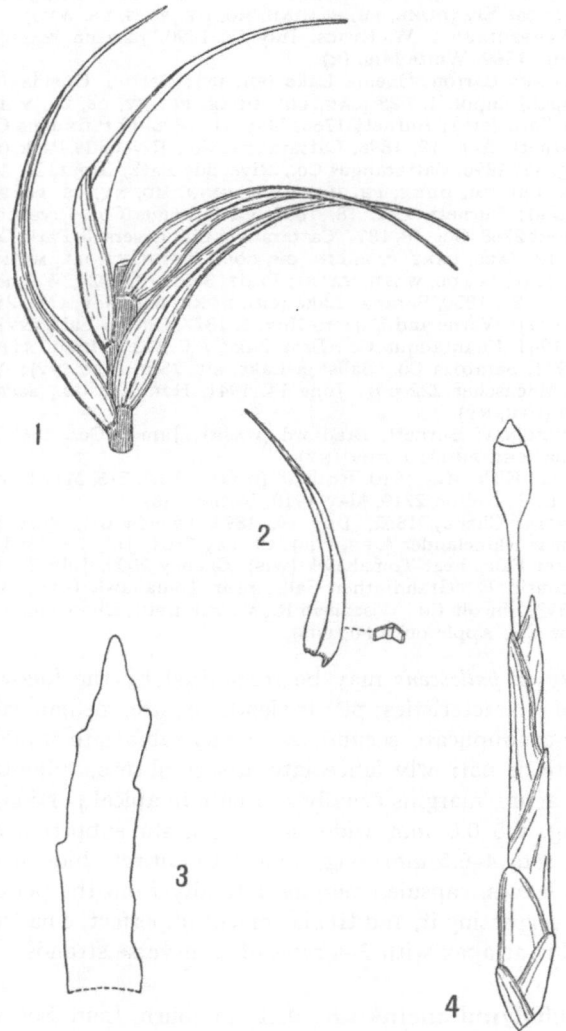

FIG. 34. – *Dichelyma uncinatum* Mitt. – 1.Portion of stem with median cauline leaves. 2.Median cauline leaf. 3.Leaf apex. 4.Perichaetium and sporophyte. Nos. 1–3 drawn from Lyall, in 1861, Oregon Boundary Commission from Fort Colville to Rocky Mts., British Columbia; no. 4 after Mitten, in Journ. Linn. Soc. 8: pl. 8.

pinnately divided; branches few to numerous, spreading, close to
distant, up to 2.5 cm long, ends of foliated stems and branches
obtuse and uncinate with falcate-secund leaves; median cauline
leaves moderately close, subimbricate, imbricate, or slightly distant
at base, bases up to 0.5 mm apart, blades firm, erect-spreading,
carinate-conduplicate, falcate-secund, narrowly lanceolate or lanceolate-subulate, margins plane in lower half or two-thirds,
usually serrulate or sinuolate in the upper one-half or one-third,
sometimes entire, width of leaves gradually decreasing from base into
narrow and long subulate apices, median cauline blades 4–5 mm long,
0.4–0.7 mm wide, 6–11 : 1; costa greenish yellow or brownish yellow,
30–45 μ in width in conduplicate leaves, long excurrent, usually
serrulate at apex and entire elsewhere; median cells of leaves linear,
ends attenuate, 5–6.5 μ wide, 12–20 : 1; alar cells indistinct, slightly
enlarged, usually quadrate or rectangular, occasionally subhexagonal,
walls yellowish or brownish yellow, auricles none, leaf bases not decurrent to very briefly so, up to 0.25 mm; median branch leaves
similar to median cauline except smaller in size, frequently falcate-
secund; perichaetial branch 5–8 mm long; perichaetium narrowly
cylindrical, 0.3–0.5 mm in diameter; perichaetial leaves ecostate, the
inner lanceolate, narrowly ovate-lanceolate, or linear, apices generally acute or acuminate, occasionally narrowly obtuse, tips usually
entire, occasionally subserrulate, blades convolute, twisted; calyptra
clasping seta by base when young but free when mature, long conical,
dimidiate, 4–5.75 mm long, 0.25–0.5 mm in diameter; capsule erect,
extending to the apices of the perichaetial leaves or surpassing them;
operculum conical, acuminate, 1–1.5 mm long, 0.5–0.75 mm in diameter; seta erect, yellowish brown, brown, orange, or reddish, usually
4–12 mm long, occasionally up to 21.5 mm; urn immersed or emergent,
subcylindrical, cylindrical, or suboval, brownish, brownish orange,
or brownish yellow, commonly straight but sometimes slightly curved,
rounded to somewhat attenuate at the base, 1–2.5 mm long, occasionally up to 3 mm in length, 0.5–0.75 mm in diameter, rarely up to 0.9
mm in diameter, 2–6 : 1, usually contracted beneath mouth when
dry, occasionally not; peristome teeth yellowish, brownish yellow,
or brownish orange, linear, subulate, 0.5–1 mm long, muricate,
lamellae 12–21; trellis yellowish, brownish yellow, or brownish orange,
perfect, generally 0.1–0.15 mm longer than the teeth, occasionally the
same length, 0.6–1 mm long, muricate, transverse bars complete,
muricate; spores yellowish green or green, 10.2–17 μ in diameter,
usually muricate, occasionally smooth, ripe in summer. [The specific
name, *uncinatum*, is significant of the uncinate ends of the stems and
branches.]

Type: In the Herbarium of the New York Botanical Garden.
Type collector: Lyall.
Type locality: British Columbia: Fort Colville.
Distribution: North America: northwestern United States and southwestern Canada.
Additional descriptions: Lesquereux and James, Man. Moss. N. Am., pp. 273 (as *D. uncinatum*), 274 (as *D. cylindricarpum*). 1884; Cardot, Mon. Font., pp. 137 (as *D. uncinatum*), 139 (as var. *cylindricarpum*). 1892; Welch, in Grout, Moss Fl. N. Am. 3: 260. 1934.
Illustrations: Mitten, in Journ. Linn. Soc. 8: pl. 8. 1865; Welch, in Grout, Moss Fl. N. Am. 3: pl. 79. 1934. – FIGURE 34.
Specimens examined: NORTH AMERICA: CANADA: BRITISH COLUMBIA: J. W. Bailey, July 29, 1901, Vancouver, Comox Lake (ABS); J. W. Bailey, Aug. 20, 1901, Vancouver, alt. 180 m (G-BOIS); J. W. Bailey, Aug. 20, 1901, Vancouver, Comox Lake, alt. 180 m, Grout, N. Am. Musci Pl. 92 (CAN, CM, DUKE, FH, ILL, MICH, MINN, MO, NY, OS, PC, S, UC, US, WASH, YALE); Brinkman 493, Oct. 12, 1911, alt. 975 m (E); A. J. Hill, Mar. 29, 1901, Ruskin, Fraser R. (NY); Lyall, Fort Colville (TYPE of *D. uncinatum*, NY; DPU); Lyall, in 1858, Oregon Boundary Commission, near the 49th parallel of latitude (K); Lyall, 1858–1859, Oregon Boundary Commission, near the 49th parallel of latitude (FH, H, K, PC, WIS); Lyall, in 1861, Oregon Boundary Commission from Fort Colville to Rocky Mts. (FH, K, NY, PC, S, WIS); Macoun, in 1872, upper Peace R., Macleod Lake (PC); Macoun 36, in 1872, Peace R. (H); Macoun, Nov. 1872, west of Rocky Mts. (CHI); Macoun, Nov. 5, 1872, Macleod Lake, Fl. Can. 262 (CAN); Macoun, Nov. 5, 1872, Macleod Lake, Can. Musci 764 (CAN); Macoun, Nov. 7, 1872, upper Peace R., Macleod Lake, Can. Fl. 764 (CAN); Macoun 194, Nov. 7, 1872, upper Peace R., Macleod Lake (MO, NY); Macoun, in 1875 (PC, S); Macoun, in 1875, Cascade Mts. (PC); Macoun, in 1875, Fl. Can. 260 (K); Macoun, in 1875, Fl. Can. 262 (K); Macoun 287, in 1875 (H); Macoun, May 11, 1875, Cascade Mts., Fl. Can. 258 (CAN); Macoun, May 17, 1875, Fraser R. (TRT); Macoun, May 17, 1875, Cascade Mts. (NY); Macoun, May 17, 1875, Cascade Mts., Fl. Can. 258 (CAN); Macoun, May 17, 1875, Fraser R. Valley, Fl. Can. 760 (CAN); Macoun, May 17, 1875, Fraser R. Valley, Fl. Can. 2155 (NY); Macoun 889, May 17, 1875, Cascade Mts. (H, PC, S); Macoun, May 29, 1875, Fraser R. Valley (TRT); Macoun, June 25, 1875, Can. Musci 225 (BM); Macoun, June 27, 1875, Macleod Lake, Fl. Can. 2157 (NY); Macoun, June 29, 1875, Macleod Lake (K, NY); Macoun, in 1890 (PC); Macoun, June 21, 1890, Pass Creek, Robson, Can. Moss. 211 (BM, CAN, FH, K, MO, MT, NY, US); Macoun, June 21, 1890, Can. Musci 234 (FH, H, K, MINN, MO, NY, PC, S, TRT, UC, WIS, YALE); Macoun, June 21, 1890, Peace Creek, Can. Musci 270 (B, CAN, S); Macoun, June 21, 1890, Can. Musci 607 (DUKE, FH, PC, US, WASH).

VANCOUVER ISLAND: Macoun, June 6, 1908, Victoria, Can. Moss. 213 (BM, CAN, FH, K, MO, MT, NY, US); Macoun 65, June 6, 1908, Victoria (S); Macoun 327, in 1913, Sidney (V); Röll 90, May 23, 1888, Victoria (B, G-BOIS, PC).

Canada without locality: Macoun, July 27, 1869, Can. Musci 761 p.p. (CAN).

UNITED STATES: IDAHO: Leiberg 231, near Lake Pend Oreille (NY); Leiberg 81, in 1888, Kootenai Co., North Fork basin (CAN, FH, K, WIS); Leiberg, Oct. 1888, Kootenai Co. (DPU, NY); Leiberg 81, Oct. 1888, North Fork basin, Prairie R. (CAN, CAS, CHI, FH, K, NY, PC, US, YALE); Leiberg 81, May 1890 (US); Leiberg 81, Aug. 1892, Kootenai Co., alt. 650 m (NY); Röll, Aug. 6, 1888, Kootenai Co., Coeur d'Alene (H, PC); Röll 1201, Aug. 6, 1888, Kootenai Co., Coeur d'Alene (NY, PC); Röll 1202, Aug. 6, 1888, Kootenai Co., Coeur d'Alene (B, NY, PC); Röll 1203, Aug. 6, 1888, Kootenai Co., Coeur d'Alene (B, NY, PC); Röll 1204, Aug. 6, 1888, Kootenai Co., Coeur d'Alene (PC); Rust 1095, Sept. 1917, Kootenai Co., Coeur d'Alene (NY); Rust 1102, Sept. 1921, Kootenai Co., Coeur d'Alene (NY); J. H. Sandberg, July 1892 (PC); J. H. Sandberg, D. T. MacDougal, and A. A. Heller 1166, Aug. 1892, Kootenai Co., Hope (Herb. ?).

MONTANA: Holzinger and Blake 43, July 21, 1898, Flathead Co., near Lake

McDonald, on Mt. Lottie Stanton, and Mt. Trilby, alt. 1050–1800 m (FH, MINN, NY, US); R. S. Williams 320, Aug. 3, 1895, head of Lake McDonald (B, CAS, CHI, K, MO, NY, S); R. S. Williams, Aug. 9, 1895, Glacier Nat. Park, head of Lake McDonald (FH, WASH); R. S. Williams 320, Aug. 9, 1895, head of Lake McDonald (NY).

OREGON: Ames, in 1886, Coast Mts. (PC); Eckfeldt, in 1878, Coast Mts. (PENN); A. S. Foster 212, June 6, 1905, near Oregon City, Magoon's Landing (WASH); A. S. Foster 1430 p.p., July 20, 1910, near Silverton (YALE); N. L. Gardner, Apr. 1883, Willamette R. (UC); E. Hall (PC); E. Hall, Salem (NY); E. Hall 74 (K); E. Hall, in 1871, Salem (CAN, CHI, FH, G-DEL, MICH, NY, PC, UC); Henderson (FH, S); Henderson 52, in 1883 (CHI); Henderson 1283 (PC); J. Howell 22, in 1876, Multnomah Co. (YALE); T. Howell, Coast Mts. (PC); T. Howell, in 1880 (NY); T. Howell, March 1880, Sauvie's Isl. (G-BOIS, S, YALE); T. Howell, Mar. 26, 1880 (CHI, FH, G-DEL, MICH, S, UC); T. Howell, May 1880, Sauvie's Isl. (YALE); T. Howell, in 1885, Sauvie's Isl. (BR, FH, PC); T. Howell, Nov. 1885, Sauvie's Isl. (FH, G-DEL); T. Howell, Nov. 1, 1885, Sauvie's Isl. (CHI); T. Howell, Nov. 7, 1885, Sauvie's Isl. (CM); T. Howell, in 1886, Coast Mts. (BM, CHI, FH, PC); Mulford, May 31, 1892, near Cornwallis (FH, MINN, MO); Pennebaker, June 24, 1947, Washington Co., lowlands of Cedar Mill Creek, near Beaverton, alt. 45–60 m (DPU); Mrs. Roy (Assumed to be TYPE or portion of Type of D. cylindri-carpum, NY; K, PC); L. Summers, McMinnville (US); L. Summers, July 1878, Coast Mts. (FH, NY, PC, US); L. Summers 2211, July 1878, Coast Mts. (E, FH, NY, UC, YALE); L. Summers, July 1879, Coast Mts., Tillamook Pass (FH).

WASHINGTON: J. W. Bailey, July 4, 1911, near Seattle, Lake Washington, Juanita Bay (ABS, H); A. S. Foster, Dec. 17, 1904, Hamilton (WASH); A. S. Foster, Jan. 20, 1905, Hamilton (DUKE); A. S. Foster, Feb. 12, 1905, Hamilton, Grout, N. Am. Musci Pl. 92a (CAN, CM, DUKE, FH, ILL, MINN, MO, NY, OS, PC, S, UC, US, WASH, YALE); A. S. Foster 150h, Mar. 18, 1905, Hamilton (DUKE); A. S. Foster, Apr. 1906, Columbia R. (YALE); A. S. Foster, May 16, 1906, Rainier (ABS); A. S. Foster 398, Aug. 5–16, 1906, Columbia R., Rainier (DUKE); A. S. Foster, Aug. 20, 1908, Aberdeen (WASH); A. S. Foster, Aug. 26, 1908, near Hoquiam (DUKE, FH, H, WASH); A. S. Foster 1926, Sept. 1911, Gate (DUKE); A. S. Foster, Oct. 1911, Gate (MT); A. S. Foster 1950, Oct. 14, 1911, Gate (NY, WASH); A. S. Foster, Feb. 25, 1912, Gate (WASH); A. S. Foster 2790, June 13, 1914, Port Angeles (WASH); T. C. Frye, Sept. 16, 1904, Renton (WASH); T. C. Frye 1029, Sept. 16, 1904, Renton (WASH); N. L. Gardner 115, Seattle (NY); N. L. Gardner, in 1897 (UC); Henderson 1915, Olympic Mts. (PC); Piper 103, Mar. 1891, Seattle (DUKE, FH, PC); Piper, Mar. 29, 1891, Seattle (WASH); Piper 103, Mar. 29, 1891, Seattle (DUKE); Piper, Apr. 6, 1891, Seattle, Lake Washington (CIN, NY, YALE); Piper, June 4, 1891, Seattle, Lake Washington (CHI, US); Piper 103, June 4, 1891, Seattle, Lake Washington (FH, NY, UC, YALE); Piper 103, Mar. 12, 1892, Seattle (UC, WIS).

WYOMING: Röll 1530, Sept. 2, 1888, Yellowstone Nat. Park (B, PC).

United States without locality: Mohr, Coast Range (NY); J. H. Sandberg 81, Aug. 1888 (MINN).

Dichelyma uncinatum may be recognized by the following combination of characteristics: plants slender in size, ends of foliated stems and branches uncinate, median cauline leaves carinate-conduplicate, falcate-secund, narrowly lanceolate or lanceolate-subulate, subimbricate to imbricate at base, bases up to 0.5 mm apart, blades 4–5 mm long, 0.4–0.7 mm wide, 6–11 : 1, and costa long excurrent, usually serrulate at apex and entire elsewhere beyond the serrulate or sinuolate margins in the upper third or half of the blades. The costa is long excurrent in leaves of both *D. uncinatum* and *D. capillaceum*, but the median cauline leaves of the former are commonly falcate-secund and 4–5 mm long, and in the latter are frequently erect-ascending, erect-spreading, or subsecund, rarely falcate, and 5–7 mm in length. When in fruit, these species are easily distinguished by capsule erect, extending to apices of perichaetial leaves or surpassing them, seta usually 4–12 mm long, occasionally up to 21.5 mm, generally extending beyond the perichaetium, and trellis perfect in *D. uncinatum*, in contrast with capsule at first immersed and later laterally emergent, seta 3–5 mm long, almost to completely hidden in the perichaetial leaves, and cilia free with the exception of apices united into trellis by 2–3 rows of transverse strands in *D. capillaceum*.

Cardot, in his monograph on Fontinalaceae, states that it is impossible to separate definitely *Dichelyma cylindricarpum* Aust. from *D. uncinatum* Mitt. on the basis of length of seta and urn, because this is a variable characteristic. However, Cardot considers *cylindricarpum* as a variety of *uncinatum* and includes under the variety those plants with setae 6–12 mm long and under the species those with setae 4–6 mm in length. In the study of the North American Fontinalaceae, the author placed Austin's species and Cardot's variety in the synonymy of *D. uncinatum* because of the variability in the seta and urn length.

In the preparation of this treatise, the writer has had the opportunity to examine many additional collections determined as *D. uncinatum*, as *D. cylindricarpum*, and as *D. uncinatum* var. *cylindricarpum*. Some of the data are included in further support of the consideration of *cylindricarpum* as a synonym of *uncinatum*.

Neither of the original descriptions of the species under consideration includes exact information concerning the sizes of the various parts of the plants. The measurements which follow have been made from the type materials. In *D. cylindricarpum* Aust., the setae vary in length from 8–12 mm and the urns from 2.5–3 mm. The setae in the type collection of *D. uncinatum* are 6–8.5 mm long, and the urns 1–1.5 mm in length. Upon this evidence it seems possible to distinguish between *D. cylindricarpum* and *D. uncinatum*.

However, in addition to the above mentioned types, numerous specimens have been examined which can not be classified as clearly as the measurements of the types seem to indicate. The setal length varies from a minimum of 4.5 mm to a maximum of 21.5 mm, although most of the setae are not longer than 12 mm. Even in a single collection, a range from 6.5–20 mm has been found in the length of the setae.

Austin, in Bryological notes, 1877, in the discussion following his Latin description of the type, states: "Nearest to *D. uncinatum* Mitt.; but readily distinguished by the capsule being twice as long and on a much longer pedicel, ..." The writer has found the urns in the type collection of *D. cylindricarpum* to be 2.5–3 mm in length and those of the type material of *D. uncinatum* to be 1–1.5 mm long. Although these data agree with the statements of Austin, additional measurements of urns in collections determined as *D. cylindricarpum* or *D. uncinatum* var. *cylindricarpum*, or *D. uncinatum* do not show such distinct limitations.

Neither are the longer urns always borne on correspondingly longer pedicels, according to the author's measurements. The writer has observed setae 7, 8, and 9 mm in length, bearing urns 1 mm long; setae 5.5 and 7 mm in length, bearing urns 1.5 mm long; setae 6.5 and 7.5 mm in length, bearing urns 2.5 mm long; setae 12 mm in length, bearing urns 3 mm long; and setae 20.5 and 21.5 mm in length, bearing urns 1.5 mm long. A very large number of the plants studied could be *cylindricarpum* if only setal length were considered. On the contrary, if the urn length were used as the criterion, these same plants would be classified as *uncinatum*.

Austin in his discussion of the type specimen states that the peristome of *D. cylindricarpum* is papillose in contrast with the smooth peristomal teeth of *D. uncinatum*. The author has found the exostome and endostome of both type collections to be papillose or muricate.

The leaves and leaf apices vary, but leaves on the same plants show comparable variations in size, width of apices, and serrulation.

The writer has been unable to find any constant characteristics by which *D. uncinatum* and *D. cylindricarpum*, or *D. uncinatum* var. *cylindricarpum*, either with or without fruit, may be distinguished definitely from each other. For this reason, *cylindricarpum* is considered as a synonym of *D. uncinatum*.

5. **Dichelyma capillaceum** (With.) Myr. emend. Br. and Schimp., Bry. Eur., Suppl. 1, Fasc. 31, 5: 1. 1846.

Fontinalis capillacea With., Syst. Arr. Br. Pl. 3: 773. 1801.[1]

---

[1] Preceding 1801, the priority date of Muscineae, Dillenius, in 1741, in Hist. Musc., p. 260, published the polynomial, *Fontinalis capillacea, calycibus stili instar cuspidatis* for the plants collected by Bartram in a lake in Pennsylvania. Linnaeus, in Fl. Suec., in 1755, p. 379, indicated the binomial, *Fontinalis capillacea*, based upon the polynomial of Dillenius. Dickson, in 1790, in Pl. Crypt. Brit., Fasc. 2, p. 1, used the name, *Fontinalis capillacea*, based upon the binomial of Linnaeus. Hedwig did not use the name, *Fontinalis capillacea*, in Species Muscorum, 1801. Withering, in Systematic Arrangement of British Plants 3:

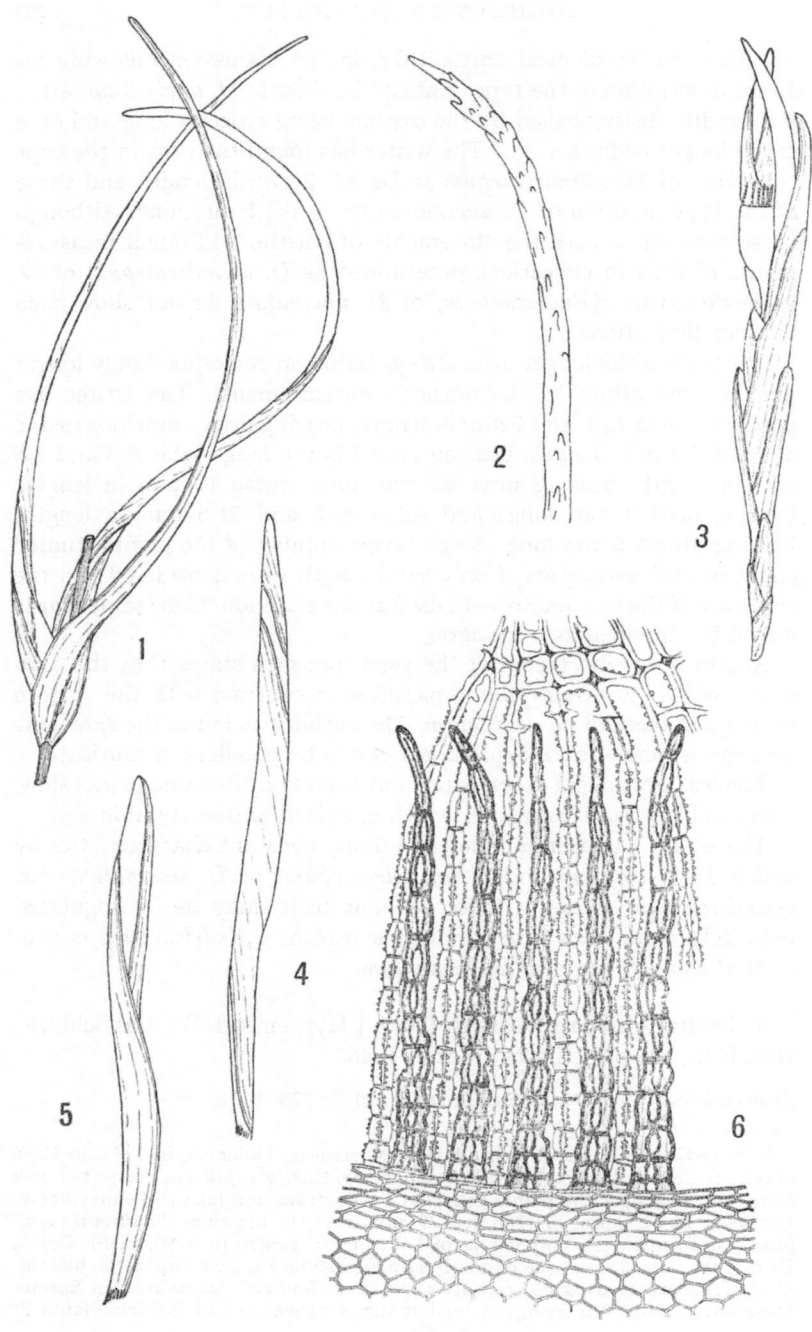

Dichelyma capillaceum Myr. var. *subulifolium* Br. and Schimp., Bry. Eur., Fasc. 16, 5: 8. 1842.
Dichelyma intermedium C. J. Hartm., Handb. Skand. Fl., p. 341. 1849.
Neckera capillacea C. Müll., Syn. Musc. 2: 144. 1850.
Dichelyma capillaceum var. *elongatum* Kindb., in Macoun, Cat. Can. Pl., Part 6: 160. 1892.

Plants slender in size, yellowish green, green, or brownish green, younger portions usually glossy, older parts commonly dull; stems subrigid to rigid, up to 20 cm in length, 0.2–0.25 mm in diameter, darker and denuded at base with age, irregularly divided; branches few to numerous, erect-spreading or spreading, close to distant, up to 4.5 cm in length, ends of foliated stems and branches obtuse, sometimes uncinate with falcate-secund leaves; median cauline leaves close to distant, bases 0.25–0.5 mm apart, blades firm, usually erect-ascending, appearing somewhat appressed, occasionally erect-spreading, carinate-conduplicate, straight to moderately curved along keel or subfalcate to falcate, generally subsecund to secund, narrowly lanceolate, width gradually decreasing from base to apex, margins entire below and entire or subserrulate above; apices long subulate or acuminate; median cauline blades 4.5–7 mm long, 0.4–0.8 mm wide, 5–11 : 1; costa brownish green or brown, 30–60 µ wide in conduplicate leaves, long excurrent, serrulate or entire at the end; median cells of leaves linear, ends attenuate, 6–7 µ wide, 10–20 : 1; alar cells indistinct, slightly enlarged, usually quadrate or rectangular, occasionally subhexagonal, walls yellowish or brownish yellow, auricles none, leaf bases not decurrent; median branch leaves similar to median cauline except smaller in size; perichaetial branch 6–8 mm long; perichaetium narrowly cylindrical, 0.35–0.65 mm in diameter; perichaetial leaves ecostate, the inner narrowly lanceolate, long acuminate, apices entire or subserrulate, blades convolute, twisted,

---

773. 1801, used the name, *Fontinalis capillacea*, based upon the *Fontinalis capillacea* of Dickson and of Dillenius. Myrin transferred the name, *Fontinalis capillacea*, to the genus, *Dichelyma*. Bruch and Schimper discovered that the plants which served as the basis of Myrin's *D. capillaceum* were not the same as those formerly known to be *Fontinalis capillacea*. According to Art. 54 of the Int. Rules of Bot. Nom. (1935), *Dichelyma capillacea* (With.) Myr. emend. Bruch and Schimper would appear to be the correct citation for this species.

FIG. 35 (*opposite*). – *Dichelyma capillaceum* (With.) Myr. emend. Br. and Schimp. – 1.Portion of stem with median cauline leaves. 2.Serrulate, excurrent costa. 3.Perichaetium and sporophyte. 4 and 5.Perichaetial leaves. 6.Upper portion of urn, bearing the outer peristome of teeth and the inner of cilia, free below and united above into an imperfect trellis by means of transverse bars, Nos. 1–6 drawn after Bruch and Schimper, Bry. Eur., Fasc. 31, 5: pl. 436.

surpassing the capsule; calyptra clasping seta by base when young, long conical, dimidiate, 1.75–2.25 mm long, 0.2–0.25 mm in diameter when immature; capsule at first immersed in the perichaetium, later laterally emergent, erect or oblique; operculum conical, acuminate, 0.5–1 mm long, 0.35–0.5 mm in diameter, beak oblique; seta erect, brownish orange or brownish yellow, 3–5 mm long, almost to completely hidden in the perichaetial leaves; urn erect or oblique, oval or oval-oblong, brownish yellow or brown when mature, rounded at base, 1–2 mm long, 0.5–1 mm in diameter, 2.3–3 : 1, frequently contracted beneath mouth when dry, sometimes not contracted; peristome teeth brownish orange, linear-lanceolate, 0.26–0.37 mm long, muricate, lamellae 10–15; cilia brownish orange, longer than the teeth, 0.45–0.6 mm in length, muricate, free almost their entire length, united into a trellis at the apex with 2–3 rows of transverse strands; spores yellowish green, green, or yellowish brown, smooth to muricate, 10–15 μ in diameter, ripe in summer. [The specific name, *capillaceum*, is indicative of the hair-like, long excurrent costa.]

Type: Not previously designated. The specimens in the J. J. Dillenius Collection in the Oxford University Herbarium, Oxford, England, first described and illustrated by Dillenius as *Fontinalis capillacea, calycibus stili instar cuspidatis,* and later known as *Fontinalis capillacea,* have been assumed to be the type.

Type collector: J. Bartram.

Type locality: Pennsylvania. Dillenius, in 1741, cites the collector and locality as, "Locis lacustribus in Pensylvania legit Jo. Bartram." Bruch and Schimper state, "In aquis Americae borealis unde e Pennsylvaniae 'locis lacustribus' a Bartramio detectore missum primus descripsit et delineavit Dillenius."

Distribution: North America and Europe: eastern United States and Canada, and western Europe; also, St. Thomas Island, West Indies.

Additional descriptions: Bridel, Musc. Recent. 2 (3): 162 (as *F. capillacea*). 1803; Smith, Fl. Brit. 3: 1337 (as *F. capillacea*). 1804; Withering, Syst. Arr. Br. Pl. 3: 969 (as *F. capillacea*). 1812; Hedwig, Schwaegrichen, Sp. Musc., Suppl. 1, 2: 307 (as *F. capillacea*). 1816; Bridel, Musc. Recent. Suppl., Part 3: 109 (as *F. capillacea*). 1817; Hedwig, Schwaegrichen, Sp. Musc., Suppl. 3, 1: pages not numbered (as *F. capillacea*). 1827; Bridel, Bry. Univ. 2: 660 (as *F. capillacea*). 1827; Swartz, Adn. Bot., p. 178 (as *F. capillacea*). 1829; Schimper, Nya Moss., p. 165. 1848; C. J. Hartman, Handb. Skand. Fl., p. 371. 1854; Wilson, Bry. Brit., p. 426. 1855; Sullivant, Musci and Hep. U. S., p. (655) 55. 1856; Schimper, Syn. Musc., pp. 460 (as *D. capillaceum*), 461 (as *D. intermedium*). 1860; Milde, Bry. Siles., p. 278. 1869; Schimper, Syn. Musc., p. 558. 1876; Lesquereux and James, Man. Moss. N. Am., p. 273. 1884; Cardot, Mon. Font., p. 140. 1892; Husnot, Musc. Gall., p. 288. 1892; Limpricht, Laubm. 2: 676. 1894; Barnes, Gen. and Sp. N. Am. Moss., p. 331 (as *D. capillaceum* var. *elongatum*). 1896; Grout, M. H. M., p. 401. 1903; Roth, Eur. Laubm. 2: 295. 1904; Warnstorf, Laubm. 2: 634. 1905; Jennings, Man. Moss. W. Penn., p. 208. 1913; Mönkemeyer, in Pascher, Süsswasserfl. 14: 108. 1914; Hj. Möller, in Ark. Bot. 17 (14): 16. 1922; Brotherus, Laubm. Fenn., p. 401. 1923; Jensen, Danm. Moss., p. 188. 1923; Mönkemeyer, Laubm. Eur. Erg.-Bd. 4: 667. 1927; Welch, in Grout, Moss Fl. N. Am. 3: 261. 1934; Jensen, Skand. Bladmossfl., p. 384. 1939; Jennings, Man. Moss. W. Penn. and Adj. Reg., p. 179. 1951.

Illustrations: Dillenius, Hist. Musc., pl. 33 (as *Fontinalis capillacea, calycibus stili instar cuspidatis*). 1741 and 1763; Hedwig, Schwaegrichen, Sp. Musc., Suppl. 3, 1: pl. 218 (as *F. capillacea*). 1827; Bruch and Schimper, Bry. Eur., Fasc. 16, 5: pl. 435, figs. of var. 1, 2, 2a (as *D. capillaceum* var. *subulifolium*). 1842, and Fasc. 31, 5: pl. 436 (as *D. capillaceum*). 1846; Schimper, Nya Moss., pl. 14. 1848; Sullivant, Musci and Hep. U. S., pl. 18. 1856; Schimper, Syn. Musc., pl. 5. 1860; Lesquereux and James, Man. Moss. N. Am., pl. 4. 1884; Husnot, Musc. Gall., pl. 81. 1892; Grout, M. H. M., fig. 219. 1903; Roth, Eur. Laubm. 2: pl. 30. 1904; Mönkemeyer, in Pascher, Süsswasserfl. 14: fig. 33. 1914; Jensen, Danm. Moss., pl. 8. 1923; Mönkemeyer, Laubm. Eur. Erg.-Bd. 4: fig. 146. 1927; Welch, in Grout, Moss Fl. N. Am. 3: pl. 79. 1934. – FIGURE 35.

Specimens examined: EUROPE: DENMARK: Jensen, Aug. 1884, Sjaelland, Helsingör (PC, S); Jensen, Aug. 19, 1884, Sjaelland, Helsingör (H).

FINLAND: H. Lindberg, June 20, 1914, Isthmus Karelicus, Muola, mouth of the Saaretjoki, V. F. Brotherus, Bryoth. Fenn. 362 (CHI, H, S, YALE).

FRANCE: V. Allorge, July 1, 1947, La Mothe, Gironde (DPU, PC); Crozals, Nov. 1, 1894 (PC).

GERMANY: Andres, Borusso-Rhenania, Köln, alt. 140 m, Krypt. Exsic. 2989 (G-BOIS, K, NY, UC, US); Brasch, Oct. 2, 1910, Rhenania, Cologne (DUKE); Brasch, Oct. 20, 1910, Rhenania, Cologne (DUKE); Freiberg, Sept. 4, 1909, West Prussia (S); Lützow, July 1881, Neustadt, West Prussia (B); Lützow, July 1882, West Prussia, Neustadt (G-BOIS, R); Lützow, Aug. 1882, West Prussia, Neustadt (B).

SCOTLAND: (See discussion.) Dickson (BM).

SWEDEN: Adlerz, Aug. 31, 1884, Närke (DPU, S); Arnell, Aug. 16, 1888, Blekinge (PC); Berggren, June 1859, Västergötland (S); Berggren, June 1860, Skåne, Vitseröd (DUKE, S); Berggren, in 1861, Skåne, Munkarp, Vitseröd (CHI); Dybeck, July 6, 1832, Västmanland (S); Dybeck, in 1833, Västmanland, Odensvi (S); Dybeck, in 1836, Västmanland, Odensvi (H, K); C. Hartman, Gästrikland (NY, PC); C. Hartman, Apr. 7, 1874, Närke, Knista, Villingsberg (DPU, S); C. Hartman, Sept. 3, 1874, Närke (S); C. Hartman, Sept. 7, 1874, Närke (S); C. Hartman, Sept. 9, 1874, Närke (WASH); C. and R. Hartman, July 3, 1848, Gästrikland, Hille, near Oslättfors (Assumed to be portion of TYPE of *D. intermedium*, H, S); C. and R. Hartman, in 1871, Gästrikland, Hille, Oslättfors (S); Hasslow, Sept. 6, 1934, Skåne (S, UC); Hovgard, Sept. 4, 1934, Skåne (CAS, S); Hovgard, Oct. 1935, Skåne, Skansen, Osby, Verdoorn, Musci Selecti et Critici 315 (BART, CHI, CM, MICH, MO, NY, WASH, WIS, YALE); Hovgard, Oct. 2, 1935, Skåne, Skansen, Osby (DPU, S); Hovgard, Oct. 2, 1935, Skåne, Skansen, Osby, Bauer, Musci Eur. et Am. Exsic. 2279 (BART); Hult, Aug. 18, 1878, Blekinge, Rödeby, Rödebyholm (S); H. Johansson, Sept. 7, 1926, Värmland (DPU, S); H. Johansson, Sept. 25, 1928, Värmland (CHI, DPU, S); Kjellmark, June 1921, Småland (S); Kjellmark, in 1923, Småland (S); Lagergren, in 1864, Värmland, Rämmen, Liljendal (S); S. O. Lindberg, Stockholm (BR, PC, US); S. O. Lindberg, in 1859, Västergötland, Hunneberg, Långvattnet, R. Hartman, Bry. Scand. Exsic. 205 (BM, K, S); S. O. Lindberg, June 1859, Västergötland (CAN, US); S. O. Lindberg, July 1859, Västergötland (S); S. O. Lindberg, July 1860, Stockholm (ILL); S. O. Lindberg, July 1860, Skåne (FH); S. O. Lindberg, July 1860, Skåne, Munkarp, Vitseröd (MINN); S. O. Lindberg, May 1862 (K); S. O. Lindberg, in 1864, Stockholm (K); S. O. Lindberg, June 30, 1864, Stockholm (CAN, DPU, MINN, MO, NY, PC, S, US, WIS, YALE); S. O. Lindberg, July 1864, Stockholm, Rabenhorst, Bryoth. Eur. 778 (FH, G-BOIS, H, NY, PC, S, WASH); S. O. Lindberg, Sept. 21, 1864, Stockholm (K); S. Medelius, July 1912, Småland, Madesjö (BART, DPU, DUKE, FH, MT, WASH); S. Medelius, July 18, 1913, Småland (S); S. Medelius, Aug. 29, 1914, Småland (H); Nyman, June 29, 1896, Uppland, Söderfors (S); Scheutz, in 1878, Småland, Växjö, Kvarnhagen (US); Scheutz, in 1879, Småland, Växjö, Kvarnhagen (B, DPU, S); Schimper, Stockholm (K); O. L. Sillén, Sept. 1870, Västmanland, Karbenning, Högfors, Musci Frond. Scand. Exsic. 117 (S); Strandmark, Sept. 1867, Småland, S. Ljunga (CHI, DPU, S);

Tufvesson, July 12, 1933, Skåne (CHI); Tufvesson, Oct. 28, 1934 (US); Waldheim, June 5, 1933, Närke (CHI, S); Westling, Aug. 1869, Linai (CHI).
    NORTH AMERICA: CANADA: CAPE BRETON ISLAND: G. E. Nichols, July 22,
1915, Barrasois R. (NY, YALE).
    LABRADOR: J. A. Allen, July 27, 1882, Esquimaux R. (YALE).
    MANITOBA: Hand 528, July 20, 1938, near Rennie (DPU).
    NEW BRUNSWICK: Fowler, Sept. 1879, Grand Lake (TRT); Moser, 675 of
Macoun (PC); Moser, Queen's Co., Canaan Forks, Macoun, Can. Musci 431 (FH,
H, K, MINN, MO, NY, TRT, UC, WIS, YALE); Moser, Queen's Co., Canaan Forks,
Macoun, Can. Musci 608 (DUKE, FH, PC, US. WASH); Moser, Aug. 1, 1889, Macoun,
Can. Musci 506 (PC); Moser, Aug. 24, 1889, Queen's Co., Macoun, Can. Moss.
212 (CAN, E, FH, K, MO, MT, NY, US); Moser, in 1894, Queen's Co., Canaan Forks (S).
    NEWFOUNDLAND: Fowler, Sept. 18, 1879, Grand Lake (CAN, PC).
    NOVA SCOTIA: M. S. Brown 530, Aug. 24, 1934, Kedji (DUKE); Macoun 175,
June 13, year ?, Yarmouth (DPU, DS); Robinson, July 17, 1903, Hartley's Falls
(ABS).
    ONTARIO: R. F. Cain, Dec. 24, 1943, Brant Co., near Hatchley (DPU, TRT);
R. F. Cain, Sept. 15, 1945, Skunk Lake, Lake Timagami (DPU, TRT); Drummond,
in 1832, Holland Landing, Musci Am. 234 p.p. (BM, CAN, K, PHIL); Hand 650,
July 22, 1938, Bruce Co., near Sable Beach (DPU); Kucyniak 42–131, Aug. 5,
1942, Carleton Co., Lake Constance (DPU, MT); Macoun, July 16, 1869, Can.
Musci 226 (BM); Macoun 288, July 27, 1869, 30 miles north of Michipicoten R.,
Lake Superior (Assumed to be TYPE of *D. capillaceum* var. *elongatum*, S).
    QUEBEC: Cléonique-Joseph 10176 p.p., July 2, 1938, St. Jean-d'Iberville Co.,
Sabrevois (DPU, MT); Dupret, Aug. 1, 1906, Oka, near Ottawa R., Grout, N.
Am. Musci Pl. 344 (CAN, CM, DUKE, FH, H, MINN, MO, NY, OS, PC, UC, US, WASH,
YALE); Dupret, Aug. 10, 1906, Oka, Ottawa R. (DUKE, FH, PC); Dupret 1365
p.p., Aug. 10, 1906, Deux-Montagnes Co., Oka, Ottawa R. (MT); Fabius 1851,
Sept. 3, 1947, Brome Lake (DPU); Marie-Anselme 1448, June 15, 1937, Shefford
Co., Waterloo (DPU); Marie-Anselme 1634, July 22, 1937, Shefford Co., Lake
Waterloo (DPU); Marie-Anselme 1861, Oct. 2, 1937 (DPU); Marie-Anselme 1908,
Oct. 23, 1937, Shefford Co., Lake Waterloo (CHI, DPU); Marie-Anselme 1910,
Oct. 23, 1937, Shefford Co., Lake Waterloo (DPU, MT); Marie-Anselme 1913 and
1914, Oct. 23, 1937, Shefford Co., Lake Waterloo (DPU); Marie-Victorin, May
1912, Chambly Co., Longueuil (MT); Marie-Victorin, Aug. 1912, Chambly Co.,
Longueuil (CHI, NY, US); Marie-Victorin, Sept. 1912, Chambly Co., Longueuil
(MT); Marie-Victorin 4 and 3726, Sept. 1912, Chambly Co., Longueuil (DPU, MT);
Marie-Victorin 9371, Aug. 1913, Chambly Co., Longueuil (DPU, MT); Marie-
Victorin 30a, July 5, 1920, Megantic Co., Lake Bécancour (DPU, MT); Marie-
Victorin and Rousseau 50846, Sept. 28, 1941, Pontiac Co., Flannagan Bay
(DPU, MT).
    Canada without locality: Drummond, North America (G-BOIS, NY, YALE);
Drummond, British North America (BM, H, K); Drummond, Canada (K);
Macoun, Can. Musci 226 (BM).
    UNITED STATES: ALABAMA: Lesquereux (PC).
    CONNECTICUT: J. A. Allen, in 1876, near New Haven (NY); Beardslee, Jan. 2,
1877 (OS); Beardslee, Apr. 19, 1877 (OS); Bishop, Feb. 8, 1880, Plainville (CHI,
FH, G-DEL, UC); Donaldson, Litchfield Co., Washington (DPU); Eaton, New
Haven (FH, UC); Eaton, May 1874 (FH); Eaton, May 11, 1874 (CAS, YALE);
Eaton, Nov. 1875, Orange (YALE); Eaton, Nov. 15, 1875, New Haven, Maltby
Park (CAS); Eaton, May 15, 1876, Middlesex Co. (YALE); Eaton, in 1879, New
Haven (PENN); Eaton, Dec. 3, 1884, East Haven (US); Eaton, Dec. 10, 1891,
East Haven, Renauld and Cardot, Musci Am. Sept. Exsic. 187 (CAN, FH, MICH,
NY, PC, YALE); A. W. Evans, Oct. 5, 1889 (CIN, US, YALE); Gleason 43–1, Oct. 3,
1943, Windham Co., Chaplin (DPU); C. B. Graves 422, New London Co., North
Stonington (YALE); C. B. Graves 462, Dec. 23, 1882, New London Co. (YALE);
Hadley, Nov. 7, 1904, South Canterbury (ABS, FH, H, WASH); Lorenz, May 25,

1909, Hartford (YALE); Lorenz, Apr. 6, 1913, Bristol (YALE); Merriam, Oct. 28, 1875, New Haven (CAS); G. E. Nichols, Oct. 11, 1906, Wilmington (YALE); G. E. Nichols, Oct. 12, 1906, Stafford (YALE); G. E. Nichols, Apr. 13, 1911, North Stonington (YALE); G. E. Nichols, Apr. 14, 1911, North Stonington (YALE); G. E. Nichols, June 24, 1912, North Colebrook (YALE); G. E. Nichols, July 1912, Pomfret (YALE); G. E. Nichols, July 1, 1912, Voluntown (YALE); G. E. Nichols, July 27, 1912, Nineveh Falls (YALE); G. E. Nichols, Aug. 1912, Pomfret (YALE); Setchell, Nov. 29, 1884, Norwich (NY, UC); R. S. Williams, Nov. 29, 1900, New Canaan (CHI); Wright, June 9, 1882 (FH); Wright, June 30, 1882, Hartford (FH).

DELAWARE: Commons, May 17, 1895 (PHIL).

DISTRICT OF COLUMBIA: Lehnert (US); Oldberg, in 1874 (MINN); Oldberg, Nov. 18, 1874 (US).

FLORIDA: Wagner 2464, May 13, 1950, Wakulla Co., near Sopchoppy (DPU).

INDIANA: Welch 9695, July 27, 1935, Porter Co., Dunes State Park (DPU).

LOUISIANA: Drummond 33, New Orleans (BM).

MAINE: M. R. Adams, July 1935, Oxford Co., Buckfield (NY); J. A. Allen, Aug. 26, 1880, Buckfield (YALE); Blake, Oct. 1878, Harrison (NY); Chamberlain 160, Aug. 23, 1898, Bristol (ABS, FH); Chamberlain 178, Aug. 31, 1898, Belgrade (FH); Crockett, Oct. 15, 1902, Camden (FH, PENN, WASH); Farlow, July 26, 1911 (FH); F. L. Harvey, Orono (FH).

MARYLAND: J. D. Smith, Aug. 10, 1876, Garrett Co. (US).

MASSACHUSETTS: Clarke, Sept. 27, 1900, Manchester (NY); Clarke, Oct. 30, 1900, Manchester (WASH); Clarke, Oct. 1, 1901, Magnolia (DUKE); Clarke, Oct. 1, 1901, Manchester (ABS); Clarke, Oct. 17, 1901 (FH); Clarke, Aug. 8, 1902, Montserrat, Beverly (MICH); Coleman, in 1875, Cheshire (US); Cummings, South Natick (FH); Cummings, in 1883, South Natick (NY); Cummings, Oct. 1883, South Natick (FH, UC, US); Cummings, Oct. 12, 1885, Wellesley (B); Cummings, Nov. 14, 1887, Wellesley (WASH); Edwards, Oct. 30, 1883, South Natick (MO, US); Edwards, Nov. 3, 1883, South Natick (CM, WIS); Edwards, Nov. 5, 1883, Wellesley (CHI); Emerson, July 12, 1904, Woods Hole (NY); Faxon 338, May 24, 1881, Dedham (NY); Faxon 339, July 3, 1883, Jamaica Plain (NY); Faxon 337, Mar. 1, 1884, Plymouth (NY); Faxon 336, Oct. 25, 1885, Jamaica Plain (NY); Gerritson, Nov. 29, 1903, Waltham (ABS, FH); Handy, Nov. 1908, Fall R. (DUKE, S); Handy, Aug. 1909, Bristol (FH); Huntington, in 1899, Essex Co., Amesbury (NY); Huntington, Nov. 15, 1899, Essex Co., Amesbury (DUKE, FH); Huntington, Dec. 15, 1899, Essex Co., Amesbury (DUKE, FH, MO, PENN, WASH); Huntington, Sept. 24, 1906, Essex Co., Amesbury (DUKE); James, Andover (MINN, PC); James, Cambridge (B, K, PC); James, July 1853, Andover (K, PC); James 93, Aug. 1853, Andover (BM); James, Dec. 1870, Cambridge (FH, S); Kennedy, Dec. 11, 1896, Westwood (FH); Kennedy, May 8, 1905, Walpole (FH); Kennedy, Sept. 20, 1907, Milton (FH); Kingman, May 1, 1912, Reading (FH); A. H. Moore 174e, Sept. 22, 1901, Andover (CHI); Dr. Nichols, June 2, 1852, Danvers (FH); Oakes, Ipswich (NY); Pringle, July 3, 1881, Jamaica Plain (DUKE, PC); Rice, Aug. 23, 1940, Nantucket (DUKE); Rice, May 27, 1942, Foxboro (DPU); Seymour, Nov. 24, 1913, Wayland (FH); Stevens, Aug. 1906, Essex Co., Manchester (FH); Stone, Worcester (FH).

MICHIGAN: Coleman, in 1874 (US); G. E. Nichols, June 1937, Marquette Co., Huron Isl. (YALE); Phinney 339, July 31, 1941, Cheboygan Co. (CHI); Schnooberger, Sept. 8, 1940, Gratiot Co., Alma, Grout, N. Am. Musci Perf. 408 (CAS, CHI, CM, DPU, MINN, MT, UC, WASH, YALE); Schnooberger 3610, Sept. 8, 1940, Gratiot Co., near Alma (DPU, MICH, NY); Alex Smith, July 2, 1930, near South Lyon (MICH).

NEW HAMPSHIRE: L. A. Carter, Belmont (ILL); Farlow, Sept. 1897, Shelburne (FH); Farlow, Sept. 1897, Shelburne (PC); Farlow, Apr. 25, 1903, Joffry (FH); Farlow, Aug. 1904, Chaurna (FH); Farlow, Sept. 29, 1908, Chocorua (FH).

NEW JERSEY: C. F. Austin, Closter (CAN, H); C. F. Austin, Oct. 1862, Bergen Co. (NY); C. F. Austin, in 1865, Closter (CAN); C. F. Austin, Dec. 1869 (NY);

C. F. Austin, Oct. 1877, Closter (NY); C. F. Austin, in 1878, Closter (PC); Eaton, in 1862 (CHI); Eaton, in 1862, Bergen Co. (YALE).

NEW YORK: C. F. Austin, Oneida (H); E. G. Britton, Nov. 14, 1886, Staten Isl. (NY); E. G. Britton, Oct. 21, 1889, Staten Isl. (NY); E. G. Britton, Nov. 1891, Staten Isl. (WASH); E. G. Britton, Nov. 14, 1891, Staten Isl. (ABS, CIN, DUKE, MO, PENN, NY, WASH); Burnett, Oct. 18, 1897, Riverside (G-BOIS); E. C. Howe, Fort Edward (MICH, NY, US); E. C. Howe, in 1865, Fort Edward (NY); E. C. Howe, in 1866, Fort Edward (NY); Latham 3717, May 4, 1914, Long Isl., Orient (ABS); Lesquereux, Catskill Mts. (PC); Muenscher, 290 of Winne, July 16, 1941, Tompkins Co., alt. 390 m (DPU, NY); Muenscher and Isely, 515 of Winne, Sept. 9, 1941, Ulster Co., Ashokan Reservoir, alt. 180 m (DPU); Nash, July 13, 1893, Cairo (DUKE, PC); Pratt, Sept. 17, 1924, Wayne Co. (WASH); Rau, Sept. 1875, Catskill Mts., South Lake (NY); Sullivant, near Bloomingdale, Musci Allegh. 151 (FH, G-BOIS, K, MICH, NY, US); H. Wheeler, July 26, 1895, Chatham (MO); Winne 313, Aug. 3, 1941, Warren Co., Lake George, alt. 96 m (DPU, NY); Winne 764 and 774, Nov. 19, 1941, Schenectady Co., Mariaville Lake, alt. 384 m (DPU, NY); Winne 858, Dec. 5, 1941, Saratoga Co., near Round Lake, alt. 48 m (DPU, NY); Winne, Shannon, and Muenscher 222 p.p., June 14, 1941, Hamilton Co., Sacandaga Lake, alt. 522 m. (DPU .

NORTH CAROLINA: Conard, Oct. 20, 1949, Green Co., Snow Hill (DPU); F. W. Gray 608b, Oct. 12–20, 1927, near Charlotte (BART); F. W. Gray M1070, June 10–20, 1928, Charlotte (ABS); F. W. Gray M1071, June 11, 1928, Charlotte (ABS, BART, PENN); F. W. Gray M1072, June 11–20, 1928, Charlotte (ABS); F. W. Gray, Mar. 3, 1931, Charlotte (WYO); F. W. Gray N.C.M. 292, Mar. 3, 1931, Charlotte (BART); F. W. Gray, Mar. 26, 1931, Charlotte, E. Bauer, Musci Eur. et Am. Exsic. 2129 (BART, NY, S).

OHIO: Bartley and Pontius, June 7, 1936, Jackson Co. (DUKE, OS); Bartley and Pontius, July 19, 1936, Ross Co. (OS); Bartley and Pontius 162, July 20, 1936, Ross Co. (NY); Bartley and Pontius, Aug. 2, 1936, Ross Co. (OS); Sterki, Apr. 1910, near Geneva (CM); Sterki, in 1911, near Geneva (CM); Sterki, Oct. 30, 1912, near Geneva (CM).

PENNSYLVANIA: J. Bartram (Assumed to be the TYPE of *Fontinalis capillacea, calycibus stili instar cuspidatis*, OXF); E. G. Britton, June 8, 1889, Pocono Summit (NY); E. G. Britton, July 4, 1893, Tobyhanna (DUKE); James, June 1850, Chester Co. (CHI, PHIL); E. T. Moul 5773, Sept. 9, 1946, Crawford Co., near Rendalls Corners, alt. 510 m (DPU, PENN); Rau, Bethlehem (CHI, FH, MICH, NY); Rau, Lehigh Co. (PHIL); Rau 132, Bethlehem (FH); Rau, June 1874, Lehigh Co. (PHIL); Rau, June 28, 1874 (NY); Rau, in 1877 (PENN); Rau, July 5, 1877, Bethlehem (OS); Rau, Nov. 1877 (PHIL); Rau, Nov. 19, 1882, Bethlehem (PC); Rau, Nov. 26, 1882 (NY); J. K. Small, Sept. 2–5, 1889, Monroe Co., Pocono Plateau (US, WIS); Wolle, Bethlehem (US); Wolle, in 1874 (NY).

RHODE ISLAND: Olney, Providence (NY).

TENNESSEE: Clebsch 962, Sept. 10, 1949, Montgomery Co., near Warfield (DPU); A. J. Sharp, May 17, 1931, Fentress Co., Buffalo Cove, alt. 480 m (CHI, DPU, WASH); A. J. Sharp 93, May 17, 1931, Fentress Co., Buffalo Cove, alt. 480 m (DUKE).

VERMONT: Blanchard, Sept. 1889, Barnet (MINN); Grout, July 26, 1900, Newfane (DUKE, G-BOIS, S); Grout, Aug. 31, 1900, Newfane (PENN); Grout, Aug. 31, 1900, Newfane, N. Am. Musci Pl. 46 (BART, CAN, CAS, CM, COLO, DUKE, FH, ILL, MINN, MO, NY, OS, PC, UC, WASH, YALE).

VIRGINIA: R. P. Carrol 85, Mar. 15, 1932, Dinwiddie Co., Curtis Pond (DPU); Leonard 17465, Sept. 21, 1935, Great Falls (US).

WEST VIRGINIA: F. W. Gray 4029, Jan. 22, 1932, Rainelle (DPU); F. W. Gray M4029, Jan. 22, 1932, Rainelle (BART).

WISCONSIN: Cheney, July 19, 1893, Oneida Co., Wisconsin R., Hat Rapids, near Rhinelander (WASH); Cheney 1857, July 19, 1893, Oneida Co., Wisconsin R., Hat Rapids, near Rhinelander (DPU, WIS); Cheney 2904, Aug. 4, 1893,

Marathon Co., Wisconsin R., Granite Heights, near Wausau (DPU, WIS); Cheney 3500, July 4, 1894, Portage Co., Wisconsin R., near Steven's Point (DPU, WIS).
United States without locality: C. F. Austin, Musci Appal. 252 (CAN, CAS, DUKE, G-BOIS, K, NY, PC, PHIL, US, WIS); Sullivant and Lesquereux, Musci Bor.-Am. Exsic 345 (CHI, CM, FH, G-BOIS, K, MICH, NY, PC, PHIL, UC, WIS, YALE); Ingraham, Sullivant and Lesquereux, Musci Bor.-Am. Exsic. 346 (B, BM, CAN, CHI, CM, FH, G-BOIS, MICH, NY, PC, PHIL, US, WIS, YALE).
WEST INDIES: [See discussion.] Krebs, St. Thomas Isl. (H).

Dichelyma capillaceum may be recognized by the following combination of characteristics: plants slender, median cauline leaves carinate-conduplicate, usually erect-ascending, appearing somewhat appressed, occasionally erect-spreading, straight to moderately curved along keel or subfalcate to falcate, generally subsecund to secund, narrowly lanceolate, bases close to distant, 0.25–0.50 mm apart, blades 4.5–7 mm long, 0.4–0.8 mm wide, 5–11 : 1; apices long subulate or acuminate, costa long excurrent, seta 3–5 mm long, almost to completely hidden in the perichaetial leaves, capsule at first immersed in perichaetium, later laterally emergent, and cilia free with exception of apices united by 2–3 rows of transverse strands.

In the British Museum of Natural History, the author examined a specimen of *D. capillaceum*, labeled as *Fontinalis capillacea*, collected by Dickson in Scotland. The plants are *D. capillaceum* without doubt. The collection was bequeathed by Robert Brown in 1858. Withering, in 1801, indicated the locality as "Mountain rivulets in Scotland." Wilson, in Bry. Brit., p. 426, in 1855, reports that for many years *D. capillaceum* has been assumed to be a dubious moss in Britain, and suggests that Dickson may have accidentally substituted a foreign specimen of *Dichelyma* for a sterile one of *Blindia acuta*, a common British moss which superficially resembles *Dichelyma*. Dixon, in Handb. Br. Moss., p. 389, 1924, states that there is much uncertainty as to the origin of this specimen and that this is the only record of this genus from the British Isles. Duncan, in Cat. Br. Moss., 1926, does not report any species of *Dichelyma*. The writer has seen no other specimen of *Dichelyma* from the British Isles.

In the Herbarium of Botanisches Museum der Universität, Helsinki, Finland, the author has studied a collection of *D. capillaceum* collected by Krebs, bearing the data, "St. Thomas ins. Antillarum." This is the only collection of Fontinalaceae from the West Indies which has been examined by the writer. The range of distribution is rather unusual for this species. No other records of collections of this family in the West Indies have come to the attention of the writer. For these reasons the authenticity of this collection seems doubtful.

FOSSIL SPECIES

Five fossil species of *Fontinalis* and one of *Dichelyma* have been noted in literature. H. N. Dixon, in W. Jongmans, Fossilium Catalogus, reported the following species, geological periods, and years of discovery.

? *Fontinalis Tournalii* (Brongn.) Schimp. (*Muscites Tournalii* Brongn.). Miocene. 1828–1838. Schimper in Traité de Pal. Végétale 1: 245. 1869, classified this moss as being pleurocarpous, with characters resembling those of a *Fontinalis* or a *Dichelyma*, and cites it as "*Fontinalis* (?) *Tournalii* (Brongn.) Schimp."

Fontinalis Sismondana Schimp. Tertiary. 1869. In Traité de Pal. Végétale 1: 245, Schimper regarded this species as being closely related to *F. antipyretica*.

? *Fontinalis pristina* Lesq. Tertiary Florissant shales. 1883. E. G. Britton and A. Hollick doubted the correctness of its reference to the genus, *Fontinalis*. F. H. Knowlton, in 1916, published the conclusive statement that *F. pristina* is a feather.

Fontinalis hypnoides C. J. Hartm. Quaternary. 1898. This fossil was first reported from Finland, by G. Andersson.

Fontinalis antipyretica Hedw. Diluvial 1914. Poland, according to A. J. Zmuda.

? *Fontinalis antipyretica* Hedw. Postglacial. 1919. France. Fragmentary material, similar to *F. antipyretica*, cited as, " *Fontinalis* ! *antipyretica* ?," by E. Gadeceau.

Fontinalis species have been reported by: W. Dawson and D. P. Penhallow, in the Pleistocene, 1890, Canada; G. Andersson, Quaternary, 1898, Finland; D. P. Penhallow, Pleistocene, 1900, Canada; H. N. Dixon, Recent, 1911, Iceland peat; and F. C. Baker, Pleistocene, 1920.

Dichelyma species have been reported but with uncertainty. *Dichelyma capillaceum* (L.) Schimp. was reported by F. C. Baker, from the Pleistocene or Glacial period, in 1920: "*Dichelyma capillaceum* (see *Distichium capillaceum*); *Distichium* (*Dichelyma*) *capillaceum*." *Dichelyma* sp. (See *Fontinalis Tournalii*.)

W. C. Steere, in Pleistocene Mosses from the Aftonian Interglacial
Deposits of Iowa, states that two different species of *Fontinalis* were
present in the collections of mosses, in a subfossil condition. The
species were not determined because the material was very fragmentary.

EXCLUDED and UNCERTAIN SPECIES

Dichelyma antarcticum C. Müll., in Engler, Bot. Jahrb. 5: 82. 1884. Plants which are assumed to be a portion of the type collection have been examined by Frances E. Wynne, who regards them as *Drepanocladus aduncus* (Hedw.) Warnst. var. *capillifolius* (Warnst.) Wynne.

Dichelyma brevinerve Kindb., in Rev. Bryol. 36: 99. 1909. Plants which are assumed to be a portion of the type collection have been examined by Wynne, who regards them as *Leptodictyum riparium* (Hedw.) Warnst.

Dichelyma californicum Aust. in herb., according to Cardot's Monograph, is not in Fontinalaceae, according to Cardot, Mont. Font., p. 144. 1892.

Dichelyma capillaceum (Brid.) Br. and Schimp., Suppl. 1, Fasc. 31. 1846, not Myrin, according to Podpěra, Consp. Musc. Eur., p. 511. 1954. Podpěra based this name upon *Fontinalis capillacea* Brid., Bry. Univ. 2: 660. 1827. Withering published the name, *F. capillacea*, in 1801.

Dichelyma capillaceum (Brid.) Hartm., according to Jensen, Skand. Bladmossfl., p. 384. 1939. Hartman, Handb. Skand. Fl., p. 433. 1843, cites *D. capillaceum* Myr. which is regarded to be a synonym of *D. pallescens* Br. and Schimp. The writer considers the authors of *D. capillaceum* to be (With.) Myr. emend. Bruch and Schimper.

Dichelyma capillaceum [Dicks.] Hartm., according to Hj. Möller, in Ark. Bot. 17 (14): 16. 1922. For reasons stated immediately above and in note 1, p. 227, the author regards this name to be *D. capillaceum* (With.) Myr. emend. Bruch and Schimper.

Dichelyma capillaceum [L.] (Smith) Schimp., according to Baehni, in Candollea 8: 189. 1941. For reasons given in note 1, p. 227, the writer regards the correct name to be *D. capillaceum* (With.) Myr. emend. Br. and Schimp.

Dichelyma capillaceum C. Müll, mss. in Musc. Röll, not Br. and Sch., is a synonym of *D. uncinatum* Mitt., fide Cardot, Mon. Font., p. 138. 1892. The writer regards this name as a manuscript name, apparently without a published description.

Dichelyma capillaceum Myr. var. *subuliferum* Br. and Schimp., cited in Podpěra, Consp. Musc. Eur., p. 511. 1954, is an error, referring to var. *subulifolium* Br. and Schimp.

Dichelyma distichum Myr., in Act. Reg. Acad. Sci. Holm. 1832, p. 282. 1833. Myrin states that the plants are very uncertain as to determination and that he regards the species as a *Dichelyma* rather than a *Fontinalis* on the basis of the dimidiate calyptra. His brief description, "capsula elongata curvata," is not applicable to any plant of *Fontinalis* or *Dichelyma* studied by the writer. Myrin gives as a synonym of his *D. distichum*, p. 283, *Fontinalis disticha* Spreng. If this is a reference to *F. disticha* Swartz, Prodr. Fl. Ind. Occ., p. 138. 1788, in one of Sprengel's publications, it is now considered as a synonym of *Neckeropsis* (*Neckera*) *disticha* (Hedw.) Kindb.

Dichelyma falcatum (Hedw.) Myr. var. *amblystegioides* Hj. Möll., in Ark. Bot. 17 (14): 16. 1922. A portion of the type has been examined by Wynne, who regards it as *Drepanocladus aduncus* (Hedw.) Warnst. var. *typicus* (Ren.) Wynne.

Dichelyma longinerve Kindb., in Macoun, in Bull. Torr. Bot. Club 16: 97. 1889, is a synonym of *Drepanocladus exannulatus* (Br. and Schimp.) Warnst. var. *Rotae* (De Not.) Grout, Moss Fl. N. Am. 3: 114. 1931, according to Wynne, in Bryol. 47: 165. 1944.

Dichelyma Novae Brunswiciae Mac., in Paris, Index Bryol. 2: 3. 1904, is considered to be an error in reference to *D. novae brunsviciae* Kindb. in litt., according to Cardot, Mon. Font., p. 143. 1892, and *D. novae-brunsviciae*, in Macoun, Cat. Can. Pl., Part 6: 159. 1892. See synonymy of *D. pallescens* Br. and Schimp.

Dichelyma sinense C. Müll., in Nuov. Giorn. Bot. Ital. 5: 190. 1898, is regarded as identical with *Ditrichum crispatissimum* (C. Müll.) Par., by Brotherus, in Engler and Prantl, Nat. Pflanz. 2: 732. 1905.

Dichelyma Swartzii Lindb. in Schimper, Syn. Musc., p. 461. 1860, was probably *Drepanocladus exannulatus* (Br. and Schimp.) Warnst. var. *typicus* (Dixon) Wynne, according to Wynne, in Bryol. 47: 75. 1944. Sullivant and Lesquereux, Musci Bor.–Am. Exsic. 344, issued as *D. Swartzii* Lindb., is *Drepanocladus exannulatus* (Br. and Schimp.) Warnst. var. *typicus* (Dixon) Wynne, according to Wynne, in Bryol. 47: 72. 1944.

Fontinalis antipyretica Hedw. subsp. *arvernica* (Ren.) Kindb., Sp. Eur. and N. Am. Bryin., Part 1: 149. 1896. Cardot published this combination in 1891.

Fontinalis antipyretica Hedw. subsp. *gracilis* (Lindb.) Giacom., in Atti Ist. Bot. Univ., Pavia, Lab. Crittog. (Ser. 5) 4 (2): 250. 1947. Kindberg published this combination in 1883.

Fontinalis antipyretica Hedw. subsp. *Kindbergii* (Ren. and Card.) Kindb., in Can. Rec. Sci. 6: 75. 1894. Cardot published this combination in 1891.

Fontinalis antipyretica Hedw. var. *laxa* Milde, in Bry. Siles., p. 276. 1869. The author's opinion is discussed in footnote 1, p. 22.

Fontinalis antipyretica Hedw. var. *minor* Roth, Eur. Laubm. 2: 279. 1904. Bridel published this combination in 1827.

Fontinalis antipyretica Hedw. var. *minor* Wahlenb., Fl. Upsal., p. 365. 1820, is *gracilis* ?, fide S. O. Lindberg, in Not. Sällsk. Fauna et Fl. Fenn. 9: 274. 1868. Lindberg included this name in his synonymy of *F. gracilis*. The writer interprets this citation to indicate Lindberg's uncertainty regarding the name, the description, or the plants. The author has been unsuccessful in locating the type specimens of *F. antipyretica* var. *minor* Wahlenb., has accepted Lindberg's uncertainty regarding this variety, and, for these reasons, is considering it as an uncertain entity.

Fontinalis antipyretica Hedw. f. *robusta* Warnst., in Verh. Bot. Ver. Brandenburg 41: 66. 1899 (nomen nudum). Cardot published this combination in 1892.

Fontinalis antipyretica Hedw. subsp. *vulgaris* (Mönkem.) Giacom. var. *danubica* Mönkem. et Warnst., in Podpěra, Consp. Musc. Eur., p. 506. 1954. Apparently the citation of Mönkemeyer with Warnstorf as an author of var. *danubica* is an error. Warnstorf in Bryol. Zeitschr. 1: 40. 1916, cites the taxon as var. *danubica* (Card.), in removing the variety from *F. fasciculata* Lindb. to *F. antipyretica* L., thus var. *danubica* (Card.) Warnst. Mönkemeyer, in Laubm. Eur., p. 665. 1927, gives the credit of the transfer to Warnstorf.

Fontinalis antipyretica subsp. *vulgaris* (Mönkem.) Giacom. f. *minor* (Roth) Podp., Consp. Musc. Eur., p. 505. 1954; *F. antipyretica* var. *minor* Roth, Eur. Laubm. 2: 279. 1904. Bridel published this combination in 1827.

Fontinalis capillacea Hook. and Wils., in Drummond, Musci Amer. 234 p.p., not Dickson, occurs in Cardot's synonymy of *Dichelyma pallescens* Br. and Schimp., Mon. Font., p. 143. 1892. The author has been unable to find additional data regarding this name. Drummond, Musci Amer. 234 is, in part, *D. pallescens* Br. and Schimp., and, in part, *D. capillaceum* (With.) Myr. emend. Br. and Schimp.

Fontinalis capillacea Schwaegr., in Hedwig, Schwaegrichen, Sp. Musc., Suppl. 1 (2): 307. 1816. The date of Withering's publication of this name is 1801.

Fontinalis dalecarlica Br. and Schimp. var. *atra* Lindb., fide Hj. Möller, in Ark. Bot. 17 (14): 57. 1922. The author has seen collections in herbaria under this name. Limpricht described *F. dale-carlica* var. *atra* in 1894.

Fontinalis dichelymoides Arn. and Nordst., not Lindb., fide Cardot, Mon. Font., p. 71. 1892 (nomen nudum). This seems to be a herbarium name.

Fontinalis Duthieae (Dixon Mss.) Sim, in Trans. Roy. Soc. S. Afr. 15: 354. 1926, is a synonym of *Wardia hygrometrica* Harv. and Hook., according to author, in Bryol. 50: 187. 1947.

Fontinalis falcata Hampe, in Paris, Index Bryol. 2: 238. 1904, is regarded as an error in citation of author as Hampe gives the name as *F. falcata* Hedw. in Linnaea 13: 45. 1839.

Fontinalis fasciculata Herb. hort. bot. Bruxell. et herb. Boissier, not Lindb., according to Cardot, Mon. Font., p. 110. 1892 (nomen nudum). This appears to be a herbarium name. Lindberg published the name, *F. fasciculata*, in 1869.

Fontinalis heterophylla Warnst., in Jaap, in Verhandl. Naturwiss. Ver. Hamb. 3 (7): 30. 1899, seems to be a name without description.

Fontinalis hypnoides C. J. Hartm. var. *Ravani* (Hy) Husn., Musc. Gall., p. 287. 1892. Cardot made *Ravani* a variety in 1891.

Fontinalis Juliana Savi, Bot. Etrusc. 3: 107. 1818, is a synonym of *Conomitrium Julianum* (Savi) Mont., in Ann. Sci. Nat. (Sér. 2) 8: 246. 1837.

Fontinalis laxa De Not., according to Cardot, Mon. Font., p. 102. 1892 (nomen nudum). Cardot refers to this as a herbarium name.

Fontinalis laxa (Milde) Warnst., Laubm., p. 631. 1905. Since Milde may be considered as having described *laxa* as a species of *Fontinalis* in 1869, the writer regards the correct citation to be *F. laxa* Milde. See footnote 1, p. 22.

Fontinalis seriata Von Klinggr., Leber- u. Laubm. West- u. Ostpreuss., p. 229. 1893, has been cited by some authors. On p. 229, Von Klinggraeff gives the authority of the name as *F. seriata* Lindb. This name was published by Lindberg in 1882.

Fontinalis squamosa Hedw. var. *dalecarlica* (Br. and Schimp.) Husn., Musc. Gall., p. 287. 1892. C. J. Hartman made this combination in 1849.

Fontinalis squamosa Hedw. subsp. *Delamarei* (Ren. and Card.) Kindb., Sp. Eur. and N. Am. Bryin., Part 1: 148. 1896. Cardot published this combination in 1891.

Fontinalis squamosa Hedw. var. *latifolia* Schimp., according to Mönkemeyer, Laubm. Eur. Erg.–Bd. 4: 662. 1927. The writer has been unsuccessful in locating the original description of *F. squamosa* var. *latifolia* Schimp. The name has been noted on herbarium labels as

F. squamosa var. *latifolia* Schimp. in litt. Luisier was author of *F. squamosa* var. *latifolia* in 1924.

Fontinalis squamosa Hedw. f. *latifolia* Card., in Rev. Bryol. 18: 82. 1891 (nomen nudum). This name seems to have been replaced by *F. squamosa* f. *latifolia* Grav. in Cardot, Mon. Font., p. 83. 1892.

Fontinalis squamosa Hedw. f. *latifolia* Schimp., fide Mönkemeyer, in Pascher, Süsswasserfl. 14: 106. 1914. The writer has not been successful in locating the original description of *F. squamosa* f. *latifolia* Schimp. According to Cardot, Mon. Font., p. 83. 1892, Gravet is author of *F. squamosa* f. *latifolia*.

Fontinalis Sullivantii Aust., not Lindb., in Cardot, in Rev. Bryol. 18: 82. 1891 (nomen nudum). This is a name apparently without description. It is cited by Cardot as a synonym of *F. Renauldi* Card. in herb. Cardot includes *F. Lescurii* Sull. var. *ramosior* Sull.? in the same synonymy. Since *F. Lescurii* Sull. var. *ramosior* Sull. is regarded by the writer as a synonym of *F. novae-angliae*, *F. Sullivantii* Aust. is assumed to be in the same synonymy. The name *F. Sullivanti* was published by Lindberg in 1869.

Fontinalis Sullivantii Card., in Rev. Bryol. 18: 85. 1891, in descriptive key, not Lindb., according to Cardot, Mon. Font., p. 121. 1892, and Cardot, in Rev. Bryol. 18: 83. 1891 (nomen nudum). However, on pp. 83 and 85. 1891, Cardot cites Lindberg as the author of *F. Sullivantii*. Cardot regarded his *F. Sullivantii* as a synonym of *F. microdonta* Ren. in litt., fide Cardot, Mon. Font., pp. 120 and 121. 1892. The writer considers *F. microdonta* Ren., collected by Lesquereux in 1883 in the United States, (cited by Cardot as *F. microdonta* Ren. in litt., Mon. Font., p. 121. 1892), to be *F. disticha* Hook. and Wils. Thus *F. Sullivantii* Card. falls into the synonymy of *F. disticha*. The author also regards *F. Sullivanti* Lindb., published in 1869, as a synonym of *F. disticha* Hook. and Wils.

Fontinalis Sullivantii Lesq. and James, Man. Moss. N. Am., p. 271. 1884, not Lindb., fide Cardot, Mon. Font., p. 117. 1892. Lesquereux and James, p. 271, cite the name as *F. Sullivantii* Lindb. The writer has determined the specimens cited with their description to be *F. novae-angliae* Sull., and regards the synonym given by Lesquereux and James, *F. Lescurii* Sull. var. *ramosior* Sull., as belonging to the synonymy of *F. novae-angliae* Sull. *F. Sullivanti* Lindb. was published in 1869.

Fontinalis tenuissima Borszczow, in Ruprecht, in Fl. Bor.-Ural., p. 44. 1854, is a synonym of *Hygrohypnum ochraceum* (Turn.) Broth. var. *tenuissimum* (Borszcz.) Savicz, according to Savicz, in Bull. Imp. Bot. Gard. Peter Great 16: 321. 1916.

Fontinalis turfacea Herz., in Biblioth. Bot. 21 (87): 106. 1916, is *Scorpidium turfaceum* Herz., according to Herzog, Bryoph. Zweit. Reise Bolivia, p. 20. 1920.

Fontinalis Uleana Broth., in Hedwigia 45: 286. 1906 (nomen nudum). According to Brotherus, in Hedwigia 45: 286. 1906 *F. Uleana* Broth. is *Potamium Uleanum* (Broth.) Broth.

EXSIC CATI CITED WITH THEIR
NUMBERS AND SPECIES

Austin, C. F. *Musci Appalachiani*. (1870)

243   *Fontinalis antipyretica* Hedw. var. *gigantea* (Sull.) Sull.
244   *Fontinalis novae-angliae* Sull.
245   *Fontinalis biformis* Sull.
246 p.p.   *Fontinalis flaccida* Ren. and Card.
246 p.p.   *Fontinalis novae-angliae* Sull.
247   *Fontinalis novae-angliae* Sull.
248   *Fontinalis novae-angliae* Sull. var. *cymbifolia* (Aust.) Welch.
249   *Fontinalis disticha* Hook. and Wils.
250   *Fontinalis filiformis* Sull. and Lesq.
251   *Fontinalis dalecarlica* Br. and Schimp.
251b   *Fontinalis neo-mexicana* Sull. and Lesq.
252   *Dichelyma capillaceum* (With.) Myr. emend. Br. and Schimp.

Austin, C. F. *Musci Appalachiani* Supplement 1. (1878)

524   *Fontinalis disticha* Hook. and Wils.
525   *Dichelyma pallescens* Br. and Schimp.

Bauer, E. *Bryotheca Bohemica*. (Centurie 1, 1898; Centurie 2, 1900; Centurie 3, 1902)

43   *Fontinalis squamosa* Hedw.
44   *Fontinalis squamosa* Hedw.
142   *Fontinalis antipyretica* Hedw.
246   *Fontinalis antipyretica* Hedw. var. *gigantea* (Sull.) Sull.
340   *Fontinalis antipyretica* Hedw. var. *gigantea* (Sull.) Sull.
341   *Fontinalis antipyretica* Hedw.
342   *Fontinalis antipyretica* Hedw. var. *gracilis* (Lindb.) Schimp.
343   *Fontinalis antipyretica* Hedw.
344   *Fontinalis squamosa* Hedw.

Bauer, E. *Musci Europaei Exsiccati*. (1900, 1909, 1915)

490   *Fontinalis antipyretica* Hedw.
491   *Fontinalis Bryhnii* Limpr.
492   *Fontinalis dalecarlica* Br. and Schimp.
493   *Fontinalis antipyretica* Hedw. var. *gracilis* (Lindb.) Schimp.
494   *Fontinalis antipyretica* Hedw. var. *gracilis* (Lindb.) Schimp.
495   *Fontinalis antipyretica* Hedw. var. *gracilis* (Lindb.) Schimp.
496   *Fontinalis antipyretica* Hedw.
497   *Fontinalis antipyretica* Hedw. var. *gracilis* (Lindb.) Schimp.
498   *Fontinalis dalecarlica* Br. and Schimp.
499   *Fontinalis squamosa* Hedw.
500   *Fontinalis squamosa* Hedw.
551   *Fontinalis antipyretica* Hedw.
552   *Fontinalis antipyretica* Hedw. var. *gracilis* (Lindb.) Schimp.

553	*Fontinalis antipyretica* Hedw.
554	*Fontinalis antipyretica* Hedw.
555	*Fontinalis antipyretica* Hedw.
556	*Fontinalis antipyretica* Hedw.
557	*Fontinalis antipyretica* Hedw.
558	*Fontinalis squamosa* Hedw.
559	*Fontinalis squamosa* Hedw.
560	*Fontinalis antipyretica* Hedw.
561	*Fontinalis dalecarlica* Br. and Schimp.
562	*Fontinalis antipyretica* Hedw.
563	*Fontinalis antipyretica* Hedw.
564	*Fontinalis hypnoides* C. J. Hartm.
565	*Fontinalis dalecarlica* Br. and Schimp.
566	*Dichelyma falcatum* (Hedw.) Myr.
1153	*Fontinalis antipyretica* Hedw.
1154	*Fontinalis antipyretica* Hedw.
1155	*Fontinalis antipyretica* Hedw.
1156	*Fontinalis hypnoides* C. J. Hartm.
1157	*Fontinalis hypnoides* C. J. Hartm.
1158	*Fontinalis hypnoides* C. J. Hartm.
1159	*Fontinalis antipyretica* Hedw.
1160	*Fontinalis antipyretica* Hedw. var. *gigantea* (Sull.) Sull.
1161	*Fontinalis dalecarlica* Br. and Schimp.
1162	*Fontinalis dalecarlica* Br. and Schimp.
1163	*Fontinalis dalecarlica* Br. and Schimp.
1164	*Fontinalis dalecarlica* Br. and Schimp.
1165	*Fontinalis squamosa* Hedw.
1395	*Fontinalis hypnoides* C. J. Hartm.
1396	*Fontinalis squamosa* Hedw.

Bauer, E. *Musci Europaei et Americani Exsiccati.*

1744	*Fontinalis antipyretica* Hedw. var. *gracilis* (Lindb.) Schimp.
1780	*Fontinalis Duriaei* Schimp.
1781	*Fontinalis novae-angliae* Sull.
1929	*Fontinalis antipyretica* Hedw.
1930	*Fontinalis Duriaei* Schimp.
1931	*Fontinalis antipyretica* Hedw. var. *gigantea* (Sull.) Sull.
1932	*Fontinalis antipyretica* Hedw.
2034	*Fontinalis neo-mexicana* Sull. and Lesq.
2129	*Dichelyma capillaceum* (With.) Myr. emend. Br. and Schimp.
2279	*Dichelyma capillaceum* (With.) Myr. emend. Br. and Schimp.
2280a	*Dichelyma falcatum* (Hedw.) Myr.
2280b	*Dichelyma falcatum* (Hedw.) Myr.

Billot, C. *Flora Galliae et Germaniae Exsiccata.*

2194	*Fontinalis antipyretica* Hedw. var. *gigantea* (Sull.) Sull.

Boros, A. *Plantae Hungariae Exsiccatae.*

1	*Fontinalis antipyretica* Hedw.
2	*Fontinalis antipyretica* Hedw.
3	*Fontinalis antipyretica* Hedw.
930	*Fontinalis Duriaei* Schimp.

Breutel. *Musci Frondosi Exsiccati.*

185	*Fontinalis squamosa* Hedw.

Brotherus, V. F. *Bryotheca Fennica.*

171	*Fontinalis hypnoides* C. J. Hartm.
257	*Dichelyma falcatum* (Hedw.) Myr.
358a	*Fontinalis antipyretica* Hedw.
358b	*Fontinalis antipyretica* Hedw.
359	*Fontinalis dalecarlica* Br. and Schimp.
360a	*Fontinalis hypnoides* C. J. Hartm.
360b	*Fontinalis hypnoides* C. J. Hartm.
361	*Fontinalis antipyretica* Hedw. var. *gracilis* (Lindb.) Schimp.
362	*Dichelyma capillaceum* (With.) Myr. emend. Br. and Schimp.

Brotherus, V. F. *Musci Fennici Exsiccati.* (1894)

22	*Fontinalis antipyretica* Hedw.
23a	*Fontinalis antipyretica* Hedw. var. *gracilis* (Lindb.) Schimp.
23b	*Fontinalis antipyretica* Hedw. var. *gracilis* (Lindb.) Schimp.
24	*Fontinalis dalecarlica* Br. and Schimp.
131a	*Fontinalis dalecarlica* Br. and Schimp.
131b	*Fontinalis dalecarlica* Br. and Schimp.
132	*Dichelyma falcatum* (Hedw.) Myr.
133	*Fontinalis antipyretica* Hedw.
199	*Fontinalis hypnoides* C. J. Hartm.
457	*Fontinalis dalecarlica* Br. and Schimp.

Dismier. *Bryotheca Gallica.*

164	*Fontinalis squamosa* Hedw.
293	*Fontinalis Duriaei* Schimp.
299	*Fontinalis antipyretica* Hedw.
309	*Fontinalis antipyretica* Hedw.
363	*Fontinalis squamosa* Hedw.

Drummond, T. *Musci Americani.* (1828) (From Canada and the Rocky Mts.)

232	*Fontinalis antipyretica* Hedw.
233	*Fontinalis dalecarlica* Br. and Schimp.
234 p.p.	*Dichelyma capillaceum* (With.) Myr. emend. Br. and Schimp.
234 p.p.	*Dichelyma pallescens* Br. and Schimp.

Drummond, T. *Musci Americani Ed. 2.* (Southern United States). (1841)

151 p.p.	*Brachelyma subulatum* (Palis. Beauv.) Schimp.
151 p.p.	*Fontinalis disticha* Hook. and Wils.
152	*Fontinalis novae-angliae* Sull. var. *cymbifolia* (Aust.) Welch
153	*Brachelyma subulatum* (Palis. Beauv.) Schimp.

Ehrhart, F. *Plantae Cryptogamae Exsiccatae.*

205	*Dichelyma falcatum* (Hedw.) Myr.

Erbario Crittogamico Italiano.

1005	*Fontinalis antipyretica* Hedw.
5(1005)	*Fontinalis antipyretica* Hedw.
1006	*Fontinalis antipyretica* Hedw. var. *gracilis* (Lindb.) Schimp.
6(1006)	*Fontinalis antipyretica* Hedw. var. *gracilis* (Lindb.) Schimp.
1103	*Fontinalis hypnoides* C. J. Hartm.

Familler, I. *Flora Exsiccata Bavarica: Bryophyta.*

26	*Fontinalis antipyretica* Hedw.
26b	*Fontinalis antipyretica* Hedw.
27	*Fontinalis squamosa* Hedw.
285	*Fontinalis antipyretica* Hedw. var. *gracilis* (Lindb.) Schimp.
678	*Fontinalis squamosa* Hedw.
794	*Fontinalis antipyretica* Hedw. var. *gracilis* (Lindb.) Schimp.

Fleischer, M. and Warnstorf, C. *Bryotheca Europaea Meridionalis.* (Centurie 1, 1896; Centurie 2, 1897; Centurie 3, 1906; Centurie 4, 1910.)

72	*Fontinalis antipyretica* Hedw.
73	*Fontinalis antipyretica* Hedw.
74	*Fontinalis Duriaei* Schimp.
171	*Fontinalis antipyretica* Hedw.
172	*Fontinalis antipyretica* Hedw.
173	*Fontinalis antipyretica* Hedw.
259	*Fontinalis hypnoides* C. J. Hartm.
364	*Fontinalis antipyretica* Hedw.
365	*Fontinalis antipyretica* Hedw. var. *gracilis* (Lindb.) Schimp.
366	*Fontinalis squamosa* Hedw.

Gandoger, M. *Flora Algeriensis Exsiccata.*

| 1801 | *Fontinalis antipyretica* Hedw. |

Gravet, F. *Bryotheca Belgica.* (Nos. 231 and 283. 1874; no. 334. 1875.)

231	*Fontinalis squamosa* Hedw.
283	*Fontinalis antipyretica* Hedw.
334	*Fontinalis squamosa* Hedw.

Grout, A. J. *North American Musci Perfecti.*

22	*Fontinalis dalecarlica* Br. and Schimp.
22a	*Fontinalis dalecarlica* Br. and Schimp.
332	*Fontinalis neo-mexicana* Sull. and Lesq.
375	*Dichelyma falcatum* (Hedw.) Myr.
408	*Dichelyma capillaceum* (With.) Myr. emend. Br. and Schimp.

Grout, A. J. *North American Musci Pleurocarpi.*

46	*Dichelyma capillaceum* (With.) Myr. emend. Br. and Schimp.
73	*Fontinalis flaccida* Ren. and Card.
84	*Fontinalis neo-mexicana* Sull. and Lesq.
91	*Fontinalis antipyretica* Hedw. var. *oreganensis* Ren. and Card.
92	*Dichelyma uncinatum* Mitt.
92a	*Dichelyma uncinatum* Mitt.
121	*Dichelyma pallescens* Br. and Schimp.
121a	*Dichelyma pallescens* Br. and Schimp.
137	*Fontinalis antipyretica* Hedw. var. *gigantea* (Sull.) Sull.
140	*Fontinalis antipyretica* Hedw. var. *oreganensis* Ren. and Card.
158	*Fontinalis patula* Card.
158x	*Fontinalis antipyretica* Hedw. var. *gigantea* (Sull.) Sull.
190	*Fontinalis Duriaei* Schimp.
210	*Fontinalis antipyretica* Hedw. var. *oreganensis* Ren. and Card.
263	*Fontinalis neo-mexicana* Sull. and Lesq.

272   *Fontinalis novae-angliae* Sull.
279   *Fontinalis missourica* Card.
280   *Fontinalis novae-angliae* Sull.
281   *Fontinalis antipyretica* Hedw. var. *gigantea* (Sull.) Sull.
286a   *Fontinalis dalecarlica* Br. and Schimp.
286a   *Fontinalis dalecarlica* Br. and Schimp.
297 p.p. *Dichelyma pallescens* Br. and Schimp.
297 p.p. *Fontinalis Duriaei* Schimp.
344   *Dichelyma capillaceum* (With.) Myr. emend. Br. and Schimp.
348   *Fontinalis novae-angliae* Sull. var. *cymbifolia* (Aust.) Welch
387   *Fontinalis Langloisii* Card.
388   *Fontinalis novae-angliae* Sull. var. *cymbifolia* (Aust.) Welch
395   *Fontinalis Allenii* Card.
396   *Dichelyma falcatum* (Hedw.) Myr.
404   *Fontinalis Duriaei* Schimp.
407   *Fontinalis flaccida* Ren. and Card.
428   *Fontinalis novae-angliae* Sull.
437   *Fontinalis antipyretica* Hedw. var. *oreganensis* Ren. and Card.

Grout, A. J. *North American Musci Pleurocarpi Supplement.*

48   *Fontinalis antipyretica* Hedw. var. *oreganensis* Ren. and Card.
56   *Fontinalis antipyretica* Hedw. var. *mollis* (C. Müll.) Welch
71   *Fontinalis biformis* Sull.

Habeeb, H. *Musci Novi Brunsvici.*

28   *Fontinalis antipyretica* Hedw. var. *gigantea* (Sull.) Sull.
51   *Fontinalis Duriaei* Schimp.
52   *Fontinalis dalecarlica* Br. and Schimp.

Hartman, R. *Bryaceae Scandinaviae Exsiccatae.* (Nos. 83–86. 1858; no. 205. 1860; nos. 411–412. 1873.)

83   *Fontinalis antipyretica* Hedw.
84   *Fontinalis dalecarlica* Br. and Schimp.
85   *Fontinalis hypnoides* C. J. Hartm.
86   *Dichelyma falcatum* (Hedw.) Myr.
205   *Dichelyma capillaceum* (With.) Myr. emend. Br. and Schimp.
411   *Fontinalis antipyretica* Hedw. var. *gracilis* (Lindb.) Schimp.
412   *Fontinalis dalecarlica* Br. and Schimp.

Husnot, T. *Genera Muscorum Europaeorum Exsiccata.*

74   *Fontinalis antipyretica* Hedw.

Husnot, T. *Musci Galliae.* (Index for nos. 651–850. 1898; for nos. 901–959. 1907.)

87   *Fontinalis antipyretica* Hedw.
88   *Fontinalis squamosa* Hedw.
673   *Fontinalis antipyretica* Hedw. var. *gigantea* (Sull.) Sull.
674   *Fontinalis dalecarlica* Br. and Schimp.
736   *Dichelyma falcatum* (Hedw.) Myr.
775   *Fontinalis squamosa* Hedw.
776   *Fontinalis hypnoides* C. J. Hartm.
832   *Fontinalis antipyretica* Hedw.
931   *Fontinalis antipyretica* Hedw. var. *gigantea* (Sull.) Sull.
932   *Fontinalis antipyretica* Hedw. var. *gracilis* (Lindb.) Schimp.
933   *Fontinalis Duriaei* Schimp.

Kavina, K. and Hilitzer, A. *Cryptogamae Čechoslovenicae Exsiccatae.*

193 *Fontinalis antipyretica* Hedw.

Kerner, A. *Flora Exsiccata Austro-Hungarica.*

1110 *Fontinalis antipyretica* Hedw.
1921 *Fontinalis antipyretica* Hedw. var. *gracilis* (Lindb.) Schimp.

Kopsch, A. *Bryotheca Saxonica.*

255 *Fontinalis antipyretica* Hedw.
256 *Fontinalis antipyretica* Hedw.
257 *Fontinalis antipyretica* Hedw.
258 *Fontinalis antipyretica* Hedw.
259 *Fontinalis squamosa* Hedw.
341 *Fontinalis squamosa* Hedw.
342 *Fontinalis squamosa* Hedw.

Kryptogamae Exsiccatae editae a Museo Palatino Vindobonensi.
(No. 297. 1898; nos. 594–595. 1900; nos. 2090–2095. 1913; nos.
2988–2989. 1926.)

297 *Fontinalis antipyretica* Hedw. var. *gracilis* (Lindb.) Schimp.
594 *Fontinalis squamosa* Hedw.
594b *Fontinalis squamosa* Hedw.
2090 *Fontinalis antipyretica* Hedw. var. *gigantea* (Sull.) Sull.
2091 *Fontinalis antipyretica* Hedw.
2092 *Fontinalis antipyretica* Hedw.
2093 *Fontinalis antipyretica* Hedw. var. *gracilis* (Lindb.) Schimp.
2094 *Fontinalis antipyretica* Hedw.
2095 *Fontinalis hypnoides* C. J. Hartm.
2989 *Dichelyma capillaceum* (With.) Myr. emend. Br. and Schimp.

Langeron, M. and Sullerot, H. *Exsiccata Muscorum Côte-d'Or.*

526 *Fontinalis antipyretica* Hedw.

Lenormand, R. *Flora Galliae et Germaniae Exsiccata.*

587 *Fontinalis squamosa* Hedw.

Limpricht, K. G. *Bryotheca Silesiaca.*

21 *Fontinalis squamosa* Hedw.
33 *Fontinalis squamosa* Hedw.
33b *Fontinalis squamosa* Hedw.
34 *Dichelyma falcatum* (Hedw.) Myr.
38 *Fontinalis antipyretica* Hedw. var. *gracilis* (Lindb.) Schimp.
234 *Fontinalis antipyretica* Hedw.
336 *Fontinalis antipyretica* Hedw. var. *gracilis* (Lindb.) Schimp.

Luisier, A. *Bryotheca Lusitanica.*

9 *Fontinalis antipyretica* Hedw.
10 *Fontinalis squamosa* Hedw.

Macoun, J. *Canadian Cryptogams.*

292 *Fontinalis Duriaei* Schimp. (June 6, 1898).

Macoun, J. *Canadian Flora*.

764   *Dichelyma uncinatum* Mitt.

Macoun, J. *Canadian Mosses*.

201 p.p.   *Fontinalis antipyretica* Hedw. (July 11, 1888.) [1]
201 p.p.   *Fontinalis antipyretica* Hedw. var. *gigantea* (Sull.) Sull. (June 13,
            1883; July 8, 1884; May 30, 1893; July 16, 1898; July 18, 1898.)
202   *Fontinalis Howellii* Ren. and Card.
203 p.p.   *Fontinalis antipyretica* Hedw.
203 p.p.   *Fontinalis neo-mexicana* Sull. and Lesq. (Apr. 6, 1889; May 30,
            1890; June 1, 1893.)
204   *Fontinalis dalecarlica* Br. and Schimp. (Aug. 4, 1898; July 3, 1899;
            July 4, 1899.)
205 p.p.   *Fontinalis antipyretica* Hedw. var. *gigantea* (Sull.) Sull. (June 12, 1899.)
205 p.p.   *Fontinalis dalecarlica* Br. and Schimp. (Aug. 26, 1883.)
205 p.p.   *Fontinalis Duriaei* Schimp. (Aug. 16, 1883).
205 p.p.   *Fontinalis Howellii* Ren. and Card. (June 15, 1891.)
205 p.p.   *Fontinalis neo-mexicana* Sull. and Lesq. (July 3, 1908.)
206 p.p.   *Fontinalis Duriaei* Schimp. (Sept. 26, 1900; Oct. 21, 1900; Oct. 27,
            1900.)
206 p.p.   *Fontinalis hypnoides* C. J. Hartm. (Sept. 26, 1900; Oct. 27, 1900.)
207   *Fontinalis hypnoides* C. J. Hartm.
208 p.p.   *Fontinalis antipyretica* Hedw. (May 8, 1893; May 15, 1893.)
208 p.p.   *Fontinalis antipyretica* Hedw. var. *oreganensis* Ren. and Card. (May
            6, 1893; May 15, 1893.)
209   *Fontinalis antipyretica* Hedw. (May 26, 1893; July 1, 1893.)
210   *Dichelyma pallescens* Br. and Schimp.
211   *Dichelyma uncinatum* Mitt.
212   *Dichelyma capillaceum* (With.) Myr. emend. Br. and Schimp.
213   *Dichelyma uncinatum* Mitt.
234   *Dichelyma uncinatum* Mitt.
432   *Fontinalis hypnoides* C. J. Hartm. (July 7, 1889; July 3, 1890.)

Macoun, J. *Canadian Musci*.

2   *Fontinalis antipyretica* Hedw. var. *oreganensis* Ren. and Card.
3   *Fontinalis antipyretica* Hedw.
4   *Fontinalis antipyretica* Hedw. (July 1, 1893; May 26, 1894.)
9   *Fontinalis antipyretica* Hedw. var. *oreganensis* Ren. and Card.
103   *Fontinalis neo-mexicana* Sull. and Lesq.
104   *Fontinalis patula* Card.
129   *Fontinalis neo-mexicana* Sull. and Lesq.
133   *Fontinalis Howellii* Ren. and Card.
206   *Fontinalis hypnoides* C. J. Hartm.
207   *Fontinalis neo-mexicana* Sull. and Lesq.
225   *Dichelyma uncinatum* Mitt.
226   *Dichelyma capillaceum* (With.) Myr. emend. Br. and Schimp.
227 p.p.   *Fontinalis antipyretica* Hedw. (Apr. 30, 1867; Apr. 30, 1887; June 6,
            1887.)
227 p.p.   *Fontinalis antipyretica* Hedw. var. *oreganensis* Ren. and Card.
227 p.p.   *Fontinalis Howellii* Ren. and Card. (June 8, 1887.)
227 p.p.   *Fontinalis novae-angliae* Sull. (Sept. 20, 1878.)

---

[1] The date of collection is given in many instances to aid in checking collections of different dates under the same number in the exsiccati. Some collections in Macoun's exsiccati were not made by him, although the majority were.

227 p.p. *Fontinalis patula* Card.
228 p.p. *Fontinalis antipyretica* Hedw. (Aug. 26, 1885.)
228 p.p. *Fontinalis antipyretica* Hedw. var. *gigantea* (Sull.) Sull. (July 26, 1865; June 13, 1883.)
229 p.p. *Fontinalis neo-mexicana* Sull. and Lesq. (June 8, 1887; Apr. 6, 1889.)
229 p.p. *Fontinalis novae-angliae* Sull.
230 *Fontinalis dalecarlica* Br. and Schimp. (June 13, 1883; July 23, 1883; June 13, 1885; in 1889.)
231 p.p. *Fontinalis dalecarlica* Br. and Schimp. (In 1889.)
231 p.p. *Fontinalis Duriaei* Schimp.
231 p.p. *Fontinalis hypnoides* C. J. Hartm.
231 p.p. *Fontinalis novae-angliae* Sull.
231 p.p. *Fontinalis novae-angliae* Sull. var. *cymbifolia* (Aust.) Welch
232 *Fontinalis hypnoides* C. J. Hartm.
233 *Fontinalis Howellii* Ren. and Card.
234 *Dichelyma uncinatum* Mitt.
235 *Dichelyma pallescens* Br. and Schimp.
251 *Fontinalis Howellii* Ren. and Card.
270 *Dichelyma uncinatum* Mitt.
368 *Fontinalis patula* Card.
431 *Dichelyma capillaceum* (With.) Myr. emend. Br. and Schimp.
432 *Fontinalis hypnoides* C. J. Hartm. (July 8, 1889; July 11, 1889; July 17, 1889.)
506 *Dichelyma capillaceum* (With.) Myr. emend. Br. and Schimp.
534 *Dichelyma pallescens* Br. and Schimp.
598 p.p. *Fontinalis antipyretica* Hedw. var. *mollis* (C. Müll.) Welch. (May 26, 1893.)
598 p.p. *Fontinalis Howellii* Ren. and Card. (July 14, 1898.)
598a *Fontinalis antipyretica* Hedw. var. *gigantea* (Sull.) Sull.
598b *Fontinalis Howellii* Ren. and Card. (June 23, 1887; June 25, 1887.)
598d p.p. *Fontinalis antipyretica* Hedw. (May 15, 1893.)
598d p.p. *Fontinalis antipyretica* Hedw. var. *oreganensis* Ren. and Card. (May 15, 1893.)
598d p.p. *Fontinalis patula* Card. (May 8, 1893; May 15, 1893.)
598e p.p. *Fontinalis antipyretica* Hedw. (May 26, 1893.)
598e p.p. *Fontinalis antipyretica* Hedw. var. *oreganensis* Ren. and Card. (May 26, 1893.)
599 *Fontinalis neo-mexicana* Sull. and Lesq.
600 *Fontinalis dalecarlica* Br. and Schimp. (June 13, 1883; Aug. 4, 1898.)
600a *Fontinalis dalecarlica* Br. and Schimp.
601 *Fontinalis novae-angliae* Sull.
602 *Fontinalis hypnoides* C. J. Hartm.
602a *Fontinalis novae-angliae* Sull.
603a *Fontinalis Duriaei* Schimp.
604 p.p. *Fontinalis Duriaei* Schimp. (Sept. 21, 1900.)
604 p.p. *Fontinalis hypnoides* C. J. Hartm. (July 18, 1880; May 31, 1890.)
604 p.p. *Fontinalis novae-angliae* Sull. (July 24, 1900.)
604a *Fontinalis hypnoides* C. J. Hartm.
605 *Fontinalis novae-angliae* Sull.
605a *Fontinalis novae-angliae* Sull.
607 *Dichelyma uncinatum* Mitt.
608 *Dichelyma capillaceum* (With.) Myr. emend. Br. and Schimp.
609 *Dichelyma pallescens* Br. and Schimp.
752 p.p. *Fontinalis Duriaei* Schimp. (June 15, 1881.)
752 p.p. *Fontinalis novae-angliae* Sull. (July 9, 1870.)
757 p.p. *Fontinalis antipyretica* Hedw. var. *gigantea* (Sull.) Sull. (July 13, 1865.)
757 p.p. *Fontinalis antipyretica* Hedw. var. *oreganensis* Ren. and Card. (May 7, 1875.)

759b   *Fontinalis Duriaei* Schimp. (May 1871; July 19, 1880.)
761 p.p. *Dichelyma falcatum* (Hedw.) Myr.
761 p.p. *Dichelyma uncinatum* Mitt.
764   *Dichelyma uncinatum* Mitt.
994   *Dichelyma pallescens* Br. and Schimp.
995   *Fontinalis antipyretica* Hedw.

Macoun, J. *Flora Canadensis.*

255   *Fontinalis dalecarlica* Br. and Schimp.
257 p.p. *Dichelyma falcatum* (Hedw.) Myr. (July 27, 1869.)
257 p.p. *Fontinalis Duriaei* (Schimp. (May 8, 1875; in 1876.)
257 p.p. *Fontinalis hypnoides* C. J. Hartm. (May 8, 1875.)
258   *Dichelyma uncinatum* Mitt. (May 11, 1875; May 17, 1875.)
259   *Fontinalis antipyretica* Hedw. var. *gigantea* (Sull.) Sull.
260   *Dichelyma uncinatum* Mitt.
261   *Dichelyma falcatum* (Hedw.) Myr.
262   *Dichelyma uncinatum* Mitt. (Nov. 5, 1872; in 1875.)
750   *Fontinalis Duriaei* Schimp.
751   *Fontinalis Howellii* Ren. and Card.
755   *Fontinalis dalecarlica* Br. and Schimp. var. *Macounii* Card.
760   *Dichelyma uncinatum* Mitt.
2155   *Dichelyma uncinatum* Mitt.
2157   *Dichelyma uncinatum* Mitt.
2222   *Fontinalis antipyretica* Hedw. var. *gigantea* (Sull.) Sull.
2223   *Fontinalis dalecarlica* Br. and Schimp.
2224   *Fontinalis dalecarlica* Br. and Schimp.
2225   *Fontinalis novae-angliae* Sull.
2226   *Fontinalis hypnoides* C. J. Hartm.
2227   *Fontinalis hypnoides* C. J. Hartm.
2615   *Fontinalis Duriaei* Schimp.

Migula, W. *Kryptogamae Germaniae, Austriae et Helvetiae Exsiccatae.*

69   *Fontinalis squamosa* Hedw.
316   *Fontinalis dalecarlica* Br. and Schimp.
435   *Fontinalis antipyretica* Hedw.
594b   *Fontinalis squamosa* Hedw.

Mikutowicz, J. *Bryotheca Baltica.*

37   *Fontinalis antipyretica* Hedw. var. *gracilis* (Lindb.) Schimp.
37a   *Fontinalis antipyretica* Hedw. var. *gracilis* (Lindb.) Schimp.
37b   *Fontinalis antipyretica* Hedw. var. *gracilis* (Lindb.) Schimp.
294   *Fontinalis antipyretica* Hedw. var. *gigantea* (Sull.) Sull.
294a   *Fontinalis antipyretica* Hedw.
295   *Fontinalis antipyretica* Hedw. var. *gracilis* (Lindb.) Schimp.
296   *Fontinalis hypnoides* C. J. Hartm.
398   *Fontinalis dalecarlica* Br. and Schimp.

Péterfi, T. M. *Flora Romaniae Exsiccata.*

715   *Fontinalis squamosa* Hedw.
1132   *Fontinalis antipyretica* Hedw.

Petrak, F. *Flora Bohemiae et Moraviae Exsiccata.*

32   *Fontinalis antipyretica* Hedw.
33   *Fontinalis antipyretica* Hedw.
117   *Fontinalis antipyretica* Hedw. var. *gigantea*(Sull.) Sull.
118   *Fontinalis antipyretica* Hedw. var. *gigantea* (Sull.) Sull.

Rabenhorst, L. *Bryotheca Europaea.*
(Nos. 401–450. 1861; nos. 601–650. 1863; nos. 751–800. 1864; nos.
901–950. 1867; nos. 1101–1150. 1871; nos. 1201–1250. 1873; nos.
1251–1300. 1875; nos. 1301–1350. 1876.)

431	*Fontinalis antipyretica* Hedw.
432	*Fontinalis squamosa* Hedw.
628	*Dichelyma falcatum* (Hedw.) Myr.
629	*Fontinalis hypnoides* C. J. Hartm.
630	*Fontinalis squamosa* Hedw.
631	*Fontinalis squamosa* Hedw.
778	*Dichelyma capillaceum* (With.) Myr. emend. Br. and Schimp.
779	*Dichelyma falcatum* (Hedw.) Myr.
926	*Fontinalis squamosa* Hedw. var. *Curnowii* Card.
927	*Fontinalis squamosa* Hedw.
1132	*Dichelyma falcatum* (Hedw.) Myr.
1179	*Fontinalis dalecarlica* Br. and Schimp.
1228	*Fontinalis hypnoides* C. J. Hartm.
1229	*Fontinalis antipyretica* Hedw. var. *gracilis* (Lindb.) Schimp.
1292	*Fontinalis antipyretica* Hedw.
1313	*Fontinalis hypnoides* C. J. Hartm.
1314	*Fontinalis squamosa* Hedw.

Renauld, F. and Cardot, J. *Musci Americae Septentrionalis Exsiccati.*
(Nos. 1–350. 1902; nos. 351–400. 1908.)

71	*Fontinalis antipyretica* Hedw. var. *gigantea* (Sull.) Sull.
72	*Fontinalis novae-angliae* Sull.
185	*Fontinalis novae-angliae* Sull.
186	*Fontinalis novae-angliae* Sull. var. *cymbifolia* (Aust.) Welch
187	*Dichelyma capillaceum* (With.) Myr. emend. Br. and Schimp.
229	*Fontinalis Langloisii* Card.
230	*Fontinalis filiformis* Sull. and Lesq.
231	*Fontinalis Langloisii* Card.
314	*Fontinalis neo-mexicana* Sull. and Lesq.
382	*Fontinalis Duriaei* Schimp.
383	*Fontinalis Duriaei* Schimp.

Renauld, F. and Cardot, J. *Musci Europaei Exsiccati.*

30	*Fontinalis antipyretica* Hedw.
185	*Fontinalis antipyretica* Hedw. var. *gracilis* (Lindb.) Schimp.
186	*Dichelyma falcatum* (Hedw.) Myr.
238	*Fontinalis antipyretica* Hedw. var. *gigantea* (Sull.) Sull.

Sillén, O. L. *Musci Frondosi Scandinaviae Exsiccati.*

117	*Dichelyma capillaceum* (With.) Myr. emend. Br. and Schimp.
118	*Fontinalis antipyretica* Hedw.

Sullivant, W. S. *Musci Alleghanienses.* (1845)

151	*Dichelyma capillaceum* (With.) Myr. emend. Br. and Schimp.
189	*Fontinalis dalecarlica* Br. and Schimp.
190	*Fontinalis disticha* Hook. and Wils.
191	*Fontinalis biformis* Sull.
192	*Fontinalis biformis* Sull.

Sullivant, W. S. and Lesquereux, L. *Musci Boreali-Americani Exsiccati.* (1856)

224   *Fontinalis antipyretica* Hedw. var. *gigantea* (Sull.) Sull.
224b   *Fontinalis neo-mexicana* Sull. and Lesq.
224c   *Fontinalis novae-angliae* Sull.
225   *Fontinalis novae-angliae* Sull.
226 p.p. *Dichelyma falcatum* (Hedw.) Myr.
226 p.p. (No specimen. Printed label bearing Latin description of *Fontinalis biformis* Sull.)
226b p.p. *Fontinalis biformis* Sull.
226b p.p. *Fontinalis novae-angliae* Sull.
226c   *Fontinalis biformis* Sull.
227   *Fontinalis disticha* Hook. and Wils.
228   *Fontinalis novae-angliae* Sull.
229 p.p. *Fontinalis antipyretica* Hedw. var. *gigantea* (Sull.) Sull.
229 p.p. *Fontinalis dalecarlica* Br. and Schimp.
229b   *Dichelyma falcatum* (Hedw.) Myr.

Sullivant, W. S. and Lesquereux, L. *Musci Boreali-Americani Exsiccati.* (1865)

333   *Fontinalis antipyretica* Hedw.
334 p.p. *Fontinalis antipyretica* Hedw.
334 p.p. *Fontinalis neo-mexicana* Sull. and Lesq.
335 p.p. *Fontinalis antipyretica* Hedw. var. *gigantea* (Sull.) Sull.
335 p.p. *Fontinalis novae-angliae* Sull.
336   *Fontinalis novae-angliae* Sull.
337 p.p. *Fontinalis biformis* Sull.
337 p.p. *Fontinalis novae-angliae* Sull.
338 p.p. *Fontinalis biformis* Sull.
338 p.p. *Fontinalis flaccida* Ren. and Card.
339 p.p. *Brachelyma subulatum* (Palis. Beauv.) Schimp.
339 p.p. *Fontinalis disticha* Hook. and Wils.
339 p.p. *Fontinalis filiformis* Sull. and Lesq.
340 p.p. *Fontinalis biformis* Sull.
340 p.p. *Fontinalis flaccida* Ren. and Card.
340 p.p. *Fontinalis novae-angliae* Sull.
341 p.p. *Fontinalis disticha* Hook. and Wils.
341 p.p. *Fontinalis novae-angliae* Sull.
342   *Fontinalis dalecarlica* Br. and Schimp.
343   *Dichelyma falcatum* (Hedw.) Myr.
345   *Dichelyma capillaceum* (With.) Myr. emend. Br. and Schimp.
346   *Dichelyma capillaceum* (With.) Myr. emend. Br. and Schimp.

Vendrely. *Flora Cryptogamae Sequaniae Exsiccata.* (1895)

138   *Fontinalis antipyretica* Hedw.

Verdoorn, F. *Bryophyta Arduennae Exsiccata.* (1927)

7   *Fontinalis squamosa* Hedw.
8   *Fontinalis squamosa* Hedw.

Verdoorn, F. *Musci Selecti et Critici.* (Series 1. 1934; Series 7. 1940.)

23   *Fontinalis antipyretica* Hedw.
315   *Dichelyma capillaceum* (With.) Myr. emend. Br. and Schimp.

Wehrhahn, R. *Bryophyta Exsiccata Hannoverana.*

102   *Fontinalis antipyretica* Hedw.
103   *Fontinalis antipyretica* Hedw.

Zmuda, A. *Bryotheca Polonica.*

132   *Fontinalis antipyretica* Hedw.

TABLE SHOWING DISTRIBUTION OF SPECIES

	AFRICA	ASIA	EUROPE	N. AMERICA	S. AMERICA
B. robustum				+	
B. subulatum				+	
D. capillaceum			+	+	
D. falcatum		+	+	+	
D. japonicum		+			
D. pallescens				+	
D. uncinatum				+	
F. Allenii				+	
F. antipyretica	+	+	+	+	
var. gigantea	+		+	+	
var. gracilis		+	+		
var. Heldreichii			+		
var. mollis				+	
var. oreganensis				+	
F. biformis				+	
F. bogotensis					+
F. Bryhnii			+		
F. chrysophylla				+	
F. dalecarlica			+	+	
var. Macounii				+	
F. disticha				+	
F. Duriaei	+	+	+	+	+
F. filiformis				+	
F. flaccida				+	
F. Howellii				+	+
F. hypnoides		+	+	+	
F. Langloisii				+	
F. Mac-Millanii				+	
F. missourica				+	
F. neo-mexicana				+	+
F. novae-angliae				+	
var. cymbifolia				+	
var. latifolia				+	
F. patula				+	
F. squamosa	+		+		
var. Curnowii			+		
TOTAL 36	4	6	12	29	4

DISTRIBUTION OF SPECIES

(*B.* = *Brachelyma*; *D.* = *Dichelyma*; *F.* = *Fontinalis*)

AFRICA
Abyssinia
  F. antipyretica
Algeria
  F. antipyretica
    var. *gigantea*
  F. Duriaei
  F. squamosa
Cape Colony
  F. antipyretica
Eritrea
  F. antipyretica
  F. Duriaei
Morocco
  F. antipyretica
  F. Duriaei
Tunisia
  F. antipyretica

ASIA
Asia Minor
  F. antipyretica
Japan
  D. japonicum
  F. antipyretica
    var. *gracilis*
  F. Duriaei
  F. hypnoides
Korea
  F. antipyretica var. *gracilis*
Mesopotamia
  F. Duriaei
Persia (Iran)
  F. antipyretica
Syria
  F. antipyretica
Union of Soviet Socialist Republics
  D. falcatum
  F. antipyretica
    var. *gracilis*
  F. Duriaei
  F. hypnoides

EUROPE
Albania
  F. antipyretica
Austria
  F. antipyretica
    var. *gigantea*
    var. *gracilis*
  F. hypnoides
Azores
  F. antipyretica
Belgium
  F. antipyretica
    var. *gigantea*
    var. *gracilis*
  F. dalecarlica
  F. squamosa
Bulgaria
  F. antipyretica
Canary Islands
  F. antipyretica
Corsica
  F. antipyretica
    var. *gigantea*
  F. Duriaei
  F. squamosa
Czechoslovakia
  D. falcatum
  F. antipyretica
    var. *gigantea*
    var. *gracilis*
  F. squamosa
Denmark
  D. capillaceum
  F. antipyretica
    var. *gigantea*
  F. dalecarlica
  F. hypnoides
England
  F. antipyretica
    var. *gigantea*
    var. *gracilis*
  F. dalecarlica
  F. hypnoides
  F. squamosa
    var. *Curnowii*

Estonia
  F. antipyretica
Faeroe Islands
  F. antipyretica
    var. *gracilis*
Finland
  D. capillaceum
  D. falcatum
  F. antipyretica
    var. *gigantea*
    var. *gracilis*
  F. dalecarlica
  F. hypnoides
France
  D. capillaceum
  F. antipyretica
    var. *gigantea*
    var. *gracilis*
  F. Duriaei
  F. hypnoides
  F. squamosa
    var. *Curnowii*
Germany
  D. capillaceum
  D. falcatum
  F. antipyretica
    var. *gigantea*
    var. *gracilis*
  F. dalecarlica
  F. Duriaei
  F. hypnoides
  F. squamosa
Giglio
  F. antipyretica
  F. Duriaei
Greece
  F. antipyretica
    var. *Heldreichii*
  F. Duriaei
Holland
  F. antipyretica
Hungary
  F. antipyretica
  F. Duriaei
Iceland
  F. antipyretica
Ireland
  F. antipyretica
    var. *gigantea*
    var. *gracilis*
  F. dalecarlica
  F. squamosa
Italy
  F. antipyretica
    var. *gigantea*
    var. *gracilis*
  F. dalecarlica

F. Duriaei
F. hypnoides
F. squamosa
Latvia (Livland)
  F. antipyretica var. *gigantea*
  F. antipyretica var. *gracilis*
  F. dalecarlica
  F. hypnoides
Lithuania
  F. antipyretica
Madeira Island
  F. antipyretica
Minorca Island
  F. Duriaei
Norway
  D. falcatum
  F. antipyretica
    var. *gracilis*
  F. Bryhnii
  F. dalecarlica
  F. hypnoides
  F. squamosa
    var. *Curnowii*
Poland
  F. antipyretica
Portugal
  F. antipyretica
    var. *gracilis*
  F. Duriaei
  F. squamosa
Rumania
  D. falcatum
  F. antipyretica
  F. squamosa
Sardinia
  F. antipyretica
    var. *gigantea*
    var. *gracilis*
  F. Duriaei
Scotland
  ? *D. capillaceum*
  F. antipyretica
    var. *gigantea*
    var. *gracilis*
  F. dalecarlica
  F. squamosa
Sicily
  F. Duriaei
Spain
  F. antipyretica
    var. *gigantea*
    var. *gracilis*
  F. Duriaei
  F. squamosa
Sweden
  D. capillaceum
  D. falcatum

F. antipyretica
  var. *gigantea*
  var. *gracilis*
F. dalecarlica
F. hypnoides
F. squamosa
Switzerland
  D. falcatum
  F. antipyretica
    var. *gigantea*
    var. *gracilis*
  F. dalecarlica
  F. squamosa
Thrace
  F. antipyretica
Union of Soviet Socialist Republics
  D. falcatum
  F. antipyretica
    var. *gigantea*
    var. *gracilis*
  F. dalecarlica
  F. Duriaei
  F. hypnoides
Wales
  F. antipyretica
    var. *gracilis*
  F. squamosa
Yugoslavia
  F. antipyretica
    var. *gracilis*

NORTH AMERICA

Alaska
  D. falcatum
  F. antipyretica
    var. *gigantea*
  F. Duriaei
  F. neo-mexicana
  F. patula
Canada
  Alberta
    D. falcatum
    F. Duriaei
    F. hypnoides
  Alberta or Saskatchewan
    F. dalecarliaca var. *Macounii*
  Anticosti Island
    D. falcatum
    F. antipyretica
    F. dalecarlica
    F. Duriaei
  British Columbia
    D. falcatum
    D. uncinatum
    F. antipyretica
      var. *gigantea*
      var. *mollis*

    var. *oreganensis*
  F. Duriaei
  F. Howellii
  F. hypnoides
  F. neo-mexicana
  F. patula
Cape Breton Island
  D. capillaceum
  F. antipyretica
    var. *gigantea*
  F. dalecarlica
  F. flaccida
  F. Howellii
  F. hypnoides
  F. novae-angliae
    var. *cymbifolia*
  F. patula
Labrador
  D. capillaceum
  D. falcatum
  F. antipyretica
    var. *gigantea*
  F. dalecarlica
Manitoba
  D. capillaceum
  D. falcatum
  F. Duriaei
  F. hypnoides
  F. Mac-Millanii
  F. novae-angliae
Miquelon
  F. novae-angliae
New Brunswick
  D. capillaceum
  D. pallescens
  F. antipyretica
    var. *gigantea*
  F. dalecarlica
  F. Duriaei
  F. novae-angliae
    var. *cymbifolia*
Newfoundland
  D. capillaceum
  D. falcatum
  D. pallescens
  F. antipyretica var. *gigantea*
  F. dalecarlica
  F. novae-angliae
    var. *cymbifolia*
Northwest Territories
  F. hypnoides
  F. patula
Nova Scotia
  D. capillaceum
  D. pallescens
  F. antipyretica var. *gigantea*
  F. dalecarlica

F. Duriaei
F. flaccida
F. Howellii
F. hypnoides
F. novae-angliae
  var. *cymbifolia*
F. patula
Ontario
  D. capillaceum
  D. falcatum
  D. pallescens
  F. antipyretica
    var. *gigantea*
  F. dalecarlica
  F. Duriaei
  F. flaccida
  F. Howellii
  F. hypnoides
  F. novae-angliae
    var. *cymbifolia*
Prince Edward Island
  F. antipyretica
    var. *gigantea*
Quebec
  D. capillaceum
  D. falcatum
  D. pallescens
  F. antipyretica
    var. *gigantea*
  F. dalecarlica
  F. Duriaei
  F. flaccida
  F. hypnoides
  F. missourica
  F. novae-angliae
    var. *cymbifolia*
    var. *latifolia*
Queen Charlotte Islands
  F. antipyretica var. *oreganensis*
  F. Duriaei
  F. neo-mexicana
Saint-Pierre & Miquelon
  F. dalecarlica
Saskatchewan
  F. dalecarlica
  F. Duriaei
  F. hypnoides
Saskatchewan or Alberta
  F. dalecarlica var. *Macounii*
Vancouver Island
  D. uncinatum
  F. antipyretica
    var. *gigantea*
    var. *oreganensis*
  F. Duriaei
  F. Howellii
  F. hypnoides

F. neo-mexicana
F. patula
Greenland
  F. antipyretica
  F. dalecarlica
United States
Alabama
  B. subulatum
  D. capillaceum
  F. disticha
  F. Duriaei
  F. flaccida
  F. novae-angliae
    var. *cymbifolia*
Arizona
  F. antipyretica
  F. Duriaei
Arkansas
  F. disticha
  F. flaccida
  F. missourica
  F. novae-angliae
    var. *latifolia*
California
  F. antipyretica
    var. *mollis*
  F. Duriaei
  F. Howellii
  F. neo-mexicana
  F. patula
Colorado
  D. falcatum
  F. antipyretica
    var. *gigantea*
  F. Duriaei
  F. hypnoides
  F. neo-mexicana
Connecticut
  D. capillaceum
  F. Allenii
  F. antipyretica var. *gigantea*
  F. dalecarlica
  F. Duriaei
  F. flaccida
  F. hypnoides
  F. novae-angliae
    var. *latifolia*
  F. patula
Delaware
  D. capillaceum
  F. antipyretica var. *gigantea*
  F. dalecarlica
  F. disticha
  F. flaccida
  F. novae-angliae
District of Columbia
  D. capillaceum

F. novae-angliae
Florida
  B. robustum
  B. subulatum
  D. capillaceum
  F. dalecarlica
  F. disticha
  F. filiformis
  F. flaccida
  F. novae-angliae
    var. *cymbifolia*
Georgia
  B. robustum
  B. subulatum
  F. dalecarlica
  F. disticha
  F. filiformis
  F. flaccida
  F. novae-angliae
Idaho
  D. uncinatum
  F. antipyretica
    var. *gigantea*
    var. *mollis*
    var. *oreganensis*
  F. Duriaei
  F. Howellii
  F. hypnoides
  F. neo-mexicana
  F. patula
Illinois
  B. subulatum
  F. biformis
  F. filiformis
  F. hypnoides
  F. missourica
  F. novae-angliae
Indiana
  D. capillaceum
  F. biformis
  F. dalecarlica
  F. disticha
  F. Duriaei
  F. novae-angliae
    var. *latifolia*
Iowa
  F. Duriaei
Kansas
  F. missourica
Kentucky
  F. biformis
  F. filiformis
  F. flaccida
  F. novae-angliae
    var. *cymbifolia*
Louisiana
  B. subulatum

D. capillaceum
  F. disticha
  F. filiformis
  F. flaccida
  F. Langloisii
  F. novae-angliae
    var. *cymbifolia*
Maine
  D. capillaceum
  D. falcatum
  D. pallescens
  F. antipyretica
    var. *gigantea*
  F. dalecarlica
  F. Duriaei
  F. filiformis
  F. flaccida
  F. Howellii
  F. hypnoides
  F. novae-angliae
    var. *cymbifolia*
    var. *latifolia*
Maryland
  D. capillaceum
  F. dalecarlica
  F. filiformis
  F. flaccida
  F. novae-angliae
Massachusetts
  D. capillaceum
  D. pallescens
  F. Allenii
  F. antipyretica
    var. *gigantea*
  F. dalecarlica
  F. disticha
  F. Duriaei
  F. flaccida
  F. Howellii
  F. hypnoides
  F. novae-angliae
    var. *cymbifolia*
    var. *latifolia*
Michigan
  D. capillaceum
  D. falcatum
  D. pallescens
  F. antipyretica
    var. *gigantea*
  F. disticha
  F. Duriaei
  F. filiformis
  F. flaccida
  F. hypnoides
  F. missourica
  F. neo-mexicana
  F. novae-angliae

var. *cymbifolia*
F. patula
Minnesota
  D. falcatum
  D. pallescens
  F. antipyretica
    var. *gigantea*
    var. *mollis*
  F. dalecarlica
  F. Duriaei
  F. hypnoides
  F. Mac-Millanii
  F. missourica
  F. novae-angliae
Mississippi
  F. Langloisii
  F. novae-angliae
    var. *cymbifolia*
Missouri
  F. Duriaei
  F. filiformis
  F. flaccida
  F. hypnoides
  F. missourica
  F. novae-angliae
    var. *cymbifolia*
    var. *latifolia*
Montana
  D. uncinatum
  F. antipyretica
    var. *gigantea*
    var. *mollis*
  F. Duriaei
  F. Howellii
  F. hypnoides
  F. neo-mexicana
  F. patula
Nevada
  F. antipyretica
    var. *mollis*
    var. *oreganensis*
New Hampshire
  D. capillaceum
  D. falcatum
  D. pallescens
  F. antipyretica
    var. *gigantea*
  F. dalecarlica
  F. flaccida
  F. Mac-Millanii
  F. novae-angliae
    var. *cymbifolia*
    var. *latifolia*
New Jersey
  D. capillaceum
  F. antipyretica var. *gigantea*
  F. dalecarlica

F. disticha
F. filiformis
F. flaccida
F. novae-angliae
  var. *cymbifolia*
New Mexico
  F. antipyretica
  F. neo-mexicana
New York
  D. capillaceum
  D. falcatum
  D. pallescens
  F. antipyretica
    var. *gigantea*
  F. dalecarlica
  F. disticha
  F. Duriaei
  F. flaccida
  F. hypnoides
  F. novae-angliae
    var. *cymbifolia*
    var. *latifolia*
North Carolina
  B. robustum
  B. subulatum
  D. capillaceum
  F. dalecarlica
  F. disticha
  F. flaccida
  F. novae-angliae
    var. *cymbifolia*
    var. *latifolia*
North Dakota
  F. Duriaei
Ohio
  D. capillaceum
  F. antipyretica var. *gigantea*
  F. biformis
  F. dalecarlica
  F. disticha
  F. Duriaei
  F. missourica
  F. novae-angliae
    var. *cymbifolia*
Oklahoma
  F. novae-angliae
Oregon
  D. uncinatum
  F. antipyretica
    var. *mollis*
    var. *oreganensis*
  F. Duriaei
  F. Howellii
  F. hypnoides
  F. neo-mexicana
  F. patula

Pennsylvania
  D. capillaceum
  D. pallescens
  F. Allenii
  F. antipyretica
    var. *gigantea*
  F. dalecarlica
  F. Duriaei
  F. flaccida
  F. novae-angliae
    var. *cymbifolia*
Rhode Island
  D. capillaceum
  F. antipyretica var. *gigantea*
  F. dalecarlica
  F. flaccida
  F. novae-angliae
    var. *cymbifolia*
    var. *latifolia*
South Carolina
  B. subulatum
  F. dalecarlica
  F. filiformis
  F. novae-angliae
    var. *cymbifolia*
South Dakota
  F. Duriaei
  F. hypnoides
Tennessee
  D. capillaceum
  F. dalecarlica
  F. disticha
  F. flaccida
  F. novae-angliae
    var. *cymbifolia*
Texas
  B. subulatum
  F. Duriaei
  F. filiformis
Utah
  D. falcatum
  F. antipyretica
    var. *mollis*
  F. Duriaei
  F. hypnoides
  F. neo-mexicana
Vermont
  D. capillaceum
  D. pallescens
  F. antipyretica var. *gigantea*
  F. dalecarlica
  F. Duriaei
  F. novae-angliae
    var. *cymbifolia*
Virginia
  B. subulatum
  D. capillaceum
  F. dalecarlica

  F. filiformis
  F. flaccida
  F. novae-angliae
    var. *cymbifolia*
    var. *latifolia*
Washington
  D. uncinatum
  F. antipyretica
    var. *gigantea*
    var. *mollis*
    var. *oreganensis*
  F. chrysophylla
  F. Duriaei
  F. Howellii
  F. neo-mexicana
  F. patula
West Virginia
  D. capillaceum
  F. dalecarlica
  F. novae-angliae
    var. *cymbifolia*
    var. *latifolia*
Wisconsin
  D. capillaceum
  D. falcatum
  D. pallescens
  F. antipyretica
    var. *gigantea*
  F. biformis
  F. dalecarlica
  F. Duriaei
  F. hypnoides
  F. novae-angliae
Wyoming
  D. falcatum
  D. uncinatum
  F. antipyretica
    var. *gigantea*
    var. *oreganensis*
  F. Duriaei
  F. hypnoides
  F. neo-mexicana
  F. patula

SOUTH AMERICA
Argentina
  F. neo-mexicana
Brazil
  F. Duriaei
Colombia
  F. bogotensis
Peru
  F. Howellii

WEST INDIES
St. Thomas Island
  ? *D. capillaceum*

LITERATURE

Adlerz, E., 1907: Bladmossflora för Sveriges lågland med särskilt avseende på arternas utbredning inom Närke. Page 26.

Allorge, Pierre, 1936: Le *Fontinalis islandica* Card. en Bretagne. Revue Bryologique et Lichénologique (Nouvelle Série) 9: 148.

Andersson, Gunnar, 1898: Studier öfver Finlands Torfmossar och fossila Kvartärflora. Bulletin de la Commission géologique de Finlande 2: 34, 71, 134, 136.

Arnell, Hampus W., 1882: Bryological Notes from the Meetings of the Society pro Fauna et Flora Fennica. Revue Bryologique 9: 85.

Austin, Coe F., 1870: Musci Appalachiani. Tickets of Specimens of Mosses collected mostly in the eastern part of North America. Pages 41–43.

Austin, Coe F., 1877: Bryological Notes. Botanical Gazette 2: 111, 143.

Austin, Coe F., 1878: Musci Appalachiani: Supplement 1. Tickets of (100) specimens of Mosses collected mostly in the eastern part of North America. Pages 13–14.

Baehni, Charles, 1941: Sur quelques Mousses originales de Dickson. Candollea 8: 189.

Baker, Frank C., 1920: The Life of the Pleistocene or Glacial Period as recorded in the deposits laid down by the great ice sheets. University of Illinois Bulletin 17: 332, 335, 377.

Barnes, Charles R., 1892: Artificial Keys to the Genera and Species of Mosses recognized in Lesquereux and James's Manual of the Mosses of North America. Transactions of Wisconsin Academy of Sciences, Arts and Letters 8: 62–63. 1888–1891.

Barnes, Charles R., 1896: Analytic Keys to the Genera and Species of North American Mosses. Bulletin of the University of Wisconsin (Science Series) 1: 224–226, 326–331.

Bauer, Ernst, 1898: Bryotheca Bohemica. Centurie 1. 1898. Botanisches Centralblatt 76: 131.

Bauer, Ernst, 1900: Bryotheca Bohemica. Centurie 2. 1899. Botanisches Centralblatt 83: 45.

Bauer, Ernst, 1904: Bryotheca Bohemica. Centurie 3. Sonderabdruck aus den Sitzungsberichten des deutschen naturwissenschaftlich-medizinischen Vereins für Böhmen "Lotos." Number 5, page 6.

Bauer, Ernst, 1909: Musci Europaei Exsiccati. Schedae zur zehnten Serie. Separate, pages 6–7.

Bauer, Ernst, 1909: Musci Europaei Exsiccati. Schedae und kritische Bemerkungen zur zwölften Serie. Separate, pages 1–3.

Bauer, Ernst, 1915: Musci Europaei Exsiccati. Schedae und Bemerkungen zur 21. bis einschliesslich 27. Serie. Separate, pages 17–19.

Bauer, Ernst, 1927: Musci Europaei et Americani Exsiccati. Schedae und Bemerkungen zur 39. Serie. Nos. 1901–1950, pages 8–9.

Bauhin, Johann et Cherler, Johann H., 1651: Historia Plantarum Universalis 3: 770.

Bescherelle, Émile, 1882: Catalogue des Mousses observées en Algérie. Pages 29–30.

Bizot, Maurice, 1946: *Fontinalis antipyretica* var. *robusta* Card. f. *subsecunda* Biz. f. nov. Revue Bryologique et Lichénologique (Nouvelle Série) 15: 166.

Bizot, Maurice et Hillier, Louis, 1945: Le Problème de *Fontinalis Durieui* Schpr. en Côte-d'Or. Revue Bryologique et Lichénologique (Nouvelle Série) 15: 70–71.

Braithwaite, Robert, 1905: The British Moss-Flora 3: 204–205, 209–215, 219, 222, 226. Plates 122–123.

Bridel-Brideri, Samuel E., 1798: Muscologia Recentiorum seu Analysis, Historia et Descriptio Methodica omnium Muscorum Frondosorum hucusque cognitorum ad normam Hedwigii. Pages 135–137; plate 4.

Bridel-Brideri, Samuel E., 1803: Muscologia Recentiorum seu Analysis, Historia et Descriptio Methodica omnium Muscorum Frondosorum hucusque cognitorum ad normam Hedwigii 2 (3): 157–165.

Bridel-Brideri, Samuel E., 1817: Muscologiae Recentiorum Supplementum seu Species Muscorum 3: 107–111.

Bridel-Brideri, Samuel E., 1827: Bryologia Universa seu Systematica ad Novam Methodum Dispositio, Historia et Descriptio omnium Muscorum Frondosorum hucusque Cognitorum cum Synonymia ex Auctoribus Probatissimis. Accedunt tabulae aeneae tredecim. Volume 2: 654–662. Plate 10.

Britton, Elizabeth G., 1896: The Water Nymphs. The Observer 7: 442–447.

Britton, Elizabeth G., 1904: Notes on Nomenclature 3. The Bryologist 7: 48–50.

Britton, Elizabeth G., 1905: Notes on Nomenclature 4. The Genus *Neckera* Hedw. The Bryologist 8: 5–6.

Britton, Elizabeth G., 1906: Notes on Nomenclature 6. The Bryologist 9: 37–38.

Britton, Elizabeth G., 1913: West Indian Mosses 1. Bulletin of the Torrey Botanical Club 40: 653, 655–657.

Britton, Elizabeth G. and Hollick, Arthur, 1907: American Fossil Mosses, with Description of a New Species from Florissant, Colorado. Bulletin of the Torrey Botanical Club 34: 139–140. Plate 9.

Brongniart, Adolphe T., 1828–1838: Histoire des végétaux renfermés dans les diverses couches du globe. Page 93; plate 10.

Brotherus, Viktor F., 1899: Neue Beiträge zur Mossflora Japans. Hedwigia 38: 225–226.

Brotherus, Viktor F., 1905: Fontinalaceae. Engler, A. und Prantl, K., Die natürlichen Pflanzenfamilien: Bryophyta 2: 722–732. Figures 544–548.

Brotherus, Viktor F., 1906: Musci Amazonici et Subandini Uleani. Hedwigia 45: 286.

Brotherus, Viktor F., 1909: Nachträge und Verbesserungen. Engler, A. und Prantl, K., Die natürlichen Pflanzenfamilien: Bryophyta 2: 1213, 1239.

Brotherus, Viktor F., 1923: Die Laubmoose Fennoskandias. Societas pro Fauna et Flora Fennica. Flora Fennica 1. Pages 392–401; figures 68–69.

Brotherus, Viktor F., 1925: Fontinalaceae. Engler, A. und Prantl, K., Die natürlichen Pflanzenfamilien: Musci (Laubmoose) 11 (2): 54–63. Figures 471–478.

Bruch, Philipp, Schimper, Wilhelm P., et Gümbel, T., 1842: *Fontinalis*. Bryologia Europaea seu Genera Muscorum Europaeorum monographice illustrata. Fascicle 16, 5: 1–6. Plates 429–430.

Bruch, Philipp, Schimper, Wilhelm P., et Gümbel, T., 1842: *Dichelyma*. Bryologia Europaea seu Genera Muscorum Europaeorum monographice illustrata. Fascicle 16, 5: 1–10. Plates 433–435.

Bruch, Philipp, Schimper, Wilhelm P., et Gümbel, T., 1846: *Fontinalis*. Bryologia Europaea seu Genera Muscorum Europaeorum monographice illustrata. Fascicle 31, 5: 1–9. Plates 431–432.

Bruch, Philipp, Schimper, Wilhelm P., et Gümbel, T., 1846: *Dichelyma*. Bryologia Europaea seu Genera Muscorum Europaeorum monographice illustrata. Supplementum 1, Fascicle 31, 5: 1–2. Plate 436.

Bryhn, N., 1907: Ad muscologiam (bryophytologiam) Norvegiae contributiones sparsae. IV. Nyt Magazin for Naturvidenskaberne grundlagt af den Physiographiske Forening I Christiana 45, 2: 123.

Cardot, Jules, 1882: Note bryologique sur les environs d'Anvers. Revue Bryologique 9: 88–89.

Cardot, Jules, 1887: F. Renauld, Énumération des Muscinées récoltées par le Dr. Delamare, à l'Île Miquelon, (Amérique septentrionale). Revue Bryologique 14: 4–6.

Cardot, Jules, 1888: Note sur une Fontinale du Rhône. Revue Bryologique 15: 13.

Cardot, Jules, 1888: F. Renauld, Notice sur quelques Mousses de l'Amérique du Nord. Revue Bryologique 15: 71.

Cardot, Jules, 1891: Tableau Méthodique et Clef Dichotomique du Genre *Fontinalis*. Revue Bryologique 18: 81–87.

Cardot, Jules, 1892: Monographie des Fontinalacées. Mémoires de la Société Nationale des Sciences Naturelles et Mathématiques de Cherbourg 28: 1–152.

Cardot, Jules, 1895: Noms de Genres à rayer de la Nomenclature Bryologique. Revue Bryologique 22: 17–18.

Cardot, Jules, 1895: Une Fontinale Nouvelle. Revue Bryologique 22: 53–54.

Cardot, Jules, 1896: Fontinales Nouvelles. Revue Bryologique 23: 67–72.

Cardot, Jules, 1897: Fontinales Japonaises. Revue Bryologique 24: 33–36.

Cardot, Jules, 1897: Mosses of the Azores and of Madeira. Annual Report of the Missouri Botanical Garden 8: 66.

Cardot, Jules, 1899: Études sur la Flore Bryologique de l'Amérique du Nord. Revision des Types d'Hedwig et de Schwaegrichen. Bulletin de l'Herbier Boissier 7: 313.

Cardot, Jules, 1903: Two new Species of *Fontinalis*. Minnesota Botanical Studies 3: 129–130. Plates 21–22.

Cardot, Jules, 1904: An Answer to Mrs. E. G. Britton's last Article "Notes on Nomenclature." The Bryologist 7: 80–81.

Cardot, Jules, 1904: Le Genre *Cryphaeadelphus*. Revue Bryologique 31: 6–8.

Cardot, Jules, 1909: Mousses Nouvelles du Japon et de Corée. Bulletin de la Société Botanique de Genève (Deuxième Série) 1: 131–132.

Cardot, Jules, 1910: .*Fontinalis maritima* et *F. mollis*. Revue Bryologique 37: 45–46.

Cardot, Jules and Thériot, Iréné, 1904: New or Unrecorded Mosses of North America 2. Botanical Gazette 37: 363–382. Plate 23.

Cardot, Jules and Thériot, Iréné, 1906: New or Unrecorded Mosses of North America. The Bryologist 9: 7–8.

Celsius, Olof, 1732: Plantarum circa Upsaliam sponte nascentium catalogus. Acta Literaria et Scientiarum Sveciae. Page 21.

Coppey, A., 1911: Études phytogéographiques sur les Mousses de la Haute-Saône. Revue Bryologique 38: 119.

Dawson, William, and Penhallow, D. P., 1890: On the Pleistocene Flora of Canada. Bulletin of the Geological Society of America 1: 315, 332, 334.

De Beck, G. und Zahlbruckner, A., 1897: Schedae ad Kryptogamas Exsiccatas. Centuria 3. Editae a Museo Palatino Vindobonensi. Annalen des Kaiserlich-Königlichen Naturhistorischen Hofmuseums 12 (2): 97.

De Heldreich, Theodor, 1883: Bericht über die Botanischen Ergebnisse einer Bereisung Thessaliens. Sitzungsberichte der Königlich-preussischen Akademie der Wissenschaften zu Berlin. Part 1: 158.

Delamare, Ernest, Renauld, Ferdinand, et Cardot, Jules, 1888: Flora Miquelonensis. Florule de l'Île Miquelon (Amérique du Nord). Énumération systématique avec notes descriptives Phanérogames, Cryptogames vasculaires, Mousses, Sphaignes, Hépatiques et Lichens. Annales Société Botanique de Lyon (1887) 15: 113.

De Necker, Noel J., 1771: Methodus Muscorum per Classes, Ordines, Genera ac Species cum Synonymis, Nominibus Trivialibus, Locis Natalibus, Observationibus Digestorum, AEneisque Figuris Illustratorum. Pages 191–192.

Dickson, James, 1790: Fasciculus Plantarum Cryptogamicarum Britanniae. Fasciculus secundus. Pages 1–2; plate 4.

Dillenius, John J., 1741: Historia Muscorum, in qua circiter sexcentae species veteres et novae ad sua genera relatae describuntur et iconibus genuinis illustrantur; cum appendice at indice synonymorum. Page 254; plate 33, figures 1, 3, and 5.

Dillenius, John, 1763: Historia Muscorum: a general history of Land and Water, etc. Mosses and Corals. Page 6; plate 33, figures 1, 3, and 5.

Dixon, Hugh N., 1896: The Student's Handbook of British Mosses. Pages 352–357; plate 48.

Dixon, Hugh N., 1897: The Student's Handbook of British Mosses. Pages 389–395; plate 48.

Dixon, Hugh N., 1904: The Student's Handbook of British Mosses. Pages 389–395; plate 48.

Dixon, Hugh N., 1911: Mosses determined by H. N. Dixon from various Strata in the Peat. Lewis, Francis J., The Plant Remains in the Scottish Peat Mosses. Part 4. Transactions of the Royal Society of Edinburgh 47: 831.

Dixon, Hugh N., 1921: Miscellanea Bryologica 7. The Journal of Botany 59: 137.

Dixon, Hugh N., 1924: The Student's Handbook of British Mosses. Pages 388–394; plates 48–49.

Dixon, Hugh N., 1927: Muscineae. Jongmans, W., Fossilium Catalogus 2. Plantae. Pars 13: 18, 49–50, 105.

Drummond, Thomas, 1828: Musci Americani; or Specimens of the Mosses collected in British North America, and chiefly among the Rocky Mountains, during the Second Land Arctic Expedition under the Command of Captain Franklin, R. N. 1: numbers 1–158; 2: numbers 159–286.

Duncan, J. B., 1926: A Census Catalogue of British Mosses, compiled for the British Bryological Society. Page 46.

Evans, Alexander W. and Nichols, George E., 1908: The Bryophytes of Connecticut. State Geological and Natural History Survey. Bulletin 11: 128–130.

Familler, Ignaz, 1911: Die Laubmoose Bayerns. Eine Zusammenstellung der bisher bekannt gewordenen Standortsangaben. Denkschriften der Königlich-bayerischen botanischen Gesellschaft in Regensburg 11: 1–3.

Familler, J., 1918: Bemerkungen über bayerische Moose. Kryptogamische Forschungen herausgegeben von der Kryptogamenkommission der Bayerischen Botanischen Gesellschaft zur Erforschung der heimischen Flora 3: 167.

Farneti, Rodolfo, 1894: Muschi della Provincia di Pavia. Atti Istituto Botanico dell' Università, Laboratorio Crittogamico, Pavia (Serie 2) 3: 68–71. Plate 24.

Farneti, Rodolfo, 1897: Briologia Insubrica Prima Contribuzione. Muschi della Provincia di Brescia. Atti Istituto Botanico dell' Università, Laboratorio Crittogamico, Pavia (Serie 2) 4: 139.

Fitzgerald, R. D., 1951: The Discovery of *Fontinalis Dixoni* Card. et Dix. in Ireland. The Irish Naturalists' Journal 10: 164–166.

Fleischer, Max, 1900: Die Musci der Flora von Buitenzorg zugleich Laubmoosflora von Java 1: xii, xviii.

Fleischer, Max, 1906: Die Musci der Flora von Buitenzorg zugleich Laubmoosflora von Java mit Berücksichtigung aller Familien und Gattungen der gesamten Laubmooswelt 3: 720–721.

Fleischer, Max und Warnstorf, Carl, 1896: Bryotheca Europaea Meridionalis. Centurie 1. Botanisches Centralblatt 65; 300.

Fleischer, Max und Warnstorf, Carl, 1897: Bryotheca Europaea Meridionalis. Centurie 1. 1896. Beiblatt zur Hedwigia 36: 74.

Fleischer, Max und Warnstorf, Carl, 1898: Bryotheca Europaea Meridionalis. Centurie 2. 1897. Beiblatt zur Hedwigia 37: 141.

Fuchsig, Heinrich, 1924: Die im Wasser wachsenden Moose des Lunzer Seengebietes. Internationale Revue der gesamten Hydrobiologie und Hydrographie 12: 200–202.

Fuchsig, Heinrich, 1926: Vergleichende anatomisch-physiologische Untersuchungen an Formen von *Fontinalis antipyretica*. Österreichische Botanische Zeitschrift 75: 114, 118–121. Figures a – b, page 119.

Gadeceau, Émile, 1919: Les Forêts submergées de Belle-Île-en-mer. Bulletin Biologique de la France et de la Belgique 53: 295.

Gaume, R., 1956: Catalogue des Muscinées de Bretagne d'après les documents inédits du Dr. F. Camus. Revue Bryologique et Lichénologique 25: 70–72.

Geheeb, Adelbert, 1872: Bryologische Notizen. Hedwigia 11: 166.

Geheeb, Adelbert, 1877: Notes sur quelques Mousses rares ou peu connues. Revue Bryologique 4: 2–3.

Geheeb, Adelbert, 1880: Bibliographie Hollandaise. Revue Bryologique 7: 30.

Geheeb, Adelbert, 1886: Bryologische Fragmente 3. C. Griechische Laubmoose. Flora 69: 343.

Geheeb, Adelbert, 1907: Neue Formen und Varietäten von Laubmoosen aus der Europäischen Flora. Beihefte zum Botanischen Centralblatt 22 (2): 100.

Giacomini, V., 1947: Syllabus Bryophytarum Italicarum. Pars Prima: Andreae-ales et Bryales. Atti Istituto Botanico della Università, Laboratorio Critto-gamico, Pavia (Serie 5) 4 (2): 249–250.

Grønlund, Chr., 1896: Tillaeg til Islands Kryptogamflora, indeholdende Liche-nes, Hepaticae og Musci. Botanisk Tidsskrift 20: 109–110.

Grout, Amos J., 1903: Mosses with Hand-lens and Microscope. Pages 395–402; plates 87–88; figures 217–220.

Grout, Amos J. (*Editor*), 1904: The Moss Flora of the Upper Minnesota River by John M. Holzinger. The Bryologist 7: 11–12.

Grout, Amos J., 1924: Mosses with a Hand-lens. Pages 219, 221–224; plates 69–70; figures 107–108.

Grout, Amos J., 1928: Classification of the Musci. The Bryologist 31: 59–61.

Győrffy, István, 1916: Beiträge zur Moosflora des Balaton (Platten)-Sees und seiner Umgebung 1. Magyar Botanikai Lapok (Ungarische Botanische Blätter) 15: 235–242. Plate 6.

Hagen, I., 1908: Mousses Nouvelles. Kongelige Norske Videnskabers-Selskabs Skrifter. Number 3: 40–43.

Hampe, Ernst, 1839: Relation über die von dem Reisenden C. Beyrich auf seiner letzten Reise in Nordamerika gesammelten Laubmoose. Linnaea 13: 45–47.

Hampe, Ernst, 1862: Beitrag zu einer Moosflora von Neu-Granada. Flora 45: 454.

Hampe, Ernst, 1865: Prodromus Florae Novo-Granatensis ou Énumération des Plantes de la Nouvelle-Grenade avec description des espèces nouvelles, par MM. J. Triana et J. E. Planchon. Musci exposuit E. Hampe. Annales des Sciences Naturelles cinquième Série. Botanique 4: 351.

Hartman, Carl J., 1820: Handbok i Skandinaviens Flora, innefattande Sveriges och Norriges Vexter, till och med Mossorna. Page 434.

Hartman, Carl J., 1843: Handbok i Skandinaviens Flora, innefattande Sveriges och Norriges Vexter, till och med Mossorna. Pages 433–434.

Hartman, Carl J., 1849: Handbok i Skandinaviens Flora, innefattande Sveriges och Norriges Vexter, till och med Mossorna. Pages 340–342.

Hartman, Carl J., 1854: Handbok i Skandinaviens Flora, innefattande Sveriges och Norriges Vexter, till och med Mossorna. Pages 370–371.

Hartman, Carl J., 1861: Handbok i Skandinaviens Flora, innefattande Sveriges och Norriges Vexter, till och med Mossorna. Page 353.

Hedwig, Joannis, 1792: Descriptio et Adumbratio microscopico-analytica Mus-corum Frondosorum nec non aliorum vegetantium e classe cryptogamica Linnaei novorum dubiisque vexatorum 3: 32–33, 57–58. Plates 12 and 24.

Hedwig, Joannis, 1801: Species Muscorum Frondosorum descriptae et tabulis aeneis coloratis illustratae. Opus Posthumum editum a Friderico Schwaeg-richen. Pages 298–300.

Hedwig, Joannis, 1816: Species Muscorum Frondosorum. Opus Posthumum. Supplementum Primum. Scriptum a Friderico Schwaegrichen. Pages 307–308.

Hedwig, Joannis, 1827: Species Muscorum Frondosorum. Opus Posthumum. Supplementum Tertium. Scriptum a Friderico Schwaegrichen. Volume 1. (Pages of descriptions not given.) Plate 218.

Herzog, Theodor, 1916: Die Bryophyten meiner zweiten Reise durch Bolivia. Bibliotheca Botanica 21 (87): 106.

Herzog, Theodor, 1920: Die Bryophyten meiner zweiten Reise durch Bolivia. Page 20.

Hill, Ellsworth J., 1915: *Fontinalis Umbachii* Cardot. The Bryologist 18: 10–12.

Hobkirk, Charles P., 1884: A Synopsis of the British Mosses. Pages 28–29, 181–182.

Holzinger, John M., 1898: Some Musci of International Boundary. Minnesota Botanical Studies 2: 43.

Holzinger, John M., 1903: The Moss Flora of the Upper Minnesota River. Minnesota Botanical Studies 3: 120–121. Plate 22.

Hooker, William J., 1841: Remarks on Drummond's Musci Americani. Hooker, William J., Journal of Botany 3: 440.

Hooker, William J. and Drummond, Thomas, 1835: Notice concerning Drummond's Collections made chiefly in the southern and western parts of the United States: described by W. J. Hooker. Hooker, William J., Companion to the Botanical Magazine 1: 21–26, 39–49, 95–101, 170–177.

Hooker, William J. and Drummond, Thomas, 1836: Notice concerning Drummond's Collections made chiefly in the southern and western parts of the United States: described by W. J. Hooker. Hooker, William J., Companion to the Botanical Magazine 2: 60–64.

Hooker, William J. and Taylor, Thomas, 1827: Muscologia Britannica, containing the mosses of Great Britain and Ireland, systematically arranged and described. Pages 140–142; plate 22.

Hudson, William, 1762: Flora Anglica. Page 468.

Husnot, T., 1892: Muscologia Gallica. Descriptions et Figures des Mousses de France et des Contrées Voisines. Deuxième Partie – Pleurocarpes. Pages 285–288; plates 80–81.

Hy, Abbé, 1882: *Fontinalis Ravani*. Mémoires de la Société Nationale d'Agriculture, Sciences et Arts d'Angers (Nouvelle Période) 24: 127–136. Plate (without number), page 137.

Iishiba, Eikichi, 1929: Nihon-san senrui sosetsu. Pages 130–132. (Catalogue of the mosses of Japan, including keys and brief descriptions).

Iishiba, Eikichi, 1932: Nihon-senrui no bunrui. Pages 28–29, 107, 120.

Irmscher, P. Edgar, 1912: Über die Resistenz der Laubmoose gegen Austrocknung und Kälte. Pages 1–63.

Jaap, Otto, 1899: Beiträge zur Moosflora der Umgegend von Hamburg. Verhandlungen des Naturwissenschaftlichen Vereins in Hamburg 3 (7): 29–30.

Jaap, Otto, 1901: Bryologische Beobachtungen in der nördlichen Prignitz aus dem Jahre 1900 und früheren Jahren. Verhandlungen des Botanischen Vereins der Provinz Brandenburg 43: 66–71.

Jaeger, Augusto, 1876: Genera et species muscorum systematice disposita seu Adumbratio florae muscorum totius orbis terrarum. Bericht über die Thätigkeit der St. Gallischen naturwissenschaftlichen Gesellschaft während des Vereinsjahres 1874–75. Pages 158–166.

Jaeger, Augusto et Sauerbeck, Fr., 1879: Supplementum ad Adumbrationem muscorum et Conspectus Systematis. Bericht über die Thätigkeit der St. Gallischen naturwissenschaftlichen Gesellschaft während des Vereinsjahres 1877–78. Pages 456–457.

Jelenc, F., 1955: Muscinées de l'Afrique du Nord (Algérie, Tunisie, Maroc, Sahara). Extrait du Bulletin de la Société de Géographie et d'Archéologie de la Province d'Oran. Tomes 72–76; fascicules 227–228, 230–232; années 1949–1953. Pages 117–119.

Jennings, Otto E., 1913: A Manual of Mosses of Western Pennsylvania. Pages 203–209; plates 30–31.

Jennings, Otto E., 1951: A Manual of Mosses of Western Pennsylvania and Adjacent Regions. Pages 172–179; plates 34, 35, 64.

Jensen, C., 1885: *Fontinalis longifolia* C. Jens. Botaniska Notiser, 1885: 83–84.

Jensen, C., 1887: Oversigt over Grønlands Mosser ved. Joh. Lange og C. Jensen. Pages 26–27.

Jensen, C., 1896: *Fontinalis thulensis* C. Jens. Grønlund, Chr. Tillaeg til Islands Kryptogamenflora indeholdende Lichenes, Hepaticae og Musci. Botanisk Tidsskrift. Fascicle 2, 20: 110.

Jensen, C., 1923: Danmarks Mosser. Pages 187–195; plates 8–9.

Jensen, C., 1939: Skandinaviens Bladmoosflora. Pages 375–384.

Jones, George N., 1930: The Moss Flora of Southeastern Washington and adjacent Idaho. Research Studies of the State College of Washington 1: 182–183.

Kaalaas, B., 1902: Zur Bryologie Norwegens 1. Nyt-Magazin for Naturvidenskaberne 40: 259–260.

Keissler, C., 1926: Schedae ad "Kryptogamas Exsiccatas" editae a Museo Historiae Naturalis Vindobonensi. Separat-Abdruck aus dem 40. Bande der Annalen des Naturhistorischen Museums in Wien. Page 148.

Kindberg, N. Conrad, 1882: Novitier för Sveriges och Norges Mossflora Botaniska Notiser. Page 146.

Kindberg, N. Conrad, 1882: Die Familien und Gattungen der Laubmoose (Bryineae) Schwedens und Norwegens. Bihang till Kongliga Svenska Vetenskaps-Akademiens Handlingar 6 (19): 10.

Kindberg, N. Conrad, 1883: Die Arten der Laubmoose (Bryineae) Schwedens und Norwegens. Bihang till Kongliga Svenska Vetenskaps-Akademiens Handlingar 7 (9): 50–51.

Kindberg, N. Conrad, 1887: Contributions à la Flore Bryologique de la Grèce. Revue Bryologique 14: 53–54.

Kindberg, N. Conrad, 1894: Check-List of European and North American Mosses (Bryineae). Canadian Record of Science 6: 75.

Kindberg, N. Conrad, 1895: New or less known Species of Pleurocarpous Mosses from North America and Europe. Revue Bryologique 22: 88.

Kindberg, N. Conrad, 1896: Species of European and North American Bryineae (Mosses). Part 1. Pleurocarpous. Pages 146–152.

Kindberg, N. Conrad, 1897: European and North American Bryineae (Mosses). Part 1: 24.

Kindberg, N. Conrad, 1909: Bryological Notes. Revue Bryologique 36: 99.

Kindberg, N. Conrad, 1910: Bryological Notes. Revue Bryologique 37: 14.

Knowlton, F. H., 1916: A Review of the Fossil Plants in the United States National Museum from the Florissant Lake Beds at Florissant, Colorado, with Description of New Species and List of Type-Specimens. Proceedings of the United States National Museum 51: 245 (footnote 1).

Kotilainen, Mauno J., 1927: Ein beachtenswerter Moosfund, *Fontinalis dichelymoides* Lindb. mit Sporogonen. Memoranda Societatis pro Fauna et Flora Fennica (1925–1926) 2: 38–39.

Krok, Th. O. B. N., 1925: Bibliotheca Botanica Suecana ab antiquissimis temporibus ad Finem Anni MCMXVIII. Svensk Botanisk Litteratur Från Äldsta Tider T.O.M. 1918. Pages 268–270.

Kucyniak, James, 1944: Le *Fontinalis disticha* Hook. and Wils. dans le Québec; une extension d'aire intéressante. (Abstract.) Les Annales de l'Association Canadienne-Française pour l'Avancement des Sciences 10: 92.

La Marck, J. B. A. P., Monnet de, 1778: La Flore Française ou description succincte de toutes les plantes, qui croissent naturellement en France 1: 64.

Lange, Johan, 1867: Flora Danica, Iconum, Fascicle 46. Icones Plantarum sponte nascentium in regno Daniae et in ducatibus Slesvici, Holsatiae et Lauenburgiae, ad illustrandum opus de Iisdem Plantis, regio jussu exarandum, Florae-danicae nomine inscriptum. Fascicle 46, 16: 17. Plate 2751. [Fascicle 46, page 17, and plate 2751 were published in 1867; volume 16 was published in 1871.]

Lange, Johan, 1887: Nomenclator "Florae Danicae." Page 105.

Lange, Johan og Jensen, C., 1887: Oversigt over Grønlands Mosser. Pages 26–27.

Lange, Johan og Jensen, C., 1887: Grønlands Mosser. Lange, Johan. Conspectus Florae Groenlandicae 1 (2): 342–343.

Leitgeb, H., 1868: Beiträge zur Entwicklungsgeschichte der Pflanzenorgane 1. Wachsthum des Stämmchens von *Fontinalis antipyretica*. Sitzungsberichte der Kaiserlichen Akademie der Wissenschaften, Mathematisch-Naturwissenschaftliche Classe 57: 308–309. Plate 4.

Lesquereux, Leo, 1868: Catalogue of Pacific Coast Mosses. Memoirs presented to California Academy of Sciences 1: 28.

Lesquereux, Leo, 1869: On Californian Mosses. Transactions of the American Philosophical Society (New Series) 13: 11, 18, 23.

Lesquereux, Leo, 1883: Enumeration and Description of the Species of Fossil Plants known from the Green River group, in Cretaceous and Tertiary Flora. Report United States Geological Survey of the Territories 8: 135, 140. Plate 21.

Lesquereux, Leo and James, Thomas P., 1884: Manual of the Mosses of North America. Pages 268–275, 413; plate 4.

Limpricht, K. Gustav, 1881: Neue Muscineen für Schlesien. Jahresbericht der Schlesischen Gesellschaft für vaterländische Cultur. 4. Botanische Section. Volume 58: 185.

Limpricht, K. Gustav, 1894: Die Laubmoose 2. Rabenhorst, Ludwig, Kryptogamen-Flora von Deutschland, Oesterreich und der Schweiz 4: 647–677. Figures 325–328.

Limpricht, K. Gustav, 1903: Die Laubmoose 3. Rabenhorst, Ludwig, Kryptogamen-Flora von Deutschland, Oesterreich und der Schweiz 4: 802–803.

Lindberg, Sextus O., 1867: Über einige Fontinalideen. Hedwigia 6: 38–41.

Lindberg, Sextus O., 1868: Musci Novi Scandinavici. Notiser ur Sällskapets pro Fauna et Flora Fennica Förhandlingar (Ny Serie) 9: 274–279.

Lindberg, Sextus O., 1869: Nya Mossor. Öfversigt af Finska Vetenskaps-Societetens Förhandlingar 12: 76–78.

Lindberg, Sextus O., 1871: Revisio critica Iconum in opere Flora Danica Muscos illustrantium. Acta Societatis Scientiarum Fennicae 10: 39, 94, 101.

Lindberg, Sextus O., 1883: Om Fyra för den Skandinaviska Mossfloran Nya Arter. (Meddelande den 3 December 1881.) Meddelanden af Societas pro Fauna et Flora Fennica 9: 127–128.

Lindberg, Sextus O. und Arnell, Hampus W., 1890: Musci Asiae Borealis Beschreibung der von den Schwedischen Expeditionen nach Sibirien in den Jahren 1875 und 1876 gesammelten Moose mit Berücksichtigung aller früheren bryologischen Angaben für das Russische Nord-Asien. Zweiter Theil. Laubmoose. Kongliga Svenska Vetenskaps-Akademiens Handlingar 23 (10): 160–162.

Linnaeus, Carl, 1745: Flora Suecica. Page 315.

Linnaeus, Carl, 1753: Species Plantarum 2: 1107–1108.

Linnaeus, Carl, 1755: Flora Suecica. Pages 379–380.

Linnaeus, Carl, 1763: Species Plantarum 2: 1571.

Linnaeus, Carl, 1761: Fauna Suecica. Page 558.

Linnaeus, Carl (filio), 1781: Methodus Muscorum illustrata. Page 368.

Loeske, Leopold, 1910: Studien zur vergleichenden Morphologie und phylogenetischen Systematik der Laubmoose. Pages 147–149.

Loeske, Leopold, 1918: Zur Bryogeographie Mitteleuropas. Bryologische Zeitschrift 1: 142–144.

Lorentz, Paul G., 1860: Beiträge zur Biologie und Geographie der Laubmoose. Page 21.

Loureiro, Juan, 1790: Flora Cochinchinensis, sistens plantas in regno Cochinchina nascentes, quibus accedunt aliae observatae in sinensi imperio, Africa orientali, Indiaeque locis variis 2: 684.

Luisier, Alphonse, 1907: Notes de Bryologie Portugaise. Annales Scientíficos da Academia Politécnica do Porto 2: 240.

Luisier, Alphonse, 1910: Bryotheca Lusitanica. Brotéria: Série Botanica 9: 68.

Luisier, Alphonse, 1916: Fragments de Bryologie Ibérique. Les débris d'une collection de mousses portugaises. Brotéria: Série Botanica 14: 39.

Luisier, Alphonse, 1924: Musci Salmanticenses. Descriptio et Distributio Specierum hactenus in Provincia geographica Salmanticensi cognitarum. Brevi addito conspectu Muscorum totius Peninsulae Ibericae. Memorias de la Real Academia de Ciencias Exactas, Físicas y Naturales de Madrid (Serie 2a) 3: 197–199.

Macoun, John, 1889: Contributions to the Bryology of Canada. Bulletin of the Torrey Botanical Club 16: 97.

Macoun, John, 1892: Catalogue of Canadian Plants. Part 6. Musci. Geological and Natural History Survey of Canada. Pages 157–160, 230, 273–274.

Macoun, John, 1902: Catalogue of Canadian Plants. Part 7. Lichens and Hepaticae. (Also, addendum to Part 6. Musci.) Geological Survey of Canada. Pages 267–269.

Maire, René et Werner, Roger G., 1934: Contribution à la Flore Cryptogamique du Maroc. Fascicule 4. Bulletin de la Société d'Histoire Naturelle de l'Afrique du Nord 25: 56.

Maskovski, Ed., 1930: Einige falsche Namenableitungen von den Gattungen Dichelyma, etc. Repertorium specierum novarum regni vegetabilis. Fedde, F., Zentralblatt für Sammlung und Veröffentlichung von Einzeldiagnosen neuer Pflanzen. Fasciculus 27: 293.

Milde, Julius, 1869: Bryologia Silesiaca. Laubmoos-Flora von Nord- und Mittel-Deutschland, unter besonderer Berücksichtigung Schlesiens und mit Hinzunahme der Floren von Jütland, Holland, der Rheinpfalz, von Baden, Franken, Böhmen, Mähren und der Umgegend von München. Pages 275–279.

Mitten, William, 1865: The "Bryologia" of the Survey of the 49th Parallel of Latitude. The Journal of the Linnean Society 8: 44. Plate 8.

Möller, Hjalmar, 1922: Lövmossornas utbredning i Sverige 7. Hookeriaceae och Fontinalaceae. Arkiv för Botanik 17 (14): 1–91. Plates 1–17; figures 1–35.

Möller, Hjalmar, 1933: Claes Gustaf Myrin's Mossherbarium. Botaniska Notiser 1933: 373–376.

Mönkemeyer, Wilhelm, 1914: Bryales. Fontinalaceae. Pascher, A., Die Süsswasserflora Deutschlands, Österreichs und der Schweiz 14: 101–108. Figures 31–33.

Mönkemeyer, Wilhelm, 1927: Die Laubmoose Europas, Ergänzungsband, 4: 653–667. Figs. 143–146. Rabenhorst, Ludwig, Kryptogamen-Flora von Deutschland, Österreich und der Schweiz.

Montagne, Camille, 1837: Monographie du genre Conomitrium, de la famille des Mousses. Annales des Sciences Naturelles (Seconde Série) 8: 246.

Müller, Carolo (Hal), 1850: Synopsis Muscorum Frondosorum omnium hucusque cognitorum 2: 143–145, 148–150.

Müller, Carolo (Hal), 1884: Die auf der Expedition S.M.S. "Gazelle" von Dr. Naumann gesammelten Laubmoose. Engler, Adolf, Botanische Jahrbücher 5: 82.

Müller, Carolo (Hal), 1887: Beiträge zur Bryologie Nord-Amerikas. Flora 70: 225.

Müller, Carolo (Hal), 1898: Bryologia Provinciae Schen-si Sinensis ex Collectione Giraldiana III. Nuovo Giornale Botanico Italiano (Nuova Serie). Memorie della Società Botanica Italiana 5: 190.

Müller, Hermann, 1867: Westfalens Laubmoose. Erster Nachtrag zur Geographie der in Westfalen beobachteten Laubmoose. Verhandlungen des Naturhistorischen Vereins der Preussischen Rheinlande und Westfalens. 24. Jahrgang, 3. Folge, 4: 138.

Myrin, Claës G., 1833: Dichelyma. Ett nytt slägte bland mossorna. Acta Regiae Academiae Scientiarum Holmiensis; or Kongliga Vetenskaps-Akademiens Handlingar för År 1832: 273–283. Plates 6–7.

Nichols, George E., 1913: Notes on Connecticut Mosses 4. Rhodora 15: 9–11.

Nicholson, W. E., 1901: New Variety of *Fontinalis antipyretica* L. The Journal of Botany 39: 427–428.

Noguchi, Akira, 1948: Notes on Japanese Musci VIII. Journal of Japanese Botany 22: 82–87. Figures 33–34.

Nordstedt, Carl F. O., 1882: Smärre Notiser. Lärda sällskaps sammanträden. Botaniska Notiser för År 1882: 26–27.

Okamura, Shutai, 1911: Neue Beiträge zur Moosflora Japans 2. The Botanical Magazine, Tokyo 25: 137–139. Figure 5.

Palisot de Beauvois, A. M. F. J., 1805: Prodrome des cinquième et sixième familles de l'AEthéogamie. Les Mousses. Les Lycopodes. Pages 30, 57–58.

Paris, E. G., 1903–1906: Index Bryologicus sive enumeratio muscorum ad diem ultimam anni 1900 cognitorum adjunctis Synonymia distributioneque geographica locupletissimis. Volumes 1–5.

Penhallow, D. P., 1900: The Canadian Pleistocene Flora of the Don Valley. Report of the Seventieth Meeting of the British Association for the Advancement of Science held at Bradford in September 1900. Page 335.

Persson, Herman, 1942: Bryophytes from the bottom of some lakes in north Sweden. Botaniska Notiser. Pages 308–324.

Persson, Herman, 1944: Existence de Mousses au fond des lacs en Suède. Revue Bryologique et Lichénologique 13: 84–88.

Philibert, Henri, 1880: Notes sur quelques espèces rares ou critiques. Revue Bryologique 7: 44.

Philibert, Henri, 1888: Études sur le Péristome. Revue Bryologique 15: 50–53.

Piré, Louis, 1871: Nouvelles recherches bryologiques. Quatrième fascicule. Bulletin Société Royale de Botanique de Belgique 10: 105.

Podpěra Josef, 1932: Výsledky bryologického výzkumu Moravy za léta 1923–1931. Zprávy komise na přírodovědecký výzkum Moravy a Slezska, Oddělení botanické č. 9: 16.

Podpěra, Josef, 1954: Conspectus Muscorum Europaeorum. Pages 504–511.

Prager, E., 1907: Neues aus der Moosflora des Riesengebirges. Allgemeine Botanische Zeitschrift 13: 124. 2 figures.

Rabenhorst, Ludwig, 1872. Bryotheca Europaeae. Fascicles 1–24. Pages 7 and 9.

Renauld, Ferdinand, 1888: Notice sur un *Fontinalis* de l'Auvergne. Revue Bryologique 15: 69.

Renauld, Ferdinand, et Cardot, Jules, 1887: Énumération des Muscinées récoltées par le Dr. Delamare, à l'Île Miquelon (Amérique Septentrionale). Revue Bryologique 14: 4–6.

Renauld, Ferdinand and Cardot, Jules, 1888: New Mosses of North America 1. Botanical Gazette 13: 200–201. Plates 18–19.

Renauld, Ferdinand et Cardot, Jules, 1888: Notice sur quelques Mousses de l'Amérique du Nord. Revue Bryologique 15: 71.

Renauld, Ferdinand et Cardot, Jules, 1888: Mousses Nouvelles de l'Amérique du Nord 1. Mémoires de la Société de Botanique de Belgique. Bulletin de la Société Royale de Botanique de Belgique 27: 133–135. Plates 8–9.

Renauld, Ferdinand and Cardot, Jules, 1889: New Mosses of North America 2. Botanical Gazette 14: 96–97. Plate 14.

Renauld, Ferdinand et Cardot, Jules, 1889: Mousses Nouvelles de l'Amérique du Nord. Bulletin de la Société Royale de Botanique de Belgique 28: 129–130. Plate 9.

Renauld, Ferdinand and Cardot, Jules, 1890: New Mosses of North America 4. Botanical Gazette 15: 58–59. Plate 9.

Renauld, Ferdinand et Cardot, Jules, 1890: Mousses Nouvelles de l'Amérique du Nord. Mémoires de la Société Royale de Botanique de Belgique. Bulletin de la Société Royale de Botanique de Belgique 29: 155. Plate 6.

Renauld, Ferdinand et Cardot, Jules, 1893: Musci Americae Septentrionalis, ex operibus novissimis recensiti et methodice dispositi. Revue Bryologique 20: 9.

Renauld, Ferdinand et Cardot, Jules, 1896: Musci Americae Septentrionalis Exsiccati. Notes sur quelques Espèces distribuées dans cette Collection. Bulletin de l'Herbier Boissier 4: 18.

Renauld, Ferdinand et Cardot, Jules, 1902: Mousses des Canaries. Bulletin de l'Herbier Boissier (Deuxième Série) 2: 450.

Rodeck, Hugo G., 1941: Distribution Problems in some Moraine Ponds. University of Colorado Studies, Series D, Physical and Biological Sciences 1: 193–201.

Röll, Julius, 1890: Vorläufige Mittheilungen über die von mir im Jahre 1888 in Nord-Amerika gesammelten neuen Arten und Varietäten der Laubmoose. Botanisches Centralblatt 44: 421–422.

Röll, Julius, 1893: Nordamerikanische Laubmoose, Torfmoose und Lebermoose, gesammelt von Dr. Julius Röll in Darmstadt. Hedwigia 32: 298–299.

Röll, Julius, 1897: Beiträge zur Moosflora von Nord-Amerika. Hedwigia 36: 61.

Roth, Georg, 1904: Die Europäischen Laubmoose 2: 275–296. Plates 30–31.

Roth, Georg, 1910: Neuere und noch weniger bekannte Europäische Laubmoose. Hedwigia 49: 220–222. Plate 8.

Roth, Georg, 1913: Neuere und noch weniger bekannte Europäische Laubmoose, über welche in meinen Büchern aus den Jahren 1904 und 1905 noch keine Zeichnungen vorhanden sind. Hedwigia 53: 128. Plate 3.

Ruprecht, Franz J., 1854: Flora Boreali-Uralensis. Über die Verbreitung der Pflanzen im nördlichen Ural. Nach den Ergebnissen der Ural-Expedition in den Jahren 1847–1848. Page 44.

Ruthe, R., 1872: Eine neue Art der Laubmoosgattung *Fontinalis*. Hedwigia 11: 166–167.

Savi, Gaetano, 1818: Botanicon etruscum 3: 107.

Savicz, Lydie, 1916: Notes concerning *Fontinalis tenuissima* Borszczow. The Bulletin of the Imperial Botanical Garden of Peter the Great 16: 312–324. 4 figures.

Sayre, Geneva, 1945: The Distribution of *Fontinalis* in a Series of Moraine Ponds. The Bryologist 48: 34–36.

Schiffner, Viktor, 1913: Bryophyta aus Mesopotamien und Kurdistan, Syrien, Rhodos, Mytilini und Prinkipo, gesammelt von Dr. Heinrich Frh. von Handel-Mazzetti. Annalen des Kaiserlich-Königlichen Naturhistorischen Hofmuseums 27: 498–499. Figures 88–91.

Schimper, Wilhelm P., 1845: Einige Bemerkungen zur Bryologia Europaea. Flora 28: 146.

Schimper, Wilhelm P., 1848: Nya Mossor, först funna under en Resa i Skandinavien År 1844. Kongliga Vetenskaps-Akademiens Handlingar för År 1846: 165–170. Plates 14–16.

Schimper, Wilhelm P., 1855: Corollarium Bryologiae Europaeae conspectum diagnosticum Familiarum, Generum et Specierum, Adnotationes Novas atque Emendationes Complectens. Pages 96–97.

Schimper, Wilhelm P., 1860: Synopsis Muscorum Europaeorum praemissa introductione de elementis bryologicis tractante. Pages 455–461.

Schimper, Wilhelm P., 1869: Traité de Paléontologie Végétale, ou la flore du monde primitif dans ses rapports avec les formations géologiques et la flore du monde actuel 1: 245–246.

Schimper, Wilhelm P., 1876: Synopsis Muscorum Europaeorum praemissa introductione de elementis bryologicis tractante. Editio secunda valde aucta et emendata. Pages 551–559.

Scopoli, Johann A., 1772: Flora Carniolica, exhibens plantas Carnioliae indigenas et distributas in classes, genera, species, varietates ordine Linneano 2: 307.

Sim, T. R., 1926: The Bryophyta of South Africa. Transactions of the Royal Society of South Africa 15: 353–355. Figures A, B, D, W, page 355.

Simmons, H. G., 1896: *Fontinalis antipyretica* L. var. *monensis* Cardot et Simmons nov. var. Botaniska Notiser. Page 222.

Smith, James E., 1804: Flora Britannica 3: 1336–1338.

Steere, William C., 1942: Pleistocene Mosses from the Aftonian Interglacial Deposits of Iowa. Papers of the Michigan Academy of Science, Arts, and Letters 27: 83, 87.

Steere, William C., 1942: Notes on Michigan Bryophytes 4. The Bryologist 45: 167.

Størmer, P., 1947: New Records of Norwegian Bryophytes. Blyttia 5: 125.

Sturm, Jakob, 1818: Deutschlands Flora in Abbildungen nach der Natur mit Beschreibungen 4 (2): 15–16. Plate 24.

Sullivant, William S., 1846: Musci Alleghanienses, sive enumeratio Muscorum atque Hepaticarum, quos in itinere a Marylandia usque ad Georgiam per tractus montium. 1843. Decerpserunt Asa Gray et W. S. Sullivant (interjectis nonnullis aliunde collectis.) Pages 38, 46–47.

Sullivant, William S., 1848: The Mosses and Liverworts. A Manual of the Botany of the Northern United States, from New England to Wisconsin and south to Ohio and Pennsylvania inclusive. Page 677.

Sullivant, William S., 1856: Description of the Mosses and Liverworts, Number 5 in Part 5. Report of the Botany of the Expedition. Explorations and Surveys for a Railroad Route from the Mississippi River to the Pacific Ocean. Route near the thirty-fifth Parallel, explored by Lieutenant A. W. Whipple, Topographical Engineers, in 1853 and 1854. Senate Reports of Explorations and Surveys to ascertain the most practicable and economical route for a railroad from the Mississippi River to the Pacific Ocean 4: 189.

Sullivant, William S., 1856: The Musci and Hepaticae of the United States east of the Mississippi River. Contributed to the second edition of Gray's Manual of Botany. Pages 654 (54)–655 (55); Additions and Corrections, pages 104–105; plate 18. (Another edition, in 1871.)

Sullivant, William S., 1864: Icones Muscorum. Pages 99–106; plates 59–66.

Sullivant, William S., 1874: Icones Muscorum. Supplement. (Posthumus.) Page 76; plate 57.

Sullivant, William S. et Lesquereux, Leo, 1856: Musci Boreali-Americani, sive Specimina Exsiccata Muscorum in Americae Rebuspublicis Foederatis detectorum. Index. (Number 226: Latin description of *Fontinalis biformis* Sull.)

Sullivant, William S. et Lesquereux, Leo, 1865: Musci Boreali-Americani, sive Specimina Exsiccata Muscorum in Americae Rebuspublicis Foederatis detectorum. Pages 56–58, 92–93.

Swartz, Olof, 1781: Methodus Muscorum Illustrata. Pages 29–30.

Swartz, Olof, 1788: Nova Genera et Species Plantarum, seu Prodromus descriptionum vegetabilium maximam partem incognitorum quae sub itinere in Indiam Occidentalem. Page 138.

Swartz, Olof, 1788: Flora Indiae Occidentalis aucta atque illustrata, sive descriptiones plantarum in Prodromo recensitarum. Page 138.

Swartz, Olof, 1791: Observationes Botanicae. Page 406.

Swartz, Olof, 1829: Adnotationes Botanicae. Pages 176–180. (See Wikström.)

Thériot, Iréné, 1927: Deux Mousses Nouvelles. Archives de Botanique 1: 67. Figure 1.

Thériot, Iréné, 1935: Jules Cardot. Revue Bryologique et Lichénologique (Nouvelle Série) 8: 5–13.

Tongiorgi, Ezio, 1938: Fontinalaceae dell' Africa Orientale Italiana dalle Collezioni del R. Erbario Coloniale di Firenze. Schedulae Bryologicae Africae Orientalis Italicae 2–3. Istituto Botanico della R. Università di Pisa. Nuovo Giornale Botanico Italiano (Nuova Serie) 45: 400–402.

Triana, José et Planchon, J. E., 1865; Prodromus Florae Novo-Granatensis ou Énumération des Plantes de la Nouvelle-Grenade avec description des espèces nouvelles. Musci exposuit E. Hampe. Annales des Sciences Naturelles, cinquième Série. Botanique 4: 351.

Velenovský, Josef, 1903: Bryologické přispěvky z Čech za rok 1901–1902. Rozpravy České Akademie Císaře Františka Josefa pro Vědy, Slovesnost a Umění 12: 13.

Villar, Dominique, 1786: Histoire des Plantes du Dauphiné 3: 919.

Voit, Johann Gottlob W., 1812: Historia Muscorum Frondosorum in Magno Ducatu Herbipolitano crescentium. Page 125.

Von Dalla Torre, K. W. und von Sarnthein, Ludwig, G., 1904: Die Moose (Bryophyta) von Tirol, Vorarlberg und Liechtenstein. Flora der Gefürsteten Grafschaft Tirol des Landes Vorarlberg und des Fürstenthumes Liechtenstein 5: 436–437.

Von Klinggraeff, Hugo, 1880: Versuch einer topographischen Flora der Provinz Westpreussen. Page 112.

Von Klinggraeff, Hugo, 1881: Zeitschrift der Naturforschenden Gesellschaft in Danzig 5: 193.

Von Klinggraeff, Hugo, 1883: Bericht über die Botanischen Reisen im Neustädter Kreise im Sommer 1882. Schriften der Naturforschenden Gesellschaft in Danzig (Neue Folge) 6: 24.

Von Klinggraeff, Hugo, 1893: Die Leber- und Laubmoose West- und Ostpreussens. Pages 226–231.

Von Wied-Neuwied, Maximilian A. P., 1841: Musci Frondosi. Verzeichnis der Pflanzen, welche der Prinz Maximilian von Wied 1833 von seiner Reise am obern Missouri mit zurückbrachte; bestimmt und beschrieben von Dr. Nees von Esenbeck. (Reise Nord-Amerika 2.) Page 27.

Wahlenberg, Göran, 1820: Flora Upsaliensis. Page 365.

Warnstorf, Carl, 1899: Neue Beiträge zur Kryptogamenflora der Mark Brandenburg. Verhandlungen des Botanischen Vereins der Provinz Brandenburg 41: 66.

Warnstorf, Carl, 1905: Laubmoose. Kryptogamenflora der Mark Brandenburg und angrenzender Gebiete, herausgegeben von dem Botanischen Verein der Provinz Brandenburg 2: 623–638. Figures 5–6.

Warnstorf, Carl, 1913: Eine Erinnerung an Dr. E. Zickendrath. Hedwigia 53: 298.

Warnstorf, Carl, 1916: Bryologische Neuigkeiten. Bryologische Zeitschrift 1: 38–41.

Weber, Friedrich und Mohr, Daniel M. H., 1807: Botanisches Taschenbuch auf das Jahr 1807. Deutschlands kryptogamische Gewächse. Page 376.

Weber, Georg H., 1778: Spicilegium Florae Gottingensis, plantas inprimis cryptogamicas Hercyniae illustrans. Page 38.

Welch, Winona H., 1934: Fontinalaceae. Grout, A. J., Moss Flora of North America North of Mexico 3: 233–262. Plates 73–79.

Welch, Winona H., 1943: The Systematic Position of the Genera *Wardia*, *Hydropogon*, and *Hydropogonella*. The Bryologist 46: 25–46. Figures 1–39.

Welch, Winona H., 1947: *Fontinalis Duthieae* (Dixon MSS.) Sim. The Bryologist 50: 187–188.

Welch, Winona H., 1948: Vegetative Propagation in *Fontinalis*. The Bryologist 51: 192–193.

Welch, Winona H., 1948: Dimorphism in the leaves of *Fontinalis biformis* Sull. The Bryologist 51: 194–197.

Werner, Roger G., 1932: Contribution à la Flore Cryptogamique du Maroc. Revue Bryologique et Lichénologique (Nouvelle Série) 5: 227.

Wikström, Johann E., 1829: Adnotationes Botanicae, quas reliquit Olavus Swartz. Post mortem Auctoris collectae, examinatae, in ordinem systematicum redactae atque notis praefatione instructae. Pages 176–180.

Williams, Robert S., 1930: Some deep-water Mosses. The Bryologist 33: 32.

Wilson, William, 1855: Bryologia Britannica. The Mosses of Great Britain and Ireland. Pages 422–426.

Withering, William, 1801: Systematic Arrangement of British Plants 3: 773.

Withering, William, 1812: Systematic Arrangement of British Plants 3: 967-969.

Wynne, Frances E., 1944: Studies in *Drepanocladus* 3. Doubtful and Excluded Names. The Bryologist 47: 72, 74, 75.

Wynne, Frances E., 1944: Studies in *Drepanocladus* 4. Taxonomy. The Bryologist 47: 165.

Zahlbruckner, Alexander, 1900: Schedae ad Kryptogamas Exsiccatas. Centuria 5–6. Editae a Museo Palatino Vindobonensi. Annalen des Kaiserlich-Königlichen Naturhistorischen Hofmuseums 15 (2): 213.

Zahlbruckner, Alexander, 1913: Schedae ad Kryptogamas Exsiccatas editae a Museo Palatino Vindobonensi. Centuria 21. Annalen des Kaiserlich-Königlichen Naturhistorischen Hofmuseums 27: 277–278.

Zmuda, Antoni J., 1914: Fossile Flora des Krakauer Diluviums. Bulletin International de l'Académie des Sciences de Cracovie. Classe des Sciences Mathématiques et Naturelles Série B: Sciences Naturelles. Pages 215, 228, 233, 249, 272.

INDEX of COLLECTIONS CITED [1]

Åberg, G.
135 *Dichelyma falcatum*
136 *D. falcatum*
137 *Fontinalis antipyretica*
138 *F. antipyretica*
139 *F. hypnoides*
140 *F. squamosa*
466 *F. antipyretica*
467 *F. hypnoides*
468 *F. squamosa*
689 *F. antipyretica*
782 *F. antipyretica*
    var. *gigantea*
841 *D. falcatum*
842 *F. antipyretica*
    var. *gracilis*
843 *F. antipyretica*
844 *F. squamosa*
1013 *D. falcatum*
1014 *D. falcatum*
1015 *D. falcatum*
1395 *F. antipyretica*
    var. *gracilis*
1396 *F. antipyretica*
    var. *gracilis*
1596 *F. antipyretica*
1597 *F. antipyretica*
1598 *F. antipyretica*
    var. *gracilis*
1599 *F. dalecarlica*
1600 *F. squamosa*
1601 *F. hypnoides*
1602 *D. falcatum*
1904 *D. falcatum*
1905 *D. falcatum*
1906 *F. antipyretica*
1907 *F. antipyretica*
1908 *F. antipyretica*
1910 *F. antipyretica*
1911 *F. dalecarlica*
1912 *F. dalecarlica*
1913 *F. dalecarlica*
3046 *D. falcatum*

3047 *F. dalecarlica*
3329 *D. falcatum*
3330 *F. antipyretica*
    var. *gracilis*
3331 *F. dalecarlica*
4696 *F. dalecarlica*
(1916) *F. antipyretica*
    var. *gracilis*
(1919) *F. antipyretica*
    var. *gracilis*
(1920) *F. squamosa*
(1923) *F. antipyretica*
s.n. *F. antipyretica*
Abramon, T.
(1910) *F. antipyretica*
    var. *gracilis*
Adam, J. C.
(1915) *F. antipyretica*
    var. *gracilis*
(1916) *F. antipyretica*
    var. *gracilis*
(1916) *F. squamosa*
Adams, Clara D.
(1937) *F. antipyretica*
    var. *gigantea*
(1937) *F. dalecarlica*
(1938) *D. falcatum*
(1938) *F. novae-angliae*
(1939) *D. falcatum*
(1940) *F. novae-angliae*
Adams, M. R.
(1935) *D. capillaceum*
Adams, Mrs. Merle R.
(1941) *F. novae-angliae*
Adlerz, E.
(1884) *D. capillaceum*
(1886) *F. antipyretica*
(1897) *D. falcatum*
(1904) *F. antipyretica*
    var. *gracilis*
(1904) *F. dalecarlica*
(1904) *F. hypnoides*
(1906) *F. hypnoides*

[1] If the collector's number is not available, the year of collection is enclosed in parentheses; and if neither the number nor the year is known, "s.n." is used.

(1908) *F. hypnoides*
Agelin, F.
  (1930) *F. antipyretica*
      var. *gigantea*
Ährling, E.
  (1855) *F. antipyretica*
  (1857) *F. antipyretica*
      var. *gracilis*
  (1861) *F. antipyretica*
      var. *gracilis*
Alcorn, Gordon
  (1934) *F. Howellii*
Alexander, E. J.
  (1934) *F. novae-angliae*
Alexandrov, T.
  (1911) *F. antipyretica*
Allard, E. C.
  18142 *F. novae-angliae*
Allard, H. A.
  4602 *F. dalecarlica*
  6478 *F. novae-angliae*
  6527 *F. novae-angliae*
  10949 *F. novae-angliae*
  10949a *F. novae-angliae*
  10951 *F. novae-angliae*
Allen, Ethan
  (1928) *F. antipyretica*
      var. *oreganensis*
Allen, J. A.
  25 *D. falcatum*
  80 *F. neo-mexicana*
  81a *F. Howellii*
  81b *F. Howellii*
  82 p.p. *F. antipyretica*
      var. *gigantea*
  82 p.p. *F. neo-mexicana*
  87 *F. neo-mexicana*
  98 *F. Howellii*
  106 *F. antipyretica*
      var. *gigantea*
  (1876) *D. capillaceum*
  (1876) *F. antipyretica*
  (1877) *F. dalecarlica*
  (1878) *F. novae-angliae*
  (1879) *F. novae-angliae*
  (1880) *D. capillaceum*
  (1880) *D. falcatum*
  (1880) *F. Allenii*
  (1880) *F. antipyretica*
      var. *gigantea*
  (1880) *F. dalecarlica*
  (1880) *F. Duriaei*
  (1880) *F. flaccida*
  (1880) *F. novae-angliae*
  (1881) *F. novae-angliae*
  (1882) *D. capillaceum*
  (1882) *D. falcatum*

(1882) *F. dalecarlica*
(1883) *F. dalecarlica*
(1885) *D. falcatum*
(1898) *F. antipyretica*
(1898) *F. antipyretica*
      var. *oreganensis*
(1898) *F. Howellii*
(1898) *F. neo-mexicana*
(1898) *F. patula*
(1912) *F. Duriaei*
(1913) *F. antipyretica*
      var. *oreganensis*
Allen, Lucy
  (1907) *F. biformis*
Allen, Oscar D.
  (1876) *F. dalecarlica*
  (1877) *F. dalecarlica*
  (1878) *F. dalecarlica*
  (1880) *F. dalecarlica*
  (1881) *F. novae-angliae*
  (1901) *F. Howellii*
  s.n. *F. antipyretica*
      var. *gigantea*
  s.n. *F. novae-angliae*
Alles, H.
  (1945) *F. antipyretica*
      var. *gigantea*
  (1945) *F. novae-angliae*
      var. *latifolia*
Allorge, Pierre
  (1924) *F. squamosa*
  (1929) *F. antipyretica*
  (1931) *F. antipyretica*
      var. *gracilis*
  (1931) *F. Duriaei*
  (1933) *F. squamosa*
Allorge, Pierre & Valentine
  (1937) *F. antipyretica*
Allorge, Valentine
  (1947) *D. capillaceum*
Allyre, Bro.
  2145 *F. novae-angliae*
Alm, Carl G.
  (1917) *F. squamosa*
  (1919) *F. antipyretica*
      var. *gracilis*
Ames, Mary E. Pulsifer
  (1886) *D. uncinatum*
  s.n. *F. neo-mexicana*
Ammann, George A.
  100 *F. neo-mexicana*
Ammons, Nelle
  (1929) *F. novae-angliae*
  (1929) *F. novae-angliae*
      var. *cymbifolia*
  (1932) *F. dalecarlica*
  (1932) *F. novae-angliae*

Arvén, A.
 (1891) *F. antipyretica*
   var. *gigantea*
 (1892) *D. falcatum*
 (1893) *F. antipyretica*
 (1900) *D. falcatum*
 (1901) *D. falcatum*
 (1903) *F. dalecarlica*
 (1910) *D. falcatum*
 (1912) *D. falcatum*
 (1913) *F. antipyretica*
 (1914) *D. falcatum*
 (1914) *F. dalecarlica*
 (1916) *D. falcatum*
 (1920) *D. falcatum*
 (1920) *F. antipyretica*
Arwidsson, Thorsten
 (1921) *F. antipyretica*
 (1931) *D. falcatum*
Atkinson, George F.
 11958 *F. novae-angliae*
Aurell, C. E.
 (1889) *D. falcatum*
Austin, Coe Finch
 (1861) *F. dalecarlica*
 (1862) *D. capillaceum*
 (1862) *F. antipyretica*
   var. *gigantea*
 (1862) *F. disticha*
 (1862) *F. flaccida*
 (1862) *F. novae-angliae*
 (1862) *F. novae-angliae*
   var. *cymbifolia*
 (1863) *F. novae-angliae*
 (1863) *F. novae-angliae*
   var. *cymbifolia*
 (1865) *D. capillaceum*
 (1865) *F. antipyretica*
   var. *gigantea*
 (1865) *F. novae-angliae*
   var. *cymbifolia*
 (1867) *F. novae-angliae*
 (1867) *F. novae-angliae*
   var. *cymbifolia*
 (1868) *F. flaccida*
 (1868) *F. novae-angliae*
 (1869) *D. capillaceum*
 (1870) *F. novae-angliae*
 (1872) *F. antipyretica*
   var. *gigantea*
 (1872) *F. dalecarlica*
 (1872) *F. novae-angliae*
 (1873) *F. dalecarlica*
 (1877) *D. capillaceum*
 (1878) *D. capillaceum*
 (1878) *F. disticha*
 (1887) *F. novae-angliae*

 s.n. *D. capillaceum*
 s.n. *F. dalecarlica*
 s.n. *F. disticha*
 s.n. *F. flaccida*
 s.n. *F. novae-angliae*
 s.n. *F. novae-angliae*
   var. *cymbifolia*
Austin, Mrs. R. M.
 (1877) *F. neo-mexicana*
 (1880) *F. Duriaei*
Axelson
 (1901) *D. falcatum*

Babet, V.
 (1914) *F. antipyretica*
   var. *gigantea*
Backman
 (1903) *D. falcatum*
Bacon, W. L.
 254 *F. dalecarlica*
 348 *F. antipyretica*
 (1932) *F. dalecarlica*
Baenitz, C.
 (1894) *D. falcatum*
 (1897) *F. antipyretica*
 (1898) *F. antipyretica*
Bailey, Harold & Virginia
 9 *F. neo-mexicana*
 148 *F. neo-mexicana*
Bailey, Harriet B.
 (1902) *D. falcatum*
Bailey, John W.
 4 *F. Howellii*
 144 *F. patula*
 1171 *F. neo-mexicana*
 (1901) *D. uncinatum*
 (1901) *F. antipyretica*
   var. *oreganensis*
 (1901) *F. neo-mexicana*
 (1901) *F. patula*
 (1904) *F. antipyretica*
   var. *oreganensis*
 (1904) *F. Howellii*
 (1904) *F. neo-mexicana*
 (1904) *F. patula*
 (1905) *F. antipyretica*
 (1907) *F. neo-mexicana*
 (1911) *D. uncinatum*
 (1930) *F. neo-mexicana*
 (1931) *F. neo-mexicana*
 (1934) *F. antipyretica*
   var. *oreganensis*
 s.n. *F. patula*
Bailey, Liberty Hyde
 82 *F. dalecarlica*
 (1886) *F. dalecarlica*

Bailey, Sarah Elizabeth
  (1939) *F. novae-angliae*
Bailey, Winona
  (1908) *F. antipyretica*
      var. *mollis*
  (1910) *F. neo-mexicana*
Bailey, W. B.
  (1882) *F. dalecarlica*
Baker, Carl F.
    1 *F. hypnoides*
    3 *F. antipyretica*
   41 *F. antipyretica*
   42 *F. Duriaei*
  631 *F. hypnoides*
 1464 *F. antipyretica*
      var. *oreganensis*
 3000 *F. antipyretica*
  (1896) *F. Duriaei*
  (1897) *F. flaccida*
Baker, J. G.
  (1870) *F. antipyretica*
Baker, M. S.
 309b *F. neo-mexicana*
  (1894) *F. antipyretica*
Baker & Nutting
  (1894) *F. antipyretica*
  (1894) *F. Duriaei*
Baldacci, Antonio
  164 *F. antipyretica*
Ballard, C.
  (1904) *F. antipyretica*
      var. *gigantea*
Bank, T. P.
  F-23 *F. neo-mexicana*
Banker, Howard J.
  (1907) *F. biformis*
  (1919) *F. antipyretica*
      var. *gigantea*
Barber
  s.n. *F. dalecarlica*
Bardell, E. M.
  (1910) *F. patula*
Barker
  s.n. *F. antipyretica*
      var. *gigantea*
Barker, T.
  (1868) *F. antipyretica*
  (1870) *F. squamosa*
  (1899) *F. antipyretica*
  (1899) *F. squamosa*
  (1900) *F. squamosa*
Barkley, Fred A.
 1640 *F. neo-mexicana*
  (1937) *F. neo-mexicana*
Barkley, Fred A. & Diettert, R. A.
 1640 *F. neo-mexicana*
 1767 *F. neo-mexicana*

Barlow, William
  (1844) *F. antipyretica*
      var. *gracilis*
Barnard, K. H.
 35628 *F. antipyretica*
Barnes, Charles R.
  (1859) *F. biformis*
  s.n. *F. Duriaei*
Barrett
  s.n. *F. neo-mexicana*
Barron, Alfred
  s.n. *D. pallescens*
  s.n. *F. antipyretica*
      var. *gigantea*
Barth, J.
  (1874) *F. antipyretica*
Bartholomew, Elizabeth
  (1939) *F. novae-angliae*
Bartley, Floyd
  (1946) *F. biformis*
  (1946) *F. novae-angliae*
      var. *cymbifolia*
Bartley, Floyd & Pontius, Leslie L.
  162 *D. capillaceum*
  172 *F. novae-angliae*
  232 *F. novae-angliae*
  531 *F. novae-angliae*
  (1935) *F. novae-angliae*
  (1936) *D. capillaceum*
  (1937) *F. novae-angliae*
  (1938) *F. biformis*
Bartram, Edwin B.
  233 *F. Duriaei*
  237 *F. novae-angliae*
  238 *F. antipyretica*
      var. *gigantea*
  543 *F. novae-angliae*
 1513 *F. missourica*
Bartram, J.
  s.n. *D. capillaceum*
Bauer
  (1839) *D. falcatum*
Bauer, D.
  88 *F. antipyretica*
 125 *F. antipyretica*
Bauer, E.
  (1897) *F. antipyretica*
  (1897) *F. squamosa*
  (1898) *F. antipyretica*
  (1898) *F. antipyretica*
      var. *gracilis*
  (1899) *F. antipyretica*
      var. *gigantea*
  (1903) *F. antipyretica*
  (1903) *F. squamosa*
  (1904) *F. squamosa*

(1911) *F. antipyretica*
  s.n. *F. squamosa*
Baumgartner, Julius
  (1909) *F. antipyretica*
  (1911) *F. antipyretica*
  (1912) *F. hypnoides*
  (1916) *F. hypnoides*
  (1925) *F. antipyretica*
    var. *gigantea*
  s.n. *F. hypnoides*
Baur, W.
  (1890) *F. squamosa*
  (1891) *F. antipyretica*
Beardslee, H. C.
  (1877) *D. capillaceum*
Beaulac, Aldéric
  416 *F. Duriaei*
  429 *F. novae-angliae*
  536 *F. novae-angliae*
  657 *D. pallescens*
  (1931) *D. pallescens*
Beauvard
  (1912) *F. antipyretica*
Becker, T.
  (1866) *F. antipyretica*
  (1867) *F. antipyretica*
Beeker, H. R.
  (1933) *F. novae-angliae*
Behm, F.
  (1865) *F. squamosa*
  (1878) *F. antipyretica*
  (1886) *F. dalecarlica*
Bell, William
  (1860) *F. antipyretica*
Bellerby, W.
  (1926) *F. antipyretica*
Benedict, Bro. A.
  2338 *F. neo-mexicana*
Bennett, James L.
  s.n. *F. antipyretica*
    var. *gigantea*
Berger, Alwin
  (1899) *F. antipyretica*
    var. *gigantea*
Berggren, S.
  (1859) *D. capillaceum*
  (1859) *F. dalecarlica*
  (1860) *D. capillaceum*
  (1861) *D. capillaceum*
  (1865) *D. falcatum*
  (1886) *F. antipyretica*
Bergman
  (1855) *F. hypnoides*
Bergner, K.
  (1921) *F. antipyretica*
Bergström, S.
  (1918) *D. falcatum*

Bergström, S. & C.
  (1913) *F. antipyretica*
  (1915) *F. antipyretica*
Bernet, H.
  729 *F. antipyretica*
    var. *gracilis*
  (1872) *F. antipyretica*
  (1881) *F. antipyretica*
    var. *gigantea*
  (1883) *F. antipyretica*
    var. *gigantea*
  (1883) *F. antipyretica*
    var. *gracilis*
  (1883) *F. dalecarlica*
  (1883) *F. squamosa*
  (1884) *F. antipyretica*
  (1886) *F. antipyretica*
  (1886) *F. antipyretica*
    var. *gracilis*
  (1886) *F. dalecarlica*
  (1886) *F. squamosa*
  (1888) *F. antipyretica*
  (1890) *F. squamosa*
  (1895) *F. antipyretica*
  s.n. *F. antipyretica*
    var. *gigantea*
Bertaud, Bro.
  207 *F. neo-mexicana*
Bertram, Pastor W.
  (1874) *F. antipyretica*
  (1874) *F. antipyretica*
    var. *gracilis*
  (1874) *F. squamosa*
  s.n. *F. antipyretica*
    var. *gracilis*
  s.n. *F. squamosa*
Berudes, Wilhelm
  (1868) *F. hypnoides*
Best, George N.
  (1890) *F. novae-angliae*
  (1892) *F. antipyretica*
    var. *gigantea*
  (1892) *F. dalecarlica*
Bethel, E.
  (1904) *F. neo-mexicana*
  (1905) *F. neo-mexicana*
Beyrich, Carl
  (1834) *F. missourica*
  s.n. *F. missourica*
  s.n. *F. novae-angliae*
Bigelow, J. W.
  (1854) *F. antipyretica*
Bimont, G.
  4304 *F. antipyretica*
Binstead, C. H.
  (1896) *F. antipyretica*
  (1896) *F. hypnoides*

s.n. *F. antipyretica*
s.n. *F. neo-mexicana*
Boldt
  (1892) *D. falcatum*
Bomansson, J. O.
  (1864) *D. falcatum*
  (1864) *F. antipyretica*
  (1872) *F. hypnoides*
  (1885) *F. hypnoides*
  (1887) *D. falcatum*
  (1890) *D. falcatum*
  s.n. *F. hypnoides*
Boner
  (1847) *F. squamosa*
Bonpland
  s.n. *F. antipyretica*
Bonser, Thomas
  (1906) *F. neo-mexicana*
Bornmüller
  (1928) *F. squamosa*
Bornmüller, J.
  3613 *F. antipyretica*
  4477 *F. antipyretica*
Boros, A.
  (1919) *F. antipyretica*
  (1920) *F. antipyretica*
  (1925) *F. antipyretica*
  (1926) *F. Duriaei*
  (1928) *F. antipyretica*
Borshchov, E.
  206 *F. antipyretica*
Boswell, H.
  (1878) *F. squamosa*
  s.n. *F. antipyretica*
Bottini, A.
  (1879) *F. antipyretica*
  (1887) *F. antipyretica*
  (1887) *F. Duriaei*
  (1888) *F. Duriaei*
  s.n. *F. Duriaei*
Boulay, Abbé
  (1873) *F. Duriaei*
  (1874) *F. hypnoides*
Bouly de Lesdain
  (1895) *F. antipyretica*
  (1896) *F. antipyretica*
  (1896) *F. antipyretica*
    var. *gigantea*
  (1897) *F. antipyretica*
  (1899) *F. antipyretica*
  (1910) *F. antipyretica*
  (1912) *F. antipyretica*
Bourgeau, E.
  (1857–1858) *F. Duriaei*
  (1859) *F. Duriaei*
  s.n. *F. Duriaei*

Bouvet
  s.n. *F. antipyretica*
Bové
  (1839) *F. Duriaei*
Boyd, O. G.
  M24 *F. antipyretica*
  M25 *F. neo-mexicana*
Boyer, A.
  50 *F. flaccida*
  99 *F. novae-angliae*
  (1906) *F. flaccida*
  (1907) *F. flaccida*
  (1909) *F. novae-angliae*
  s.n. *F. flaccida*
  s.n. *F. novae-angliae*
Bradwell
  (1798) *F. antipyretica*
  s.n. *F. antipyretica*
Braithwaite, Robert
  (1891) *F. antipyretica*
  s.n. *F. antipyretica*
Braithwaite & Beesley
  (1903) *F. antipyretica*
    var. *gracilis*
Brandegee, T. S.
  (1874–1878) *F. Duriaei*
  (1874–1878) *F. neo-mexicana*
  (1880) *F. neo-mexicana*
  (1882) *F. neo-mexicana*
  (1882) *F. patula*
  (1883) *F. antipyretica*
  (1885) *F. antipyretica*
  s.n. *F. antipyretica*
  s.n. *F. dalecarlica*
  s.n. *F. Duriaei*
  s.n. *F. neo-mexicana*
Branik
  (1876) *F. antipyretica*
Brann, G.
  (1877) *F. squamosa*
  (1878) *F. squamosa*
Brasch
  (1910) *D. capillaceum*
Braun, Alexander
  (1824) *F. squamosa*
  (1833) *F. squamosa*
  (1855) *F. hypnoides*
  (1899) *F. hypnoides*
  s.n. *F. hypnoides*
  s.n. *F. squamosa*
Braun, E. Lucy
  (1928) *F. dalecarlica*
  (1931) *F. novae-angliae*
  (1932) *F. biformis*
  (1937) *F. novae-angliae*
Brébisson (See de Brébisson.)

Breidler
  (1870) *F. antipyretica*
       var. *gigantea*
  (1870) *F. squamosa*
Brenckle, J. F.
  (1910) *F. Duriaei*
  s.n. *F. Duriaei*
Brendel, Fred
  (1844) *F. antipyretica*
Brendle, Dr.
  (1869) *F. novae-angliae*
  s.n. *F. novae-angliae*
Brenner, Lewis G., Jr.
  (1940) *F. novae-angliae*
Breutel
  41 *F. antipyretica*
  184 *F. antipyretica*
  s.n. *F. antipyretica*
  s.n. *F. squamosa*
Breutel & Spohrleder
  s.n. *F. squamosa*
Brevière, L.
  s.n. *F. squamosa*
Brewer, William H.
  952 *F. antipyretica*
  (1858) *F. Duriaei*
Brinker, Robert
  498 *Brachelyma robustum*
Brinkman, A. H.
  68 *F. neo-mexicana*
  125 *F. neo-mexicana*
  140 *F. antipyretica*
      var. *gigantea*
  145 *F. Duriaei*
  219 *F. hypnoides*
  231 *F. hypnoides*
  246 *F. Howellii*
  247 *F. hypnoides*
  442 *F. antipyretica*
  444 *F. patula*
  493 *D. uncinatum*
  614 *F. Howellii*
  674 *F. hypnoides*
  719 *F. hypnoides*
  4309 *F. Duriaei*
  (1908) *F. hypnoides*
  (1909) *F. neo-mexicana*
Briquet, J.
  (1906) *F. antipyretica*
Britton, Elizabeth Gertrude
  237 *F. novae-angliae*
  238 *F. dalecarlica*
  240 *F. dalecarlica*
  (1882) *F. novae-angliae*
  (1886) *D. capillaceum*
  (1886) *F. novae-angliae*
  (1889) *D. capillaceum*

(1889) *F. antipyretica*
    var. *gigantea*
(1889) *F. dalecarlica*
(1891) *D. capillaceum*
(1892) *F. novae-angliae*
(1893) *D. capillaceum*
(1893) *F. antipyretica*
    var. *gigantea*
(1894) *F. dalecarlica*
(1896) *F. antipyretica*
    var. *gigantea*
(1896) *F. dalecarlica*
(1897) *F. antipyretica*
    var. *gigantea*
(1897) *F. dalecarlica*
(1897) *F. novae-angliae*
(1898) *F. antipyretica*
    var. *gigantea*
(1898) *F. dalecarlica*
(1899) *F. antipyretica*
    var. *gigantea*
(1899) *F. dalecarlica*
(1900) *F. antipyretica*
(1900) *F. antipyretica*
    var. *gigantea*
(1900) *F. novae-angliae*
(1901) *F. antipyretica*
    var. *gigantea*
Britton, E. G. & A. M. S.
(1901) *F. antipyretica*
    var. *gigantea*
Britton, Nathaniel L.
(1882) *F. novae-angliae*
(1885) *F. antipyretica*
    var. *gigantea*
(1886) *F. novae-angliae*
(1901) *F. antipyretica*
    var. *gigantea*
Britton, N. L. & E. G. & Timmerman,
Millie
(1889) *F. novae-angliae*
    var. *cymbifolia*
Brocas, F. Y.
(1851) *F. antipyretica*
(1853) *F. antipyretica*
Brotherus, A. H. & V. F.
  44 *F. antipyretica*
(1881) *F. antipyretica*
Brotherus, Victor F.
  78 *F. dalecarlica*
  180 *F. antipyretica*
    var. *gracilis*
  181 *F. dalecarlica*
  182 *F. antipyretica*
  442 *F. dalecarlica*
  490 *F. antipyretica*
(1868) *D. falcatum*

(1868) *F. antipyretica*
(1869) *D. falcatum*
(1869) *F. antipyretica*
    var. *gracilis*
(1869) *F. dalecarlica*
(1869) *F. hypnoides*
(1870) *F. dalecarlica*
(1871) *F. antipyretica*
(1871) *F. antipyretica*
    var. *gracilis*
(1872) *F. antipyretica*
(1872) *F. antipyretica*
    var. *gracilis*
(1872) *F. dalecarlica*
(1873) *F. dalecarlica*
(1877) *F. antipyretica*
(1881) *F. antipyretica*
(1883) *D. falcatum*
(1887) *D. falcatum*
(1887) *F. antipyretica*
(1887) *F. antipyretica*
(1887) *F. antipyretica*
    var. *gracilis*
(1887) *F. dalecarlica*
(1896) *F. antipyretica*
(1910) *F. dalecarlica*
(1916) *D. falcatum*
(1916) *F. antipyretica*
    var. *gracilis*
s.n. *F. antipyretica*
    var. *gracilis*
Brouard, Bro. Arsène (See Arsène, Bro.)
Brown, A., Hogg, Thomas, Vail, Anna Murray, Timmerman, Millie, and Britton, N. L. & E. G.
(1890) *F. dalecarlica*
Brown, A. C. and Floyd
(1949) *F. novae-angliae*
Brown, D. M.
    M39 *F. novae-angliae*
Brown, H. E.
    32 *F. antipyretica*
    432 *F. antipyretica*
(1897) *F. patula*
Brown, H. H.
    117 *F. dalecarlica*
    821 *F. Duriaei*
    887 *F. hypnoides*
Brown, Margaret S.
    530 *D. capillaceum*
    568 *F. antipyretica*
    var. *gigantea*
    569 *F. dalecarlica*
    570b *F. novae-angliae*
    571a *F. novae-angliae*
(1907) *F. novae-angliae*

(1923) *F. dalecarlica*
(1923) *F. novae-angliae*
(1923) *F. novae-angliae*
    var. *cymbifolia*
(1924) *F. novae-angliae*
(1924) *F. novae-angliae*
    var. *cymbifolia*
(1931) *F. novae-angliae*
Brown, R. G.
(1946) *F. novae-angliae*
Browne, Carlotta H.
(1907) *F. dalecarlica*
Brückner, Ad.
(1903) *F. antipyretica*
    var. *gracilis*
Brun
(1910) *F. squamosa*
Brunard
(1909) *F. antipyretica*
Brundin
(1889) *D. falcatum*
(1892) *D. falcatum*
Bryhn, N.
(1876) *D. falcatum*
(1878) *D. falcatum*
(1879) *F. antipyretica*
    var. *gracilis*
(1885) *F. Bryhnii*
(1885) *F. dalecarlica*
(1887) *F. antipyretica*
    var. *gracilis*
(1887) *F. Bryhnii*
(1889) *F. squamosa*
(1892) *F. Bryhnii*
(1892) *F. dalecarlica*
(1899) *D. falcatum*
(1899) *F. Bryhnii*
(1900) *F. antipyretica*
    var. *gracilis*
(1900) *F. Bryhnii*
(1900) *F. dalecarlica*
(1901) *D. falcatum*
(1901) *F. Bryhnii*
(1902) *F. dalecarlica*
(1902) *F. squamosa*
(1904) *D. falcatum*
(1904) *F. Bryhnii*
(1906) *F. antipyretica*
    var. *gracilis*
(1907) *F. squamosa*
Buch, Hans
(1904) *D. falcatum*
(1904) *F. hypnoides*
(1911) *D. falcatum*
(1911) *F. antipyretica*
(1911) *F. hypnoides*

Bullard, Charles
  (1938) *F. dalecarlica*
Bumstead, F. J.
  s.n. *F. antipyretica*
    var. *gigantea*
Bunett, P. G.
  s.n. *F. neo-mexicana*
Bureau, Emile
  293 *F. Duriaei*
  (1894) *F. Duriaei*
  (1895) *F. Duriaei*
Bureau, E. & Camus, F.
  (1891) *F. hypnoides*
  (1892) *F. antipyretica*
  (1894) *F. antipyretica*
  s.n. *F. Duriaei*
Burgess, T. J. W.
  (1881) *F. novae-angliae*
    var. *cymbifolia*
Burlingame, G. W.
  11152 *F. dalecarlica*
Burnett, D. A.
  496 *F. antipyretica*
    var. *gigantea*
  496a *F. antipyretica*
    var. *gigantea*
  497 *F. novae-angliae*
  1768 *D. pallescens*
  2065 *D. pallescens*
  2540 *F. antipyretica*
    var. *gigantea*
  2768 *D. pallescens*
  (1893) *F. novae-angliae*
  (1896) *D. pallescens*
  (1896) *F. antipyretica*
    var. *gigantea*
  (1896) *F. novae-angliae*
  (1897) *D. capillaceum*
  (1897) *D. pallescens*
  s.n. *D. pallescens*
  s.n. *F. antipyretica*
    var. *gigantea*
Burnett, D. A. & Gates
  1704 *F. antipyretica*
    var. *gigantea*
Burnham, S. H.
  (1896) *F. antipyretica*
    var. *gigantea*
  (1915) *F. antipyretica*
    var. *gigantea*
Burrage, Walter L.
  (1883) *F. dalecarlica*
Burrell, W. H.
  (1921) *F. antipyretica*
    var. *gracilis*
Burrell, W. H. & C. A. C.
  (1920) *F. antipyretica*
    var. *gracilis*

Bush, B. F.
  187 *F. missourica*
  1454 *F. novae-angliae*
  5414 *F. flaccida*
  5414a *F. flaccida*
  (1894) *F. hypnoides*
  (1897) *F. disticha*
  (1897) *F. Duriaei*
  (1898) *F. disticha*
  (1898) *F. Duriaei*
  (1899) *F. Duriaei*
  (1899) *F. hypnoides*
  (1899) *F. missourica*
  (1899) *F. novae-angliae*
  (1905) *F. missourica*
Butler, Abigail
  (1919) *F. antipyretica*
    var. *gigantea*
  (1927) *F. antipyretica*
    var. *gigantea*
Byxbee, E. S.
  (1895) *F. antipyretica*

Cahn, A. R.
  720 *F. Duriaei*
  740 *F. Duriaei*
Cain, Roy F.
  (1931) *F. novae-angliae*
  (1932) *F. novae-angliae*
  (1939) *D. falcatum*
  (1939) *F. antipyretica*
    var. *gigantea*
  (1939) *F. dalecarlica*
  (1939) *F. novae-angliae*
  (1939) *F. novae-angliae*
    var. *cymbifolia*
  (1940) *D. falcatum*
  (1940) *F. dalecarlica*
  (1940) *F. novae-angliae*
  (1941) *D. pallescens*
  (1941) *F. antipyretica*
    var. *gigantea*
  (1941) *F. dalecarlica*
  (1941) *F. Duriaei*
  (1941) *F. novae-angliae*
  (1942) *F. Duriaei*
  (1943) *D. capillaceum*
  (1943) *F. Duriaei*
  (1944) *F. antipyretica*
    var. *gigantea*
  (1944) *F. dalecarlica*
  (1944) *F. novae-angliae*
  (1945) *D. capillaceum*
  (1945) *D. falcatum*
  (1945) *D. pallescens*
  (1945) *F. novae-angliae*

(1946) *D. pallescens*
(1946) *F. novae-angliae*
Cain, Stanley A.
   11 *F. novae-angliae*
   48 *F. novae-angliae*
  632 *F. novae-angliae*
Cain, Stanley A. & Sharp, Aaron J.
  254 *F. novae-angliae*
      var. *cymbifolia*
California State Survey
  4603 *F. antipyretica*
Campbell, M. S.
  (1938) *F. antipyretica*
      var. *gigantea*
Camus, F.
  (1878) *F. squamosa*
  (1881) *F. squamosa*
  (1884) *F. antipyretica*
      var. *gracilis*
  (1889) *F. antipyretica*
      var. *gracilis*
  (1890) *F. antipyretica*
  (1890) *F. Duriaei*
  (1891) *F. hypnoides*
  (1892) *F. antipyretica*
  (1892) *F. antipyretica*
      var. *gracilis*
  (1894) *F. antipyretica*
  (1894) *F. Duriaei*
  (1895) *F. antipyretica*
  (1895) *F. squamosa*
  (1896) *F. squamosa*
  (1898) *F. squamosa*
  (1899) *F. antipyretica*
      var. *gigantea*
  (1900) *F. antipyretica*
  (1900) *F. squamosa*
  (1901) *F. antipyretica*
  (1901) *F. squamosa*
  (1904) *F. antipyretica*
  (1910) *F. antipyretica*
      var. *gracilis*
  (1910) *F. squamosa*
Camus, F. & Bureau, E.
  s.n. *F. Duriaei*
Canepsa
  (1867) *F. Duriaei*
Cardot, Jules
  (1883) *F. squamosa*
  (1884) *F. antipyretica*
  (1884) *F. squamosa*
  (1889) *F. squamosa*
  (1917) *F. Duriaei*
Carle, Robert G.
  (1908) *F. antipyretica*
      var. *gigantea*

Carle, Rosie M.
  3584 *F. dalecarlica*
  3585 *F. dalecarlica*
Carlson, G. W. F.
  116 *F. antipyretica*
  (1901) *F. antipyretica*
      var. *gracilis*
Carroll, Gladys
  (1940) *F. novae-angliae*
  (1941) *F. dalecarlica*
  (1941) *F. novae-angliae*
Carroll, I.
  (1851) *F. squamosa*
Carroll, R. P.
  85 *D. capillaceum*
  86 *B. subulatum*
  172 *F. novae-angliae*
Carter, Alice
  (1888) *F. antipyretica*
      var. *gigantea*
Carter, Laura A.
  (1898) *F. antipyretica*
      var. *gigantea*
  (1901) *F. antipyretica*
      var. *gigantea*
  (1901) *F. dalecarlica*
  (1901) *F. novae-angliae*
  (1904) *F. dalecarlica*
  s.n. *D. capillaceum*
  s.n. *F. antipyretica*
      var. *gigantea*
Carter, Laura A. & W. A.
  (1901) *F. dalecarlica*
Carter, W. A.
  (1901) *F. dalecarlica*
Casares-Gil, A.
  24 *F. antipyretica*
  (1902) *F. Duriaei*
  (1909) *F. squamosa*
  s.n. *F. Duriaei*
Caspary, R.
  (1876) *F. dalecarlica*
  (1877) *F. dalecarlica*
  (1880) *F. dalecarlica*
  (1881) *F. hypnoides*
  (1882) *F. hypnoides*
  (1883) *F. hypnoides*
  (1884) *F. dalecarlica*
  s.n. *F. dalecarlica*
Cassebeer, J. H.
  40 *F. antipyretica*
Catcheside, David G.
  4759 *F. antipyretica*
  4760 *F. antipyretica*
  47186 *F. dalecarlica*
  (1924) *F. antipyretica*
  (1924) *F. squamosa*

(1925) *F. antipyretica*
(1926) *F. antipyretica*
(1927) *F. antipyretica*
(1929) *F. antipyretica*
(1929) *F. antipyretica*
    var. *gracilis*
(1929) *F. squamosa*
(1932) *F. antipyretica*
(1940) *F. antipyretica*
(1940) *F. squamosa*
Cedergren, Gösta R.
(1910) *F. antipyretica*
(1916) *F. antipyretica*
    var. *gracilis*
(1920) *F. antipyretica*
Celestin
    1025 *F. novae-angliae*
        var. *cymbifolia*
(1892) *F. Langloisii*
Chamberlain, Edward B.
    160 *D. capillaceum*
    178 *D. capillaceum*
    368 *D. capillaceum*
    3318 *F. dalecarlica*
    3588 *F. antipyretica*
        var. *gigantea*
    3589 *F. novae-angliae*
(1907) *F. antipyretica*
        var. *gigantea*
Chamberlain, Edward B. & Bartram, Edwin B.
(1924) *F. antipyretica*
        var. *gigantea*
Chamberlain, E. B. & Collins, J. F.
    1684 *F. novae-angliae*
Chamberlain, E. B. and Norton, A. H.
    3588 *F. antipyretica*
        var. *gigantea*
Chapman, A. W.
    39 *B. robustum*
    s.n. *B. robustum*
    s.n. *B. subulatum*
    s.n. *F. dalecarlica*
Chapman, F. S.
(1920) *F. dalecarlica*
(1921) *F. novae-angliae*
(1932) *F. antipyretica*
        var. *gigantea*
(1932) *F. dalecarlica*
Chapman, Stuart
    2064 *F. dalecarlica*
(1923) *F. dalecarlica*
Charrier, J.
(1935) *F. squamosa*
Chassagne
(1912) *F. antipyretica*
(1913) *F. antipyretica*

Chatterton, F. W.
(1890) *F. dalecarlica*
Cheetham, C. A.
(1912) *F. antipyretica*
    var. *gracilis*
Cheever, Clarence A.
(1904) *F. antipyretica*
    var. *gigantea*
Chen, P. C.
(1937) *F. antipyretica*
Chenevard, P.
(1911) *F. antipyretica*
Cheney, L. S.
    161 *F. antipyretica*
        var. *gigantea*
    162 *F. antipyretica*
        var. *gigantea*
    196a *F. Duriaei*
    204 *F. antipyretica*
        var. *gigantea*
    205 *F. antipyretica*
        var. *gigantea*
    638 *F. Duriaei*
    639 *F. Duriaei*
    640 *F. Duriaei*
    796 *F. antipyretica*
        var. *gigantea*
    796a *F. Duriaei*
    841 *F. Duriaei*
    875 *F. Duriaei*
    880 *F. Duriaei*
    881 *F. Duriaei*
    966 *F. antipyretica*
        var. *gigantea*
    976 *F. Duriaei*
    1533 *F. antipyretica*
        var. *gigantea*
    1730 *F. Duriaei*
    1853 *F. Duriaei*
    1857 *D. capillaceum*
    1862 *D. pallescens*
    1867 *F. Duriaei*
    1927 *F. antipyretica*
    2015 *F. Duriaei*
    2058 *F. antipyretica*
        var. *gigantea*
    2504 *D. pallescens*
    2505 *F. Duriaei*
    2509 *D. pallescens*
    2554 *D. falcatum*
    2690 *F. novae-angliae*
    2707 *F. antipyretica*
        var. *gigantea*
    2818 *D. pallescens*
    2843 *F. Duriaei*
    2844 *F. dalecarlica*
    2848 *F. dalecarlica*

2849 *F. Duriaei*
2862 *D. falcatum*
2885 *D. fulcatum*
2904 *D. capillaceum*
2916 *F. antipyretica*
   var. *gigantea*
2967 *D. falcatum*
3203 *F. novae-angliae*
3325 *F. dalecarlica*
3375 *F. Duriaei*
3376 *F. antipyretica*
   var. *gigantea*
3477 *F. Duriaei*
3500 *D. capillaceum*
4066 *F. Duriaei*
4835 *F. antipyretica*
5440 *D. falcatum*
5943 p.p. *F. antipyretica*
5943 p.p. *F. antipyretica*
   var. *gigantea*
5943a *F. antipyretica*
   var. *gigantea*
5944 *D. falcatum*
5945 *D. falcatum*
6313 *F. antipyretica*
   var. *gigantea*
6841 *F. antipyretica*
   var. *gigantea*
7346 *F. antipyretica*
   var. *gigantea*
7565 *D. falcatum*
7993 *F. Duriaei*
9595 *F. antipyretica*
   var. *gigantea*
9635 *D. falcatum*
9721 *F. Duriaei*
9798 *F. antipyretica*
   var. *gigantea*
9932 *D. falcatum*
9935 *F. antipyretica*
   var. *gigantea*
9941 *F. antipyretica*
   var. *gigantea*
10273 *F. Duriaei*
10273a *F. Duriaei*
10737 *F. biformis*
10737a *F. biformis*
10893 *F. Duriaei*
10920 *F. Duriaei*
10927 *F. Duriaei*
11133 *F. novae-angliae*
12957 *F. novae-angliae*
13086 *F. Duriaei*
13130 *F. Duriaei*
13143 *F. Duriaei*
13144 *F. Duriaei*
13170 *F. Duriaei*

13171 *F. antipyretica*
   var. *gigantea*
13207a *F. Duriaei*
13207b *F. Duriaei*
13227 *F. antipyretica*
   var. *gigantea*
(1891) *F. antipyretica*
   var. *gigantea*
(1893) *D. capillaceum*
(1893) *F. dalecarlica*
(1893) *F. Duriaei*
(1894) *F. dalecarlica*
(1894) *F. novae-angliae*
(1896) *F. antipyretica*
(1896) *F. Duriaei*
(1912) *D. falcatum*
Chernayev
(1836) *F. antipyretica*
Christ, J.
47 *F. neo-mexicana*
Clapp, Ida D.
(1904) *F. novae-angliae*
s.n. *F. novae-angliae*
Clark, B. Hartwell
(1931) *F. dalecarlica*
(1931) *F. hypnoides*
Clarke, Cora H.
(1900) *D. capillaceum*
(1900) *F. dalecarlica*
(1901) *D. capillaceum*
(1902) *D. capillaceum*
(1905) *F. dalecarlica*
(1907) *F. dalecarlica*
Clarke, S. S.
(1895) *F. dalecarlica*
Clason, I. G.
(1831) *F. dalecarlica*
s.n. *F. dalecarlica*
Clebsch, Alfred
958 *F. novae-angliae*
962 *D. capillaceum*
(1944) *F. dalecarlica*
(1945) *F. dalecarlica*
(1947) *F. novae-angliae*
Cleland, Ralph E. & Taylor, William R.
(1917) *F. antipyretica*
   var. *gigantea*
Cléonique-Joseph, Bro.
5989 *F. antipyretica*
7929 *F. missourica*
8700 *F. antipyretica*
   var. *gigantea*
9223 *D. pallescens*
10176 p.p. *D. capillaceum*
10176 p.p. *D. pallescens*
10177 *F. hypnoides*
10660 *F. dalecarlica*

292 THE FONTINALACEAE

Cleve, P. T.
(1862) *F. antipyretica*
var. *gracilis*
Clinton, G. W.
s.n. *F. antipyretica*
var. *gigantea*
s.n. *F. Duriaei*
s.n. *F. hypnoides*
s.n. *F. novae-angliae*
Clover, J.
(1907) *F. antipyretica*
var. *gigantea*
Cohoe, Edith & Jotter, Mary L.
(1937) *F. neo-mexicana*
Cole, D. L.
(1943) *F. novae-angliae*
Cole, Lilian A.
1017 p.p. *F. antipyretica*
var. *gigantea*
1017 p.p. *F. novae-angliae*
(1918) *F. antipyretica*
var. *gigantea*
Coleman, N.
(1874) *D. capillaceum*
(1875) *D. capillaceum*
Collander, Runar
(1938) *F. antipyretica*
Collin, Otto
(1878) *F. antipyretica*
(1883) *D. falcatum*
Collinder, E.
(1873) *D. falcatum*
Collins, J. Franklin
833 *F. dalecarlica*
885 *F. novae-angliae*
var. *cymbifolia*
1840 *F. dalecarlica*
2368 *F. dalecarlica*
2429 *F. novae-angliae*
2545 *F. dalecarlica*
3163 *F. dalecarlica*
4310 *F. dalecarlica*
5897 *F. flaccida*
5900 *F. novae-angliae*
Collins, Fernald, & Pease
3788 *F. antipyretica*
var. *gigantea*
Commons, A.
(1874) *F. antipyretica*
var. *gigantea*
(1890) *F. flaccida*
(1890) *F. novae-angliae*
(1895) *D. capillaceum*
Compton, Mary
(1883) *F. neo-mexicana*
Conard, Henry S.
22 *F. Duriaei*

26 *F. novae-angliae*
27 *F. Duriaei*
237 *F. novae-angliae*
8–122 *F. hypnoides*
40–1222 *F. novae-angliae*
40–1249 *F. novae-angliae*
40–1263 *F. novae-angliae*
40–1267 *F. novae-angliae*
40–1285 *F. novae-angliae*
42–12 *F. Duriaei*
46–119 *F. antipyretica*
var. *gigantea*
46–122 *F. dalecarlica*
46–127 *F. dalecarlica*
47–5 *F. Duriaei*
47–6 *F. Duriaei*
47–7 *F. Duriaei*
47–9 *F. Duriaei*
47–28 *F. Duriaei*
47–29 *F. Duriaei*
47–30 *F. Duriaei*
47–31 *F. Duriaei*
47–32 *F. Duriaei*
47–52 *F. Duriaei*
47–54 *F. Duriaei*
47–59 *F. Duriaei*
47–62 *F. Duriaei*
47–65 *F. Duriaei*
47–68 *F. Duriaei*
47–153 *F. missourica*
48–271 *F. neo-mexicana*
48–352 *F. patula*
(1924) *F. Duriaei*
(1932) *F. antipyretica*
(1933) *F. novae-angliae*
(1934) *F. novae-angliae*
(1934) *F. novae-angliae*
var. *cymbifolia*
(1935) *F. antipyretica*
(1935) *F. dalecarlica*
(1935) *F. squamosa*
(1937) *F. antipyretica*
var. *gigantea*
(1938) *F. novae-angliae*
(1941) *F. hypnoides*
(1946) *F. dalecarlica*
(1947) *F. antipyretica*
var. *oreganensis*
(1947) *F. Duriaei*
(1949) *B. subulatum*
(1949) *D. capillaceum*
(1950) *B. subulatum*
Congdon, Joseph W.
(1858) *F. novae-angliae*
Conradi, F. E.
(1899) *D. falcatum*
(1899) *F. antipyretica*
var. *gracilis*

(1899) *F. dalecarlica*
Conti, I.
    (1894) *F. antipyretica*
Cook, O. F.
    (1888) *F. Duriaei*
Cooper, William S. & Andrews, Frances E.
    80 *F. antipyretica*
        var. *gigantea*
Coppey, A.
    (1909) *F. antipyretica*
    (1910) *F. antipyretica*
Corbière, L.
    (1886) *F. antipyretica*
    s.n. *F. antipyretica*
Cornet, Arthur
    96 *F. squamosa*
    (1899) *F. antipyretica*
    (1903) *F. antipyretica*
        var. *gigantea*
    (1904) *F. antipyretica*
        var. *gigantea*
    (1904) *F. antipyretica*
        var. *gracilis*
    (1904) *F. squamosa*
    (1910) *F. antipyretica*
    (1921) *F. antipyretica*
        var. *gigantea*
    (1921) *F. antipyretica*
        var. *gracilis*
    (1921) *F. squamosa*
Correll, Don S.
    4439 *F. novae-angliae*
    8658 *F. novae-angliae*
        var. *latifolia*
    04489 *F. novae-angliae*
    11562 *F. filiformis*
    11567 *B. subulatum*
Correll, D. S. & H. B.
    8143 *F. novae-angliae*
        var. *cymbifolia*
    9982 *F. novae-angliae*
    11076 *F. antipyretica*
        var. *gigantea*
Correll, D. S. & McDowell, G. W.
    10722 *F. novae-angliae*
Courbon
    (1846) *F. antipyretica*
Coville, Frederick V. & Applegate, Elmer I.
    1142 *F. Duriaei*
Coville, Frederick V. & Leiberg, John B.
    229 *F. antipyretica*
Cowles, Henry C.
    782 *F. patula*
Craft, James H.
    832 *F. Duriaei*

Croall, A.
    470 *F. antipyretica*
        var. *gracilis*
    569 *F. antipyretica*
    (1849) *F. squamosa*
    (1853) *F. antipyretica*
        var. *gracilis*
Croasdale, Hannah T.
    48 *F. novae-angliae*
    49 *F. antipyretica*
        var. *gigantea*
    (1948) *F. novae-angliae*
Crockett, Alice L.
    (1900) *F. antipyretica*
        var. *gigantea*
    (1900) *F. dalecarlica*
    (1902) *D. capillaceum*
    (1903) *D. pallescens*
Crossland, C. & Needham
    (1904) *F. squamosa*
Crozals, A.
    (1894) *D. capillaceum*
    (1902) *F. antipyretica*
    (1902) *F. Duriaei*
    s.n. *F. antipyretica*
        var. *gigantea*
    s.n. *F. squamosa*
Cuervo, Rev.
    s.n. *F. bogotensis*
Cummings, Clara E.
    (1878) *F. dalecarlica*
    (1879) *F. dalecarlica*
    (1883) *D. capillaceum*
    (1883) *D. pallescens*
    (1884) *F. antipyretica*
        var. *gigantea*
    (1884) *F. dalecarlica*
    (1885) *D. capillaceum*
    (1887) *D. capillaceum*
    s.n. *D. capillaceum*
    s.n. *D. pallescens*
Curnow, W.
    (1865) *F. squamosa*
    (1865) *F. squamosa*
        var. *Curnowii*
Cussac, J.
    (1857) *F. antipyretica*
        var. *gigantea*
Daily, W. A.
    (1940) *F. novae-angliae*
Darker, G. D.
    6692 *F. antipyretica*
        var. *gigantea*
Darlington, H. T.
    346 *F. Duriaei*
Daubenmire, R. F.
    4243 *F. patula*
    4433 *F. neo-mexicana*

4464 *F. neo-mexicana*
45230 *F. antipyretica*
Daugherty, A. M.
  (1921) *F. antipyretica*
    var. *oreganensis*
Davidson, D. W.
  9 *F. dalecarlica*
Davidsson, Olafur
  (1898) *F. antipyretica*
  (1903) *F. antipyretica*
Davis, John
  (1915) *F. Duriaei*
Davy, Joseph B. & Blasdale, Walter C.
  5225 *F. antipyretica*
Dawson
  (1888) *F. Howellii*
Dawson, Ray F.
  (1941) *F. novae-angliae*
Deam, Charles C.
  29527 *F. dalecarlica*
  49117 *F. Duriaei*
  55177 *F. Duriaei*
  57125 *F. Duriaei*
  58343 *F. novae-angliae*
  59827 *F. Duriaei*
Deardon, Ruth
  (1944) *F. novae-angliae*
De Brébisson, L. Alphonse
  51 *F. antipyretica*
  52 *F. squamosa*
  s.n. *F. squamosa*
Degener, Otto
  16920 *F. antipyretica*
De Gruchy, James
  20 *F. novae-angliae*
Delamare, Ernest
  119 *F. novae-angliae*
  (1881) *F. novae-angliae*
  (1886) *F. antipyretica*
    var. *gigantea*
  (1887) *F. novae-angliae*
  (1892) *F. novae-angliae*
  (1894) *F. antipyretica*
    var. *gigantea*
  (1896) *F. novae-angliae*
  s.n. *F. antipyretica*
    var. *gigantea*
  s.n. *F. novae-angliae*
Delestre
  (1845) *F. antipyretica*
Delisle
  s.n. *F. filiformis*
Delogne
  s.n. *F. antipyretica*
    var. *gigantea*
Delogne, C.
  30 *F. squamosa*
  (1868) *F. squamosa*

Demaree, Delzie
  22722 *F. novae-angliae*
    var. *latifolia*
  26541 *F. disticha*
  (1942) *F. novae-angliae*
    var. *latifolia*
Demaret
  1546a *F. antipyretica*
Demetrio, C. H.
  172 *F. missourica*
  (1884) *F. missourica*
  (1886) *F. filiformis*
  (1894) *F. missourica*
  (1896) *F. missourica*
  (1907) *F. missourica*
  (1910) *F. missourica*
  s.n. *F. missourica*
Dennings, Frank
  1091 *F. antipyretica*
  1092 *F. Duriaei*
De Notaris, Giuseppe
  (1838) *F. antipyretica*
  (1839) *D. falcatum*
Deschner, A.
  (1899) *F. antipyretica*
Déséglise, A.
  (1855) *F. antipyretica*
Desmazières, J. B. H. J.
  97 *F. antipyretica*
  533 *F. squamosa*
  1133 *F. squamosa*
Dickson, James
  25 p.p. *F. antipyretica*
    var. *gracilis*
  25 p.p. *F. squamosa*
Dickson, M.
  s.n. *D. capillaceum*
Dietrich, D.
  s.n. *F. antipyretica*
Diettrich-Kolkhoff
  (1911) *F. antipyretica*
Dismier
  (1896) *F. squamosa*
  (1897) *F. antipyretica*
  (1909) *F. squamosa*
  (1922) *F. Duriaei*
  (1927) *F. antipyretica*
  (1930) *F. squamosa*
Dixon, George
  (1852) *F. antipyretica*
  s.n. *F. antipyretica*
Dixon, Hugh N.
  (1888) *F. squamosa*
  (1894) *F. dalecarlica*
  (1895) *F. antipyretica*
  (1895) *F. antipyretica*
    var. *gracilis*

(1896) *F. antipyretica*
(1897) *F. squamosa*
(1901) *F. antipyretica*
(1901) *F. squamosa*
(1905) *F. antipyretica*
    var. *gracilis*
  s.n. *F. antipyretica*
Dobbin, Frank
(1907) *F. novae-angliae*
Doctvovsky, W.
(1908) *F. antipyretica*
    var. *gracilis*
Dodge, Charles K.
(1917) *F. antipyretica*
Donaldson, H. H.
  s.n. *D. capillaceum*
Doty, Maxwell S.
  1221 *F. neo-mexicana*
  2947 *F. Howellii*
  4256 *F. antipyretica*
  5260 *F. antipyretica*
  6625 *F. flaccida*
  6632 *F. flaccida*
Douglas, D.
(1835) *F. neo-mexicana*
Douret, Aug.
(1884) *F. antipyretica*
Drew, W. B.
  7438 *F. Duriaei*
  14738 *F. novae-angliae*
Drexler, R. V.
  120 *F. novae-angliae*
  133 *F. antipyretica*
    var. *gigantea*
  158 *F. novae-angliae*
  188 *D. falcatum*
  376 *F. antipyretica*
    var. *gigantea*
  461 *F. novae-angliae*
  491 *F. novae-angliae*
  655 *F. novae-angliae*
  847 *F. neo-mexicana*
  993 *F. neo-mexicana*
  1667 *F. novae-angliae*
  1706 *D. pallescens*
  3781 *F. antipyretica*
    var. *mollis*
(1939) *F. novae-angliae*
(1948) *F. Duriaei*
Drolet, Madeleine, Raymond, M., &
  Kucyniak, J.
  44–3 *F. novae-angliae*
  44–4 *F. novae-angliae*
Drouet, Francis
  3614 *F. novae-angliae*
Drouet, Francis, Patrick, Ruth, &
  Hodge, Charles
  3599 *F. flaccida*

Drummond, Thomas
  23 *F. disticha*
  33 *D. capillaceum*
  61 *F. disticha*
  62 *B. subulatum*
(1832) *D. capillaceum*
  s.n. *B. subulatum*
  s.n. *D. capillaceum*
  s.n. *D. pallescens*
  s.n. *F. disticha*
Dudley, M. G.
(1938) *F. hypnoides*
Dufour, Léon
(1914) *F. antipyretica*
  s.n. *F. antipyretica*
Duncan, J. B.
(1902) *F. antipyretica*
    var. *gracilis*
(1933) *F. squamosa*
Duncan, W. H.
  3301 *F. dalecarlica*
Dunham, Elizabeth M.
(1905) *F. flaccida*
(1906) *F. dalecarlica*
  s.n. *D. falcatum*
Dunham, H. C.
(1905) *F. Allenii*
Dupret, H.
  46 *F. Duriaei*
  57 *F. dalecarlica*
  118 *F. novae-angliae*
  123 *F. novae-angliae*
  449 *F. dalecarlica*
  1364 *F. novae-angliae*
  1365 p.p. *D. capillaceum*
  1365 p.p. *F. Duriaei*
  1752 *F. novae-angliae*
  1758 *F. Duriaei*
  3705 *F. Duriaei*
  4107 *D. pallescens*
  10470 *F. dalecarlica*
(1905) *F. antipyretica*
(1905) *F. antipyretica*
    var. *gigantea*
(1905) *F. dalecarlica*
(1905) *F. Duriaei*
(1906) *D. capillaceum*
(1906) *D. pallescens*
(1906) *F. antipyretica*
    var. *gigantea*
(1906) *F. Duriaei*
(1906) *F. novae-angliae*
(1906) *F. novae-angliae*
    var. *cymbifolia*
(1907) *D. pallescens*
(1907) *F. novae-angliae*
(1910) *F. dalecarlica*

(1911) *F. novae-angliae*
(1913) *F. Duriaei*
(1915) *F. dalecarlica*
(1915) *F. Duriaei*
(1916) *D. pallescens*
(1924) *D. pallescens*
(1925) *F. dalecarlica*
(1929) *F. novae-angliae*
    var. *cymbifolia*
  s.n. *F. Duriaei*
Durand, E. J.
  12097 *F. dalecarlica*
  12117 *F. novae-angliae*
Durieu de Maisonneuve
  134 *F. squamosa*
  144 *F. squamosa*
  (1835) *F. squamosa*
  (1840) *F. antipyretica*
  (1840) *F. Duriaei*
  s.n. *F. antipyretica*
  s.n. *F. Duriaei*
  s.n. *F. squamosa*
Dusén, K. F.
  (1870) *D. falcatum*
  (1872) *F. antipyretica*
    var. *gracilis*
Dusén, P.
  (1889) *F. antipyretica*
    var. *gracilis*
Dutilly, LePage, & O'Neill
  22110 *F. neo-mexicana*
Dutton, D. Lewis
  324 *F. dalecarlica*
  404 *F. dalecarlica*
  420 *F. novae-angliae*
  439 *F. Duriaei*
  741 *F. Duriaei*
  900 *F. Duriaei*
  965 *F. antipyretica*
    var. *gigantea*
  1307 *F. dalecarlica*
  2061 *F. antipyretica*
    var. *gigantea*
  2173 *F. novae-angliae*
  (1909) *F. dalecarlica*
  (1911) *F. Duriaei*
  (1923) *F. antipyretica*
    var. *gigantea*
DuVall, Robert H.
  (1934) *F. antipyretica*
    var. *mollis*
  (1934) *F. patula*
  (1935) *F. antipyretica*
  (1935) *F. antipyretica*
    var. *mollis*
  (1935) *F. Howellii*
  (1935) *F. neo-mexicana*

Dybeck
  (1832) *D. capillaceum*
  (1833) *D. capillaceum*
  (1833) *D. falcatum*
  (1836) *D. capillaceum*
Dyer
  (1867) *F. antipyretica*

Eastwood, Alice
  (1896) *F. antipyretica*
Eastwood, Sidney K.
  (1935) *F. antipyretica*
    var. *gigantea*
  (1935) *F. dalecarlica*
  (1935) *F. novae-angliae*
Eaton, Daniel C.
  57 *F. novae-angliae*
  (1855) *F. dalecarlica*
  (1855) *F. flaccida*
  (1855) *F. novae-angliae*
  (1856) *D. falcatum*
  (1856) *F. antipyretica*
    var. *gigantea*
  (1856) *F. novae-angliae*
  (1862) *D. capillaceum*
  (1864) *F. antipyretica*
    var. *gigantea*
  (1866) *F. dalecarlica*
  (1869) *F. antipyretica*
    var. *gigantea*
  (1873) *F. antipyretica*
    var. *gigantea*
  (1874) *D. capillaceum*
  (1874) *F. antipyretica*
    var. *gigantea*
  (1874) *F. novae-angliae*
  (1875) *D. capillaceum*
  (1875) *F. novae-angliae*
  (1876) *D. capillaceum*
  (1877) *F. dalecarlica*
  (1879) *D. capillaceum*
  (1884) *D. capillaceum*
  (1884) *F. antipyretica*
    var. *gigantea*
  (1890) *F. dalecarlica*
  (1891) *D. capillaceum*
  s.n. *D. capillaceum*
  s.n. *F. novae-angliae*
Eckfeldt, John W.
  (1868) *D. falcatum*
  (1870) *F. antipyretica*
    var. *gigantea*
  (1878) *D. uncinatum*
  (1879) *F. novae-angliae*
Eckham
  (1873) *F. novae-angliae*
    var. *cymbifolia*

Edwards, William
  (1883) *D. capillaceum*
Eggers
  (1880) *F. neo-mexicana*
Ehlers, J. H.
    136 *F. Duriaei*
    2072 *F. Duriaei*
    2074 *F. Duriaei*
    2075 *F. Duriaei*
    (1931) *F. Duriaei*
    (1937) *F. disticha*
Ehlers, J. & Blinks, L.
  (1920) *F. Duriaei*
Ehrhart, F.
    s.n. *D. falcatum*
Eiben, C. E.
    34 *F. antipyretica*
Ekstrand, E. V.
    s.n. *F. antipyretica*
Ekstrand, G. N.
  (1878) *F. hypnoides*
Eleinke, Aug.
  (1900) *F. antipyretica*
Elftman, Arthur H.
  (1897) *F. antipyretica*
      var. *gigantea*
Elfving
  (1873) *D. falcatum*
Eliás
    12 *F. antipyretica*
Elliott, R.
  (1889) *F. dalecarlica*
Elmer, A. D.
  (1899) *F. neo-mexicana*
Elmqvist, F.
  (1873) *D. falcatum*
  (1874) *F. antipyretica*
  (1879) *D. falcatum*
Emerson, J. T.
  (1904) *D. capillaceum*
Emmitt, Ronald W.
    1060a *F. disticha*
    1613 *F. Duriaei*
Engelmann, Georg
  (1837) *F. flaccida*
    s.n. *F. flaccida*
Engleman, J.
  (1933) *F. novae-angliae*
Enwald, R.
  (1880) *F. antipyretica*
      var. *gracilis*
Eriksson, Jakob
  (1873) *F. antipyretica*
      var. *gigantea*
Erlandsson, Stellan
  (1936) *F. antipyretica*
Ervideira, A.
  (1924) *F. Duriaei*

Evans, Alexander W.
  (1889) *D. capillaceum*
  (1889) *F. antipyretica*
      var. *gigantea*
  (1889) *F. novae-angliae*
  (1890) *F. dalecarlica*
  (1891) *F. dalecarlica*
  (1891) *F. novae-angliae*
  (1907) *F. novae-angliae*
Evans, William
  (1902) *F. antipyretica* .
  (1902) *F. antipyretica*
      var. *gracilis*
Everken
  (1869) *F. squamosa*
Eyerdam, W. J.
    580 *F. patula*
    581 *F. patula*
    582 *F. antipyretica*
    874 *F. neo-mexicana*

Fabius, Bro.
    H–62 *F. Duriaei*
    1033 *F. novae-angliae*
    1055 *F. Duriaei*
    1056 *F. Duriaei*
    1085 *F. novae-angliae*
    1086 *F. novae-angliae*
    1256 *F. antipyretica*
      var. *gigantea*
    1263 *F. antipyretica*
      var. *gigantea*
    1271 *F. novae-angliae*
    1273 *F. novae-angliae*
    1280 *F. novae-angliae*
    1516 *F. dalecarlica*
    1522 *F. novae-angliae*
    1531 *F. novae-angliae*
    1538 *F. novae-angliae*
    1582 *F. dalecarlica*
    1591 *F. dalecarlica*
    1707 *F. antipyretica*
      var. *gigantea*
    1737 *F. novae-angliae*
    1749 *F. novae-angliae*
    1781 *F. novae-angliae*
      var. *cymbifolia*
    1851 *D. capillaceum*
    2262 *F. novae-angliae*
    2266 *F. antipyretica*
      var. *gigantea*
    2268 *F. novae-angliae*
    2334 *F. dalecarlica*
    2342 *F. novae-angliae*
    2467 *F. novae-angliae*
    2601 *F. dalecarlica*
    2612 *F. dalecarlica*
    3125 *F. novae-angliae*

3126 *F. antipyretica*
      var. *gigantea*
3475 *F. novae-angliae*
3477 *F. dalecarlica*
3493 *F. novae-angliae*
3496 *F. novae-angliae*
3499 *F. antipyretica*
      var. *gigantea*
3630 *F. antipyretica*
      var. *gigantea*
Fahr
   s.n. *F. antipyretica*
Falk, Kurt
(1920) *F. antipyretica*
      var. *gracilis*
Familler, Ignaz
(1907) *F. squamosa*
(1911) *F. antipyretica*
Farlow, W. G.
 579 *D. falcatum*
(1882) *F. antipyretica*
      var. *gigantea*
(1882) *F. dalecarlica*
(1897) *D. capillaceum*
(1897) *D. falcatum*
(1899) *D. falcatum*
(1903) *D. capillaceum*
(1903) *F. flaccida*
(1903) *F. novae-angliae*
(1904) *D. capillaceum*
(1904) *F. dalecarlica*
(1905) *D. falcatum*
(1906) *F. novae-angliae*
      var. *cymbifolia*
(1908) *D. capillaceum*
(1911) *D. capillaceum*
Fasset, Norman C.
8770 *F. antipyretica*
8772 *F. antipyretica*
8773 *F. antipyretica*
8775 *F. antipyretica*
Faurie
 374 *F. antipyretica*
      var. *gracilis*
3049 *D. japonicum*
3104 *F. antipyretica*
3203 *F. Duriaei*
8621 *F. antipyretica*
      var. *gracilis*
8623 *F. antipyretica*
      var. *gracilis*
8691 *F. hypnoides*
9073 *F. Duriaei*
9073a *F. Duriaei*
9073b *F. Duriaei*
12305 *F. Duriaei*
12645 *F. Duriaei*

(1892) *F. antipyretica*
      var. *gracilis*
   s.n. *F. Duriaei*
Faxon, C. E.
(1886) *F. disticha*
Faxon, Edwin
 320 *F. antipyretica*
      var. *gigantea*
 321 *F. dalecarlica*
 322 *F. dalecarlica*
 323 *F. dalecarlica*
 325 *F. dalecarlica*
 326 *F. novae-angliae*
 327 *F. Duriaei*
 328 *F. Duriaei*
 329 *F. novae-angliae*
 330 *F. novae-angliae*
 331 *F. novae-angliae*
 332 *F. flaccida*
 334 *F. novae-angliae*
 336 *D. capillaceum*
 337 *D. capillaceum*
 338 *D. capillaceum*
 339 *D. capillaceum*
 473 *F. novae-angliae*
 554 *F. dalecarlica*
(1891) *F. antipyretica*
      var. *gigantea*
(1891) *F. dalecarlica*
(1891) *F. Mac-Millanii*
(1891) *F. novae-angliae*
Feddersen, A.
(1884) *F. antipyretica*
(1886) *F. antipyretica*
Fedtschenko
 899 *F. antipyretica*
3081 *F. antipyretica*
(1869) *F. antipyretica*
(1902) *F. antipyretica*
Ferguson, J.
(1876) *F. antipyretica*
      var. *gracilis*
(1880) *D. falcatum*
(1880) *F. antipyretica*
      var. *gracilis*
   s.n. *F. antipyretica*
Fergusson, I.
(1877) *F. antipyretica*
      var. *gigantea*
Fernald, M. L.
 139 *D. pallescens*
 182 *F. novae-angliae*
Fernald, M. L., Bartram, E. B., &
Long, Bayard
 736 *F. novae-angliae*
Fernald, Sarah
3631 *F. neo-mexicana*

Fhurarlet
  (1912) *F. antipyretica*
      var. *gracilis*
Fioro, Adriano
  (1909) *F. Duriaei*
Fischer, G.
  583d *F. antipyretica*
      var. *gracilis*
  587a *F. antipyretica*
Fisher, A. K.
  (1918) *F. antipyretica*
Fisher, H. L.
  (1896) *F. novae-angliae*
  (1897) *F. novae-angliae*
Fitzgerald
  126 *F. dalecarlica*
  (1880) *F. novae-angliae*
      var. *cymbifolia*
  (1881) *F. novae-angliae*
  s.n. *F. novae-angliae*
Flagey, C.
  (1880) *F. antipyretica*
  s.n. *F. antipyretica*
Fleiden & Berneek
  (1843) *F. antipyretica*
Fleischer, Max
  (1894) *F. antipyretica*
  (1894) *F. Duriaei*
  (1895) *F. antipyretica*
  (1896) *F. antipyretica*
  (1908) *F. antipyretica*
      var. *gracilis*
  (1908) *F. squamosa*
  s.n. *F. Duriaei*
Fleischer & Warnstorf
  s.n. *F. squamosa*
Fleming, A.
  s.n. *F. antipyretica*
Flett, J. B.
  2068 *F. neo-mexicana*
  (1901) *F. antipyretica*
      var. *oreganensis*
  (1902) *F. antipyretica*
  (1903) *F. patula*
  (1914) *F. Howellii*
  (1919) *F. antipyretica*
      var. *oreganensis*
  (1933) *F. Howellii*
Florin, R.
  (1923) *F. antipyretica*
  (1924) *D. falcatum*
  (1927) *F. antipyretica*
  (1929) *F. antipyretica*
  (1931) *D. falcatum*
  (1932) *D. falcatum*
  (1933) *F. antipyretica*
Flowers, Seville
  806 *F. antipyretica*

1705 *D. falcatum*
2139 *D. falcatum*
2323 *F. antipyretica*
2324 *F. hypnoides*
2522 *F. antipyretica*
2524 *F. Duriaei*
2525 *F. Duriaei*
2526 *F. antipyretica*
2527 *F. Duriaei*
2528 *F. Duriaei*
  (1934) *F. novae-angliae*
Fogg, J. M., Jr.
  (1923) *F. dalecarlica*
Ford, E. S. & Redfearn, P. L.
  179 *F. disticha*
Foster, Adriance S.
  150 H *D. uncinatum*
  212 *D. uncinatum*
  398 *D. uncinatum*
  439 *F. neo-mexicana*
  439b *F. antipyretica*
      var. *mollis*
  539 *F. neo-mexicana*
  747 *F. antipyretica*
      var. *oreganensis*
  883 *F. antipyretica*
  883 or 906b *F. antipyretica*
  884 *F. neo-mexicana*
  907b *F. antipyretica*
      var. *oreganensis*
  907c *F. antipyretica*
      var. *oreganensis*
  967 *F. neo-mexicana*
  1045 *F. Howellii*
  1046 *F. antipyretica*
      var. *oreganensis*
  1098 *F. antipyretica*
  1213a p.p. *F. neo-mexicana*
  1213a p.p. *F. patula*
  1295 *F. neo-mexicana*
  1426 *F. antipyretica*
      var. *oreganensis*
  1427 *F. antipyretica*
      var. *oreganensis*
  1428 p.p. *F. antipyretica*
  1428 p.p. *F. antipyretica*
      var. *oreganensis*
  1428 p.p. *F. Howellii*
  1429 *F. antipyretica*
      var. *oreganensis*
  1430 p.p. *D. uncinatum*
  1430 p.p. *F. antipyretica*
  1430 p.p. *F. Howellii*
  1431 *F. Howellii*
  1529 *F. antipyretica*
      var. *oreganensis*
  1821 p.p. *F. antipyretica*
  1821 p.p. *F. Duriaei*

1901 F. neo-mexicana
1921 F. neo-mexicana
1921a F. neo-mexicana
1926 D. uncinatum
1950 D. uncinatum
2171 F. antipyretica
2396 F. neo-mexicana
2684 F. antipyretica
      var. oreganensis
2790 D. uncinatum
2792 F. antipyretica
      var. oreganensis
(1904) D. uncinatum
(1904) F. antipyretica
      var. mollis
(1904) F. neo-mexicana
(1905) D. uncinatum
(1905) F. antipyretica
(1905) F. neo-mexicana
(1906) D. uncinatum
(1907) F. antipyretica
(1907) F. antipyretica
      var. oreganensis
(1907) F. neo-mexicana
(1907) F. patula
(1908) D. uncinatum
(1908) F. antipyretica
(1908) F. antipyretica
      var. oreganensis
(1908) F. neo-mexicana
(1909) F. antipyretica
(1909) F. antipyretica
      var. oreganensis
(1909) F. neo-mexicana
(1909) F. patula
(1910) F. antipyretica
(1910) F. antipyretica
      var. oreganensis
(1910) F. Howellii
(1910) F. neo-mexicana
(1911) D. uncinatum
(1911) F. antipyretica
(1911) F. antipyretica
      var. oreganensis
(1911) F. neo-mexicana
(1912) D. uncinatum
(1912) F. antipyretica
(1912) F. patula
(1914) F. antipyretica
      var. mollis
(1920) F. patula
Foster, C. F.
(1905) F. neo-mexicana
Foster, H.
(1896) F. neo-mexicana
(1896) F. patula
Fowler
   89 F. dalecarlica

Fowler, James
(1865) F. dalecarlica
(1868) F. dalecarlica
(1869) F. antipyretica
      var. gigantea
(1869) F. dalecarlica
(1873) F. antipyretica
(1873) F. dalecarlica
(1878) D. pallescens
(1879) D. capillaceum
(1879) D. pallescens
(1879) F. antipyretica
      var. gigantea
(1890) D. pallescens
(1892) F. dalecarlica
(1901) F. novae-angliae
(1902) F. antipyretica
      var. gigantea
s.n.   F. antipyretica
      var. gigantea
Fox, W. B.
(1939) F. dalecarlica
(1940) F. novae-angliae
Freiberg, Wilhelm
(1909) D. capillaceum
(1909) F. antipyretica
(1909) F. dalecarlica
(1909) F. hypnoides
(1910) F. antipyretica
(1910) F. dalecarlica
(1911) F. dalecarlica
(1911) F. hypnoides
(1912) F. antipyretica
(1912) F. dalecarlica
(1912) F. hypnoides
(1913) F. antipyretica
(1913) F. dalecarlica
(1914) F. dalecarlica
(1915) F. squamosa
(1916) F. squamosa
(1920) D. falcatum
(1933) F. squamosa
s.n.   F. hypnoides
Freye, J. B.
(1904) F. antipyretica
Fries, Th.
(1863) F. dalecarlica
Fries, Th. & Hensehen, R.
(1863) F. dalecarlica
Fristedt & Looin
(1853) F. antipyretica
      var. gracilis
Fritze, R.
(1865) D. falcatum
Frost, C. C.
(1856) F. novae-angliae
(1858) F. antipyretica
(1858) F. dalecarlica

(1858) *F. novae-angliae*
(1859) *F. novae-angliae*
s.n. *F. antipyretica*
var. *gigantea*
s.n. *F. dalecarlica*
s.n. *F. novae-angliae*
Frye, Theodore C.
1029 *D. uncinatum*
3200 *F. Howellii*
3254 *F. neo-mexicana*
(1904) *D. uncinatum*
(1904) *F. antipyretica*
(1904) *F. antipyretica*
var. *oreganensis*
(1904) *F. patula*
(1906) *F. neo-mexicana*
(1907) *F. antipyretica*
(1907) *F. neo-mexicana*
(1907) *F. patula*
(1908) *F. antipyretica*
(1908) *F. antipyretica*
var. *oreganensis*
(1908) *F. neo-mexicana*
(1909) *F. Howellii*
(1909) *F. neo-mexicana*
(1911) *F. patula*
(1914) *F. antipyretica*
var. *mollis*
(1915) *F. Howellii*
(1917) *F. antipyretica*
(1921) *F. antipyretica*
(1923) *F. antipyretica*
(1923) *F. antipyretica*
var. *oreganensis*
(1925) *F. neo-mexicana*
(1925) *F. patula*
(1928) *F. antipyretica*
var. *oreganensis*
(1928) *F. Duriaei*
(1929) *F. antipyretica*
var. *mollis*
(1929) *F. antipyretica*
var. *oreganensis*
(1929) *F. Duriaei*
(1929) *F. Howellii*
(1929) *F. neo-mexicana*
(1929) *F. patula*
(1930) *F. neo-mexicana*
(1931) *F. antipyretica*
var. *mollis*
(1931) *F. neo-mexicana*
(1932) *F. antipyretica*
var. *oreganensis*
(1932) *F. neo-mexicana*
(1932) *F. patula*
(1933) *F. neo-mexicana*
(1934) *F. antipyretica*

(1934) *F. antipyretica*
var. *oreganensis*
(1934) *F. Duriaei*
(1934) *F. neo-mexicana*
(1937) *F. antipyretica*
var. *oreganensis*
(1937) *F. Howellii*
(1937) *F. neo-mexicana*
s.n. *F. neo-mexicana*
Frye, Wilbert
(1933) *F. novae-angliae*
var. *latifolia*
Fuckel
(1894) *F. antipyretica*
var. *gigantea*
Fulford, Margaret
(1933) *F. novae-angliae*
(1936) *F. antipyretica*
var. *gigantea*
(1939) *F. dalecarlica*
(1939) *F. novae-angliae*
Funck, Heinrich C.
76 *F. antipyretica*
77 *F. squamosa*
116 *F. antipyretica*
117 *F. squamosa*
s.n. *F. squamosa*

Gaimard, Paul & Robert
(1835) *F. antipyretica*
Galen, James
s.n. *F. dalecarlica*
Gandoger, M.
(1880) *F. antipyretica*
Garber, A. P.
(1868) *F. antipyretica*
var. *gigantea*
(1868) *F. dalecarlica*
Gardiner, William
(1844) *F. antipyretica*
(1845) *F. antipyretica*
var. *gracilis*
Gardner, G.
(1835) *F. antipyretica*
Gardner, N. L.
26 *F. antipyretica*
27 *F. patula*
28 *F. neo-mexicana*
115 *D. uncinatum*
122 *F. antipyretica*
1741 *F. neo-mexicana*
(1883) *D. uncinatum*
(1883) *F. antipyretica*
var. *oreganensis*
(1897) *D. uncinatum*
(1927) *F. antipyretica*
s.n. *F. antipyretica*

s.n. *F. antipyretica*
    var. *oreganensis*
s.n. *F. neo-mexicana*
s.n. *F. patula*
Garrigus
  (1849) *F. antipyretica*
Gasilien, Bro.
   379 *F. antipyretica*
     var. *gigantea*
  (1901) *F. antipyretica*
     var. *gracilis*
Gattefossé, J.
  (1932) *F. antipyretica*
  (1932) *F. Duriaei*
Gaudefroy, E.
  (1869) *F. antipyretica*
Gauthier, Roger
   416 *F. dalecarlica*
  2146 *F. antipyretica*
     var. *gigantea*
  2162 *F. dalecarlica*
  2163 *F. novae-angliae*
  2220 *F. antipyretica*
  2247 *F. novae-angliae*
 11175 *F. antipyretica*
     var. *gigantea*
 11204 *F. dalecarlica*
 11218 *F. novae-angliae*
 11241 *F. dalecarlica*
 11304 *F. dalecarlica*
 11308 *F. antipyretica*
     var. *gigantea*
 11332 *F. antipyretica*
     var. *gigantea*
 11412 *F. dalecarlica*
 11533 *F. antipyretica*
     var. *gigantea*
 11603 *F. antipyretica*
     var. *gigantea*
Gäumann
  (1907) *F. antipyretica*
Geheeb, Adelbert
  (1871) *F. antipyretica*
     var. *gracilis*
  (1872) *F. antipyretica*
  (1872) *F. antipyretica*
     var. *gracilis*
  (1873) *F. antipyretica*
     var. *gracilis*
  (1875) *F. antipyretica*
     var. *gracilis*
  (1877) *F. antipyretica*
     var. *gracilis*
  (1878) *F. antipyretica*
     var. *gracilis*
  (1886) *F. antipyretica*
     var. *gracilis*

s.n. *F. antipyretica*
s.n. *F. antipyretica*
     var. *gracilis*
Gennari, Patrizio
  (1851) *F. antipyretica*
  (1862) *F. antipyretica*
     var. *gracilis*
Gerard, W. R.
  s.n. *F. novae-angliae*
Gerritson, Walter
  (1903) *D. capillaceum*
  (1906) *F. novae-angliae*
Gibbs, Gertrude
   181 *F. neo-mexicana*
Gibetti, G.
  (1876) *F. Duriaei*
Gier, L. J.
   701a *F. missourica*
  2652 *F. missourica*
  2660 *F. missourica*
  4495 *F. novae-angliae*
     var. *cymbifolia*
Giles, George H.
  (1939) *F. antipyretica*
Gilkey, H. M.
    8 *F. neo-mexicana*
Gillett, J. M.
  2293 *F. Duriaei*
Gillman, Henry
  (1866) *F. Duriaei*
  (1867) *F. antipyretica*
     var. *gigantea*
Giraudias
  (1884) *F. antipyretica*
Girod, Louis-André
   727 *F. antipyretica*
  727b *F. antipyretica*
  (1893) *F. antipyretica*
  (1901) *F. antipyretica*
  (1903) *F. antipyretica*
  (1907) *F. antipyretica*
     var. *gracilis*
Gist, E. B.
   109 *F. novae-angliae*
Githens, Thomas S.
   236 *F. dalecarlica*
   242 *F. novae-angliae*
   244 *F. dalecarlica*
  1050 *F. hypnoides*
Gleason, Henry A., Jr.
  43–1 *D. capillaceum*
  2337 *F. Duriaei*
  2409 *F. antipyretica*
  2448 *F. novae-angliae*
  3077 *F. novae-angliae*
     var. *cymbifolia*
  3204 *F. novae-angliae*

3380 *F. novae-angliae*
47078 *F. novae-angliae*
48022 *F. novae-angliae*
48087 *F. novae-angliae*
(1941) *B. subulatum*
Glowacki
(1897) *F. antipyretica*
Glowenke, Stanley L.
598 *F. novae-angliae*
877 *F. dalecarlica*
907 *F. dalecarlica*
908 *F. novae-angliae*
915 *F. novae-angliae*
941 *F. antipyretica*
var. *gigantea*
Godman, F. D.
s.n. *F. antipyretica*
Goessl, Charles
7507 *F. antipyretica*
Goldmann, A.
(1906) *F. antipyretica*
(1906) *F. antipyretica*
var. *gigantea*
(1907) *F. antipyretica*
(1908) *F. antipyretica*
(1908 & 1909) *F. antipyretica*
(1910) *F. antipyretica*
s.n. *F. antipyretica*
Goodding, Leslie N.
(1903) *F. neo-mexicana*
Gordiagin, A.
16 *F. hypnoides*
74 *F. antipyretica*
var. *gracilis*
Gordon
(1929) *F. novae-angliae*
Gorodkov, B. N.
(1913) *F. antipyretica*
var. *gracilis*
(1913) *F. hypnoides*
(1914) *F. hypnoides*
(1915) *F. Duriaei*
Gorum, Cavin
s.n. *F. antipyretica*
var. *gracilis*
Gräbner
(1893) *F. antipyretica*
Graef, Hugo
(1884) *F. antipyretica*
var. *gigantea*
Graham, Edward H.
7981 *F. antipyretica*
Grant, J. M.
6510 *F. patula*
(1916) *F. antipyretica*
var. *oreganensis*

(1917) *F. antipyretica*
var. *oreganensis*
(1918) *F. antipyretica*
var. *oreganensis*
(1919) *F. antipyretica*
(1920) *F. neo-mexicana*
(1920) *F. patula*
(1922) *F. antipyretica*
(1922) *F. antipyretica*
var. *oreganensis*
(1923) *F. antipyretica*
(1926) *F. neo-mexicana*
(1926) *F. patula*
(1927) *F. neo-mexicana*
(1927) *F. patula*
(1932) *F. antipyretica*
(1932) *F. neo-mexicana*
s.n. *F. patula*
Grape, A.
(1875) *D. falcatum*
(1880) *F. dalecarlica*
(1900) *F. antipyretica*
var. *gracilis*
Grassl, C. O.
7676 *F. Duriaei*
Graves, C. B.
422 *D. capillaceum*
462 *D. capillaceum*
466 *F. novae-angliae*
468 *F. antipyretica*
var. *gigantea*
Graves, J. A.
(1897) *F. novae-angliae*
Gravet, Frédéric
(1868) *F. antipyretica*
(1869) *F. squamosa*
(1871) *F. squamosa*
(1872) *F. antipyretica*
(1874) *F. antipyretica*
(1881) *F. squamosa*
(1886) *F. antipyretica*
(1886) *F. squamosa*
Gray, Fred W.
608b *D. capillaceum*
1053 *F. novae-angliae*
1553 *F. novae-angliae*
4029 *D. capillaceum*
M607 *B. subulatum*
M957 *F. novae-angliae*
M1070 *D. capillaceum*
M1071 *D. capillaceum*
M1072 *D. capillaceum*
M1210 *F. novae-angliae*
M1552 *F. novae-angliae*
M1636 *F. dalecarlica*
M1641 *F. novae-angliae*
M4029 *D. capillaceum*

M4036 *F. novae-angliae*
M4041 *F. novae-angliae*
M4055 *F. novae-angliae*
N.C.M. 292 *D. capillaceum*
(1931) *D. capillaceum*
Grebe, C.
  (1884) *F. hypnoides*
  (1895) *F. antipyretica*
    var. *gracilis*
  (1899) *F. antipyretica*
  (1900) *F. antipyretica*
  (1904) *F. antipyretica*
  s.n. *F. antipyretica*
Green, Traill
  (1857) *F. antipyretica*
    var. *gigantea*
  (1857) *F. dalecarlica*
  (1859) *F. dalecarlica*
  (1863) *F. antipyretica*
    var. *gigantea*
  (1865) *F. disticha*
  s.n. *F. dalecarlica*
Green, William
  (1878) *F. antipyretica*
Greenalch, Wallace
  (1898) *D. falcatum*
  (1898) *F. dalecarlica*
Gresino, P.
  (1922) *F. antipyretica*
  (1924) *F. antipyretica*
  (1925) *F. antipyretica*
  (1926) *F. antipyretica*
  (1929) *F. antipyretica*
Greve, Jan
  (1890) *F. antipyretica*
  (1900) *F. dalecarlica*
Greville
  s.n. *F. antipyretica*
  s.n. *F. squamosa*
Griffin
  (1909) *F. novae-angliae*
Grigoletto
  (1923) *F. antipyretica*
Grønlund, C.
  (1876) *F. antipyretica*
Gross, Catherine
  (1935) *F. antipyretica*
    var. *gigantea*
  (1937) *F. antipyretica*
    var. *gigantea*
  (1937) *F. dalecarlica*
  (1937) *F. novae-angliae*
Gross, H.
  (1921) *F. dalecarlica*
Grosvenor, J. W.
  s.n. *F. novae-angliae*

Grotenfelt
  (1875) *F. antipyretica*
    var. *gigantea*
Grout, A. J.
  99 *F. dalecarlica*
  (1893) *F. antipyretica*
    var. *gigantea*
  (1893) *F. novae-angliae*
  (1896) *F. dalecarlica*
  (1897) *F. novae-angliae*
  (1898) *F. dalecarlica*
  (1899) *F. novae-angliae*
  (1900) *D. capillaceum*
  (1900) *F. dalecarlica*
  (1900) *F. novae-angliae*
  (1900) *F. novae-angliae*
    var. *cymbifolia*
  (1901) *F. antipyretica*
    var. *gigantea*
  (1902) *F. novae-angliae*
  (1902) *F. patula*
  (1903) *F. novae-angliae*
  (1904) *F. dalecarlica*
  (1906) *F. antipyretica*
    var. *gigantea*
  (1906) *F. dalecarlica*
  (1906) *F. novae-angliae*
  (1907) *F. dalecarlica*
  (1909) *F. Duriaei*
  (1909) *F. novae-angliae*
  (1910) *F. novae-angliae*
    var. *cymbifolia*
  (1913) *F. dalecarlica*
  (1913) *F. novae-angliae*
  (1914) *F. antipyretica*
  (1914) *F. novae-angliae*
  (1918) *F. dalecarlica*
  (1918) *F. novae-angliae*
  (1921) *F. dalecarlica*
  (1926) *F. novae-angliae*
  (1927) *F. novae-angliae*
    var. *cymbifolia*
  (1929) *F. antipyretica*
    var. *gigantea*
  (1929) *F. novae-angliae*
  (1931) *F. dalecarlica*
  s.n. *F. novae-angliae*
Guinet, Auguste
  (1890) *F. squamosa*
  (1894) *F. antipyretica*
    var. *gigantea*
  (1904) *F. antipyretica*
    var. *gigantea*
  (1906) *F. antipyretica*
  (1917) *F. antipyretica*

Habeeb, Herbert
  174 *F. antipyretica*
    var. *gigantea*
  275 *F. dalecarlica*
  449 *F. Duriaei*
  449a *F. Duriaei*
  (1944) *F. antipyretica*
    var. *gigantea*
  (1944) *F. Duriaei*
  (1948) *F. novae-angliae*
Habeeb, Herbert & Davidson, D. W.
  13 *D. pallescens*
  (1944) *F. dalecarlica*
Hadley, Sarah B.
  (1904) *D. capillaceum*
  (1907) *F. novae-angliae*
  s.n. *F. novae-angliae*
Hagen, I.
  (1887) *F. Bryhnii*
  (1889) *F. antipyretica*
  (1891) *D. falcatum*
  (1891) *F. antipyretica*
  (1891) *F. dalecarlica*
  (1894) *D. falcatum*
  (1896) *F. dalecarlica*
  (1900) *D. falcatum*
Haines, Mary P.
  (1877) *F. antipyretica*
    var. *gigantea*
Hale
  s.n. *F. flaccida*
Hale, T. J.
  (1861) *F. antipyretica*
  s.n. *F. antipyretica*
    var. *gigantea*
Halin, M.
  (1898) *F. dalecarlica*
  (1899) *F. dalecarlica*
  (1904) *F. dalecarlica*
Hall, C. C.
  41 *F. antipyretica*
Hall, Elihu
  74 *D. uncinatum*
  (1860) *F. hypnoides*
  (1861) *F. hypnoides*
  (1862) *F. neo-mexicana*
  (1870) *F. missourica*
  (1871) *D. uncinatum*
  s.n. *D. uncinatum*
  s.n. *F. antipyretica*
  s.n. *F. neo-mexicana*
  s.n. *F. novae-angliae*
    var. *latifolia*
Hall, E. & Harbour, J. P.
  (1862) *F. neo-mexicana*
Halle, T. G.
  (1925) *F. antipyretica*

(1928) *D. falcatum*
(1928) *F. antipyretica*
(1936) *D. falcatum*
Hallier, J. G.
  (1892) *F. antipyretica*
Hällström, Ed. Y.
  (1905) *F. antipyretica*
    var. *gracilis*
Hamilton, W. P.
  (1888) *F. squamosa*
  (1895) *F. squamosa*
  (1897) *F. antipyretica*
  (1898) *F. antipyretica*
  (1904) *F. antipyretica*
  s.n. *F. antipyretica*
Hamnström, C. O.
  (1873) *F. squamosa*
  (1875) *D. falcatum*
Hampe, Ernst
  (1846) *D. falcatum*
  s.n. *F. squamosa*
Hand, Clarence H.
  466 *F. antipyretica*
    var. *gigantea*
  468 *F. hypnoides*
  469 *F. novae-angliae*
  527 *F. novae-angliae*
  528 *D. capillaceum*
  586 *F. Duriaei*
  594 *F. novae-angliae*
  611 *F. hypnoides*
  612 *D. pallescens*
  617 *F. dalecarlica*
  646 *F. Duriaei*
  649 *F. dalecarlica*
  650 *D. capillaceum*
  738 *F. antipyretica*
    var. *gigantea*
  739 *F. novae-angliae*
  741 *F. novae-angliae*
  742 *F. antipyretica*
    var. *gigantea*
  743 *F. novae-angliae*
  744 *D. falcatum*
  750 *F. Duriaei*
  753 *F. dalecarlica*
  754 *F. dalecarlica*
  755 *F. dalecarlica*
  756 *F. dalecarlica*
  1021 *F. Duriaei*
Handy, Mrs. J. B.
  (1908) *D. capillaceum*
  (1909) *D. capillaceum*
  (1909) *F. novae-angliae*
  (1910) *F. novae-angliae*
Hapeman, H.
  (1892) *F. neo-mexicana*

Haradjian, Manoog
  3916 *F. antipyretica*
Hardin, S.
  (1833) *F. antipyretica*
Hardy, Robert M.
  68 *F. neo-mexicana*
Harger, E. B.
  336 *F. antipyretica*
    var. *gigantea*
Haring, Inez M.
  134 *F. novae-angliae*
    var. *latifolia*
  3259 *F. antipyretica*
  4500 *F. dalecarlica*
  (1936) *F. dalecarlica*
Harper, Edward T. & Susan A.
  (1895) *F. antipyretica*
    var. *gigantea*
  (1895) *F. dalecarlica*
  (1906) *F. antipyretica*
Harper, Roland M.
  104 *B. subulatum*
  1377a *B. robustum*
  1808a *F. dalecarlica*
  1919a *B. robustum*
  2111a *B. robustum*
  2142a *B. robustum*
  2151a *F. filiformis*
  2185b *F. flaccida*
  3065 *B. subulatum*
  (1903) *B. robustum*
  (1905) *F. novae-angliae*
  (1933) *B. subulatum*
Harrington, M. W.
  (1872) *F. antipyretica*
Harris, S. K.
  (1932) *F. antipyretica*
    var. *gigantea*
Harris, Wilson P.
  494 *F. antipyretica*
  877 *F. novae-angliae*
    var. *cymbifolia*
  (1900) *F. antipyretica*
    var. *gigantea*
  (1900) *F. novae-angliae*
Harrison, Arthur K.
  42 *F. antipyretica*
    var. *gigantea*
Harrison, B. F.
  10765 *F. antipyretica*
Harrison, Charles
  (1939) *F. Howellii*
  (1939) *F. neo-mexicana*
Hartman
  (1842) *F. hypnoides*
  (1850) *F. hypnoides*
  (1851) *D. falcatum*

(1859) *D. falcatum*
(1873) *D. falcatum*
(1891) *F. dalecarlica*
(1891) *F. hypnoides*
s.n. *D. falcatum*
s.n. *F. hypnoides*
Hartman, C.
(1852) *F. hypnoides*
(1870) *D. falcatum*
(1872) *F. antipyretica*
(1873) *D. falcatum*
(1874) *D. capillaceum*
s.n. *D. capillaceum*
s.n. *D. falcatum*
s.n. *F. antipyretica*
s.n. *F. dalecarlica*
Hartman, C. & R.
(1842) *F. hypnoides*
(1848) *D. capillaceum*
(1871) *D. capillaceum*
Hartman, C. & Strömbäck, E. A.
(1845) *F. hypnoides*
Hartman, C. J.
(1830) *D. falcatum*
Hartman, R.
(1842) *F. hypnoides*
(1847) *F. hypnoides*
(1852) *F. hypnoides*
(1854) *F. hypnoides*
(1858) *D. falcatum*
(1858) *F. antipyretica*
(1858) *F. dalecarlica*
(1858) *F. hypnoides*
(1859) *D. falcatum*
(1859) *F. hypnoides*
(1870) *D. falcatum*
(1871) *F. hypnoides*
s.n. *F. hypnoides*
Harvey, F. L.
(1890) *F. antipyretica*
    var. *gigantea*
s.n. *D. capillaceum*
s.n. *F. filiformis*
Harvey, W. H.
(1831) *F. antipyretica*
Harvill, A. M.
5144 *F. novae-angliae*
    var. *cymbifolia*
5177 *F. novae-angliae*
6841 *F. flaccida*
7287 *F. novae-angliae*
    var. *cymbifolia*
Harvill, A. M. & Crawford, L. C.
6051 *F. novae-angliae*
Hässler, C. Aron
(1921) *F. antipyretica*
    var. *gracilis*

(1921) *F. squamosa*
Hasslow, Olof J.
(1921) *F. dalecarlica*
(1934) *D. capillaceum*
(1934) *F. antipyretica*
Hatakeyama, Hisashige
(1909) *D. japonicum*
Hatcher, Raymond
113:9 *F. missourica*
Haussknecht, C.
(1869) *F. antipyretica*
Haw
(1891) *F. neo-mexicana*
Hay, G. N.
(1884) *F. antipyretica*
var. *gigantea*
Hazslinszky, F.
s.n. *F. antipyretica*
Hedley, A., Raymond, M., & Kucyniak, J.
45–82 *F. dalecarlica*
Hegelmaier, F.
(1873) *F. Duriaei*
(1878) *F. Duriaei*
Heimerl, A.
73 *F. antipyretica*
(1934) *F. antipyretica*
Heimerl, A. & Conard, H. S.
(1935) *F. antipyretica*
var. *gracilis*
Heldreich, Th. de
38 *F. antipyretica*
var. *Heldreichii*
1000 *F. antipyretica*
var. *Heldreichii*
(1882 & 1883) *F. antipyretica*
var. *Heldreichii*
(1883) *F. antipyretica*
var. *Heldreichii*
(1887) *F. antipyretica*
var. *Heldreichii*
Hellbom, P. J.
(1867) *D. falcatum*
Heller, A. A. & E. Gertrude
3450 *F. antipyretica*
Hellsing, Gustaf
72 *F. antipyretica*
74 *F. antipyretica*
105 *F. squamosa*
107 *D. falcatum*
110 *F. antipyretica*
160 *F. antipyretica*
194 *D. falcatum*
211 *F. squamosa*
267 *D. falcatum*
278 *F. dalecarlica*
279 *F. dalecarlica*

298 *F. dalecarlica*
303 *F. dalecarlica*
306 *F. dalecarlica*
335 *D. falcatum*
370 *F. dalecarlica*
397 *F. antipyretica*
398 *F. dalecarlica*
466 *F. antipyretica*
(1893) *D. falcatum*
(1893) *F. dalecarlica*
(1893) *F. hypnoides*
(1895) *F. antipyretica*
(1895) *F. antipyretica*
var. *gracilis*
(1896) *F. hypnoides*
(1898) *F. dalecarlica*
(1899) *D. falcatum*
(1899) *F. hypnoides*
(1912) *F. antipyretica*
var. *gracilis*
(1918) *D. falcatum*
(1918) *F. dalecarlica*
Henderson, Louis F.
52 *D. uncinatum*
1282 *F. Howellii*
1283 *D. uncinatum*
1741 *F. neo-mexicana*
1742 *F. patula*
1744 *F. Howellii*
1907 *F. chrysophylla*
1915 *D. uncinatum*
2886 *F. neo-mexicana*
3313 *F. antipyretica*
var. *gigantea*
12067 *F. antipyretica*
(1883) *F. antipyretica*
var. *oreganensis*
(1890) *F. chrysophylla*
s.n. *D. uncinatum*
s.n. *F. Howellii*
s.n. *F. neo-mexicana*
s.n. *F. patula*
Henrotay, L. A.
(1863) *F. antipyretica*
Henry, L. K.
(1925) *F. novae-angliae*
(1930) *F. novae-angliae*
Hepner, C. M.
(1932) *F. antipyretica*
var. *gigantea*
(1933) *F. antipyretica*
var. *gigantea*
(1933) *F. novae-angliae*
Hepp
(1816) *F. squamosa*
Héribaud, Bro. R. F.
(1887) *F. antipyretica*

(1888) *F. antipyretica*
(1892) *F. antipyretica*
(1896) *F. antipyretica*
(1902) *F. antipyretica*
(1904) *F. antipyretica*
    var. *gigantea*
(1906) *F. antipyretica*
Héribaud, Biélawski, & Gonod d'Artemare
(1891) *F. antipyretica*
Hermann, F. J.
(1936) *F. Duriaei*
Herzog, Th.
(1906) *F. antipyretica*
    var. *gigantea*
Hespe, Th.
    242a *F. antipyretica*
Hesselbo, August
(1906) *F. hypnoides*
(1908) *F. hypnoides*
(1909) *F. antipyretica*
(1912) *F. antipyretica*
Hill, Albert J.
    101 *F. antipyretica*
        var. *oreganensis*
    789 *F. antipyretica*
(1901) *D. uncinatum*
(1901) *F. patula*
(1903) *F. antipyretica*
(1903) *F. antipyretica*
    var. *oreganensis*
(1904) *F. antipyretica*
(1904) *F. antipyretica*
    var. *oreganensis*
    s.n. *F. antipyretica*
Hill, Ellsworth J.
    216. 1880 *F. neo-mexicana*
    98. 1882 *F. novae-angliae*
    157. 1882 *F. Duriaei*
    158. 1882 *F. dalecarlica*
    123. 1883 *F. dalecarlica*
    92. 1909 *F. novae-angliae*
    9. 1911 *F. missourica*
    19. 1911 *F. missourica*
(1911) *F. missourica*
Hintze, F.
(1911) *F. dalecarlica*
(1912) *F. dalecarlica*
Hise
(1862) *D. falcatum*
Hjärne, Carl E.
(1900) *F. antipyretica*
(1900) *F. antipyretica*
    var. *gracilis*
(1901) *F. antipyretica*
(1928) *F. antipyretica*
(1930) *F. antipyretica*

(1934) *F. antipyretica*
Hjelt, A. H.
(1874) *D. falcatum*
Hjelt, A. H. & Hult, R.
(1877) *D. falcatum*
(1877) *F. dalecarlica*
Hjertman, Ernst
(1912) *F. antipyretica*
Hobson, Dean
(1939) *F. neo-mexicana*
Hobson, E.
    s.n. *F. antipyretica*
Högman, Laura
(1906) *D. falcatum*
Holler
    350 *F. antipyretica*
Holmberg, Otto R.
(1893) *F. antipyretica*
    var. *gigantea*
(1896) *F. antipyretica*
Holmgren, H.
(1867) *D. falcatum*
(1867) *F. squamosa*
(1868) *D. falcatum*
    s.n. *F. antipyretica*
        var. *gracilis*
Holt, W. P.
(1905) *D. falcatum*
Holton, I. F.
(1852) *F. bogotensis*
Holzinger, J. M.
    77 *F. Duriaei*
(1891) *F. novae-angliae*
(1892) *F. novae-angliae*
(1894) *F. hypnoides*
(1894) *F. novae-angliae*
(1896) *F. Duriaei*
(1897) *D. pallescens*
(1897) *F. Duriaei*
(1897) *F. missourica*
(1897) *F. novae-angliae*
(1901) *F. Duriaei*
(1902) *D. falcatum*
(1902) *F. Duriaei*
(1907) *F. dalecarlica*
(1909) *F. dalecarlica*
    s.n. *F. novae-angliae*
Holzinger, J. M. & Blake, J. B.
    42 *F. antipyretica*
        var. *gigantea*
    43 *D. uncinatum*
(1898) *F. antipyretica*
    var. *gigantea*
Holzinger, J. M. & Elftman, A. H.
(1897) *F. antipyretica*
    var. *gigantea*
(1897) *F. hypnoides*

(1897) *F. missourica*
Hommey, J.
(1889) *F. antipyretica*
Hone, Daisy
  225 *F. neo-mexicana*
  (1903) *F. neo-mexicana*
Hood, S. C.
  (1911) *F. flaccida*
  (1912) *F. flaccida*
  s.n. *F. flaccida*
Hopkins, L. S.
  (1909) *F. biformis*
  (1910) *F. biformis*
  (1912) *F. biformis*
Hornemann, Jens Wilken
  s.n. *F. dalecarlica*
Hovgard, Åke
  (1926) *D. falcatum*
  (1929) *F. antipyretica*
      var. *gracilis*
  (1930) *F. antipyretica*
      var. *gracilis*
  (1934) *D. capillaceum*
  (1934) *F. antipyretica*
  (1934) *F. antipyretica*
      var. *gracilis*
  (1935) *D. capillaceum*
  (1935) *F. antipyretica*
  (1936) *F. antipyretica*
  (1936) *F. hypnoides*
  (1938) *F. antipyretica*
Howe, E. C.
  (1865) *D. capillaceum*
  (1866) *D. capillaceum*
  (1866) *F. antipyretica*
      var. *gigantea*
  (1866) *F. novae-angliae*
  s.n. *D. capillaceum*
  s.n. *F. novae-angliae*
Howe, Marshall A.
  103 *F. neo-mexicana*
  103b *F. neo-mexicana*
  (1894) *F. Howellii*
  (1894) *F. neo-mexicana*
Howell, Joseph
  22 *D. uncinatum*
Howell, Thomas
  19 *F. antipyretica*
      var. *oreganensis*
  33 *F. Howellii*
  34 *F. neo-mexicana*
  44/no. 20 *F. Howellii*
  (1880) *D. uncinatum*
  (1882) *F. neo-mexicana*
  (1883) *F. neo-mexicana*
  (1884) *F. Howellii*
  (1885) *D. uncinatum*

(1885) *F. antipyretica*
(1885) *F. patula*
(1886) *D. uncinatum*
(1886) *F. patula*
s.n. *D. uncinatum*
s.n. *F. Howellii*
Howie, C.
  (1870) *F. antipyretica*
      var. *gracilis*
Hübener, J. W. P.
  s.n. *D. falcatum*
Hull, Edwin D.
  (1934) *F. Duriaei*
Hülphers, A.
  (1917) *F. antipyretica*
      var. *gracilis*
  (1918) *F. antipyretica*
      var. *gigantea*
  (1919) *F. antipyretica*
      var. *gracilis*
  (1920) *F. antipyretica*
  (1920) *F. antipyretica*
      var. *gigantea*
  (1921) *F. antipyretica*
      var. *gigantea*
  (1921) *F. antipyretica*
      var. *gracilis*
  (1922) *F. antipyretica*
      var. *gracilis*
  (1923) *F. antipyretica*
  (1925) *F. antipyretica*
      var. *gracilis*
  (1926) *F. antipyretica*
      var. *gracilis*
  (1927) *F. antipyretica*
  (1927) *F. antipyretica*
      var. *gracilis*
Hulst, George D.
  (1898) *F. novae-angliae*
  (1899) *F. dalecarlica*
  (1899) *F. novae-angliae*
      var. *cymbifolia*
Hult, R.
  (1878) *D. capillaceum*
  (1880) *D. falcatum*
  (1880) *F. hypnoides*
Hult, R. & Kihlman, O.
  229 *F. antipyretica*
      var. *gracilis*
  230 *F. dalecarlica*
Hultén, Eric
  5211 *F. neo-mexicana*
Hunt, George E.
  (1803) *F. antipyretica*
  (1865) *F. squamosa*
  (1867) *F. squamosa*
  (1869) *F. antipyretica*

(1869) *F. antipyretica*
var. *gracilis*
s.n. *F. squamosa*
Huntington, J. W.
(1899) *D. capillaceum*
(1899) *F. flaccida*
(1900) *F. flaccida*
(1903) *F. novae-angliae*
(1904) *F. novae-angliae*
(1905) *F. novae-angliae*
(1906) *D. capillaceum*
(1908) *F. novae-angliae*
s.n. *F. flaccida*
Husnot, T.
s.n. *F. antipyretica*
s.n. *F. antipyretica*
var. *gigantea*
Husnot, T. & Ménager
(1891) *F. squamosa*
Hutchinson, Ethel P.
(1946) *F. novae-angliae*
(1948) *F. novae-angliae*
Hy, F.
(1882) *F. hypnoides*
(1884) *F. hypnoides*
(1885) *F. hypnoides*
(1888) *F. hypnoides*
s.n. *F. hypnoides*
Hy & Rechin
(1894) *F. hypnoides*

Ibbatson, H.
(1847) *D. pallescens*
Ikenberry, Gilford J.
(1935) *F. antipyretica*
var. *mollis*
(1936) *F. antipyretica*
var. *gigantea*
(1937) *F. neo-mexicana*
Iltis, Hugh H.
3794 *F. novae-angliae*
3798 *F. novae-angliae*
Indebetou, Conrad
(1875) *F. dalecarlica*
(1877) *D. falcatum*
(1879) *D. falcatum*
(1880) *D. falcatum*
(1880) *F. dalecarlica*
(1882) *F. dalecarlica*
Ingham, W.
(1901) *F. antipyretica*
Ingram, Douglas C.
s.n. *F. antipyretica*

Jaap, O.
(1900) *F. hypnoides*
(1901) *F. hypnoides*

Jäderholm, (Axel) Elof
(1885) *D. falcatum*
(1893) *D. falcatum*
(1893) *F. antipyretica*
var. *gracilis*
(1893) *F. dalecarlica*
(1894) *F. hypnoides*
(1909) *D. falcatum*
(1910) *F. dalecarlica*
(1911) *F. antipyretica*
var. *gracilis*
(1911) *F. dalecarlica*
(1914) *D. falcatum*
(1916) *D. falcatum*
(1920) *F. antipyretica*
var. *gracilis*
Jaeger, A.
(1864) *F. squamosa*
Jakenchi, R.
(1867) *F. hypnoides*
James, Thomas P.
43 *D. falcatum*
59 *F. antipyretica*
59–6 *F. dalecarlica*
60–1 *D. falcatum*
85 *F. dalecarlica*
86 *F. novae-angliae*
87 *F. dalecarlica*
88 *F. dalecarlica*
90 *F. novae-angliae*
91 *F. disticha*
93 *D. capillaceum*
(1850) *D. capillaceum*
(1850) *F. dalecarlica*
(1851) *F. disticha*
(1851) *F. flaccida*
(1852) *D. falcatum*
(1852) *F. antipyretica*
(1852) *F. dalecarlica*
(1853) *D. capillaceum*
(1853) *F. dalecarlica*
(1853) *F. flaccida*
(1855) *D. falcatum*
(1855) *F. antipyretica*
var. *gigantea*
(1856) *F. novae-angliae*
(1857) *D. falcatum*
(1857) *F. antipyretica*
var. *gigantea*
(1858) *F. antipyretica*
var. *gigantea*
(1859) *F. dalecarlica*
(1860) *F. novae-angliae*
(1866) *D. falcatum*
(1869) *D. pallescens*
(1870) *D. capillaceum*
s.n. *D. capillaceum*

s.n. *D. falcatum*
s.n. *D. pallescens*
s.n. *F. antipyretica*
    var. *gigantea*
s.n. *F. dalecarlica*
s.n. *F. flaccida*
s.n. *F. neo-mexicana*
s.n. *F. novae-angliae*
Jao, C. C.
  281 *F. patula*
Jardin, E.
  (1861) *F. flaccida*
  (1865) *F. antipyretica*
  (1887) *F. novae-angliae*
  s.n. *F. novae-angliae*
Jeanpert, Ed.
  (1893) *F. antipyretica*
  (1894) *F. antipyretica*
  (1896) *F. antipyretica*
  (1899) *F. antipyretica*
  (1901) *F. antipyretica*
  (1901) *F. squamosa*
Jennings, Otto E.
  (1909) *F. dalecarlica*
  (1909) *F. Duriaei*
Jennings, Otto E. & Grace K.
  (1909) *F. novae-angliae*
  (1925) *F. dalecarlica*
Jennison, H. M.
  (1936) *F. flaccida*
Jennison & Wilson
  (1934) *F. novae-angliae*
Jensen, C.
  (1884) *D. capillaceum*
  (1884) *F. antipyretica*
  (1896) *F. antipyretica*
  (1896) *F. antipyretica*
    var. *gracilis*
  (1900) *F. antipyretica*
    var. *gigantea*
Jensen, C. & Arnell, H. W.
  (1902) *D. falcatum*
  (1902) *F. antipyretica*
    var. *gracilis*
Jepson, W. L.
  67 *F. antipyretica*
Johanson, C. J.
  (1877) *F. dalecarlica*
Johansson, Harald E.
  (1918) *D. falcatum*
  (1926) *D. capillaceum*
  (1926) *D. falcatum*
  (1927) *D. falcatum*
  (1928) *D. capillaceum*
  (1928) *D. falcatum*
Johansson, K.
  (1900) *F. antipyretica*

Johnson
  (1859) *F. antipyretica*
Johnston, Henry Halcro
  (1912) *F. antipyretica*
Jonas
  534 *F. antipyretica*
Jones, D. A.
  450 *F. squamosa*
  (1902) *F. squamosa*
  (1907) *F. squamosa*
  (1912) *F. squamosa*
  (1913) *F. squamosa*
Jones, D. A., Owen, S. J., & Duncan,
J. B.
  (1900) *F. squamosa*
Jones, D. A. & Rhodes, J.
  (1911) *F. squamosa*
Jones, George N.
  2074 *F. antipyretica*
  (1930) *F. patula*
  (1933) *F. antipyretica*
    var. *oreganensis*
Jones, Marcus E.
  6626 *F. neo-mexicana*
  (1886) *F. Duriaei*
  (1899) *F. neo-mexicana*
Jonsson
  s.n. *F. antipyretica*
Jonsson, Enar
  (1906) *F. antipyretica*
Jörgensen, E.
  (1895) *F. dalecarlica*
  (1902) *F. dalecarlica*
Jungling, Paul
  (1868–1870) *F. antipyretica*

Kaalaas, B.
  (1881) *F. antipyretica*
  (1882) *D. falcatum*
  (1886) *D. falcatum*
  (1886) *F. squamosa*
  (1889) *F. squamosa*
  (1890) *D. falcatum*
  (1892) *D. falcatum*
  (1892) *F. squamosa*
  (1894) *F. antipyretica*
  (1895) *D. falcatum*
  (1895) *F. hypnoides*
  (1896) *F. antipyretica*
  (1898) *F. squamosa*
    var. *Curnowii*
  (1900) *F. antipyretica*
  (1900) *F. squamosa*
  (1902) *F. antipyretica*
  (1903) *D. falcatum*
  (1907) *F. Bryhnii*
  (1909) *F. antipyretica*

(1910) *F. squamosa*
(1911) *F. antipyretica*
(1914) *D. falcatum*
(1916) *F. antipyretica*
Kaiser, George B.
(1910) *F. antipyretica*
var. *gigantea*
(1910) *F. dalecarlica*
(1914) *F. flaccida*
Kalmus
(1906) *F. antipyretica*
var. *gracilis*
(1907) *F. hypnoides*
Karshner & Foster
(1908) *F. antipyretica*
var. *oreganensis*
Kaulfuss, J.
(1891) *F. antipyretica*
Kaurin
(1894) *F. dalecarlica*
Keller, B.
356 *F. antipyretica*
var. *gracilis*
Kelley, A. P.
541 *F. hypnoides*
(1926) *F. antipyretica*
Kendall, W. C., Goldsborough, E. L.,
& Doolittle, A.
(1904) *F. antipyretica*
var. *gigantea*
(1904) *F. dalecarlica*
Kennedy, George G.
(1894) *F. antipyretica*
var. *gigantea*
(1896) *D. capillaceum*
(1898) *F. antipyretica*
var. *gigantea*
(1898) *F. Duriaei*
(1899) *F. filiformis*
(1899) *F. novae-angliae*
var. *cymbifolia*
(1900) *D. pallescens*
(1901) *D. falcatum*
(1905) *D. capillaceum*
(1905) *F. Duriaei*
(1907) *D. capillaceum*
Kern
(1877) *D. falcatum*
(1885) *D. falcatum*
(1886) *D. falcatum*
(1886) *F. antipyretica*
(1886) *F. antipyretica*
var. *gracilis*
(1886) *F. squamosa*
(1888) *F. hypnoides*
Kiaer, F.
40 *D. falcatum*

194 *F. antipyretica*
(1866) *D. falcatum*
(1867) *D. falcatum*
(1869) *D. falcatum*
(1878) *D. falcatum*
(1884) *D. falcatum*
(1884) *F. antipyretica*
(1884) *F. dalecarlica*
(1890) *F. antipyretica*
var. *gracilis*
(1891) *D. falcatum*
Kidston, R.
(1896) *F. antipyretica*
var. *gracilis*
Kiener, Walter
5696 *F. neo-mexicana*
Kienholz, Raymond
(1929) *F. Howellii*
Kihlman, A. O.
520 *F. dalecarlica*
521 *F. dalecarlica*
(1878) *F. hypnoides*
(1881) *F. hypnoides*
(1887) *D. falcatum*
(1887) *F. antipyretica*
(1891) *F. antipyretica*
(1892) *F. antipyretica*
(1892) *F. dalecarlica*
(1897) *F. hypnoides*
Killip, Ellsworth P.
12928 *F. novae-angliae*
Kincaid, T. C. D.
(1898) *F. neo-mexicana*
(1899) *F. neo-mexicana*
Kindberg, N. Conrad
(1853) *D. falcatum*
(1869) *F. hypnoides*
(1878) *F. antipyretica*
var. *gracilis*
(1879) *D. falcatum*
(1897) *D. falcatum*
Kingman, Chester C.
(1908) *F. antipyretica*
var. *gigantea*
(1908) *F. Howellii*
(1908) *F. novae-angliae*
(1909) *F. antipyretica*
var. *gigantea*
(1911) *F. novae-angliae*
var. *cymbifolia*
(1912) *D. capillaceum*
(1912) *F. antipyretica*
var. *gigantea*
(1912) *F. dalecarlica*
(1912) *F. novae-angliae*
Kirillov, E.
(1914) *F. Duriaei*

Labbé
(1945) *F. antipyretica*
Lachenaud, G.
(1898) *F. antipyretica*
(1898) *F. antipyretica*
var. *gracilis*
(1898) *F. squamosa*
(1899) *F. antipyretica*
var. *gigantea*
(1899) *F. squamosa*
(1900) *F. squamosa*
(1901) *F. antipyretica*
var. *gigantea*
(1901) *F. squamosa*
Lackström, E. F.
(1867) *F. dalecarlica*
(1868) *F. antipyretica*
var. *gracilis*
(1868) *F. dalecarlica*
(1871) *D. falcatum*
(1873) *F. dalecarlica*
s.n. *F. antipyretica*
var. *gracilis*
s.n. *F. dalecarlica*
Lacoste, Sande
1842–43 *F. antipyretica*
1842–53 *F. antipyretica*
(1846) *F. antipyretica*
s.n. *F. antipyretica*
Laestadius, C. P.
(1874) *D. falcatum*
Lagergren
(1864) *D. capillaceum*
Lamb, I. Mackenzie
(1946) *F. squamosa*
Lampton
617 *F. antipyretica*
var. *gigantea*
Landberg, J. H.
(1892) *F. neo-mexicana*
Lång, Gosta
(1898) *F. antipyretica*
(1898) *F. hypnoides*
(1899) *F. hypnoides*
Langdon, Fanny E.
18 *F. dalecarlica*
Lange, H.
(1925) *F. squamosa*
Lange, Joh.
(1849) *D. falcatum*
(1884) *F. dalecarlica*
Lange, Z. M. T.
(1854) *F. hypnoides*
Langeron, M. & Sullerot, H.
(1898) *F. antipyretica*

Langlois, A. B.
23c *F. novae-angliae*
var. *cymbifolia*
25 *F. flaccida*
31 *F. novae-angliae*
var. *cymbifolia*
33 *F. novae-angliae*
var. *cymbifolia*
140 *F. flaccida*
262 *F. novae-angliae*
var. *cymbifolia*
518 *F. Langloisii*
648 *F. Langloisii*
745 *F. novae-angliae*
var. *cymbifolia*
745a *F. novae-angliae*
var. *cymbifolia*
745b p.p. *F. novae-angliae*
var. *cymbifolia*
745b p.p. *F. Langloisii*
746 *F. filiformis*
750 *F. filiformis*
751 *F. Langloisii*
752 *F. novae-angliae*
var. *cymbifolia*
753 *F. novae-angliae*
var. *cymbifolia*
754 *F. Langloisii*
755 *F. novae-angliae*
var. *cymbifolia*
756 *F. novae-angliae*
var. *cymbifolia*
757 *F. novae-angliae*
var. *cymbifolia*
758 *F. novae-angliae*
var. *cymbifolia*
814 *F. Langloisii*
815 *F. Langloisii*
816 *F. Langloisii*
816[2] *F. Langloisii*
818 *F. Langloisii*
925 *F. flaccida*
1003 *B. subulatum*
(1882) *F. novae-angliae*
var. *cymbifolia*
(1884) *F. novae-angliae*
var. *cymbifolia*
(1886) *F. flaccida*
(1891) *F. flaccida*
(1891) *F. novae-angliae*
var. *cymbifolia*
(1892) *F. filiformis*
(1892) *F. Langloisii*
(1892) *F. novae-angliae*
var. *cymbifolia*
(1893) *B. subulatum*
(1893) *F. filiformis*

(1893) *F. flaccida*
(1898) *F. filiformis*
  s.n. *F. flaccida*
Lapham, I. A.
  57 *F. biformis*
(1859). *F. biformis*
(1862) *F. antipyretica*
    var. *gigantea*
  s.n. *F. biformis*
Larsen, Earl W. & Mercer, W. H.
(1945) *F. Duriaei*
Larsson, P. A.
(1920) *F. antipyretica*
    var. *gracilis*
(1921) *F. antipyretica*
(1921) *F. dalecarlica*
Latham, Roy
  54 *F. flaccida*
  1736 *F. antipyretica*
    var. *gigantea*
  3717 *D. capillaceum*
  7850 *F. flaccida*
Laurent, Vivi
(1920) *F. antipyretica*
Lawson
(1859) *F. antipyretica*
Lawton, Elva
  1014 *F. novae-angliae*
    var. *latifolia*
(1950) *F. novae-angliae*
Lazarenko, A. S.
(1928) *F. antipyretica*
Le Gallo, C.
  30 *F. dalecarlica*
Lehnert
(1884) *F. antipyretica*
(1884) *F. patula*
  s.n. *D. capillaceum*
  s.n. *F. novae-angliae*
  s.n. *F. squamosa*
Leiberg, J. B.
  81 *D. uncinatum*
  85 *F. patula*
  88 & 114 *F. neo-mexicana*
  114 *F. neo-mexicana*
  114=230 p.p. *F. antipyretica*
  114=230 p.p. *F. antipyretica*
    var. *oreganensis*
  114=230 p.p. *F. Duriaei*
  137 *F. hypnoides*
  137ML *F. hypnoides*
  231 *D. uncinatum*
  256 *F. hypnoides*
(1888) *D. uncinatum*
(1888) *F. hypnoides*
(1890) *D. uncinatum*
(1891) *F. neo-mexicana*

(1892) *D. uncinatum*
  s.n. *F. hypnoides*
Leiner
(1858) *F. antipyretica*
Lemmon, J. G.
(1874) *F. antipyretica*
  s.n. *F. neo-mexicana*
Lenormand, R.
(1841) *F. squamosa*
(1852) *F. squamosa*
  s.n. *F. squamosa*
  s.n. *F. squamosa*
    var. *Curnowii*
Lénström, C. A. E.
(1870) *F. antipyretica*
(1875) *D. falcatum*
Leonard, Emery C.
  1867 *F. novae-angliae*
  2291 *F. novae-angliae*
  17465 *D. capillaceum*
  18142 *F. novae-angliae*
  18143 *F. novae-angliae*
  19738 *F. flaccida*
Leonard, Emery C. & Killip, Ellsworth P.
  845 *F. filiformis*
Leonard, Emery C. & Stewart, Robert E.
  20321 *F. novae-angliae*
  20333 *F. novae-angliae*
Lepage, Ernest
  649 *F. novae-angliae*
  1316 *F. novae-angliae*
  1318 *F. novae-angliae*
  1783 *F. antipyretica*
    var. *gigantea*
  1870 *F. Duriaei*
  1879 *F. Duriaei*
  1882 *F. Duriaei*
  1922 *F. Duriaei*
  2005 *F. hypnoides*
  2239 *F. antipyretica*
    var. *gigantea*
  2251 *F. dalecarlica*
  2253 *F. dalecarlica*
  2254 *F. antipyretica*
    var. *gigantea*
  2260 *F. antipyretica*
    var. *gigantea*
  2778 *F. novae-angliae*
  3008 *F. dalecarlica*
  3232 *F. antipyretica*
    var. *gigantea*
  3233 *F. Duriaei*
  3383 *F. Duriaei*
  3959 *F. Duriaei*
  6194 *F. antipyretica*

9870 F. dalecarlica
Lepage, Ernest & Dutilly, Arthème
    4312 F. dalecarlica
    4366 F. novae-angliae
    4377 F. novae-angliae
    4379 F. dalecarlica
    4445 D. falcatum
    4553 F. dalecarlica
    6199 F. Duriaei
    6434 F. Duriaei
    6721 F. Duriaei
    8956 F. Duriaei
    9974 F. novae-angliae
   12147 F. Duriaei
   12151 F. Duriaei
Leroy, E.
  (1933) F. antipyretica
        var. gigantea
Lesdain (See Bouly de Lesdain.)
Lesquereux, Leo
     42 F. neo-mexicana
  (1857) F. filiformis
  (1872) F. antipyretica
        var. gigantea
  (1883) F. disticha
  s.n. D. capillaceum
  s.n. F. biformis
  s.n. F. filiformis
  s.n. F. flaccida
  s.n. F. neo-mexicana
  s.n. F. novae-angliae
Leth, Th.
  (1870) F. antipyretica
Letterman
  (1908) F. missourica
Levier, E.
     50 F. antipyretica
  (1878) F. antipyretica
  (1879) F. antipyretica
Levisky, A.
  (1909) F. antipyretica
        var. gracilis
Lewis, John F.
  (1927) F. antipyretica
        var. gigantea
  (1927) F. dalecarlica
Libert
  (1814) F. antipyretica
Lickenberger, E.
    246 F. squamosa
Lidforss, Bengt
  (1883) F. antipyretica
        var. gigantea
Liljedahl, A.
  (1920) F. antipyretica
        var. gracilis

Lillefosse, Torkel
  (1911) F. antipyretica
Lillieroth, Carl Gustaf
  (1934) F. antipyretica
  (1934) F. dalecarlica
Limpricht, (K.) Gustav
  (1860) D. falcatum
  (1864) F. squamosa
  (1865) D. falcatum
  (1866) D. falcatum
  (1866) F. squamosa
  (1867) F. squamosa
  (1868) D. falcatum
  (1868) F. antipyretica
  (1870) F. antipyretica
        var. gracilis
  s.n. F. antipyretica
        var. gracilis
Lindberg, Harald
  (1881) F. antipyretica
        var. gracilis
  (1884) D. falcatum
  (1884) F. antipyretica
        var. gracilis
  (1884) F. dalecarlica
  (1891) F. hypnoides
  (1893) D. falcatum
  (1894) D. falcatum
  (1894) F. antipyretica
        var. gracilis
  (1898) F. antipyretica
        var. gigantea
  (1898) F. hypnoides
  (1900) D. falcatum
  (1900) F. antipyretica
  (1900) F. antipyretica
        var. gracilis
  (1900) F. dalecarlica
  (1901) D. falcatum
  (1914) D. capillaceum
Lindberg, Sextus O.
  (1852) F. hypnoides
  (1853) F. hypnoides
  (1854) D. falcatum
  (1854) F. antipyretica
  (1854) F. dalecarlica
  (1854) F. hypnoides
  (1859) D. capillaceum
  (1859) D. falcatum
  (1859) F. dalecarlica
  (1860) D. capillaceum
  (1862) D. capillaceum
  (1862) F. hypnoides
  (1864) D. capillaceum
  (1865) F. antipyretica
        var. gigantea

(1866) *F. antipyretica*
var. *gracilis*
(1867) *F. dalecarlica*
(1868) *D. falcatum*
(1868) *F. antipyretica*
(1868) *F. antipyretica*
var. *gracilis*
(1868) *F. dalecarlica*
(1873) *F. antipyretica*
var. *gracilis*
(1873) *F. dalecarlica*
(1874) *F. antipyretica*
(1877) *D. falcatum*
(1893) *F. hypnoides*
s.n. *D. capillaceum*
s.n. *D. falcatum*
s.n. *F. antipyretica*
var. *gracilis*
s.n. *F. dalecarlica*
s.n. *F. hypnoides*
Lindblom
(1826) *F. dalecarlica*
Linder, Mary
s.n. *F. Duriaei*
Lindgren
s.n. *F. antipyretica*
Lindig, A.
(1860) *F. bogotensis*
Lindman, C.
(1871) *D. falcatum*
Linkola, K.
(1906) *D. falcatum*
(1928) *D. falcatum*
Linnaean Club of University of Minnesota
67 *F. Duriaei*
Linnaniemi, Walter M.
(1907) *D. falcatum*
Lippincott, C. D.
(1898) *F. antipyretica*
var. *gigantea*
Livingstone, L.
(1928) *D. pallescens*
Lochenies, G.
(1886) *F. antipyretica*
var. *gigantea*
Löddeström
(1873) *F. antipyretica*
var. *gracilis*
Loeske
(1892) *F. squamosa*
(1901) *F. antipyretica*
(1901) *F. squamosa*
(1910) *F. antipyretica*
var. *gigantea*
Loew
s.n. *F. neo-mexicana*

Löfvander, K.
(1905) *F. antipyretica*
Loitlesberger, K.
(1906) *F. antipyretica*
Lokolov, Th. W.
(1909) *F. antipyretica*
var. *gracilis*
Loleirol
s.n. *F. antipyretica*
var. *gigantea*
Long, Bayard
121 *F. antipyretica*
var. *gigantea*
143 *F. novae-angliae*
2867 *F. novae-angliae*
var. *latifolia*
Lönnkvist, Fr.
(1880) *F. dalecarlica*
Lörenson, L.
(1918) *F. antipyretica*
Lorenz, Annie
(1902) *F. hypnoides*
(1906) *F. novae-angliae*
(1907) *F. dalecarlica*
(1907) *F. novae-angliae*
(1909) *D. capillaceum*
(1913) *D. capillaceum*
(1913) *D. falcatum*
(1913) *F. flaccida*
(1913) *F. novae-angliae*
Loser
(1860) *F. antipyretica*
Low, A. P.
53 *F. dalecarlica*
Lowe
(1858) *F. antipyretica*
(1863) *F. antipyretica*
Lowe, Rachel L. (Mrs. Frank E.)
(1930) *F. novae-angliae*
(1934) *F. hypnoides*
(1935) *F. Howellii*
(1939) *F. dalecarlica*
(1939) *F. novae-angliae*
(1940) *F. novae-angliae*
(1940) *F. novae-angliae*
var. *latifolia*
Lucien, F.
1292 *F. novae-angliae*
var. *cymbifolia*
Luisier, Alfonso
9 *F. antipyretica*
10 *F. squamosa*
(1907) *F. squamosa*
(1908) *F. Duriaei*
(1909) *F. antipyretica*
(1912) *F. squamosa*

Lundgvist, P. F.
   (1873) *F. antipyretica*
      var. *gracilis*
Lundström, K. A. & Palmén
   (1865) *F. antipyretica*
      var. *gracilis*
   (1865) *F. dalecarlica*
Lundström, K. A. & Palus
   (1865) *F. dalecarlica*
Luse, Th.
   (1903) *F. antipyretica*
   (1906) *F. antipyretica*
   (1911) *F. antipyretica*
Lützow, C.
   (1880) *F. dalecarlica*
   (1881) *D. capillaceum*
   (1881) *F. dalecarlica*
   (1882) *D. capillaceum*
   (1882) *F. dalecarlica*
   (1884) *F. dalecarlica*
   (1885) *F. dalecarlica*
Lyall
   (1858) *D. uncinatum*
   (1858) *F. antipyretica*
   (1858–1859) *D. uncinatum*
   (1858–1859) *F. antipyretica*
   (1858–1859) *F. patula*
   (1861) *D. uncinatum*
   (1861) *F. hypnoides*
   (1861) *F. neo-mexicana*
      s.n. *D. uncinatum*
      s.n. *F. antipyretica*
         var. *oreganensis*
      s.n. *F. neo-mexicana*

MacDougal, D. T.
   210 *F. neo-mexicana*
MacFadden, Mrs. Fay
   727 *F. Howellii*
   1116 *F. neo-mexicana*
   8910 *F. neo-mexicana*
   9793 *F. neo-mexicana*
   17010 *F. antipyretica*
   17400 *F. antipyretica*
   19038 *F. antipyretica*
      var. *mollis*
   19297 *F. antipyretica*
      var. *mollis*
   19298 *F. antipyretica*
   19299 *F. patula*
   (1922) *F. hypnoides*
   (1925) *F. hypnoides*
   (1926) *D. falcatum*
   (1926) *F. antipyretica*
   (1926) *F. Duriaei*
   (1926) *F. Howellii*
   (1926) *F. neo-mexicana*

   (1928) *F. neo-mexicana*
   (1934) *F. antipyretica*
   (1934) *F. neo-mexicana*
   (1940) *F. antipyretica*
      s.n. *F. hypnoides*
      s.n. *F. neo-mexicana*
MacGoron, Philip T.
   (1941) *F. novae-angliae*
      var. *cymbifolia*
MacMillan, Conway
   (1895) *F. Mac-Millanii*
MacMillan, Brand, & Lyon
   (1901) *F. Duriaei*
   (1901) *F. hypnoides*
MacMillan & Lyon
   (1901) *D. falcatum*
   (1901) *F. novae-angliae*
MacMullan
      s.n. *F. dalecarlica*
McAndrew, J.
   (1892) *F. dalecarlica*
   (1898) *F. dalecarlica*
   (1901) *F. squamosa*
   (1904) *F. antipyretica*
McArdle, D.
   (1901) *F. squamosa*
McArdle, D. C.
   (1928) *F. neo-mexicana*
McArdle, R. E.
   (1929) *F. patula*
McCall
   (1925) *F. antipyretica*
McCaskey, Lois
   200 *F. neo-mexicana*
   201 *F. Duriaei*
   202 *D. falcatum*
McCorkle, Marjorie
   (1950) *F. novae-angliae*
McFarlin, James B.
   1694 *B. robustum*
McGaw, W. R.
   453 *F. antipyretica*
      var. *gracilis*
McIntosh, A. C.
   (1926) *F. Duriaei*
      s.n. *F. hypnoides*
McIntosh, C.
   (1902) *F. dalecarlica*
McKay, A. H.
   73 *F. novae-angliae*
      var. *cymbifolia*
Mackaness (See Pennebaker.)
Mackenzie, Elisabeth C.
   102 *F. Howellii*
   154 *F. Howellii*
   341 *F. neo-mexicana*

352 F. antipyretica
    var. mollis
(1930) F. Howellii
Macoun, John
  10 F. neo-mexicana
  26 F. Mac-Millanii
  36 D. uncinatum
  65 D. uncinatum
  71 D. falcatum
  98 F. antipyretica
    var. gigantea
  99 F. Duriaei
 100 F. novae-angliae
 101 F. dalecarlica
 102 F. novae-angliae
 118 F. Duriaei
 119 F. antipyretica
    var. gigantea
 122 F. dalecarlica
 155 F. antipyretica
    var. oreganensis
 163 F. Howellii
 165 F. antipyretica
 175 D. capillaceum
 186a F. antipyretica
    var. oreganensis
 189 F. novae-angliae
 190 F. Duriaei
 190 (= 190a) F. dalecarlica
 191? = 190b F. dalecarlica
 192 F. flaccida
 193 D. falcatum
 194 D. uncinatum
 209 F. antipyretica
    var. mollis
 223 p.p. F. Duriaei
 223 p.p. F. novae-angliae
 224 D. falcatum
 227 D. pallescens
 247 D. pallescens
 256 F. Duriaei
 264 F. Duriaei
 287 D. uncinatum
 288 D. capillaceum
 311 F. Duriaei
 327 D. uncinatum
 341 F. novae-angliae
 553 F. flaccida
 604 F. hypnoides
 687 F. antipyretica
    var. gigantea
 829 F. antipyretica
 832 F. novae-angliae
 874 F. Duriaei
 889 D. uncinatum
 890 D. falcatum
 987 F. Duriaei

2613 F. Duriaei
(1864) F. Duriaei
(1865) F. antipyretica
    var. gigantea
(1865) F. dalecarlica
(1866) F. Duriaei
(1866) F. novae-angliae
(1867) F. antipyretica
(1867) F. dalecarlica
(1868) F. antipyretica
    var. gigantea
(1868) F. dalecarlica
(1868) F. Duriaei
(1868) F. novae-angliae
(1868) F. novae-angliae
    var. cymbifolia
(1869) D. capillaceum
(1869) D. falcatum
(1869) D. uncinatum
(1869) F. flaccida
(1869) F. novae-angliae
(1870) F. antipyretica
    var. gigantea
(1870) F. Duriaei
(1870) F. hypnoides
(1870) F. novae-angliae
(1871) F. Duriaei
(1871) F. hypnoides
(1872) D. uncinatum
(1872) F. dalecarlica
(1872) F. dalecarlica
    var. Macounii
(1872) F. Duriaei
(1872) F. hypnoides
(1872) F. Mac-Millanii
(1872) F. novae-angliae
(1872) F. patula
(1874) D. falcatum
(1874) F. antipyretica
    var. gigantea
(1874) F. dalecarlica
(1875) D. falcatum
(1875) D. uncinatum
(1875) F. antipyretica
(1875) F. antipyretica
    var. gigantea
(1875) F. antipyretica
    var. oreganensis
(1875) F. dalecarlica
(1875) F. dalecarlica
    var. Macounii
(1875) F. Duriaei
(1875) F. Howellii
(1875) F. hypnoides
(1875) F. neo-mexicana
(1875) F. patula
(1876) F. Duriaei

(1878) *F. novae-angliae*
(1879) *F. dalecarlica*
    var. *Macounii*
(1879) *F. Duriaei*
(1880) *F. Duriaei*
(1880) *F. hypnoides*
(1881) *F. Duriaei*
(1883) *F. antipyretica*
(1883) *F. antipyretica*
    var. *gigantea*
(1883) *F. dalecarlica*
(1883) *F. Duriaei*
(1883) *F. flaccida*
(1883) *F. novae-angliae*
(1883) *F. novae-angliae*
    var. *cymbifolia*
(1884) *D. pallescens*
(1884) *F. antipyretica*
(1884) *F. antipyretica*
    var. *gigantea*
(1884) *F. Duriaei*
(1885) *F. antipyretica*
(1885) *F. dalecarlica*
(1885) *F. neo-mexicana*
(1886) *F. flaccida*
(1887) *D. pallescens*
(1887) *F. antipyretica*
(1887) *F. antipyretica*
    var. *oreganensis*
(1887) *F. Howellii*
(1887) *F. neo-mexicana*
(1887) *F. patula*
(1888) *F. antipyretica*
(1888) *F. antipyretica*
    var. *gigantea*
(1889) *D. pallescens*
(1889) *F. dalecarlica*
(1889) *F. Duriaei*
(1889) *F. hypnoides*
(1889) *F. neo-mexicana*
(1889) *F. patula*
(1890) *D. uncinatum*
(1890) *F. antipyretica*
    var. *oreganensis*
(1890) *F. hypnoides*
(1890) *F. neo-mexicana*
(1891) *F. dalecarlica*
(1891) *F. Howellii*
(1893) *F. antipyretica*
(1893) *F. antipyretica*
    var. *gigantea*
(1893) *F. antipyretica*
    var. *mollis*
(1893) *F. antipyretica*
    var. *oreganensis*
(1893) *F. neo-mexicana*
(1893) *F. patula*

(1894) *F. antipyretica*
(1894) *F. antipyretica*
    var. *mollis*
(1895) *F. dalecarlica*
(1895) *F. Duriaei*
(1896) *D. pallescens*
(1896) *F. dalecarlica*
(1896) *F. novae-angliae*
    var. *cymbifolia*
(1897) *F. Duriaei*
(1898) *F. antipyretica*
    var. *gigantea*
(1898) *F. dalecarlica*
(1898) *F. Howellii*
(1898) *F. novae-angliae*
(1899) *F. antipyretica*
    var. *gigantea*
(1899) *F. dalecarlica*
(1899) *F. flaccida*
(1900) *F. antipyretica*
(1900) *F. antipyretica*
    var. *gigantea*
(1900) *F. dalecarlica*
(1900) *F. Duriaei*
(1900) *F. hypnoides*
(1900) *F. novae-angliae*
(1900) *F. novae-angliae*
    var. *cymbifolia*
(1901) *F. antipyretica*
    var. *oreganensis*
(1905) *F. Duriaei*
(1905) *F. Howellii*
(1905) *F. neo-mexicana*
(1906) *F. hypnoides*
(1907) *F. hypnoides*
(1908) *D. uncinatum*
(1908) *F. Howellii*
(1908) *F. neo-mexicana*
(1909) *F. antipyretica*
(1909) *F. antipyretica*
    var. *oreganensis*
(1911) *F. Howellii*
(1912) *F. antipyretica*
    var. *oreganensis*
(1912) *F. Howellii*
(1912) *F. patula*
(1913) *D. uncinatum*
(1916) *F. antipyretica*
s.n. *D. pallescens*
s.n. *F. antipyretica*
    var. *gigantea*
s.n. *F. antipyretica*
    var. *oreganensis*
s.n. *F. dalecarlica*
s.n. *F. Duriaei*
s.n. *F. hypnoides*
s.n. *F. novae-angliae*

s.n. *F. novae-angliae*
    var. *cymbifolia*
s.n. *F. patula*
Madiot, V.
  (1882) *F. antipyretica*
Magnusson, A. H.
  (1908) *F. antipyretica*
    var. *gracilis*
Maguire, Bassett
  414 *F. antipyretica*
Maguire, Bassett & Piranian, George
  5326 *F. antipyretica*
  5330 *F. Duriaei*
  5331 *F. antipyretica*
  5339 *F. antipyretica*
    var. *gigantea*
Maire, R.
  (1848) *F. antipyretica*
  s.n. *F. antipyretica*
Malmberg, A. J.
  (1866) *F. antipyretica*
    var. *gracilis*
  (1866) *F. hypnoides*
Malmström, C.
  (1924) *D. falcatum*
Malts, N.
  (1918) *F. antipyretica*
    var. *gracilis*
Mann, Horace
  (1863) *F. novae-angliae*
  s.n. *F. antipyretica*
    var. *gigantea*
  s.n. *F. novae-angliae*
Mansion, Arthur
  53 *F. squamosa*
  200a *F. squamosa*
  306 *F. antipyretica*
  455 *F. squamosa*
  (1890) *F. antipyretica*
  (1893) *F. antipyretica*
Marcel, Brûlé, P., & Lorenzo, Fr.
  35–198 *F. antipyretica*
    var. *gigantea*
Marcucci
  (1866) *F. antipyretica*
  (1866) *F. Duriaei*
Marie-Anselme, Bro.
  1 *F. dalecarlica*
  11 *F. flaccida*
  63 *F. dalecarlica*
  66 *F. Duriaei*
  112 *D. pallescens*
  131 *F. antipyretica*
    var. *gigantea*
  194 *F. dalecarlica*
  197 *F. dalecarlica*
  204 *F. dalecarlica*

210 *F. dalecarlica*
211 *F. dalecarlica*
222 *F. novae-angliae*
    var. *latifolia*
224 *F. antipyretica*
    var. *gigantea*
385 *D. falcatum*
420 *F. novae-angliae*
427 *F. dalecarlica*
430 *F. novae-angliae*
433 *F. novae-angliae*
500 *D. pallescens*
501 *F. antipyretica*
    var. *gigantea*
503 *F. dalecarlica*
508 *F. dalecarlica*
691 *F. novae-angliae*
    var. *latifolia*
692 *F. dalecarlica*
705 *F. dalecarlica*
706 *D. falcatum*
725 *F. dalecarlica*
726 *F. dalecarlica*
727 *F. flaccida*
742 *F. novae-angliae*
759 *F. novae-angliae*
761 *F. novae-angliae*
    var. *latifolia*
762 *F. dalecarlica*
763 *F. novae-angliae*
764 *F. dalecarlica*
782 *F. novae-angliae*
792 *F. novae-angliae*
793 *D. falcatum*
838 *F. antipyretica*
    var. *gigantea*
839 p.p. *F. antipyretica*
    var. *gigantea*
839 p.p. *F. novae-angliae*
840 *F. antipyretica*
    var. *gigantea*
841 *F. dalecarlica*
842 *F. novae-angliae*
843 p.p. *F. Duriaei*
843 p.p. *F. novae-angliae*
912 *F. Duriaei*
1446 *D. pallescens*
1448 *D. capillaceum*
1582 *F. novae-angliae*
1583 *F. novae-angliae*
1584 *F. novae-angliae*
1585 *F. novae-angliae*
1586 *F. novae-angliae*
1599 *F. novae-angliae*
1624 *F. Duriaei*
1625 *F. Duriaei*
1626 *F. novae-angliae*

1627 F. novae-angliae
1632 F. novae-angliae
1634 D. capillaceum
1638 F. novae-angliae
1643 F. Duriaei
1708 D. falcatum
1758 F. novae-angliae
1789 F. novae-angliae
1816 D. pallescens
1861 D. capillaceum
1908 D. capillaceum
1909 D. pallescens
1910 D. capillaceum
1911 F. novae-angliae
1912 D. pallescens
1913 D. capillaceum
1914 D. capillaceum
3228 F. dalecarlica
3757 F. novae-angliae
3784 F. antipyretica
4026 F. novae-angliae
4034 F. novae-angliae
4073 F. antipyretica
       var. gigantea
4074 F. novae-angliae
4075 F. dalecarlica
4076 F. antipyretica
4186 F. antipyretica
       var. gigantea
(1934) D. pallescens
(1935) D. falcatum
(1935) D. pallescens
(1935) F. dalecarlica
(1936) F. dalecarlica
(1936) F. novae-angliae
(1939) F. antipyretica
       var. gigantea
(1940) F. dalecarlica
Marie-Jean-Eudes, Sister
86 F. novae-angliae
       var. cymbifolia
2986 F. antipyretica
       var. gigantea
3021 F. antipyretica
3021a F. novae-angliae
(1933) F. novae-angliae
Marie-Victorin, Bro.
1 F. Duriaei
2 F. dalecarlica
4 p.p. D. capillaceum
4 p.p. F. antipyretica
       var. gigantea
26 F. dalecarlica
30 F. novae-angliae
30a D. capillaceum
65 F. Duriaei
66 F. Duriaei

1335 F. Duriaei
1740 F. Duriaei
1749 F. antipyretica
       var. gigantea
3571 p.p. D. falcatum
3571 p.p. D. pallescens
3726 D. capillaceum
3751 F. Duriaei
7600 F. novae-angliae
9327 F. antipyretica
       var. gigantea
9328 F. novae-angliae
9329 F. novae-angliae
       var. cymbifolia
9330 F. Duriaei
9342 F. novae-angliae
9371 D. capillaceum
18274 F. novae-angliae
19081 D. falcatum
19086 (14) F. Duriaei
19164 F. Duriaei
19228 F. novae-angliae
       var. cymbifolia
(1906) F. antipyretica
(1912) D. capillaceum
(1913) F. antipyretica
       var. gigantea
(1916) F. novae-angliae
Marie-Victorin & Rolland-Germain,
Bros.
33636 F. novae-angliae
45630 F. antipyretica
       var. gigantea
49574 F. dalecarlica
(1933) F. antipyretica
       var. gigantea
(1940) F. Duriaei
(1943) F. dalecarlica
(1948) F. dalecarlica
Marie-Victorin, Rolland-Germain, &
Boivin, Bernard
10999 F. Duriaei
Marie-Victorin, Rolland-Germain, &
Brunel, Jules B.
45306 F. Duriaei
Marie-Victorin, Rolland-Germain,
Brunel, Jules B., & Rousseau, Z.
18312 F. antipyretica
       var. gigantea
Marie-Victorin & Rousseau, Jacques
50843 D. pallescens
50846 D. capillaceum
Marr, J.
M348 F. antipyretica
Martianov, N.
(1882) F. hypnoides
(1888) F. antipyretica

Martianov, N. M.
  (1894–1898) *F. antipyretica*
Martin, W. R.
  596 *F. novae-angliae*
Mason, H. L.
  (1930) *F. antipyretica*
Matouschek, Franz
  (1902) *F. squamosa*
  s.n. *F. antipyretica*
  s.n. *F. squamosa*
Maxon, William R.
  (1909) *F. novae-angliae*
Medelius, L.
  (1920) *F. antipyretica*
    var. *gracilis*
Medelius, Sigfrid
  (1909) *D. falcatum*
  (1911) *F. antipyretica*
  (1912) *D. capillaceum*
  (1913) *D. capillaceum*
  (1913) *F. antipyretica*
  (1914) *D. capillaceum*
  (1920) *F. antipyretica*
    var. *gigantea*
  (1920) *F. dalecarlica*
Meeker, Grace R.
  (1892) *F. missourica*
  (1893) *F. missourica*
  (1894) *F. missourica*
  (1900) *F. missourica*
Mehner, A.
  (1906) *F. antipyretica*
  (1913) *F. Howellii*
Meldrum, R. H.
  (1887) *F. antipyretica*
  (1903) *F. antipyretica*
Merriam, C. H.
  (1875) *D. capillaceum*
  (1875) *F. antipyretica*
    var. *gigantea*
  (1875) *F. novae-angliae*
Merrill, Elmer D.
  67 *F. dalecarlica*
  68 *F. antipyretica*
    var. *gigantea*
  69 p.p. *F. antipyretica*
    var. *gigantea*
  69 p.p. *F. novae-angliae*
  70 *F. novae-angliae*
  (1897) *F. dalecarlica*
  (1898) *F. antipyretica*
    var. *gigantea*
Merrill, G. K.
  141 *F. novae-angliae*
  170 *F. dalecarlica*
  184 *F. antipyretica*
    var. *gigantea*

  185 *F. novae-angliae*
  186 *F. dalecarlica*
Merrill, H. D.
  117 *D. pallescens*
Metcalf, F. P.
  889 *F. missourica*
  s.n. *F. missourica*
Meyer, F. G.
  80 *F. neo-mexicana*
Michigan Dept. of Conservation
  83 *F. novae-angliae*
  100 *F. novae-angliae*
  106 *F. novae-angliae*
  124 *F. novae-angliae*
    var. *cymbifolia*
  141 *F. flaccida*
  163 *F. novae-angliae*
Migault, Bureau, & Camus
  (1891) *F. hypnoides*
Mikutowicz, Joh.
  15297 *F. hypnoides*
  15299 *F. hypnoides*
  (1906) *F. antipyretica*
    var. *gracilis*
  (1907) *F. antipyretica*
    var. *gigantea*
  (1908) *F. antipyretica*
    var. *gracilis*
Milde, Julius
  (1860) *D. falcatum*
  (1860) *F. squamosa*
  (1866) *D. falcatum*
  (1866) *F. antipyretica*
    var. *gracilis*
  (1868) *D. falcatum*
  s.n. *D. falcatum*
  s.n. *F. antipyretica*
    var. *gracilis*
  s.n. *F. Duriaei*
  s.n. *F. squamosa*
Mille, Ida
  10 *F. antipyretica*
    var. *gigantea*
  137 *F. novae-angliae*
  151 *F. antipyretica*
    var. *gigantea*
  225 *F. novae-angliae*
  320 *F. novae-angliae*
  341 *F. novae-angliae*
    var. *cymbifolia*
  345 *F. antipyretica*
    var. *gigantea*
  394 *F. novae-angliae*
    var. *cymbifolia*
  401 *F. novae-angliae*
  410 *F. novae-angliae*
  429 *F. novae-angliae*

Miller, Mary F.
(1898) *F. antipyretica*
var. *gigantea*
(1900) *F. antipyretica*
var. *gigantea*
(1900) *F. dalecarlica*
Miller, Mary Louise
(1935) *F. antipyretica*
(1935) *F. neo-mexicana*
Millican, H. A.
(1925) *F. Howellii*
Minkevičius, A.
200 *F. antipyretica*
Mitmann, Charles F.
(1936) *F. novae-angliae*
var. *cymbifolia*
Mitten, William
(1867) *F. squamosa*
(1877) *F. antipyretica*
(1877) *F. squamosa*
(1880) *F. antipyretica*
s.n. *F. antipyretica*
Miyabe, Kingo
263 *F. antipyretica*
333 *F. hypnoides*
359 *F. antipyretica*
var. *gracilis*
Mohr, Charles
(1886) *F. novae-angliae*
(1889) *F. novae-angliae*
s.n. *B. subulatum*
s.n. *D. uncinatum*
s.n. *F. neo-mexicana*
Moir, William
(1906) *F. novae-angliae*
Moldenke, Harold N.
9526 *F. antipyretica*
var. *gigantea*
10392 *F. novae-angliae*
Molendo, Ludwig
(1867) *F. antipyretica*
s.n. *F. antipyretica*
Moller, A.
s.n. *F. Duriaei*
Möller, Fabricus
(1863) *F. antipyretica*
Möller, Hjalmar
(1883) *F. antipyretica*
var. *gigantea*
(1885) *F. antipyretica*
var. *gigantea*
(1890) *F. antipyretica*
var. *gigantea*
(1893) *F. antipyretica*
var. *gigantea*
(1897) *F. antipyretica*
var. *gracilis*

(1906) *F. antipyretica*
(1907) *F. antipyretica*
(1908) *D. falcatum*
(1908) *F. antipyretica*
(1909) *D. falcatum*
(1909) *F. antipyretica*
(1909) *F. antipyretica*
var. *gracilis*
(1909) *F. dalecarlica*
(1910) *D. falcatum*
(1910) *F. antipyretica*
(1910) *F. hypnoides*
(1911) *D. falcatum*
(1911) *F. antipyretica*
(1911) *F. antipyretica*
var. *gracilis*
(1912) *D. falcatum*
(1912) *F. antipyretica*
(1912) *F. antipyretica*
var. *gracilis*
(1913) *D. falcatum*
(1913) *F. antipyretica*
(1914) *D. falcatum*
(1914) *F. antipyretica*
(1915) *F. antipyretica*
(1916) *D. falcatum*
(1916) *F. antipyretica*
(1916) *F. antipyretica*
var. *gigantea*
(1917) *F. antipyretica*
var. *gigantea*
(1918) *D. falcatum*
(1919) *D. falcatum*
(1919) *F. antipyretica*
(1919) *F. dalecarlica*
(1920) *F. antipyretica*
var. *gracilis*
(1920) *F. hypnoides*
(1921) *D. falcatum*
(1921) *F. antipyretica*
(1921) *F. antipyretica*
var. *gracilis*
(1921) *F. dalecarlica*
(1921) *F. hypnoides*
(1922) *F. antipyretica*
var. *gracilis*
(1923) *D. falcatum*
(1924) *F. antipyretica*
(1926) *D. falcatum*
s.n. *F. squamosa*
Monguillas
(1888) *F. antipyretica*
var. *gracilis*
Mönkemeyer, Wilhelm
(1902) *F. antipyretica*
(1903) *F. antipyretica*

(1904) *F. antipyretica*
    var. *gigantea*
(1904) *F. antipyretica*
    var. *gracilis*
(1905) *F. antipyretica*
(1906) *F. antipyretica*
(1906) *F. antipyretica*
    var. *gracilis*
(1910) *F. antipyretica*
   s.n. *F. antipyretica*
    var. *gracilis*
Moore, Albert Hanford
(1901) *D. capillaceum*
(1901) *F. antipyretica*
    var. *gigantea*
(1903) *F. antipyretica*
    var. *gigantea*
(1903) *F. novae-angliae*
Moore, George
(1938) *F. novae-angliae*
Moore, W. G.
(1936) *F. antipyretica*
    var. *gigantea*
Moricand, Moise Etienne
   s.n. *F. antipyretica*
Morrison, Adrian & Dawson, Ray F.
(1933) *F. novae-angliae*
    var. *latifolia*
Morsé
   1365 *F. antipyretica*
Moseley, E. S.
(1911) *F. Duriaei*
Mosely
(1909) *F. antipyretica*
Mosén, Hj.
(1864) *F. antipyretica*
    var. *gigantea*
(1868) *D. falcatum*
Moser, J.
(1889) *D. capillaceum*
(1891) *F. antipyretica*
    var. *gigantea*
(1892) *D. pallescens*
(1894) *D. capillaceum*
(1894) *D. pallescens*
   s.n. *D. capillaceum*
Mosier, C. A.
(1892) *F. neo-mexicana*
Mosier, C. F.
(1892) *F. neo-mexicana*
Mosseray, R. & Lebrum, J.
(1933) *F. antipyretica*
Mougeot
(1814) *F. hypnoides*
Mougeot & Nestler
   238 *F. antipyretica*
   430 *F. squamosa*

(1812) *F. antipyretica*
Moul, Edwin T.
   2490 *F. antipyretica*
    var. *gigantea*
   2692 *F. novae-angliae*
   2692a *F. novae-angliae*
   2693 *F. dalecarlica*
   2798 *F. novae-angliae*
    var. *cymbifolia*
   2859 *F. dalecarlica*
   2881 *F. novae-angliae*
   3834 *F. novae-angliae*
   3856 *F. novae-angliae*
   4386 *F. dalecarlica*
   4418 *F. dalecarlica*
   4457 *F. antipyretica*
    var. *gigantea*
   4784 *F. novae-angliae*
   5192 *F. novae-angliae*
   5773 *D. capillaceum*
   5790 *F. dalecarlica*
   5791 *F. novae-angliae*
   5798 *F. novae-angliae*
   5799 *F. novae-angliae*
Moul, E. T. & H. L.
   4840 *F. dalecarlica*
   4881 *F. dalecarlica*
   4999 *F. antipyretica*
    var. *gigantea*
Moulton, E.
(1886) *F. novae-angliae*
Moxley, E. A.
   1213 *F. dalecarlica*
   7096 *F. dalecarlica*
(1925) *F. antipyretica*
(1926) *F. antipyretica*
(1927) *F. hypnoides*
(1928) *F. Duriaei*
(1929) *F. Duriaei*
(1935) *F. Duriaei*
(1936) *F. novae-angliae*
Moyle, J. B.
   2627 *F. Duriaei*
   3330 *F. Duriaei*
   3335 *F. Duriaei*
   3336 *F. Duriaei*
   3586 *F. Duriaei*
   3613 *F. antipyretica*
    var. *gigantea*
   3618 *F. Duriaei*
   3619 *F. antipyretica*
   3620 *F. antipyretica*
   3621 *F. antipyretica*
   3622 *F. novae-angliae*
   3623 *F. antipyretica*
    var. *gigantea*
   3624 *F. hypnoides*

3625 F. antipyretica
3626 F. antipyretica
         var. gigantea
3627 F. antipyretica
         var. gigantea
3628 F. antipyretica
         var. gigantea
3629 F. antipyretica
         var. gigantea
3632 F. antipyretica
         var. gigantea
3636 F. antipyretica
3644 F. antipyretica
3645 F. antipyretica
         var. gigantea
3646 F. antipyretica
         var. gigantea
3647 F. antipyretica
         var. gigantea
Muenscher, W. C.
   (1941) D. capillaceum
   (1941) F. novae-angliae
Muenscher, W. C. & Isely, Duane
   (1941) D. capillaceum
   (1941) F. antipyretica
         var. gigantea
   (1941) F. dalecarlica
   (1941) F. novae-angliae
   (1941) F. novae-angliae
         var. cymbifolia
Mulford, A. Isabel
   (1892) D. uncinatum
Müller, H.
    27 F. antipyretica
   378 F. antipyretica
Müller, J.
   (1857) F. antipyretica
Müller, L.
   (1931) F. antipyretica
   (1931) F. antipyretica
         var. gracilis
Murrill, W. Alphonso
   (1896) F. novae-angliae
   (1938) F. flaccida
Musgrave, Paul R.
   (1932) F. dalecarlica
Myers, Frank J.
   (1941) F. dalecarlica
   (1941) F. novae-angliae
Myers, James
   (1937) F. novae-angliae
Myrin, C. G.
   (1833) D. falcatum

Nakamura, M.
   992 F. antipyretica
         var. gracilis

Nash, G. V.
   (1893) D. capillaceum
Navaschin, L.
   (1888) F. antipyretica
Naveau, R.
   (1904) F. antipyretica
         var. gigantea
   (1908) F. antipyretica
Naylor, J. P.
   (1906) F. biformis
   (1907) F. biformis
Needham
   (1897) F. squamosa
Nelson, Aven
    69 F. Duriaei
    70 F. antipyretica
  2248 F. Duriaei
  2371 p.p. F. antipyretica
  2371 p.p. F. neo-mexicana
  2506 F. neo-mexicana
  2619 F. antipyretica
         var. oreganensis
  6293 F. neo-mexicana
  7540 F. neo-mexicana
  7708 F. antipyretica
  8001 F. antipyretica
  8002 F. hypnoides
  8521 p.p. F. antipyretica
  8521 p.p. F. antipyretica
         var. gigantea
  9669 F. antipyretica
  9670 F. antipyretica
         var. gigantea
  9671 F. hypnoides
   (1897) F. Duriaei
Nelson, Aven & Elias
  6293 F. neo-mexicana
  6320 F. neo-mexicana
Nelson, N. L. T.
  1020 F. missourica
  2132 F. neo-mexicana
  2341 D. capillaceum
  2398 F. patula
  2399b F. novae-angliae
  2523 F. antipyretica
  2661 F. dalecarlica
  2719 D. pallescens
  2769 F. neo-mexicana
  2817 F. novae-angliae
  2828 F. antipyretica
   (1905) F. missourica
Nevius, R. D.
   (1887) F. neo-mexicana
   s.n. F. antipyretica
   s.n. F. neo-mexicana
Newberry, J. S.
   (1883) F. neo-mexicana

Newman, S. M.
  3586 *F. dalecarlica*
  (1905) *F. dalecarlica*
Nichols
  (1852) *D. capillaceum*
Nichols, George E.
  215 *F. novae-angliae*
  459 *F. novae-angliae*
  460 *F. novae-angliae*
  682 *F. antipyretica*
  1428 *F. antipyretica*
  (1903) *F. dalecarlica*
  (1905) *F. antipyretica*
    var. *gigantea*
  (1905) *F. dalecarlica*
  (1906) *D. capillaceum*
  (1906) *F. antipyretica*
    var. *gigantea*
  (1906) *F. dalecarlica*
  (1907) *F. dalecarlica*
  (1907) *F. novae-angliae*
  (1908) *F. antipyretica*
    var. *gigantea*
  (1908) *F. dalecarlica*
  (1908) *F. novae-angliae*
  (1908) *F. novae-angliae*
    var. *latifolia*
  (1909) *F. dalecarlica*
  (1909) *F. novae-angliae*
  (1911) *D. capillaceum*
  (1911) *F. Allenii*
  (1911) *F. antipyretica*
    var. *gigantea*
  (1911) *F. dalecarlica*
  (1911) *F. Duriaei*
  (1911) *F. flaccida*
  (1911) *F. novae-angliae*
  (1911) *F. novae-angliae*
    var. *latifolia*
  (1912) *D. capillaceum*
  (1914) *F. antipyretica*
  (1914) *F. dalecarlica*
  (1914) *F. flaccida*
  (1914) *F. hypnoides*
  (1914) *F. novae-angliae*
  (1914) *F. novae-angliae*
    var. *cymbifolia*
  (1914) *F. patula*
  (1915) *D. capillaceum*
  (1915) *F. antipyretica*
  (1915) *F. antipyretica*
    var. *gigantea*
  (1915) *F. novae-angliae*
  (1915) *F. novae-angliae*
    var. *cymbifolia*
  (1915) *F. patula*
  (1920) *F. Duriaei*

  (1921) *F. Duriaei*
  (1926) *F. Duriaei*
  (1930) *F. Duriaei*
  (1934) *D. falcatum*
  (1934) *D. pallescens*
  (1934) *F. antipyretica*
    var. *gigantea*
  (1937) *D. capillaceum*
  (1937) *D. pallescens*
  (1937) *F. antipyretica*
    var. *gigantea*
  (1937) *F. Duriaei*
  (1937) *F. novae-angliae*
  (1937) *F. patula*
  s.n. *F. antipyretica*
    var. *gigantea*
  s.n. *F. dalecarlica*
  s.n. *F. novae-angliae*
Nichols, George E. & Steere, William C.
  (1935) *D. falcatum*
  (1935) *D. pallescens*
  (1935) *F. antipyretica*
  (1935) *F. antipyretica*
    var. *gigantea*
  (1935) *F. Duriaei*
  (1935) *F. missourica*
Nicholson, William Edward
  (1900) *D. falcatum*
  (1900) *F. antipyretica*
  (1901) *F. antipyretica*
  (1910) *F. antipyretica*
Niclasen, E.
  (1911) *F. antipyretica*
Niklander
  (1855) *D. falcatum*
Noble, James
  (1923) *F. antipyretica*
Nolander, Arvid
  (1903) *F. antipyretica*
Nordénström, H.
  (1885) *F. antipyretica*
Nordstedt, O.
  (1869) *F. antipyretica*
    var. *gracilis*
  (1887) *F. antipyretica*
    var. *gracilis*
  (1888) *F. antipyretica*
    var. *gracilis*
  (1891) *F. antipyretica*
    var. *gracilis*
  (1895) *F. antipyretica*
    var. *gracilis*
  (1898) *F. antipyretica*
    var. *gracilis*
  (1899) *F. antipyretica*

(1899) *F. antipyretica*
  var. *gigantea*
(1900) *F. antipyretica*
s.n. *F. antipyretica*
  var. *gracilis*
Nordström, K. B.
(1915) *F. antipyretica*
  var. *gracilis*
(1915) *F. dalecarlica*
Norman, Jim
(1947) *F. novae-angliae*
Norrlin
(1863) *F. hypnoides*
(1864) *F. hypnoides*
(1868) *F. dalecarlica*
Northrop, Mrs. A. R.
(1902) *F. novae-angliae*
Norton, A. H.
(1907) *F. flaccida*
(1907) *F. novae-angliae*
(1908) *F. novae-angliae*
(1918) *F. novae-angliae*
(1923) *D. pallescens*
(1923) *F. antipyretica*
  var. *gigantea*
(1933) *F. flaccida*
(1937) *F. novae-angliae*
(1937) *F. novae-angliae*
  var. *cymbifolia*
(1940) *F. antipyretica*
  var. *gigantea*
(1940) *F. dalecarlica*
Norton, A. H. & Fanning, J. F.
(1938) *F. dalecarlica*
Norton, J. B. S.
(1940) *F. novae-angliae*
Norton & Haven
(1927) *F. dalecarlica*
Notaris (See de Notaris.)
Novopokrovsky, T. W.
(1908) *F. antipyretica*
  var. *gracilis*
Nowell, J.
(1856) *F. antipyretica*
Nutting, Franklin P.
(1894) *F. antipyretica*
Nyberg, B. Ax.
(1864) *F. antipyretica*
  var. *gracilis*
Nylander, Fr.
(1844) *F. antipyretica*
  var. *gracilis*
Nylander, W.
s.n. *F. dalecarlica*
Nyman, Erik
(1884) *F. antipyretica*
  var. *gracilis*

(1885) *F. antipyretica*
  var. *gracilis*
(1886) *F. antipyretica*
(1888) *F. antipyretica*
(1888) *F. antipyretica*
  var. *gracilis*
(1888) *F. dalecarlica*
(1890) *D. falcatum*
(1891) *D. falcatum*
(1893) *D. falcatum*
(1893) *F. antipyretica*
(1893) *F. dalecarlica*
(1893) *F. squamosa*
(1896) *D. capillaceum*
(1896) *D. falcatum*
(1896) *F. hypnoides*

Oakes
1846 *F. antipyretica*
  var. *gigantea*
s.n. *D. capillaceum*
s.n. *F. antipyretica*
  var. *gigantea*
Oakes, Tuckerman, & James
s.n. *F. antipyretica*
  var. *gigantea*
Oakes, Hellen E.
224 *F. novae-angliae*
(1946) *D. falcatum*
(1946) *F. novae-angliae*
Ohio Field Crew of Conservation Dept.
(1937) *F. missourica*
Ohlsson, P.
s.n. *D. falcatum*
Öhrstedt, G.
(1930) *F. antipyretica*
Okamura, Shutai
805 *F. antipyretica*
959 *F. Duriaei*
960 *F. Duriaei*
Oldberg, Rud.
(1868) *D. falcatum*
(1868) *F. dalecarlica*
(1872) *F. novae-angliae*
(1873) *F. novae-angliae*
(1874) *D. capillaceum*
(1874) *F. novae-angliae*
(1877) *F. filiformis*
s.n. *F. antipyretica*
Olney
s.n. *D. capillaceum*
Olossohn, C.
(1912) *F. squamosa*
Olsson, P.
(1855) *D. falcatum*
(1856) *D. falcatum*
(1856) *F. dalecarlica*

(1857) *F. antipyretica*
(1873) *F. antipyretica*
    var. *gracilis*
(1878) *F. dalecarlica*
Ordway, D. W.
   (1936) *F. novae-angliae*
    var. *latifolia*
Östman, Magnus
   159 *F. antipyretica*
   (1895) *F. antipyretica*
   (1895) *F. squamosa*
   (1904) *D. falcatum*
   (1909) *D. falcatum*
Otis, I. C.
   1403 *F. neo-mexicana*
Ouach
   (1888) *F. antipyretica*
    var. *gigantea*
Over, W. H.
   15457 *F. Duriaei*
Owens, Anna Belle
   (1938) *F. dalecarlica*
   (1948) *F. novae-angliae*

Painter, W. H.
   (1895) *F. squamosa*
   (1902) *F. antipyretica*
    var. *gracilis*
Palisot de Beauvois, A. M. F. J.
   s.n. *B. subulatum*
Palmatier, E. A.
   (1941) *F. novae-angliae*
    var. *latifolia*
Palmén
   (1865) *F. antipyretica*
   (1865) *F. antipyretica*
    var. *gracilis*
   (1865) *F. dalecarlica*
   (1865) *F. hypnoides*
Palmer, E.
   (1855) *F. neo-mexicana*
Palmer, E. J.
   2260 *F. Duriaei*
Palmer, J. E.
   (1915) *F. antipyretica*
Palmgren, O.
   (1917) *F. antipyretica*
    var. *gigantea*
Pammel, L. H.
   256 *F. neo-mexicana*
Parish, Samuel B.
   3425 *F. antipyretica*
Parish, S. B. & W. F.
   1684 *F. antipyretica*
   1685 *F. patula*
Parker, Adella M.
   (1892) *F. neo-mexicana*

Parker, C. F.
   (1864) *F. antipyretica*
    var. *gigantea*
Parker, Dorothy
   (1938) *F. dalecarlica*
Parks, H. E.
   2842 *F. antipyretica*
Parlin, J. C.
   (1932) *F. novae-angliae*
   (1934) *F. dalecarlica*
   (1936) *F. novae-angliae*
   (1938) *D. falcatum*
   (1938) *D. pallescens*
   (1938) *F. antipyretica*
    var. *gigantea*
   (1938) *F. novae-angliae*
   (1939) *F. dalecarlica*
   (1946) *F. antipyretica*
    var. *gigantea*
   (1946) *F. dalecarlica*
Patchin, E.
   (1908) *F. antipyretica*
Patterson, Paul M.
   28 *F. dalecarlica*
   137 *F. novae-angliae*
    var. *cymbifolia*
   170 *F. antipyretica*
    var. *gigantea*
   1179 *F. novae-angliae*
   1700 *F. flaccida*
   1882 *F. flaccida*
   R–215 *F. dalecarlica*
   R–508 *F. novae-angliae*
   R–509 *F. dalecarlica*
   R–581 *F. novae-angliae*
   R–741 *F. novae-angliae*
   (1931) *F. novae-angliae*
   (1931) *F. novae-angliae*
    var. *cymbifolia*
   (1937) *F. novae-angliae*
Paul, Mrs. W.
   (1894) *F. antipyretica*
    var. *gigantea*
Payot, V. & Bernet, H.
   (1884) *F. antipyretica*
Pease, F. N.
   (1875) *F. patula*
Peck, Charles H.
   (1844) *F. novae-angliae*
   (1864) *F. antipyretica*
    var. *gigantea*
   (1864) *F. dalecarlica*
   (1864) *F. novae-angliae*
   (1865) *D. falcatum*
   (1865) *F. dalecarlica*
   (1865) *F. novae-angliae*
   s.n. *F. disticha*

Pedersen, J. P.
　(1888) *F. antipyretica*
Pelvet, A.
　s.n. *F. squamosa*
Penard, E.
　444 *F. neo-mexicana*
Pendleton, George M.
　(1909) *F. antipyretica*
　(1912) *F. neo-mexicana*
　(1913) *F. neo-mexicana*
Penfound, William T.
　25 *F. flaccida*
　37–3 *F. flaccida*
　105 *F. flaccida*
　109 *F. flaccida*
Penfound, William T. & Pennebaker,
　Faith
　(1940) *F. flaccida*
Pennebaker, Faith
　(1939) *B. subulatum*
　(1939) *F. disticha*
　(1940) *B. subulatum*
　(1940) *F. filiformis*
　(1940) *F. novae-angliae*
　(1947) *D. uncinatum*
Persson, Erik
　(1888) *D. falcatum*
Persson, Herman
　(1936) *F. squamosa*
Persson, J.
　(1913) *F. antipyretica*
Pesola, Vilho
　(1915) *D. falcatum*
Péterfi, T. M.
　(1915) *F. squamosa*
　(1916) *F. antipyretica*
Petrak, F.
　(1911) *F. antipyretica*
　(1913) *F. antipyretica*
　　var. *gigantea*
Philibert, Henri
　(1875) *F. squamosa*
　(1877) *F. antipyretica*
　　var. *gracilis*
　(1877) *F. Duriaei*
Phinney, Harry K.
　321 *F. Duriaei*
　339 *D. capillaceum*
　1002 *F. novae-angliae*
Pickard, J. T.
　(1909) *F. antipyretica*
　　var. *gigantea*
Pickett, F. L.
　222 *F. neo-mexicana*
Pier, Amy M.
　(1902) *F. novae-angliae*
　　var. *latifolia*

Pierce, M. E.
　(1936) *F. novae-angliae*
Pilous, Zdeněk
　(1933) *F. antipyretica*
　(1934) *F. antipyretica*
　　var. *gracilis*
　(1947) *F. squamosa*
Piper, Charles V.
　3a *F. antipyretica*
　　var. *oreganensis*
　37 *F. patula*
　82 *F. neo-mexicana*
　83 *F. neo-mexicana*
　103 *D. uncinatum*
　133 *F. antipyretica*
　　var. *oreganensis*
　224 p.p. *F. antipyretica*
　224 p.p. *F. antipyretica*
　　var. *mollis*
　347 *F. neo-mexicana*
　1099 *F. patula*
　(1889) *F. neo-mexicana*
　(1890) *F. neo-mexicana*
　(1891) *D. uncinatum*
　(1892) *D. uncinatum*
　(1892) *F. antipyretica*
　(1892) *F. antipyretica*
　　var. *mollis*
　(1892) *F. antipyretica*
　　var. *oreganensis*
　(1892) *F. neo-mexicana*
　(1895) *F. antipyretica*
Piré, Louis
　(1885) *F. antipyretica*
Pitard, Charles Joseph (Pitard-Briau)
　55 *F. antipyretica*
　56 *F. antipyretica*
　280 *F. Duriaei*
Podpěra, Josef
　(1906) *F. antipyretica*
　　var. *gracilis*
　(1908) *F. antipyretica*
Pontius, Leslie L. & Bartley, Floyd
　36 *F. novae-angliae*
Porter
　s.n. *D. pallescens*
Porter, Cedric L.
　656 *F. Duriaei*
　883 *F. neo-mexicana*
　1072 *F. Duriaei*
　1370 *F. Duriaei*
　1437 *F. antipyretica*
　1704 *F. antipyretica*
Porter, J.
　(1874) *F. dalecarlica*
Porter, Lillian
　(1922) *F. patula*

Porter, Thomas C.
(1852) *F. dalecarlica*
(1858) *F. antipyretica*
var. *gigantea*
(1859) *F. novae-angliae*
(1862) *F. antipyretica*
var. *gigantea*
(1862) *F. dalecarlica*
(1863) *F. Duriaei*
(1867) *F. dalecarlica*
(1873) *F. antipyretica*
s.n. *F. antipyretica*
var. *gigantea*
s.n. *F. dalecarlica*
s.n. *F. Duriaei*
Postian
141 *F. antipyretica*
(1891) *F. antipyretica*
Potter, David
11 *F. dalecarlica*
(1937) *F. antipyretica*
var. *gigantea*
Potter, G. A.
s.n. *F. Howellii*
Povah, A. H.
129 *F. novae-angliae*
Prager, E.
(1886) *F. antipyretica*
(1887) *F. antipyretica*
(1904) *F. squamosa*
(1908) *F. antipyretica*
(1908) *F. squamosa*
Prasch, W.
(1920) *F. antipyretica*
Pratt, C. L.
(1924) *D. capillaceum*
(1924) *D. pallescens*
Prince, A. R.
6187 *F. dalecarlica*
6444 *F. antipyretica*
var. *gigantea*
6458 *F. dalecarlica*
Pringle, C. G.
26c *F. neo-mexicana*
516 *F. antipyretica*
var. *oreganensis*
(1880) *D. pallescens*
(1880) *F. antipyretica*
var. *gigantea*
(1880) *F. dalecarlica*
(1880) *F. Duriaei*
(1880) *F. novae-angliae*
(1881) *D. capillaceum*
(1881) *F. Duriaei*
Progel, A.
(1851) *F. antipyretica*
(1887) *F. squamosa*

Purpus, C. A.
749 *F. antipyretica*
(1891) *F. hypnoides*
(1893) *F. antipyretica*

Rabenhorst, L.
s.n. *F. squamosa*
Rainer, C.
(1874) *F. squamosa*
Ramaley, Francis
907 *F. neo-mexicana*
Ramaley, Francis, Dodds, G. S., &
Robbins, W. W.
2961 *F. neo-mexicana*
Rancken, H.
(1907) *F. antipyretica*
var. *gracilis*
Rand, E. L.
(1898) *F. dalecarlica*
Rand, E. L. & Robinson, B. L.
(1898) *F. dalecarlica*
Rappleye, R. D.
(1948) *F. novae-angliae*
Rau, Eugene A.
132 *D. capillaceum*
134 *F. novae-angliae*
var. *cymbifolia*
135 *F. Duriaei*
137 *F. dalecarlica*
138 *F. antipyretica*
var. *gigantea*
(1872) *F. Allenii*
(1872) *F. antipyretica*
var. *gigantea*
(1873) *F. novae-angliae*
(1873) *F. novae-angliae*
var. *cymbifolia*
(1874) *D. capillaceum*
(1874) *F. novae-angliae*
(1875) *D. capillaceum*
(1875) *F. novae-angliae*
(1877) *D. capillaceum*
(1877) *F. antipyretica*
var. *gigantea*
(1880) *D. pallescens*
(1882) *D. capillaceum*
s.n. *D. capillaceum*
s.n. *F. antipyretica*
var. *gigantea*
s.n. *F. dalecarlica*
s.n. *F. Duriaei*
s.n. *F. novae-angliae*
Rauschenberg, Mathilde
(1861) *F. antipyretica*
Ravain, J. R.
(1882) *F. hypnoides*

Raymond, Marcel & Kucyniak, James
   46–163a *F. novae-angliae*
Rechinger, K. & L.
   s.n. *F. antipyretica*
Redfearn, Paul L.
   2023 *F. filiformis*
Redfield, John N.
   (1889) *F. antipyretica*
         var. *gigantea*
   (1890) *F. dalecarlica*
Regel, A.
   3289 *F. antipyretica*
   3339 *F. antipyretica*
   3352 *F. antipyretica*
   3354 *F. antipyretica*
   3356 *F. antipyretica*
   (1878) *F. antipyretica*
Reif, C.
   (1936) *F. novae-angliae*
   (1938) *F. novae-angliae*
Reineck, E. M.
   (1898) *F. Duriaei*
   (1914) *F. squamosa*
Reinhardt & Fried
   (1862) *F. antipyretica*
Reinsch, Paulus
   (1872) *D. falcatum*
   s.n. *F. antipyretica*
   s.n. *F. squamosa*
Reiss, J. P.
   (1881) *F. antipyretica*
Remy, J.
   (1855) *F. antipyretica*
         var. *mollis*
Renauld, Ferdinand
   (1886) *F. antipyretica*
   (1887) *F. Duriaei*
   s.n. *F. antipyretica*
Renauld, F. & Paillot, J.
   (1873) *F. antipyretica*
Renou, F.
   (1857) *F. antipyretica*
Renqvist, H. W.
   (1875) *F. antipyretica*
Reuter
   (1841) *F. Duriaei*
   (1847) *F. antipyretica*
Rhéole, Bro.
   (1917) *F. Duriaei*
Rhodes, J.
   (1911) *F. squamosa*
Rhodes, P. G. M.
   (1909) *F. antipyretica*
   (1910) *F. squamosa*
Rice, Mabel A.
   (1940) *D. capillaceum*
   (1941) *F. novae-angliae*

(1942) *D. capillaceum*
(1942) *F. novae-angliae*
(1943) *F. antipyretica*
         var. *gigantea*
Richards, Paul M.
   (1947) *F. squamosa*
Richter
   1110 *F. antipyretica*
   s.n. *F. antipyretica*
Riedeman, Robert G.
   337 *F. novae-angliae*
         var. *latifolia*
Riehmer, E.
   (1921) *F. squamosa*
Ritchie, J. C.
   1834 *D. falcatum*
Robert
   (1835) *F. antipyretica*
Roberts, C. M.
   (1924) *F. dalecarlica*
   (1925) *F. Howellii*
   (1925) *F. neo-mexicana*
Robinson, C. B.
   (1903) *D. capillaceum*
Rodeck, H. G.
   (1938) *F. Duriaei*
   (1938) *F. hypnoides*
Roemer, C.
   (1865) *F. hypnoides*
   (1876) *F. squamosa*
   s.n. *F. antipyretica*
   s.n. *F. dalecarlica*
Roeper
   (1829) *F. antipyretica*
Rogers, E.
   (1878) *F. squamosa*
         var. *Curnowii*
   s.n. *F. antipyretica*
         var. *gigantea*
Röll, Julius
   81 *F. antipyretica*
         var. *oreganensis*
   82 *F. Howellii*
   83 *F. patula*
   84 *F. Howellii*
   86 *F. Howellii*
   87 *F. patula*
   89 p.p. *F. Howellii*
   89 p.p. *F. patula*
   90 *D. uncinatum*
   207 *F. Howellii*
   292 *F. antipyretica*
         var. *mollis*
   409 *F. neo-mexicana*
   453 *F. antipyretica*
         var. *oreganensis*

490–491 *F. neo-mexicana*
492 *F. neo-mexicana*
660 *F. neo-mexicana*
661 *F. neo-mexicana*
662 *F. neo-mexicana*
663 *F. neo-mexicana*
665–666 p.p. *F. antipyretica*
          var. *oreganensis*
665–666 p.p. *F. Howellii*
667 *F. Howellii*
668 *F. Howellii*
668a *F. Howellii*
683 *F. antipyretica*
          var. *oreganensis*
821–822 *F. Howellii*
823 *F. Howellii*
917 *F. neo-mexicana*
918 *F. neo-mexicana*
1125 *F. neo-mexicana*
1196 *F. patula*
1197 *F. patula*
1200 *F. patula*
1201 *D. uncinatum*
1202 *D. uncinatum*
1203 *D. uncinatum*
1204 *D. uncinatum*
1242 *F. hypnoides*
1289 *F. neo-mexicana*
1432 *F. hypnoides*
1433 *F. hypnoides*
1434 *F. hypnoides*
1502 *F. hypnoides*
1503 *F. hypnoides*
1529 *F. antipyretica*
1530 *D. uncinatum*
1554 *F. hypnoides*
1582 *F. hypnoides*
1583 *F. hypnoides*
(1882) *F. antipyretica*
(1886) *F. antipyretica*
          var. *gigantea*
(1888) *D. uncinatum*
(1888) *F. antipyretica*
          var. *mollis*
(1888) *F. antipyretica*
          var. *oreganensis*
(1888) *F. Howellii*
(1888) *F. neo-mexicana*
(1888) *F. patula*
Rome, Jacques
(1886) *F. antipyretica*
Römer, C.
(1879) *F. antipyretica*
          var. *gracilis*
(1880) *F. antipyretica*
          var. *gracilis*

Rood, Almon
(1935) *F. dalecarlica*
Rose, Frank H. & Forbes, Maude
4274 *F. patula*
Rose, Lewis S.
36282 *F. antipyretica*
Rosendahl, C. O. & Dahlberg, R. C.
(1918) *F. antipyretica*
          var. *gigantea*
Roth, Georg
(1882) *F. antipyretica*
(1883) *F. antipyretica*
(1884) *F. squamosa*
(1889) *F. antipyretica*
          var. *gracilis*
(1891) *F. antipyretica*
(1902) *F. antipyretica*
(1903) *F. antipyretica*
(1905) *F. antipyretica*
          var. *gigantea*
(1907) *F. antipyretica*
(1908) *F. antipyretica*
s.n. *F. antipyretica*
s.n. *F. antipyretica*
          var. *gracilis*
Roussel
(1817) *F. antipyretica*
Routien, John B.
829 *F. novae-angliae*
Rowell
(1938) *F. antipyretica*
          var. *gigantea*
Roy
54 *F. Duriaei*
Roy, Mrs. Jessie
s.n. *D. uncinatum*
Roze & Bescherelle
93 *F. antipyretica*
Rubenson, Alb.
(1902) *F. dalecarlica*
Rudd, John
(1935) *F. patula*
Ruiz, Hipolito & Pavón, José
s.n. *F. Howellii*
Russell, Colton
54 *F. novae-angliae*
55 *F. novae-angliae*
Rust, Henry J.
1095 *D. uncinatum*
1096 *F. neo-mexicana*
1102 *D. uncinatum*
Ruthe, R.
(1870) *F. hypnoides*
(1872) *F. antipyretica*
(1872) *F. hypnoides*
(1873) *F. antipyretica*
(1873) *F. hypnoides*
(1874) *F. antipyretica*

(1874) *F. hypnoides*
(1875) *F. hypnoides*
(1876) *F. antipyretica*
(1876) *F. hypnoides*
(1879) *F. antipyretica*
(1879) *F. hypnoides*
(1887) *F. hypnoides*
  s.n. *F. hypnoides*
Ryan, E.
(1886) *D. falcatum*
(1886) *F. antipyretica*
(1887) *F. antipyretica*
(1888) *D. falcatum*
(1888) *F. antipyretica*
(1888) *F. hypnoides*
(1894) *D. falcatum*
(1896) *F. antipyretica*
  var. *gracilis*
Rydberg, P. A. & Bessey, Ernst A.
(1897) *F. neo-mexicana*

Sahlberg, John
(1869) *D. falcatum*
(1869) *F. dalecarlica*
(1870) *F. antipyretica*
  var. *gracilis*
(1870) *F. dalecarlica*
(1876) *F. Duriaei*
Saint-Vincent
(1829) *F. Duriaei*
Sakurai, K.
  399 *D. japonicum*
Samuelson, G.
(1917) *F. antipyretica*
  var. *gracilis*
Sandberg, Carl
(1921) *F. dalecarlica*
(1936) *F. antipyretica*
Sandberg, J. H.
  81 *D. uncinatum*
(1888) *F. neo-mexicana*
(1892) *D. uncinatum*
(1892) *F. antipyretica*
(1892) *F. Howellii*
Sandberg, J. H., MacDougal, D. T., &
Heller, A. A.
  1165 p.p. *F. neo-mexicana*
  1165 p.p. *F. patula*
  1166 *D. uncinatum*
Sanford, S. N. F.
(1910) *F. antipyretica*
  var. *gigantea*
(1910) *F. novae-angliae*
Sanio
(1886) *F. hypnoides*
Sartwell, H. P.
  s.n. *F. Duriaei*

Saunders, Jas.
(1882) *F. antipyretica*
(1886) *F. antipyretica*
(1896) *F. antipyretica*
  s.n. *F. antipyretica*
Savage, T. E.
(1898) *F. patula*
Savage, T. E., Cameron, J. E., Lenocker, F. E.
(1898) *F. antipyretica*
(1898) *F. patula*
Sayre, Geneva
  346 *F. neo-mexicana*
  677 *F. antipyretica*
  697 *F. neo-mexicana*
  801 *D. falcatum*
Sbarbaro, C.
(1921) *F. antipyretica*
(1924) *F. antipyretica*
  s.n. *F. dalecarlica*
Schaeffer, R. L., Jr.
(1940) *F. antipyretica*
  var. *gigantea*
Schaffner, John H.
(1895) *F. Duriaei*
(1895) *F. missourica*
Schallert, P. O.
  46 *F. dalecarlica*
  341 *F. dalecarlica*
  514 *B. subulatum*
  523 *F. dalecarlica*
  718 *F. neo-mexicana*
  2780 *F. dalecarlica*
  2987 *F. novae-angliae*
(1921) *F. dalecarlica*
(1921) *F. flaccida*
(1923) *F. novae-angliae*
(1925) *F. novae-angliae*
(1926) *B. robustum*
(1926) *B. subulatum*
(1926) *F. dalecarlica*
(1926) *F. Duriaei*
(1926) *F. novae-angliae*
(1929) *F. dalecarlica*
(1929) *F. novae-angliae*
(1931) *F. dalecarlica*
(1932) *F. dalecarlica*
(1934) *F. dalecarlica*
(1934) *F. novae-angliae*
  var. *cymbifolia*
(1935) *F. antipyretica*
  var. *gigantea*
Schekovnikov, A.
(1916) *F. antipyretica*
Schellenberg, G.
(1908) *F. antipyretica*

Schemmann, W.
(1895) *F. antipyretica*
(1909) *F. antipyretica*
Schenk, Ferdinand
(1931) *F. antipyretica*
Scheutz, N. J.
(1865) *F. dalecarlica*
(1869) *D. falcatum*
(1871) *F. dalecarlica*
(1878) *D. capillaceum*
(1878) *F. antipyretica*
(1879) *D. capillaceum*
(1879) *D. falcatum*
(1885) *F. antipyretica*
var. *gracilis*
s.n. *D. falcatum*
s.n. *F. antipyretica*
var. *gracilis*
s.n. *F. dalecarlica*
Schiffner, D.
(1886) *D. falcatum*
(1887) *F. squamosa*
(1889) *F. squamosa*
Schimper, Wilhelm Philipp
185 *D. falcatum*
1055 *F. antipyretica*
(1837) *F. antipyretica*
var. *gracilis*
(1840) *F. antipyretica*
(1840) *F. squamosa*
(1844) *D. falcatum*
(1844) *F. antipyretica*
(1844) *F. dalecarlica*
(1844) *F. hypnoides*
(1845) *D. falcatum*
(1845) *F. antipyretica*
(1845) *F. dalecarlica*
(1845) *F. squamosa*
(1856) *D. falcatum*
(1856) *F. hypnoides*
(1865) *F. squamosa*
s.n. *D. capillaceum*
s.n. *D. falcatum*
s.n. *F. antipyretica*
s.n. *F. dalecarlica*
s.n. *F. hypnoides*
s.n. *F. squamosa*
Schipchinsky, N.
(1914) *F. antipyretica*
Schischkin, T.
(1924) *F. antipyretica*
Schmidt, A.
(1885) *F. squamosa*
(1900) *F. antipyretica*
var. *gigantea*
(1911) *F. antipyretica*
var. *gigantea*

Schneck, J.
(1881) *F. filiformis*
Schneeberger, Ed.
(1929) *F. antipyretica*
Schnooberger, Irma
1373a *F. hypnoides*
1960 *F. filiformis*
1960a *F. Duriaei*
3610 *D. capillaceum*
3178 *D. pallescens*
3732 *F. Duriaei*
(1940) *D. capillaceum*
Schnooberger, Irma & Wynne, Frances E.
3178 *F. novae-angliae*
3181 *F. novae-angliae*
5011 *F. novae-angliae*
5015 *F. novae-angliae*
5037 *F. novae-angliae*
Schofield, W. B.
(1949) *F. dalecarlica*
Schofield, W. B. & M. B.
7176 *F. antipyretica*
var. *gigantea*
Schornherst, Ruth O.
872 *F. disticha*
2503 *B. robustum*
2513 *F. disticha*
Schott, Arthur
(1857) *F. novae-angliae*
Schrader, A.
(1862) *F. biformis*
(1863) *F. biformis*
Schuh, R. E.
(1891) *F. antipyretica*
var. *gigantea*
Schultz
238 *F. antipyretica*
240 *F. antipyretica*
274 *F. antipyretica*
var. *gracilis*
(1887) *F. antipyretica*
var. *gigantea*
(1899) *F. antipyretica*
Schulze
(1839) *D. falcatum*
(1868) *D. falcatum*
(1878) *D. falcatum*
(1878) *F. antipyretica*
Schwab, A.
(1900) *F. squamosa*
Schwab, A. & Familler, Ig.
(1900) *F. antipyretica*
Scott, William
73 *D. pallescens*
Scully, F. J.
1321 *F. flaccida*

1349　*F. flaccida*
Searle, B.
　(1938)　*D. falcatum*
Sendtner, Otto
　(1838)　*D. falcatum*
　(1838)　*F. squamosa*
Seringe, Nicolas C.
　　s.n.　*F. squamosa*
Setchell, William A.
　(1884)　*D. capillaceum*
　(1884)　*F. antipyretica*
　　　　　var. *gigantea*
　(1884)　*F. novae-angliae*
　(1885)　*F. dalecarlica*
　(1894)　*F. dalecarlica*
　(1916)　*F. neo-mexicana*
Seth, K. A. Th.
　　27　*F. antipyretica*
　(1879)　*D. falcatum*
Seymour, A. B.
　(1889)　*F. antipyretica*
　　　　　var. *gigantea*
　(1889)　*F. dalecarlica*
　(1901)　*F. novae-angliae*
　(1908)　*F. antipyretica*
　　　　　var. *gigantea*
　(1913)　*D. capillaceum*
　(1913)　*F. Allenii*
Shacklette, H. T.
　1760　*F. filiformis*
　1957　*F. novae-angliae*
　2026　*F. Duriaei*
　2043　*F. filiformis*
　2316　*F. antipyretica*
　2411　*F. antipyretica*
Shacklette, H. T. & Harvill, A. M.
　2175　*F. novae-angliae*
Shalosubov, N. L.
　(1906)　*F. hypnoides*
Sharp, A. J.
　　12　*F. novae-angliae*
　　33　*F. dalecarlica*
　　88　*F. dalecarlica*
　　93　*D. capillaceum*
　107　*F. dalecarlica*
　108　*F. novae-angliae*
　124　*F. novae-angliae*
　　　　　var. *cymbifolia*
　125　*F. dalecarlica*
　126　*F. dalecarlica*
　133　*F. dalecarlica*
　341　*F. flaccida*
　3721　*F. dalecarlica*
　4061　*F. novae-angliae*
　4744　*F. novae-angliae*
　4751　*F. novae-angliae*

　4761　*F. novae-angliae*
　　　　　var. *cymbifolia*
　4763　*F. novae-angliae*
　5013　*F. disticha*
　34103　*F. novae-angliae*
　34348　*F. novae-angliae*
　34377　*F. novae-angliae*
　34562　*F. novae-angliae*
　　　　　var. *cymbifolia*
　34915　*F. novae-angliae*
　　　　　var. *cymbifolia*
　34955　*F. novae-angliae*
　　　　　var. *cymbifolia*
　35223　*F. novae-angliae*
　38107　*F. novae-angliae*
　341116　*F. novae-angliae*
　(1930)　*F. dalecarlica*
　(1931)　*D. capillaceum*
　(1931)　*F. novae-angliae*
　　　　　var. *cymbifolia*
　(1932)　*F. antipyretica*
　　　　　var. *gigantea*
　(1932)　*F. dalecarlica*
　(1933)　*F. novae-angliae*
　(1937)　*F. dalecarlica*
Sharp, A. J. & Clebsch, E.
　(1947)　*F. novae-angliae*
　(1947)　*F. novae-angliae*
　　　　　var. *cymbifolia*
Sharp, Ward
　(1932)　*F. novae-angliae*
　　　　　var. *latifolia*
Sheldon, Edmund P.
　9028　*F. neo-mexicana*
Sheldon, John L.
　N143　*F. antipyretica*
　　　　　var. *gigantea*
　C406　*F. novae-angliae*
　3282　*F. dalecarlica*
　(1905)　*F. antipyretica*
　　　　　var. *gigantea*
Sherrin, W. R.
　(1918)　*F. dalecarlica*
　(1919)　*F. antipyretica*
Shields, Randolph
　(1934)　*F. flaccida*
Shiota, Kenzo
　(1921)　*F. Duriaei*
Shoop, Cora
　536　*F. novae-angliae*
　572　*F. novae-angliae*
Sickenberger, E.
　246　*F. squamosa*
Silén, Fr.
　(1861)　*D. falcatum*

Sillén, O. Leopold
   45 *F. antipyretica*
      var. *gracilis*
   46 *F. antipyretica*
      var. *gracilis*
   47 *F. hypnoides*
   48 *F. hypnoides*
   50 p.p. *F. antipyretica*
      var. *gracilis*
   50 p.p. *F. dalecarlica*
  (1834) *D. falcatum*
  (1870) *D. capillaceum*
  (1875) *F. antipyretica*
      var. *gracilis*
  (1876) *F. antipyretica*
      var. *gracilis*
  s.n. *F. antipyretica*
Simkovics
  (1874) *D. falcatum*
Simmonds, N. W.
  (1943) *F. squamosa*
Simmons, Emory G.
  1756 *F. patula*
Simmons, H. G.
   471 *F. antipyretica*
   509 *F. antipyretica*
      var. *gracilis*
   536 *F. antipyretica*
  (1893) *F. antipyretica*
      var. *gigantea*
  (1897) *F. antipyretica*
      var. *gracilis*
  s.n. *F. antipyretica*
      var. *gigantea*
Simon, J.
   56 *F. antipyretica*
      var. *gigantea*
Sioda, K.
  (1921) *F. hypnoides*
Slerardrov, P.
  (1911) *F. Duriaei*
Small, A. M.
  (1900) *F. antipyretica*
      var. *gigantea*
Small, John K.
   26 *F. novae-angliae*
   32 *F. dalecarlica*
  5068 *B. subulatum*
  5079 *B. subulatum*
  9663 *F. flaccida*
  S14 *F. dalecarlica*
  S43/9663 *F. flaccida*
  S184/5079 *B. subulatum*
  S377 *F. dalecarlica*
  S527 *F. dalecarlica*
  (1889) *D. capillaceum*

(1889) *F. antipyretica*
      var. *gigantea*
(1890) *F. dalecarlica*
(1892) *F. dalecarlica*
(1892) *F. novae-angliae*
(1893) *F. novae-angliae*
(1894) *F. flaccida*
(1931) *F. flaccida*
Small, John K. & Wherry, Edgar T.
  11660 *F. disticha*
Smirnov, V.
  (1911) *F. antipyretica*
      var. *gracilis*
Smith, Alexander H.
  (1930) *D. capillaceum*
  (1934) *F. antipyretica*
      var. *gigantea*
Smith, A. M.
  (1901) *F. antipyretica*
      var. *gigantea*
Smith, Catherine
  (1923) *F. Howellii*
Smith, Harry
  (1920) *F. antipyretica*
      var. *gigantea*
  (1920) *F. dalecarlica*
Smith, John Donnell
  (1876) *D. capillaceum*
  (1876) *F. novae-angliae*
  (1878) *F. dalecarlica*
  (1878) *F. novae-angliae*
Smith, J. S.
  (1856) *F. antipyretica*
      var. *gigantea*
Smith, W.
  (1886) *F. antipyretica*
Sola, A. A.
  (1904) *F. hypnoides*
Solheim, William
  (1939) *F. antipyretica*
Solms, H. Gr. Z.
   99 *F. antipyretica*
Sommier, S. & Levier, E.
   246 *F. antipyretica*
   662 *F. antipyretica*
Sörensen, S.
  (1911) *D. falcatum*
  (1911) *F. dalecarlica*
  (1911) *F. hypnoides*
Söyrinki, Nülo
  (1934) *D. falcatum*
Spencer, H. L.
  (1938) *F. Duriaei*
Spindler, M.
  (1906) *F. squamosa*
  (1907) *F. squamosa*

Spreadborough, W.
  (1910) *F. antipyretica*
    var. *oreganensis*
  (1910) *F. Duriaei*
  (1910) *F. neo-mexicana*
Sprules, W.
  (1938) *F. dalecarlica*
  (1939) *F. dalecarlica*
Stair, Leslie D.
  4888 *F. patula*
  5029 *F. Duriaei*
  5029f *F. Duriaei*
Standley, Paul C.
  13151 *F. novae-angliae*
  16031 *F. antipyretica*
    var. *gigantea*
  16979 *F. Howellii*
  17681 *F. patula*
  17964 *F. patula*
  18554 *F. antipyretica*
    var. *gigantea*
  (1908) *F. neo-mexicana*
Standley, P. C. & Bollman, H. C.
  10123 *F. dalecarlica*
  12026 *F. antipyretica*
    var. *gigantea*
Stapf, D. O.
  (1885) *F. antipyretica*
Stauffer, J.
  s.n. *F. dalecarlica*
Stebbins, F. L.
  C680 *F. dalecarlica*
Steenis, J. H.
  4573 *F. neo-mexicana*
  4584 *F. neo-mexicana*
Steere, William C.
  702 *D. pallescens*
  3199 *D. pallescens*
  (1931) *F. dalecarlica*
  (1934) *D. pallescens*
  (1935) *F. Duriaei*
  (1935) *F. missourica*
  (1935) *F. novae-angliae*
  (1936) *F. antipyretica*
    var. *gigantea*
Steffen, Hans
  (1921) *F. dalecarlica*
Stenholm, Carl
  (1918) *D. falcatum*
  (1919) *F. dalecarlica*
  (1920) *F. antipyretica*
  (1920) *F. dalecarlica*
  (1922) *F. antipyretica*
  (1923) *F. antipyretica*
  (1923) *F. dalecarlica*
  (1926) *D. falcatum*
  (1931) *F. antipyretica*

(1931) *F. dalecarlica*
(1932) *F. antipyretica*
(1932) *F. antipyretica*
  var. *gigantea*
(1932) *F. dalecarlica*
  s.n. *F. antipyretica*
    var. *gracilis*
Sterki, Victor
  (1910) *D. capillaceum*
  (1910) *F. Duriaei*
  (1911) *D. capillaceum*
  (1911) *F. antipyretica*
    var. *gigantea*
  (1912) *D. capillaceum*
  (1912) *F. Duriaei*
Sterner, Rikard
  (1917) *F. antipyretica*
    var. *gigantea*
Steudner
  (1862) *F. antipyretica*
Stevens, Mary L.
  (1906) *D. capillaceum*
  (1906) *F. dalecarlica*
Steyermark, Julian A.
  83 *F. novae-angliae*
    var. *latifolia*
  4226 *F. antipyretica*
    var. *gigantea*
  13616 *F. novae-angliae*
  14419 *F. missourica*
  26914 *F. Duriaei*
  40110 *F. missourica*
  40112 *F. missourica*
Stirling, J.
  (1902) *F. antipyretica*
Stone, G. E.
  s.n. *D. capillaceum*
Strandmark, J. E.
  (1867) *D. capillaceum*
Strang, Mabel
  (1904) *F. antipyretica*
    var. *oreganensis*
Strömbäck, C. & E. A. & Hartman, C., Jr.
  (1842) *F. hypnoides*
Strömbäck, E. A.
  (1844) *F. hypnoides*
  (1852) *F. hypnoides*
Stübner
  (1844) *D. falcatum*
Studhalter, R. A.
  1162 *F. neo-mexicana*
  2532 *F. neo-mexicana*
Suksdorf, Wilhelm N.
  30 *F. patula*
  (1889) *F. antipyretica*
  (1890) *F. Howellii*

(1890) *F. patula*
Sullivant, William S.
25 *F. biformis*
33 *F. biformis*
34 *F. biformis*
47 *F. biformis*
62 *F. biformis*
63 *F. biformis*
90 *F. biformis*
(1842) *F. biformis*
(1845) *F. disticha*
(1851) *F. biformis*
s.n. *B. subulatum*
s.n. *D. capillaceum*
s.n. *F. biformis*
s.n. *F. disticha*
Summers, Lucia (Mrs. R. W.)
1380 *F. neo-mexicana*
1863 *F. antipyretica*
2211 *D. uncinatum*
(1875) *F. neo-mexicana*
(1878) *D. uncinatum*
(1878) *F. antipyretica*
(1879) *D. uncinatum*
(1880) *F. neo-mexicana*
s.n. *D. uncinatum*
Summers, R. W. & Lucia
(1863) *F. antipyretica*
Sundén, G. W.
(1855) *F. hypnoides*
Sundvik, O.
(1904) *F. hypnoides*
Sutton, A.
(1917) *F. antipyretica*
Svedberz, A.
(1921) *F. antipyretica*
Svenson, S.
(1920) *F. antipyretica*
Svensson, Gösta
(1934) *F. dalecarlica*
Svihla, Ruth D.
706 *F. antipyretica*
778 *F. patula*
2076 *F. antipyretica*
2132 *F. neo-mexicana*
2176 *F. patula*
2177 *F. patula*
2206 *F. Howellii*
(1931) *F. antipyretica*
(1939) *F. antipyretica*
(1943) *F. neo-mexicana*
(1943) *F. patula*
Swails, L. F.
28 *B. subulatum*
Swartz, Olof
801 *F. hypnoides*
(1795) *D. falcatum*

(1804) *F. antipyretica*
var. *gracilis*
(1807) *D. falcatum*
(1807) *F. dalecarlica*
(1807) *F. hypnoides*
Sydow, Hans
s.n. *F. antipyretica*
Sydow, P.
(1903) *F. antipyretica*

Tameki, S.
1532 *D. japonicum*
(1912) *D. japonicum*
Tärnlund, C. Alb.
(1889) *F. antipyretica*
var. *gracilis*
(1890) *D. falcatum*
(1905) *F. squamosa*
(1918) *F. antipyretica*
var. *gracilis*
(1919) *F. antipyretica*
(1921) *F. antipyretica*
var. *gracilis*
(1923) *F. hypnoides*
(1927) *F. antipyretica*
(1929) *F. antipyretica*
(1933) *F. antipyretica*
(1934) *F. dalecarlica*
(1935) *F. antipyretica*
var. *gracilis*
(1936) *F. antipyretica*
var. *gracilis*
(1937) *F. antipyretica*
Tarzwell, C. M.
253 *F. Duriaei*
Taylor
s.n. *F. antipyretica*
Taylor, M. E.
(1896) *F. antipyretica*
Taylor, Norman
361 *F. antipyretica*
var. *gigantea*
Taylor, William Randolph
(1916) *F. dalecarlica*
(1917) *F. novae-angliae*
(1920) *F. antipyretica*
var. *gigantea*
(1922) *F. antipyretica*
var. *gigantea*
(1922) *F. dalecarlica*
(1922) *F. novae-angliae*
(1922) *F. novae-angliae*
var. *cymbifolia*
(1923) *F. antipyretica*
var. *gigantea*
Tedd, H. Griffith
K47 *F. antipyretica*

K206 *F. antipyretica*
Tees, Grace Mary
218 *F. novae-angliae*
Terracciano, A. & Pappi, A.
375 *F. antipyretica*
4663 *F. antipyretica*
Thedenius, Hugo
(1863) *F. dalecarlica*
(1863) *F. hypnoides*
(1885) *F. hypnoides*
(1888) *D. falcatum*
(1888) *F. hypnoides*
s.n. *F. antipyretica*
var. *gracilis*
s.n. *F. squamosa*
Thedenius, K. F.
(1834) *D. falcatum*
(1834) *F. antipyretica*
(1834) *F. dalecarlica*
(1837) *D. falcatum*
(1838) *F. dalecarlica*
(1873) *F. antipyretica*
var. *gigantea*
s.n. *D. falcatum*
s.n. *F. dalecarlica*
Thériot, Iréné
(1889) *F. antipyretica*
Thiry, Arzt
(1859) *F. antipyretica*
Thompson, E. Ward
(1934) *F. antipyretica*
var. *gigantea*
(1934) *F. dalecarlica*
(1934) *F. novae-angliae*
(1935) *F. dalecarlica*
(1935) *F. novae-angliae*
Thompson, James
(1843) *F. antipyretica*
(1843) *F. antipyretica*
var. *gracilis*
Thompson, J. W. & Jacobson, J. R.
5199 *F. antipyretica*
var. *gigantea*
Thorne, Robert F. & Muenscher, W.C.
9173a *F. filiformis*
Thorpe, F. J.
71 *F. antipyretica*
var. *gigantea*
Thurow, F. W.
27 *F. filiformis*
(1892) *F. filiformis*
Tilden, Josephine E.
s.n. *F. hypnoides*
Timm
(1905) *F. antipyretica*
Timofeyev, A. F.
(1907) *F. antipyretica*

(1907) *F. antipyretica*
var. *gracilis*
Tinney, F.
(1933) *F. Duriaei*
Toba, Y.
(1927) *F. antipyretica*
var. *gracilis*
Toepffer, A.
(1892) *F. antipyretica*
var. *gigantea*
(1894) *F. antipyretica*
Tohle, R.
(1913) *F. Duriaei*
Tolf, Robert
(1770) *D. falcatum*
(1885) *F. antipyretica*
var. *gracilis*
(1888) *D. falcatum*
(1891) *F. antipyretica*
Tolmatschev
(1908) *F. antipyretica*
Tomin, M.
(1908) *F. antipyretica*
var. *gracilis*
(1910) *F. antipyretica*
Tordal, L. H.
429 *F. antipyretica*
Torrey, John
(1865) *F. neo-mexicana*
s.n. *F. dalecarlica*
Trabut
(1889) *F. antipyretica*
var. *gigantea*
(1902) *F. Duriaei*
(1907) *F. Duriaei*
s.n. *F. antipyretica*
s.n. *F. Duriaei*
Trask, Habeeb, & Grout
39 *F. novae-angliae*
184 *F. dalecarlica*
201 *F. dalecarlica*
250 *F. novae-angliae*
var. *cymbifolia*
Trautman, W.
(1918) *F. antipyretica*
Treboux, J.
(1907) *F. antipyretica*
var. *gracilis*
Trelease, William
1367 *F. antipyretica*
2368 *F. antipyretica*
(1894) *F. antipyretica*
Troch, Pierre
(1885 & 1886) *F. antipyretica*
Trulsson, Åke
(1918) *F. antipyretica*

Tufvesson, Elsa
(1933) *D. capillaceum*
(1933) *F. antipyretica*
(1934) *D. capillaceum*
Tullgren, A.
(1891) *F. dalecarlica*
(1904) *F. dalecarlica*
Turyev, A.
  284 *F. antipyretica*
    var. *gracilis*

Umbach, L. M.
   58 *F. antipyretica*
  604 *F. missourica*
  777 *F. patula*
  867 *F. antipyretica*
  869 *F. neo-mexicana*
 1104 *F. missourica*
(1898) *F. missourica*
(1901) *F. neo-mexicana*
(1903) *F. antipyretica*
(1903) *F. neo-mexicana*
(1903) *F. patula*
(1909) *F. missourica*
Underwood, Lucien M.
(1887) *F. antipyretica*
    var. *gigantea*
(1889) *F. antipyretica*
    var. *gigantea*
Urban, D.
(1858) *D. pallescens*
(1858) *F. antipyretica*

Vahl, Jens
(1829) *F. antipyretica*
(1829) *F. dalecarlica*
Vail, Anna Murray
(1892) *F. dalecarlica*
Vail, Anna Murray & Britton, Elizabeth G.
   83 *F. antipyretica*
    var. *gigantea*
  211 *F. novae-angliae*
  237 *F. novae-angliae*
  238 *F. dalecarlica*
  241 *F. novae-angliae*
(1892) *F. dalecarlica*
(1892) *F. novae-angliae*
Vandenbroeck, Hn.
  462 *F. antipyretica*
    var. *gigantea*
(1881) *F. antipyretica*
(1882) *F. antipyretica*
Van Schaack
 953a *F. neo-mexicana*
Van Vlesk, J.
  s.n. *F. flaccida*

Van Wert, Mary C.
(1922) *F. neo-mexicana*
(1923) *F. neo-mexicana*
Vasey, George R.
(1868) *D. falcatum*
(1868) *F. neo-mexicana*
(1875) *F. neo-mexicana*
  s.n. *F. antipyretica*
  s.n. *F. biformis*
  s.n. *F. neo-mexicana*
Venturi
(1882) *F. antipyretica*
    var. *gracilis*
Verdoorn, Frans
(1927) *F. squamosa*
Vermont Bureau of Fisheries
(1909) *F. Duriaei*
Vesterlund, Otto
(1910) *F. antipyretica*
    var. *gracilis*
(1912) *F. antipyretica*
    var. *gracilis*
(1915) *F. antipyretica*
    var. *gracilis*
(1916) *F. antipyretica*
Vetterhall, Erland
(1874) *D. falcatum*
(1874) *F. antipyretica*
    var. *gracilis*
(1878) *F. hypnoides*
Victor, Bro. Jules
(1937) *F. novae-angliae*
Vigener, A.
(1859) *F. antipyretica*
(1874) *F. squamosa*
Vigineix, G.
(1855) *F. antipyretica*
Vinette, R.
   41 *F. dalecarlica*
Voigt, Alwin
(1874) *F. antipyretica*
  s.n. *D. falcatum*
Von Bock
  212 *F. antipyretica*
    var. *gigantea*
  341 *F. antipyretica*
    var. *gigantea*
  357 *F. antipyretica*
    var. *gracilis*
  434 *F. antipyretica*
    var. *gracilis*
  480 *F. antipyretica*
    var. *gracilis*
  908 *F. antipyretica*
    var. *gracilis*
(1907) *F. antipyretica*
    var. *gigantea*

(1907) *F. antipyretica*
var. *gracilis*
(1908) *F. antipyretica*
var. *gigantea*
(1908) *F. antipyretica*
var. *gracilis*
(1909) *F. antipyretica*
var. *gigantea*
(1909) *F. antipyretica*
var. *gracilis*
(1909) *F. squamosa*
(1910) *F. antipyretica*
var. *gracilis*
(1911) *F. antipyretica*
var. *gigantea*
(1911) *F. antipyretica*
var. *gracilis*
(1913) *F. antipyretica*
var. *gracilis*
s.n. *F. antipyretica*
var. *gracilis*
Von Handel-Mazzetti, Heinrich
348 *F. Duriaei*
821 *F. antipyretica*
Von Klinggraeff
(1861) *D. falcatum*
Von Post, Hampus A.
(1847) *D. falcatum*
Von Schrenk, Hermann
(189?) *F. dalecarlica*
Von Wied-Neuwied, Maximilian A. P.
(Von Wied)
130 *B. subulatum*
(1832) *B. subulatum*
s.n. *B. subulatum*
Voronichin
(1916) *F. Duriaei*
Voronov, G. N.
(1923) *F. antipyretica*
Vrang, Erik P.
(1896) *D. falcatum*

Waddell, C. H.
(1898) *F. antipyretica*
var. *gracilis*
Waghorne, Arthur C.
3 *F. dalecarlica*
25 *D. falcatum*
108 *D. falcatum*
(1883) *F. novae-angliae*
(1890) *F. antipyretica*
var. *gigantea*
(1890) *F. dalecarlica*
(1890) *F. novae-angliae*
var. *cymbifolia*
(1891) *D. falcatum*
(1891) *F. dalecarlica*

(1892) *D. falcatum*
(1892) *F. novae-angliae*
var. *cymbifolia*
(1893) *F. dalecarlica*
(1893) *F. novae-angliae*
(1893) *F. novae-angliae*
var. *cymbifolia*
(1894) *D. falcatum*
(1894) *F. antipyretica*
(1894) *F. antipyretica*
var. *gigantea*
(1894) *F. dalecarlica*
Wagner
s.n. *F. antipyretica*
var. *gracilis*
Wagner, Kenneth A.
1582 *F. dalecarlica*
1974 *F. dalecarlica*
1986 *F. dalecarlica*
2464 *D. capillaceum*
Wahlenberg
s.n. *D. falcatum*
Wahlstedt, L. J.
(1866) *F. squamosa*
(1867) *D. falcatum*
(1867) *F. dalecarlica*
(1869) *F. antipyretica*
Waite, M. B.
(1890) *F. novae-angliae*
Waldburg-Zeil
(1876) *F. Duriaei*
Wälde, A.
(1900) *F. squamosa*
(1902) *F. squamosa*
Waldheim
(1933) *D. capillaceum*
(1937) *F. dalecarlica*
Waldron, L. R.
23 *F. neo-mexicana*
(1901) *F. antipyretica*
(1901) *F. patula*
Walker
(1884) *F. antipyretica*
Walker, E. H.
(1928) *F. antipyretica*
var. *gigantea*
Walker, Mr. & Mrs. E. P.
902 *F. antipyretica*
Wallace, E. C.
(1926) *F. antipyretica*
(1938) *F. antipyretica*
(1948) *F. antipyretica*
Wallace, J. B.
(1933) *F. flaccida*
Wallgren
(1865) *F. antipyretica*
var. *gigantea*

Walters, Maurice B.
  45 *F. novae-angliae*
Walther, Alexander
  s.n. *F. antipyretica*
Ward, R. H.
  s.n. *F. antipyretica*
    var. *gigantea*
Wareham, Richard T.
  (1936) *F. novae-angliae*
  (1937) *F. dalecarlica*
  (1940) *F. novae-angliae*
  (1942) *F. novae-angliae*
  (1946) *F. dalecarlica*
  (1946) *F. novae-angliae*
Warne, H. A.
  s.n. *F. dalecarlica*
Warne, H. A. & Barron, Alfred
  (1877) *D. pallescens*
Warnstorf, C.
  218 *F. squamosa*
  (1873) *F. antipyretica*
  (1875) *F. squamosa*
  (1880) *F. antipyretica*
  (1883) *F. antipyretica*
  (1895) *F. antipyretica*
    var. *gigantea*
  (1899) *F. antipyretica*
  (1906) *F. antipyretica*
    var. *gracilis*
Warnstorf, Joh.
  (1908) *F. squamosa*
Watson, Hewett C.
  (1842) *F. antipyretica*
Watson, Sereno
  1452 *F. antipyretica*
  (1880) *F. Duriaei*
  (1880) *F. neo-mexicana*
Watson, W.
  451d *F. antipyretica*
Weatherby, C. A.
  (1907) *F. novae-angliae*
Weber, J.
  (1878) *D. falcatum*
  (1878) *F. antipyretica*
Weber, William A.
  2653 *F. antipyretica*
  2697 *F. neo-mexicana*
  (1945) *F. antipyretica*
Webster, G. L. & Wilbur, R. L.
  744 *F. novae-angliae*
    var. *cymbifolia*
Weed, Walter H.
  (1889) *F. hypnoides*
Wegelius, A.
  (1913) *F. antipyretica*
Wehrhahn, R.
  (1887) *F. antipyretica*

  (1898) *F. antipyretica*
Weir, J.
  284 *F. bogotensis*
Welch, Winona H.
  711 *F. antipyretica*
    var. *gigantea*
  712 *F. novae-angliae*
  713 *F. dalecarlica*
  1084 *F. novae-angliae*
    var. *latifolia*
  1086 *F. novae-angliae*
    var. *cymbifolia*
  1087 *F. novae-angliae*
    var. *latifolia*
  1088 *F. novae-angliae*
    var. *latifolia*
  2152 *F. novae-angliae*
  2220 *F. dalecarlica*
  2221 *F. dalecarlica*
  2222 *F. dalecarlica*
  2223 *F. dalecarlica*
  2224 *F. novae-angliae*
  2225 *F. novae-angliae*
  2226 *F. novae-angliae*
  2970 *F. novae-angliae*
  5050 *F. novae-angliae*
  5051 *F. novae-angliae*
  5052 *F. novae-angliae*
  5312 *D. pallescens*
  5313 *F. novae-angliae*
  5567 *F. antipyretica*
  5568 *F. antipyretica*
  5660 *F. antipyretica*
  5662 *F. squamosa*
  5666 *F. antipyretica*
  5696 *F. squamosa*
  5816 *F. squamosa*
  6368 *F. novae-angliae*
  6971 *F. novae-angliae*
  6972 *F. novae-angliae*
  6973 *F. novae-angliae*
  6974 *F. novae-angliae*
  6975 *F. novae-angliae*
  6976 *F. novae-angliae*
  6977 *F. novae-angliae*
  6978 *F. novae-angliae*
  6979 *F. novae-angliae*
  7391 *F. biformis*
  7392 *F. biformis*
  7393 *F. biformis*
  7394 *F. biformis*
  7395 *F. biformis*
  7396 *F. biformis*
  7397 *F. biformis*
  7398 *F. biformis*
  7399 *F. biformis*
  7400 *F. biformis*

7401 *F. biformis*
7402 *F. biformis*
7403 *F. biformis*
7515 *F. biformis*
9133 *F. biformis*
9134 *F. biformis*
9135 *F. biformis*
9136 *F. biformis*
9137 *F. biformis*
9138 *F. biformis*
9139 *F. biformis*
9140 *F. biformis*
9141 *F. biformis*
9191 *F. disticha*
9691 *F. novae-angliae*
9695 *D. capillaceum*
9697 *F. hypnoides*
10360 *F. hypnoides*
10419 *F. hypnoides*
11908 *F. neo-mexicana*
15074 *F. patula*
15081 *F. patula*
15082 *F. patula*
15083 *F. patula*
15084 *F. patula*
15091 *F. patula*
16057 *F. patula*
16058 *F. patula*
16059 *F. patula*
16060 *F. patula*
Wells
  (1906 or 1907) *F. antipyretica*
Welwitsch, F.
  7 *F. antipyretica*
  256 *F. Duriaei*
  260 *F. squamosa*
  (1847) *F. antipyretica*
Wennersten, O.
  (1892) *F. antipyretica*
Wentworth, M.
  (1923) *F. Howellii*
Werner, Ernst
  (1909) *F. dalecarlica*
Wesenberg-Lund & Jensen, C.
  (1908) *F. dalecarlica*
West, G. T.
  (1904) *F. antipyretica*
West, W.
  (1885) *F. squamosa*
Westerberg, F. O.
  (1897) *D. falcatum*
  (1907) *D. falcatum*
Westling, P. A.
  196 *F. hypnoides*
  (1863) *D. falcatum*
  (1863) *F. dalecarlica*
  (1863) *F. hypnoides*

  (1865) *F. antipyretica*
  (1866) *F. antipyretica*
      var. *gracilis*
  (1869) *D. capillaceum*
Wetherby, A. G.
  (1896) *F. disticha*
Wheeler, Harriet
  (1895) *D. capillaceum*
Wheeler, Louis C.
  2962 *F. antipyretica*
      var. *mollis*
  2963 *F. neo-mexicana*
  3212 *F. antipyretica*
      var. *mollis*
Whitehouse, Eula
  22139 *F. patula*
  23109 *F. Duriaei*
  27488 *F. patula*
  27834 *B. subulatum*
Whitham, J. C.
  1494 *F. antipyretica*
      var. *mollis*
  1495 *F. neo-mexicana*
Whitham, J. C. & Wetzel, W. W.
  317 *F. antipyretica*
Wicht, C. L.
  (1944) *F. antipyretica*
Wickes, Mildred L.
  25 *F. dalecarlica*
  29 *F. antipyretica*
  45 *F. dalecarlica*
  48 *F. antipyretica*
  64 *F. antipyretica*
  (1938) *F. dalecarlica*
Wiggins, Ira L.
  C54 *F. neo-mexicana*
  (1938) *F. antipyretica*
Wikström, J. E.
  s.n. *F. hypnoides*
Wilkens, Hans
  (1928) *F. dalecarlica*
Williams, H. L. & R. S.
  (1933) *F. dalecarlica*
Williams, Robert S.
  82 *F. Duriaei*
  284 *F. hypnoides*
  320 *D. uncinatum*
  361 *F. Howellii*
  362 *F. patula*
  391 *F. antipyretica*
      var. *gigantea*
  417 *F. antipyretica*
  692 *D. falcatum*
  (1886) *F. antipyretica*
  (1895) *D. uncinatum*
  (1898) *D. falcatum*
  (1900) *D. capillaceum*

Williams, Wayland
  (1946) *F. dalecarlica*
Willis
  (1883) *F. neo-mexicana*
Wilms, D.
  (1882) *F. antipyretica*
Wilson
  442 *F. antipyretica*
  443 *F. squamosa*
Wilson, A.
  (1917) *F. dalecarlica*
Wilson, Leonard R.
  100 *F. antipyretica*
  260 *F. antipyretica*
        var. *gigantea*
  2041 *F. Duriaei*
  2078 *F. antipyretica*
        var. *gigantea*
Wilson, Percy
  150 *F. novae-angliae*
Wilson, W.
  (1826) *F. antipyretica*
  (1828) *F. antipyretica*
  (1828) *F. squamosa*
  (1842) *F. antipyretica*
  (1843) *F. squamosa*
  (1844) *F. antipyretica*
  (1856) *F. antipyretica*
  (1865) *F. squamosa*
  (1865) *F. squamosa*
        var. *Curnowii*
Windle, F.
  (1905) *F. novae-angliae*
Winne, W. T.
  256 *F. novae-angliae*
  265 *F. antipyretica*
        var. *gigantea*
  271 *F. novae-angliae*
  278 *F. novae-angliae*
        var. *cymbifolia*
  279 *F. novae-angliae*
  281 *F. antipyretica*
        var. *gigantea*
  293 *F. dalecarlica*
  309 *F. Duriaei*
  313 *D. capillaceum*
  314 *F. novae-angliae*
        var. *cymbifolia*
  336 *F. Duriaei*
  401 *F. novae-angliae*
        var. *cymbifolia*
  418 *F. novae-angliae*
  477 *F. novae-angliae*
  507 *F. dalecarlica*
  564 *F. novae-angliae*
        var. *cymbifolia*
  575 *F. dalecarlica*

576 *F. dalecarlica*
578 *F. dalecarlica*
583 *D. falcatum*
589 *F. antipyretica*
      var. *gigantea*
592 *F. novae-angliae*
595 *F. novae-angliae*
596 *F. novae-angliae*
601 *F. dalecarlica*
605 *F. novae-angliae*
612 *F. novae-angliae*
616 *F. dalecarlica*
618 *F. dalecarlica*
620 *F. novae-angliae*
621 *F. dalecarlica*
624 *F. dalecarlica*
628 *F. antipyretica*
      var. *gigantea*
629 *F. novae-angliae*
630 *F. dalecarlica*
641 *F. dalecarlica*
649 *F. dalecarlica*
650 *F. antipyretica*
      var. *gigantea*
651 *F. novae-angliae*
660 *F. novae-angliae*
662 *F. novae-angliae*
668 *F. novae-angliae*
670 *F. dalecarlica*
675 *F. dalecarlica*
678 *F. antipyretica*
      var. *gigantea*
679 *F. novae-angliae*
680 *F. dalecarlica*
688 *F. dalecarlica*
691 *F. dalecarlica*
700 *F. dalecarlica*
706 *F. antipyretica*
      var. *gigantea*
707 *F. dalecarlica*
714 *F. antipyretica*
      var. *gigantea*
720 *F. novae-angliae*
      var. *latifolia*
724 *F. Duriaei*
741 *F. novae-angliae*
757 *F. novae-angliae*
764 *D. capillaceum*
768 *F. novae-angliae*
      var. *cymbifolia*
774 *D. capillaceum*
776 *F. novae-angliae*
823 *D. pallescens*
840 *F. Duriaei*
843 *D. pallescens*
858 *D. capillaceum*
864 *F. Duriaei*

866 *F. Duriaei*
867 *F. Duriaei*
868 *F. novae-angliae*
    var. *latifolia*
Winne, W. T. & Andrews, LeRoy
  129 *F. novae-angliae*
Winne, W. T. & Muenscher, W. C.
  249 *F. Duriaei*
  396 *F. dalecarlica*
  397 *F. dalecarlica*
  398 *F. novae-angliae*
Winne, W. T., Muenscher, W. C., &
  Isely, Duane
  468 *F. antipyretica*
    var. *gigantea*
Winne, W. T., Muenscher, W. C., &
  Shannon, E. L.
  171 *F. antipyretica*
    var. *gigantea*
  215 *F. novae-angliae*
    var. *cymbifolia*
  223 *F. novae-angliae*
  228 *F. novae-angliae*
  241 *F. novae-angliae*
Winne, W. T., Shannon, E. L., &
  Muenscher, W. C.
  190 *F. dalecarlica*
  202 *F. novae-angliae*
    var. *cymbifolia*
  222 p.p. *D. capillaceum*
  222 p.p. *D. pallescens*
  230 *F. dalecarlica*
  240 *F. novae-angliae*
Winne, W. T., Whitford, N.,
  & Chase, S.
  141 *F. novae-angliae*
Winslow, E. J.
  (1912) *F. dalecarlica*
Winter, D.
  (1901) *D. falcatum*
Winter, E.
  (1922) *F. antipyretica*
Winter, F.
  (1873) *F. novae-angliae*
Winter, Ferd
  (1865) *F. squamosa*
  (1869) *F. squamosa*
  s.n. *D. falcatum*
Wirtgen, Phil.
  (1858) *F. antipyretica*
  s.n. *F. antipyretica*
Witz, W. R.
  (1937) *F. dalecarlica*
Wolden, B. O.
  (1925) *F. Duriaei*
Wolf
  s.n. *F. biformis*

s.n. *F. hypnoides*
Wolf, J.
  s.n. *F. filiformis*
Wolle, F.
  (1873) *F. antipyretica*
    var. *gigantea*
  (1874) *D. capillaceum*
  (1874) *F. antipyretica*
  (1874) *F. dalecarlica*
  (1874) *F. novae-angliae*
  (1874) *F. novae-angliae*
    var. *cymbifolia*
  s.n. *D. capillaceum*
Wolle & Rau
  s.n. *F. Duriaei*
Wood
  (1816) *F. squamosa*
Wood, H. C.
  (1860) *F. dalecarlica*
Wood, J. B.
  (1878) *F. squamosa*
Woodward, C. H. & Beals, A. T.
  (1938) *F. dalecarlica*
Worthington, Ruth
  (1877) *F. Duriaei*
Wright, Charles
  (1882) *D. capillaceum*
  (1882) *F. antipyretica*
    var. *gigantea*
  (1882) *F. hypnoides*
  s.n. *F. antipyretica*
  s.n. *F. neo-mexicana*
Wright, C. A.
  (1883) *F. antipyretica*
Wulfsberg, N.
  (1867) *F. antipyretica*
  (1867) *F. dalecarlica*
  (1867) *F. squamosa*
  (1870) *D. falcatum*
Wynne, Frances E.
  1310 *F. Duriaei*
  1898 *F. novae-angliae*
    var. *cymbifolia*
  2797 *F. antipyretica*
    var. *gigantea*
  2798 *F. novae-angliae*
Wynne, Frances E. & Schnooberger,
  Irma
  204 *F. novae-angliae*

Yasuda, A.
  493 *F. antipyretica*
    var. *gracilis*
  661 *F. antipyretica*
    var. *gracilis*

Young, A. H.
  (1874) *F. novae-angliae*
Yuncker, T. G.
  10557 *F. dalecarlica*
  10559 *F. dalecarlica*
Yuncker, T. G. & E. C.
  7106 *F. patula*
  7118 *F. neo-mexicana*
  7119 *F. patula*
  7120 *F. antipyretica*
  10905 *F. Duriaei*
  12369 *D. falcatum*
  12370 *F. Duriaei*
  12371 *F. Duriaei*

Zametzer, Phil.
  s.n. *F. antipyretica*
    var. *gracilis*
Zetterstedt, J. E.
  174a *D. falcatum*
  175a *F. dalecarlica*
  175b *F. dalecarlica*
  176 *F. antipyretica*
  451 *F. antipyretica*
    var. *gigantea*
  (1853) *D. falcatum*
  (1855) *D. falcatum*
  (1859) *F. antipyretica*
    var. *gracilis*
  (1859) *F. hypnoides*
  (1865) *D. falcatum*
  (1865) *F. dalecarlica*

(1868) *D. falcatum*
(1870) *D. falcatum*
(1872) *D. falcatum*
(1872) *F. antipyretica*
  var. *gigantea*
(1873) *F. antipyretica*
  var. *gracilis*
(1875) *D. falcatum*
(1878) *D. falcatum*
(1878) *F. dalecarlica*
(1879) *F. antipyretica*
  var. *gracilis*
Zetterstedt, J. E. & Wickbom, J. A. O.
  (1870) *D. falcatum*
  (1870) *F. dalecarlica*
Zeuker, J. C. & Dietrich, F. D.
  s.n. *F. antipyretica*
Zickendrath, Ernst
  1150 *F. antipyretica*
    var. *gracilis*
  (1895) *F. antipyretica*
    var. *gracilis*
  (1895) *F. hypnoides*
Zimmermann
  (1865) *D. falcatum*
  (1868) *D. falcatum*
Zmuda, A.
  (1912) *F. antipyretica*
Zodda
  444 *F. Duriaei*
  (1907) *F. Duriaei*

INDEX of NAMES [1]

---

[1] All valid entities are in Roman type; the others are italicized.

    f. *funiculata* Hj. Möll. 24.
subsp. *gigantea* (Sull.) Kindb. 55.
  var. gigantea (Sull.) Sull. 19, 55–67, 57*.
    f. *gigantea* (Sull.) Mönkem. 56.
subsp. *gothica* (Card. and Arn.) Podp. 46.
               f. *dimorphophylla* (Hj. Möll.) Podp. 46.
               f. *stagnalis* (Kaal.) Podp. 25.
    f. *gothica* (Card. and Arn.) Perss. 46.
subsp. *gracilis* (Lindb.) Giacom. 239.
subsp. *gracilis* (Lindb.) Kindb. 46.
     var. *Grebeana* Roth 25.
     var. *patens* Bryhn 46.
  var. gracilis (Lindb.) Schimp. 20, 46–55, 48*.
  var. Heldreichii (C. Müll.) Ruthe 19, 71–73, 71*.
    f. *imbricata* Card. 23.
subsp. *islandica* (Card.) Kindb. 23.
subsp. *Kindbergii* (Ren. and Card.) Card. 89.
       f. *gracilior* Card. 91.
       f. *robustior* Card. 91.
subsp. *Kindbergii* (Ren. and Card.) Kindb. 240.
subsp. *Lachenaudi* (Card.) Podp. 26.
  var. *Lachenaudii* (Card.) Warnst. 24.
    f. *lacustris* Fuchs. 24.
  var. *latifolia* Milde 55.
    f. *latifolia* (Milde) Mönkem. 56.
    f. *latifolia-tenuis* Fuchs. 24.
    f. *latifolia-vulgaris* Fuchs. 56.
  var. *laxa* Milde 22, 240.
  var. *laxa* (Milde) Limpr. 23.
       f. *paroica* Mönkem. 24.
       f. *robustior* Fleisch. and Warnst. 23.
    f. *laxa* (Milde) Mönkem. 24.
    f. *laxior* Schimp. 23.
  var. *ligurica* Fleisch. 23.
  var. *livonica* (Roth and Von Bock) Mönkem. 56.
    f. *livonica* (Roth and Von Bock) Mönkem. 56.
  var. *macrophylla* Warnst. 55.
  var. *minor* Brid. 46.
  var. *minor* Roth 240.
  var. *minor* Wahlenb. 240.
    f. *minor* Schimp. 23.
  var. mollis (C. Müll.) Welch 19, 67–70, 68*.
  var. *mollissima* Warnst. 24.
  var. *monensis* Card. and Simm. 55.
    f. *monensis* (Card. and Simm.) Jens. 56.
  var. *montana* H. Müll. 22.
    f. *montana* (H. Müll.) Mönkem. 24.
subsp. *neo-mexicana* (Sull. and Lesq.) Card. 77.
  var. *occidentalis* Card. 26.
  var. oreganensis Ren. and Card. 20, 73–77, 74*
  var. *oregonensis* Ren. and Card. 73.
  var. *patens* Ren. and Card. 67.

[1] *Fontinalis missouriensis* Card., collected in Missouri, and cited by V. F. Brotherus, in Engler und Prantl, Die natürlichen Pflanzenfamilien: Bryophyta 2: 728. 1905, and in the 1925 edition, Musci 11 (2): 60, is regarded by the author as an error, in reference to *F. missourica* Card.